TRENDS IN ANTARCTIC TERRESTRIAL AND LIMNETIC ECOSYSTEMS

TRENDS IN ANTARCTIC TERRESTRIAL AND LIMNETIC ECOSYSTEMS

Antarctica as a Global Indicator

Edited by

D.M. BERGSTROM

Australian Government Antarctic Division, Kingston, Australia

P. CONVEY

British Antarctic Survey, Natural Environment Research Council, Cambridge, United Kingdom

A.H.L. HUISKES

Unit for Polar Ecology, NIOO-KNAW, Yerseke, The Netherlands

 Springer

A C.I.P. Catalogue record for this book is available from the Library of Congress.

ISBN-10 1-4020-5276-6 (HB)
ISBN-13 978-1-4020-5276-7 (HB)
ISBN-10 1-4020-5277-4 (e-book)
ISBN-13 978-1-4020-5277-4 (e-book)

Published by Springer,
P.O. Box 17, 3300 AA Dordrecht, The Netherlands.

www.springer.com

Printed on acid-free paper

Cover Illustration: Back ground photo: Nunataks in the Behrendt Mountains, Ellsworth Land, continental Antarctica, host some of the simplest terrestrial faunal communities known on the planet. *Photograph: Pete Convey.*

Insert photo: Eight invasive mammals are currently established on subantarctic islands. Other than rodents, the remainder were originally deliberately introduced by humans. Cats are responsible for drastic reductions in some seabird populations. This photograph shows a cat in a king penguin colony in the Iles Kerguelen. *Photograph: Jean-Louis Chapuis.*

Steven Chown (South Africa) has rolled across a cushion plant vegetation on
Heard Island (subantarctic, Indian Ocean sector) as to not damage it, in order to
sample arthropods from *Pringlea antiscorbutica* (Kerguelen cabbage).
Note: there are no footprints. *Photograph Dana Bergstrom.*

TABLE OF CONTENTS

PREFACE

Motivated by the Northern Hemisphere International Tundra Experiment (ITEX), Dana Bergstrom instigated a workshop in Brisbane, Australia, in 1998 to discuss the concept for a Southern Hemisphere ITEX using a series of networked sites in and around Antarctica. Instead of following the ITEX model of looking at variations in organismal performance with longitude, the new program would be based around the premise that latitude and altitude could act as proxies or predictors for future climate change. Using the power of the internet, these concepts and intentions were disseminated to researchers who formed various discussion groups. There were instant replies from Ad Huiskes and Burkhard Schroeter in Europe, who exclaimed that they had conceived a similar concept while tent-bound in a blizzard on the Antarctic Peninsula.

Thus began the SCAR program, Regional Sensitivity to Climate Change in Antarctic Terrestrial and Limnetic Ecosystems (RiSCC). An initial workshop was hosted by Antonio Quesada and Leo Sancho, with support from the Spanish Antarctic Program, to further develop the concept. This was followed by a science planning workshop, hosted by Steven Chown in South Africa. These workshops defined the foundations for the RiSCC Program, which were to study changes and patterns in species diversity and organismal performance around Antarctica. The data collected were to be linked with latitude along the "Antarctic Environmental Gradient", which extends over 40° of latitude from Marion Island at 47°S to the Transantarctic Mountains at 87°S, and includes a range of climatic zones present on the cool temperate oceanic islands to the frigid and arid Antarctic continent.

RiSCC ran for just over five years (2000–2005), before being absorbed within the new framework of SCAR international scientific programs, and contributing to the development of its successor program "Evolution and Biodiversity in Antarctica". Without a doubt, RiSCC has been a very successful enterprise and catalyst when seen from a number of perspectives: almost 200 peer-reviewed scientific publications have emerged from the scientific activities associated with the program, a strong international science network has developed with the active involvement of scientists from over 18 nations, and RiSCC proved the nexus for several successful international expeditions.

RiSCC has helped to improve our understanding of the interactions and linkages among climate change, indigenous and alien species, and the ways in which ecosystems function. This volume is a culmination of the efforts and findings of the RiSCC community. We would like to thank all our authors for their efforts, and the members of the RiSCC program in general whose scientific findings are reported here. We thank all referees for their prompt efforts to provide sage advice on all chapters in this volume. We would also like to thank Eric Woehler for his enormous efforts in helping us get the book to publication. Eric has filled numerous roles including sub–editor, illustrator and tireless supporter of the project, and has dedicated five months of his life to this cause.

We dedicate the book to one of the original RiSCC 'Three Musketeers', Burkhard Schroeter, who while no longer pursuing his passion for Antarctic field research, now perhaps has a far more important role in life, that of teaching and inspiring the next generation of scientists.

Dana Bergstrom, Pete Convey and Ad Huiskes (eds).

LIST OF CONTRIBUTORS

D.M. BERGSTROM
*Department of Environment
and Heritage
Australian Government Antarctic
Division
203 Channel Highway
Kingston, Tasmania 7050, Australia
dana.bergstrom@agad.gov.au*

L. BEYENS
*Universiteit Antwerpen (CMI)
Departement Biologie/PLP
Groenenborgerlaan 171
B-2020 Antwerp, Belgium
louis.beyens@ua.ac.be*

M. BÖLTER
*Institute for Polar Ecology
University of Kiel
Wischhofstr. 1-3
24148 Kiel, Germany
mboelter@ipoe.uni-kiel.de*

J.L. CHAPUIS
*Muséum National d'Histoire Naturelle
Département Ecologie et Gestion de la
Biodiversité, Paris, France
chapuis@mnhn.fr*

S.L. CHOWN
*Centre for Invasion Biology
Department of Botany and Zoology
Stellenbosch University
Private Bag X1Matieland 7602,
South Africa
slchown@sun.ac.za*

P. CONVEY
*British Antarctic Survey
Natural Environment Research Council
High Cross, Madingley Road
Cambridge CB3 0ET, United Kingdom
p.convey@bas.ac.uk*

G. COPSON
*Wildlife Management Branch,
Department of Primary Industries
and Water
GPO Box 44, Hobart, Tasmania
Geoff.Copson@dpiw.tas.gov.au*

H.J.G. DARTNALL
*Department of Biological Sciences
Macquarie University
Sydney, NSW 2109, Australia*

Y. FRENOT
*UMR 6553 CNRS-Université de Rennes
& French Polar Institute (IPEV)
Station Biologique
F-35380 Paimpont, France
yves.frenot@univ-rennes1.fr*

J.A.E. GIBSON
*Institute of Antarctic and Southern
Ocean Studies
University of Tasmania, Private Bag 77
Hobart, Tasmania 7001, Australia
John.Gibson@utas.edu.au*

xi

N. GREMMEN
Data-Analyse Ecologie
Hesselsstraat 11
7981 CD Diever, The Netherlands
gremmen@wxs.nl

F. HENNION
Impact des Changements Climatiques
UMR 6553, Centre National
de la Recherche
Scientifique – Université de Rennes 1
Campus de Beaulieu
F-35042 Rennes cedex, France
Francoise.Hennion@univ-rennes1.fr

J.E. HOBBIE
The Ecosystems Center
Marine Biological Laboratory
Woods Hole, MA 02543, USA
jhobbie@mbl.edu

D.A. HODGSON
British Antarctic Survey
Natural Environment Research Council
High Cross, Madingley Road
Cambridge CB3 0ET, United Kingdom
daho@bas.ac.uk

I.D. HOGG
Centre for Biodiversity and Ecology
Research
University of Waikato, Private Bag 3105
Hamilton, New Zealand
hogg@waikato.ac.nz

K.A. HUGHES
British Antarctic Survey
Natural Environment Research Council
High Cross, Madingley Road
Cambridge CB3 0ET, United Kingdom
kehu@bas.ac.uk

A.H.L. HUISKES
Unit for Polar Ecology
Netherlands Institute of Ecology

(NIOO-KNAW)
POB 140, 4400 AC Yerseke,
The Netherlands
a.huiskes@nioo.knaw.nl

B.B. HULL
Department of Environment
and Heritage
Australian Government Antarctic
Division
203 Channel Highway
Kingston, Tasmania 7050, Australia
bruce.hull@agad.gov.au

E. KAUP
Institute of Geology at Tallinn
University of Technology, Estonia pst 7
10143 Tallinn, Estonia
kaup@gi.ee

I. LAURION
Institut national de la recherche
scientifique
Centre Eau, Terre et Environnement
and Centre d'études nordiques
490 de la Couronne
Québec, G1K 9A9, Canada
Isabelle_Laurion@ete.inrs.ca

J. LAYBOURN-PARRY
Natural Sciences, Keele University
Keele, Straffordshire ST5 5BG, UK
j.laybourn-parry@natsci.keele.ac.uk

W.B. LYONS
Byrd Polar Research Center
The Ohio State University
Columbus, OH 43210-1002 USA
lyons.142@osu.edu

S. OTT
Botanisches Institut
Heinrich-Heine Universität Düsseldorf
Universitätsstr. 1
D-40225 Düsseldorf, Germany
otts@uni-duesseldorf.de

R. PIENITZ
Département de Géographie
and Centre d'études nordiques
Université Laval, Sainte-Foy
Québec, G1K 7P4, Canada
reinhard.pienitz@cen.ulaval.ca

J.C. PRISCU
Department of Land Resources and
Environmental Sciences
Montana State University
Bozeman, MT 59717 USA
jpriscu@montana.edu

A. QUESADA
Departamento de Biología
Universidad Autónoma de Madrid
28049 Madrid, Spain
antonio.quesada@uam.es

S. ROBINSON
Institute for Conservation Biology
University of Wollongong
Northfields Avenue
Wollongong, NSW 2522, Australia
sharonr@uow.edu.au

P.M. SELKIRK
Department of Biological Sciences
Macquarie University
Sydney, NSW 2109,Australia
pselkirk@rna.bio.mq.edu.au

M.L. SKOTNICKI
Genomic Interactions Group
Research School of Biological Sciences
Institute of Advanced Studies
Australian National University
Canberra, ACT 2601, Australia
skotnicki@rsbs.anu.edu.au

M.I. STEVENS
Allan Wilson Centre for Molecular
Ecology and Evolution, Massey
University
Private Bag 11-222
Palmerston North, New Zealand &
Department of Genetics, La Trobe
University
Bundoora, 3083 Victoria, Australia
M.I.Stevens@massey.ac.nz

A. TATON
Centre d'ingénierie des protéines
Institut de Chimie B6, Université
de Liège
B-4000 Liège, Belgium
A.Taton@student.ulg.ac.be

W.F. VINCENT
Département de Biologie
and Centre d'études nordiques
Université Laval, Sainte-Foy
Québec, G1K 7P4, Canada
warwick.vincent@bio.ulaval.ca

B. VAN DE VIJVER
National Botanic Garden of Belgium
Department of Cryptogamy
Domein van Bouchout
B-1860 Meise, Belgium
vandevijver@br.fgov.be

J. WASLEY
Department of Environment
and Heritage
Australian Government Antarctic
Division
203 Channel Highway
Kingston, Tasmania 7050, Australia
jane.wasley@agad.gov.au

K.A. WELCH
Byrd Polar Research Center
The Ohio State University
Columbus, OH 43210-1002 USA
welch.189@osu.edu

J. WHINAM
Biodiversity Conservation Branch
Department of Primary Industries
and Water
GPO Box 44, Hobart, Tasmania
Jennie.Whinam@dpiw.tas.gov.au

A. WILMOTTE
Centre d'ingénierie des protéines
Institut de Chimie B6, Université
de Liège
B-4000 Liège, Belgium
awilmotte@ulg.ac.be

1. TRENDS IN ANTARCTIC TERRESTRIAL AND LIMNETIC ECOSYSTEMS: ANTARCTICA AS A GLOBAL INDICATOR

A.H.L. HUISKES
Unit for Polar Ecology
Netherlands Institute of Ecology (NIOO-KNAW)
POB 140, 4400 AC Yerseke, The Netherlands
a.huiskes@nioo.knaw.nl

P. CONVEY
British Antarctic Survey, Natural Environment Research Council
High Cross, Madingley Road
Cambridge CB3 0ET, United Kingdom
p.convey@bas.ac.uk

D. M. BERGSTROM
Department of Environment and Heritage
Australian Government Antarctic Division
203 Channel Highway
Kingston, Tasmania 7050, Australia
dana.bergstrom@agad.gov.au

Introduction

The Antarctic provides a suite of environments and scenarios that give key opportunities to improve understanding of the consequences of climate change on terrestrial and limnetic biota. These result both from the considerable differences in the rates of contemporary change in the region, and the marked natural environmental gradients present in terrestrial and limnetic ecosystems. Antarctic communities range from polar deserts and permanently frozen lakes to lush,

1

D.M. Bergstrom et al. (eds.), Trends in Antarctic Terrestrial and Limnetic Ecosystems, 1–13.
© 2006 *Springer.*

eutrophic grasslands and nutrient-enriched ponds. Such gradients, including latitudinal, altitudinal and environmental examples, provide useful analogues for the predictions of future change trajectories. Regional differences in the nature of the changes currently experienced add a further, useful, layer of complexity. For instance, strong seasonal differences in the rate of temperature change are seen in the Antarctic Peninsula region, with autumn and winter temperatures having increased substantially more than those in the spring and summer. Precipitation rates in the Peninsula area are also predicted to increase, whereas they have been decreasing dramatically on several subantarctic islands.

The international research programme RiSCC (Regional Sensitivity to Climate Change in Antarctic Terrestrial and Limnetic Ecosystems), sponsored by the Scientific Committee on Antarctic Research, has been investigating these scenarios with the goals of (1) understanding the likely response of Antarctic biotas to changing climates and (2) contributing to the development of broadly applicable theory applying to interactions among climate change, indigenous and introduced species and ecosystem functioning. The purpose of this volume is to provide a picture of the current state of knowledge of the Antarctic terrestrial and limnetic ecosystems and a synthesis of the known and likely effects of climate change on Antarctic terrestrial and limnetic ecosystems, based on data gathered by the RiSCC programme and from the wider literature and, thereby, to contribute to their management and conservation. In doing so, we also highlight the global significance of developing understanding of climate change consequences in the Antarctic. Rates of change currently seen in this region are amongst the greatest documented worldwide, while overall knowledge of the relatively simple ecosystems is often more developed, allowing the region to act as a "canary in the coalmine" for planet Earth (Convey et al. 2003).

Antarctic opportunities

Antarctica[1] has long been described as the most remote continent on Earth, the last true wilderness, unspoilt by humans, and isolated from external environmental influences by the Polar Vortex in the upper atmosphere and the Antarctic Polar Front in the Southern Ocean. To the uninitiated, Antarctica is a hostile place, cold and windy, covered in ice and inhabited only by seabirds and mammals. It is familiar to most through media reports and television programmes, made by those few who venture there.

However, in the last few decades scientists have come to appreciate that the continent is fundamental to global processes. Furthermore, the continental icecap, averaging over 2km and up to 5km deep, contains an unique archive of the atmospheric composition at least over the last 800 000 years (EPICA 2004), trapped in air bubbles in the ice. This archive is of vital importance in the reconstruction of

---

[1] We use the term 'Antarctica' to refer to the continent in a geographical sense, and "the Antarctic" in a biogeographical sense, ie including the subantarctic islands.

climate evolution and climate fluctuation and allows the reconstruction of past climate fluctuations and refining of scenarios for future climate change. On even longer timescales, the geology of the continent provides information on past climate evolution and on the causes, mechanisms and timescales of the fragmentation of the supercontinent Gondwana.

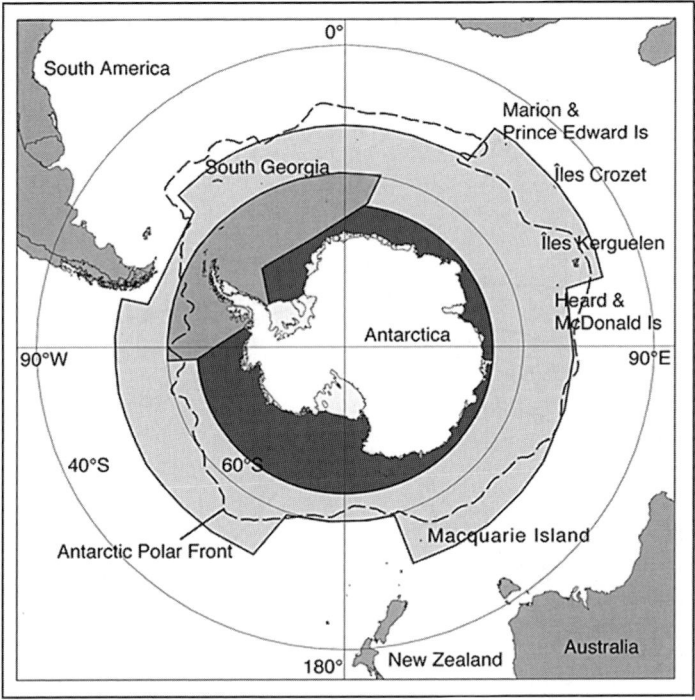

Figure 1. Map of Antarctica with the surrounding subantarctic islands. Dark shading indicates continental Antarctica, medium shading the maritime Antarctic and light shading the subantarctic region.

Looking inwards towards the Antarctic, external influences on the Antarctic environment and its biome have long been considered negligible or at most only of local importance. As late as 1985 Phillpot advocated "...caution in suggesting the possibility of 'climatic change'..." in connection with Antarctica (Phillpot 1985). Twenty years later, we are fully aware of the fact that Antarctica is not the remote continent we once thought. Anthropogenic chlorofluorocarbons (CFCs) and CO_2 and other greenhouse gases, released into the atmosphere mainly in the Northern Hemisphere, exert a profound influence on the global atmospheric environment including that of Antarctica, giving rise respectively to stratospheric ozone depletion and the Antarctic ozone hole and to global climatic warming, seen at its greatest rate in the region of the Antarctic Peninsula.

To biologists, the Antarctic has always been an important natural laboratory for the study of life under extreme conditions. The terrestrial and limnetic biota may experience drastic environmental contrasts on timescales varying from seasonal to diurnal and spatial scales varying between millimetres and thousands of kilometres - from prolonged darkness in winter to continuous daylight in summer, from a dark and isolated hypolimnion in winter to a stratified lake in summer; from nutrient poor circumstances in isolated nunataks or proglacial lakes to extremely eutrophic situations at the edges of vertebrate colonies and in lakes frequented by seals, from moist subzero circumstances on rock outcrops in the early morning hours, to dry 30°C+ situations later in the day. These environmental circumstances differ profoundly across the Antarctic, which results in significant differences in biogeography in the Antarctic biome.

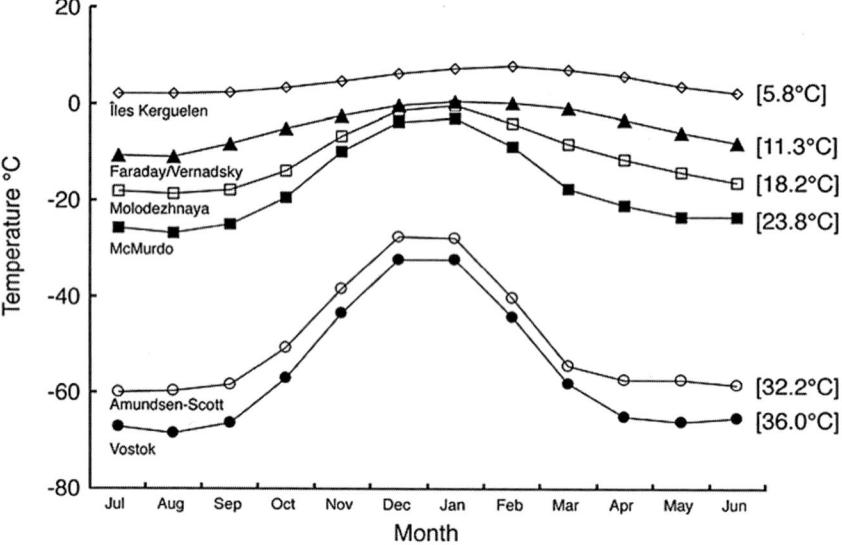

Figure 2. Mean monthly temperatures and annual ranges (°C) at selected Antarctic research stations. Iles Kerguelen represent the subantarctic annual temperature range, Faraday/Vernadsky the maritime Antarctic annual temperature range and remaining stations the continental Antarctic range. There is a clear contrast between the temperature ranges seen at coastal stations in the continental Antarctic (Molodezhnaya and McMurdo) and that of stations on the high plateau (Amundsen-Scott and Vostok) (data from Stonehouse 1989).

Biogeographic zonation

Commonly, three biogeographical zones are distinguished in the Antarctic (Smith 1984), referred to as the subantarctic, the maritime Antarctic and the continental

Antarctic zones (Fig. 1). These zones differ considerably from each other in climatic conditions and, consequently, in species diversity (Smith 1984, Convey 2001b, Chown and Convey this volume, Gibson et al. this volume).

Table 1. Number of native terrestrial invertebrate taxa reported from the three Antarctic biogeographical zones (updated from Convey 2001b); note that inadequate data are available for many smaller groups, both from a lack of taxonomic knowledge and incomplete survey coverage. * Protozoan taxa for the Subantarctic and maritime Antarctic combined.

Group	Subantarctic	Maritime Antarctic	Continental Antarctic
Protozoa		83 *	33
Rotifera	>29	>48	13
Tardigrada	>29	24	18
Nematoda	22	29	14
Annelida (Oligochaeta)	23	3	0
Mollusca	5	0	0
Crustacea (terrestrial)	6	0	0
Crustacea (non-marine but including meromictic lakes)	48	11	11
Insecta (total)	210	35	49
Collembola	>92	12	12
Mallophaga	61	25	34
Diptera	44	2	0
Coleoptera	40	0	0
Arachnida (total)	167	36	29
Araneida	20	0	0
Acarina	140	36	29
Myriapoda	3	0	0

The subantarctic zone comprises a series of remote islands and small archipelagos encircling the continent, and includes South Georgia, Marion and Prince Edwards Islands, Iles Crozet, Iles Kerguelen, Heard and McDonald Islands and Macquarie Island. Temperatures in the subantarctic are on average above freezing point year-round (Fig. 2) and precipitation is between 2000 and 3000mm per year (Gremmen 1981). These islands have in common that they are close to or just north of the Antarctic Polar Front (Fig. 1). Large and globally significant populations of seabirds and mammals breed on almost all these islands, contributing significantly to the input of nutrients. With the exception of two duck and one passerine species, indigenous terrestrial vertebrates are not present and the terrestrial fauna consists of invertebrate species (Convey 2001b, Table 1). The subantarctic flora comprises species of nearly all major taxonomic divisions (Stonehouse 1989, Table 2), although is distinctive from that of northern polar and tundra regions through the lack of woody plants and insect-pollinated flowering species.

The maritime Antarctic comprises the western coastal region of the Antarctic Peninsula to about 72°S (Smith 1996), South Shetland, South Orkney and South Sandwich archipelagos and the isolated islands of Bouvetøya and Peter I Øy. The seasonal differences in temperatures experienced are greater than in the subantarctic (Fig. 2), with summer temperatures on average around or slightly above freezing point for two to four months and mean annual precipitation is 400-500mm in the northern maritime Antarctic, decreasing to 50-100mm in the southern Antarctic Peninsula (Walton 1984, Øvstedal and Smith 2001). In summer, there is considerable seasonal snowmelt, which is directly responsible for most liquid water availability in terrestrial ecosystems. Nutrients originate from several sources, including marine vertebrates breeding, moulting or resting on land, transport in melt water, from sea spray aerosols carried inland by wind action and, to a smaller extent, from decomposition of organic material and rock weathering.

Table 2. Number of native plant species in the three biogeographical zones of the Antarctic (after Stonehouse 1989 and Convey 2001b).

Locality	Macrofungi	Lichens	Mosses	Liverworts	Ferns	Angiosperms
Subantarctic	70	>300	250	150	16	56
Maritime Antarctic	30	150	75	25	0	2
Continental Antarctic	0	125	30	1	0	0

Compared to the subantarctic, biodiversity is much reduced. The number of species in the various groups of invertebrates is significantly lower, except for the Protozoa, Rotifera, Nematoda and Tardigrada (Table 1). However, some of these apparent differences are likely merely to reflect the current lack of taxonomic and distributional data available across the Antarctic for these groups, rather than underlying biogeographical features. With respect to the flora, knowledge of species diversity is much more comprehensive (except, perhaps, the macrofungi). Again, species diversity is much lower in the maritime Antarctic - there are only two native angiosperms species (*Deschampsia antarctica* (Poaceae) and *Colobanthus quitensis* (Caryophyllaceae), no ferns and the numbers of lichen, moss, liverwort and macrofungal species are less then 50% of those found in the subantarctic (Table 2). Unlike the isolate island nature of the subantarctic, the maritime Antarctic is an integral element of the Antarctic continent. Although small in area in comparison with the continental Antarctic, it is the region of the continent in which most biodiversity is found.

In terms of area, the continental Antarctic is by far the largest biogeographical zone of the Antarctic biome. It comprises the entire landmass of the continent with the exception of the west coast of the Antarctic Peninsula north of 72°S and the

Balleny Islands. Although the coastal area of the continental Antarctic has a relatively milder climate than the interior of the continent, comparable with that of much of the maritime Antarctic, even in the coastal areas the temperature rarely rises above freezing and then only for short periods with positive mean monthly temperatures experienced for 0-1 months of the year (Walton 1984, Convey 1996a). Annual precipitation is 30-70 mm (Fogg 1998). At inland locations such as exposed nunatak summits and ablation areas such as the Victoria Land Dry Valleys, air temperatures remain permanently and often considerably below zero, while precipitation may be negligible with much of the continental area technically a frigid desert. Nevertheless, microhabitat warming allows liquid water to become available intermittently even in the most southern terrestrial habitats of the Transantarctic Mountains at 84 - 86ºS. The flora and fauna of the continental Antarctic has to be able to function by making use of short and relatively unpredictable windows of opportunity when (microhabitat) temperature are above physiological and biochemical thresholds and moisture becomes available. Only those organisms occur in this zone that are able to withstand prolonged periods of cold and desiccation stress. In the most extreme circumstances, no visible vegetation can be found, with life being limited to endolithic communities of algae, cyanobacteria, fungi, bacteria and lichens that survive in porous rocks such as certain sandstones (Friedmann 1982, 1993).

Key environmental factors

Solar radiation, temperature and moisture availability are key environmental variables governing most processes in terrestrial ecosystems, not least the physiological activity and life history of the individual constituent organisms (Block 1984, Kennedy 1993, Convey 1996b). Together, these physical factors also exert a controlling influence on the release and availability of nutrients. Irradiation may be further modulated by the dynamics of ice and snow cover – the importance of this is clear in limnetic ecosystems (Quesada et al. this volume) but also applies, often on a much smaller physical scale, in the terrestrial environment (Cockell et al. 2002, Arnold et al. 2003). The patterns of variation experienced in these factors, their interactions, optimum levels required in the context of biological activity and the thresholds at which organisms are still able to function are all important determinants of the structure and functioning of the ecosystem as a whole. Temperature (intimately linked with water availability) and solar radiation are the driving forces. Any variation in these factors is likely to have clear and identifiable consequences for Antarctic terrestrial and limnetic ecosystems, with little chance of these being obscured or buffered by the complicated inter-specific relationships existing in the more elaborate ecosystems typical of lower latitudes (Convey this volume).

Isolation

The Antarctic biome has become isolated from biomes at lower latitudes, both because of its obvious geographical isolation from the nearest Southern Hemisphere continents, the existence of the atmospheric Polar Vortex and oceanic Antarctic Polar Front and its history of almost complete glaciation during periods of glacial maxima (Convey 2001a, b, Clarke et al. 2005, Bergstrom et al. this volume). Furthermore, Antarctic terrestrial ecosystems can only be present in areas that are seasonally or permanently snow- and ice-free. These are predominantly present in the coastal zone and at inland nunataks and other oases of exposed ground. These areas, separated on scales of metres to at least hundreds of kilometres by hostile environments of permanent snow, ice or ocean are effectively "islands" of habitat. This isolation has immediate consequences for evolutionary and population processes and genetics, through inducing the isolation of species and populations (Bergstrom et al. this volume, Skotnicki and Selkirk this volume, Stevens and Hogg this volume). Colonisation of newly formed snow- and ice-free areas, once of the most obvious predicted consequences of climate change processes, is dependent largely on the composition of adjacent communities (Hughes et al. this volume, Skotnicki and Selkirk this volume). Moreover, direct human impact on the ecosystems is currently relatively minor, the subantarctic excepted (Frenot et al. 2005, Convey et al. this volume).

Global atmospheric change and the Antarctic climate

The anthropogenic release of chlorofluorocarbons (CFCs) used in various industrial and domestic processes and appliances into the atmosphere, mainly in the Northern Hemisphere, resulted in a decrease in the stratospheric ozone concentration, as CFCs chemically react with ozone under the influence of sunlight. As CFCs concentrate in the atmosphere of the higher latitudes during periods of darkness, a rapid decrease in stratospheric ozone concentration occurs in spring, when the sun comes above the horizon again. The resulting 'ozone hole' was first described by Farman et al. (1985). Since then, in most years the Antarctic ozone hole has become 'deeper' (more depleted), wider and more prolonged, and any improvements consequential on the adoption of the internationally agreed Montreal Protocol aimed at reducing CFC emissions have yet to become apparent. The biological consequences of ozone depletion, to which the Antarctic is specifically exposed through the general position of the ozone hole being determined by atmospheric circulation processes, are described by Hennion et al. (this volume).

Since the start of modern industrialisation in the Northern Hemisphere in the 19th Century, large amounts of infrared-absorbing gases have been released into the atmosphere. These so-called greenhouse gases have induced a general increase in mean annual atmospheric temperatures, and subsequently in the annual mean surface temperatures and most recently in sea temperatures. The present view is that a global mean temperature increase of approximately 2°C and attendant changes in precipitation regimes, are realistic expectations over the next 50 years (IPCC 2001). Moreover, because large-scale extinctions and movements of species ranges are known to have resulted from major climate changes in the past (Coope 1995, Roy et al. 1996), an ability to predict the consequences to biodiversity of climate change is becoming increasingly important (Mc Neeley et al. 1995).

Over the last 50 years, air temperatures and moisture regimes in parts of the Antarctic have changed markedly (Frenot et al. 1997 Weller 1998, Smith et al. 1996, King and Harangozo 1998, King et al. 2003, Convey this volume), with most recently parallel changes in near-surface sea temperatures being reported (Meredith and King 2005), and the indications are that these trends will continue. This increase is especially apparent in the Antarctic Peninsula region, where summer temperatures have increased since the 1950s by about 2°C and winter temperatures by about 5°C (King et al. 2003). However, in other areas of the Antarctic, temperature increases have been much lower and some areas have recorded regional cooling (Turner et al. 2005).

Invasive species

Human trade and movement across the planet have led to accidental or deliberate transport and introductions of virtually all groups of biota. Until the last two centuries or so, Antarctica's isolation has protected it from such invasions, but this protection has been rapidly overcome, particularly in the Subantarctic (Frenot et al. 2005, Convey et al. this volume). Here, deliberate introductions of vertebrates include reindeer, cattle, sheep, cats and rabbits, while rats and mice were introduced inadvertently. Such introductions have a profound influence on the native biota (eg Vogel et al. 1984, Chapuis et al. 1994, Smith 1996, Chown and Block 1997, Chown et al. 1998). A range of plant species have been introduced both deliberately and inadvertently to almost every subantarctic island, while many reports also exist on the introduction of invertebrate species (Frenot et al. 2005, Convey et al. this volume). About half of the recorded vascular flora of South Georgia comprises plant species introduced from South America and Europe (Convey 2001a), with some of these displacing the native flora (Smith 1996). On Marion Island, the native flora is out-competed by invasive species such as *Agrostis stolonifera* (Gremmen 1997, Gremmen and Smith 1999, Convey et al. this volume). However, establishment of alien species to the maritime Antarctic are few to date, including a dipteran and an enchytraeid worm known to have been introduced to Signy Island (Convey and Block 1996) and the grass *Poa pratensis* to the northern Antarctic Peninsula (Smith 1996, Convey et al. this volume).

While introductions are not of necessity directly connected with climate change, any increase in temperature, especially in connection with an increase in water availability, may both lower the barriers to the long distance transport stage of the colonisation process and provide opportunities for invading species to establish after arrival and survive better the adverse season, allowing them subsequently to expand at the expense of the native fauna and flora (Bergstrom and Chown 1999, Frenot et al. 2005). Increased human activity and freedom to travel separately further lowers these barriers, through bypassing the need for survival of long distance atmospheric or oceanic transport.

Responses of the biota

In many areas of the Antarctic, the biota has already shown pronounced direct and indirect responses to both climate change and the invasions of alien species (Chown and Smith 1993, Fowbert and Smith 1994, Kennedy 1995, Ellis-Evans et al. 1997, Frenot et al. 2005). Currently processes in both terrestrial (Convey this volume) and in limnetic ecosystems (Lyons et al. this volume) and the links between the two (Quesada et al. this volume) are being studied by many groups of Antarctic ecologists.

Custodians

Humanity's role in the Antarctic has changed since the early explorers' visits to the region, progressing from that of an explorer and hunter to that of an investigator and, increasingly, as a custodian. Uniquely on the planet, people's activities in Antarctica are governed by international agreements under the Antarctic Treaty System. Within this System, many regulations have been agreed by the various member states over the last decades, amongst which are clear recognition of the fragility of the Antarctic and stipulation on the need and means for protection. Understanding the effects of climate change on the Antarctic environment has fundamental implications for the way in which we protect this environment (Whinam et al. this volume, Hull and Bergstrom this volume). Our increasing knowledge of the region and its biome only serves to emphasise that this exceptional environment must be treated with care, not only as part of our legacy for future generations but also because the Antarctic is intricately linked with, and plays a fundamental role in, global processes and, hence, the environment we ourselves inhabit.

References

Arnold, R.J., Convey, P., Hughes, K.A. and Wynn-Williams, D.D. (2003) Seasonal periodicity of physical factors, inorganic nutrients and microalgae in Antarctic fellfields, *Polar Biology* **26**, 396-403.

Bergstrom, D. and Chown, S.L. (1999) Life at the front: history, ecology and change on Southern Ocean islands, *Trends in Ecology and Evolution* **14**, 472-477.

Bergstrom, D.M., Hodgson, D.A. and Convey, P. (2006) The physical setting of the Antarctic, in D.M. Bergstrom, P. Convey, and A.H.L. Huiskes (eds.), *Trends in Antarctic Terrestrial and Limnetic Ecosystems: Antarctica as a Global Indicator*, Springer, Dordrecht (this volume).

Block, W. (1984) Terrestrial Microbiology, Invertebrates and Ecosystems, in R.M. Laws (ed.), *Antarctic Ecology Volume 1*. Academic Press, London, pp. 163-236.

Chapuis, J.L., Bousses, P. and Barnard, G. (1994) Alien mammals, impact and management in the French subantarctic islands, *Biological Conservation* **67**, 97-104.

Chown, S.L. and Block, W. (1997) Comparative nutritional ecology of grass-feeding in a sub-Antarctic beetle: the impact of introduced species on *Hydromedion sparsutum* from South Georgia, *Oecologia* **111**, 216-224.

Chown, S.L. and Convey, P. (2006) Biogeography, in D.M. Bergstrom, P. Convey, and A.H.L. Huiskes (eds.), *Trends in Antarctic Terrestrial and Limnetic Ecosystems: Antarctica as a Global Indicator*. Springer, Dordrecht (this volume).

Chown, S.L. and Smith, V.R. (1993) Climate change and the short-term impact of feral house mice at the sub-antarctic Prince Edward Islands, *Oecologica* **96**, 508-518.

Chown, S.L., Gremmen, N.J.M. and Gaston, K.J. (1998) Ecological biogeography of Southern Ocean islands: Species-area relationships, human impacts, and conservation, *American Naturalist* **152**, 562-575.

Clarke, A., Barnes, D.K.A. and Hodgson, D.A. (2005) How isolated is Antarctica? *Trends in Ecology and Evolution* **20**, 1-3.

Cockell, C.S., Rettberg, P., Horneck, G., Wynn-Williams, D.D., Scherer, K. and Gugg-Helminger, A. (2002) Influence of ice and snow covers on the UV exposure of terrestrial microbial communities: dosimetric studies, *Journal of Photochemistry and Photobiology B: Biology* **68**, 23-32.

Convey P. (1996a) Overwintering strategies of terrestrial invertebrates in Antarctica - the significance of flexibility in extremely seasonal environments, *European Journal of Entomology* **93**, 489-505.

Convey, P. (1996b) The influence of environmental characteristics on life history attributes of Antarctic terrestrial biota, *Biological Reviews* **71**, 191-225.

Convey, P. (2001a) Terrestrial ecosystem response to climate changes in the Antarctic. In G.-R. Walther, C.A. Burga and P.J. Edwards (eds.), *"Fingerprints" of climate change - adapted behaviour and shifting species ranges*, Kluwer, New York, pp 17-42.

Convey, P. (2001b) Antarctic Ecosystems, in Levin, S. (ed.), *Encyclopedia of Biodiversity* Vol. 1. Elsevier/Academic Press, San Diego, pp. 171-184.

Convey, P. (2006) Antarctic climate change and its influences on terrestrial ecosystems, in D.M. Bergstrom, P. Convey, and A.H.L. Huiskes, (eds.), *Trends in Antarctic Terrestrial and Limnetic Ecosystems: Antarctica as a Global Indicator*, Springer, Dordrecht (this volume).

Convey, P. and Block, W. (1996) Antarctic Diptera: Ecology, physiology and distribution, *European Journal of Entomology* **93**, 1-13.

Convey, P., Scott, D. and Fraser, W.R. (2003) Biophysical and habitat changes in response to climate alteration in the Arctic and Antarctic, in T.E. Lovejoy and L. Hannah (eds.), *Climate Change and Biodiversity: Synergistic Impacts, International Conservation Center for Applied Biodiversity Science, Advances in Applied Biodioversity Science* **4**, pp. 79-84.

Convey, P., Frenot, Y., Gremmen, N. and Bergstrom, D.M. (2006) Biological invasions, in D.M. Bergstrom, P. Convey, and A.H.L. Huiskes (eds.), *Trends in Antarctic Terrestrial and Limnetic Ecosystems: Antarctica as a Global Indicator*, Springer, Dordrecht (this volume).

Coope, G.R. (1995) Insect faunas in ice age environments: why so little extinction? In, J.H. Lawton and R.M. May (eds.), *Extinction Rates*, Oxford University Press, Oxford, pp. 55-74.

Ellis-Evans, J.C., Laybourn-Parry, J., Bayliss, P. and Perriss, S. (1997) Human impact on an oligotrophic lake in the Larsemann Hills, in B. Battaglia, J. Valencia, and D.W.H. Walton (eds.), *Antarctic Communities: Species, Structure and Survival*, Cambridge University Press, Cambridge, pp. 396-404.

EPICA, (2004) Eight glacial cycles from an Antarctic ice core, *Nature,* **429**, 623-628.

Farman, J.C., Gardiner, B.G. and Shanklin, J.D. (1985) Large losses of total ozone in Antarctica reveal seasonal ClO_x/NO_x interaction, *Nature* **315**, 207-210.

Fogg, G.E. (1998) *The Biology of Polar Habitats,* Oxford University Press, Oxford, 263 pp.

Fowbert, J.A. and Smith, R.I.L. (1994) Rapid population increases in native vascular plants in the Argentine Islands, Antarctic Peninsula, *Arctic and Alpine Research* **26**, 290-296.

Frenot, Y., Gloaguen, J.-C., and Tréhen, P. (1997b) Climate change in Kerguelen Islands and colonization of recently deglaciated areas by *Poa kerguelensis* and *P. annua* in B. Battaglia, J. Valencia and D.W.H. Walton (eds.), *Antarctic Communities: Species, Structure and Survival*, Cambridge University Press, Cambridge, UK, pp. 358-366.

Frenot, Y., Chown, S.L., Whinam, J., Selkirk, P., Convey, P., Skotnicki, M. and Bergstrom, D. (2005) Biological invasions in the Antarctic: extent, impacts and implications, *Biological Reviews* **80**, 45-72.

Friedmann, E.I. (1982) Endolithic microorganisms in the Antarctic cold desert, *Science* **215**, 1045-1053.

Friedmann, E.I. (1993) *Antarctic Microbiology*, John Wiley and Sons, New York, 634 pp.

Gibson, J.A.E., Wilmotte, A., Taton, A., Van De Vijver, B., Beyens, L. and Dartnall, H.J.G (2006) Biogeographic trends in Antarctic lake communities, in D.M. Bergstrom, P. Convey, and A.H.L. Huiskes, (eds.) *Trends in Antarctic Terrestrial and Limnetic Ecosystems: Antarctica as a Global Indicator,* Springer, Dordrecht (this volume).

Gremmen, N.J.M. (1981) *The vegetation of the Subantarctic islands Marion and Prince Edward*, Dr. W. Junk Publishers, The Hague, 149 pp.

Gremmen, N.J.M. (1997) Changes in the vegetation of subantarctic Marion Island resulting from introduced vascular plants, in B. Battaglia, J. Valencia and D.W.H. Walton (eds.), *Antarctic Communities: Species, Structure and Survival,* Cambridge University Press, Cambridge, pp. 417-423.

Gremmen, N.J.M. and Smith, V.R. (1999) New records of alien vascular plants from Marion and Prince Edward Islands, sub-Antarctic, *Polar Biology* **21**, 401-409.

Hennion, F., Huiskes, A.H.L., Robinson, S. and Convey, P. (2006) Physiological traits of organisms in a changing environment, in D.M. Bergstrom, P. Convey, and A.H.L. Huiskes (eds.), *Trends in Antarctic Terrestrial and Limnetic Ecosystems: Antarctica as a Global Indicator*, Springer, Dordrecht (this volume).

Hughes, K.A., Ott, S., Bölter, M. and Convey, P. (2006) Colonisation processes, in D.M. Bergstrom, P. Convey, and A.H.L. Huiskes (eds.), *Trends in Antarctic Terrestrial and Limnetic Ecosystems: Antarctica as a Global Indicator*, Springer, Dordrecht (this volume).

Hull, B.B. and Bergstrom, D.M. (2006) Antarctic terrestrial and limnetic conservation and management, in D.M. Bergstrom, P. Convey, and A.H.L. Huiskes (eds.), *Trends in Antarctic Terrestrial and Limnetic Ecosystems: Antarctica as a Global Indicator*, Springer, Dordrecht (this volume).

Inter-governmental Panel on Climate Change (IPCC) (2001) *Climate change 2001: impacts, adaptation and vulnerability*, in J.J. McCarthy, O.F. Canziani, N.A. Leary, D.J. Dokken and K.S. White (eds.), Contribution of Working Group 2 to the Third Assessment Report of the IPCC, Cambridge University Press, Cambridge, UK.

Kennedy, A.D. (1993) Water as a limiting factor in the Antarctic terrestrial environment: a biogeographical synthesis, *Arctic and Alpine Research* **25**, 308-315.

Kennedy, A.D. (1995) Antarctic terrestrial ecosystem response to global environmental change, *Annual Review of Ecology and Systematics* **26**, 683-704.

King, J.C. and Harangozo, S.A. (1998) Climate change in the western Antarctic Peninsula since 1945: observations and possible causes, *Annals of Glaciology* **27**, 571-575.

King, J.C., Turner, J., Marshall, G.J., Connally, W.M. and Lachlan-Cope, T.A. (2003) Antarctic Peninsula climate variability and its causes as revealed by analysis of instrumental records, in E. Domack, A. Burnett, A. Leventer, P. Convey, M. Kirby and R. Bindschadler (eds.), *Antarctic Peninsula Climate Variability: Historical and Palaeoenvironmental Perspectives, Antarctic Research Series,* Vol. 79, American Geophysical Union, Washington, D.C., pp. 17-30.

Longton, R.E. (1988) *The biology of polar bryophytes and lichens,* Cambridge University Press, Cambridge, 391 pp.

Lyons, W.B., Laybourn-Parry, J., Welch, K.A. and Priscu, J.C. (2006) Antarctic lake systems and climate change, in D.M. Bergstrom, P. Convey, and A.H.L. Huiskes (eds.), *Trends in Antarctic Terrestrial and Limnetic Ecosystems: Antarctica as a Global Indicator*, Springer, Dordrecht (this volume).

McNeely, J.A., Gadgil, M., Leveque, C. and Redford, K. (1995) Human influences on Biodiversity, in V.H. Heywood, and R.T. Watson (eds.), *Global Biodiversity Assessment,* Cambridge University Press, Cambridge, pp. 711- 821.

Meredith, M.P. and King, J.C. (2005) Rapid climate change in the ocean to the west of the Antarctic Peninsula during the second half of the 20th century, *Geophysical Research Letters* **32**, doi:10.1029/2005gl02042.

Øvstedal, D.O. and Lewis Smith, R.I. (2001) *Lichens of Antarctica and South Georgia A Guide to their Identification and Ecology,* Cambridge University Press, Cambridge, 411 pp.

Phillpot, H.R. (1985) Physical Geography – Climate, in W.N. Bonner, and D.W.H. Walton, (eds.) *Key Environments, Antarctica,* Pergamon Press, Oxford, pp. 23-38.

Quesada, A., Vincent, W.F. Kaup, E., Hobbie, J.E., Laurion, I., Pienitz, R., López-Martínez, J. and Durán, J.-J. (2006) Landscape control of high-latitude lakes in a changing climate, in D.M. Bergstrom, P. Convey, and A.H.L. Huiskes (eds.), *Trends in Antarctic Terrestrial and Limnetic Ecosystems: Antarctica as a Global Indicators*, Springer, Dordrecht (this volume).

Roy, K., Valentine, J.W. Jablonski, D. and Kidwell, S.M. (1996) Scales of climatic variability and time averaging in Pleistocene biotas: implications for ecology and evolution, *Trends in Ecology and Evolution* **11**, 458-462.

Skotnicki, M.L. and Selkirk, P.M. (2006) Plant biodiversity in an extreme environment: genetic studies of origins, diversity and evolution in the Antarctic, in D.M. Bergstrom, P. Convey, and A.H.L. Huiskes (eds.), *Trends in Antarctic Terrestrial and Limnetic Ecosystems: Antarctica as a Global Indicator.* Springer, Dordrecht (this volume).

Smith, R.C., Stammerjohn, S.E. and Baker, K.S. (1996) Surface air temperature variations in the Western Antarctic Peninsula region, *Antarctic Research Series* **70**, 105-122.

Smith, R.I.L. (1984) Terrestrial Plant Biology of the Subantarctic and the Antarctic, in R.M. Laws (ed.), *Antarctic Ecology* Volume 1, Academic Press, London, pp. 61-162.

Smith, R.I.L. (1996) Terrestrial and freshwater biotic components of the western Antarctic Peninsula, *Antarctic Research Series* **70**, 15-59.

Stevens, M.I. and Hogg, I.D. (2006) The molecular ecology of Antarctic terrestrial and limnetic invertebrates and microbes, in D.M. Bergstrom, P. Convey, and A.H.L. Huiskes (eds.), *Trends in Antarctic Terrestrial and Limnetic Ecosystems: Antarctica as a Global Indicators*, Springer, Dordrecht (this volume).

Stonehouse, B. (1989) *Polar Ecology,* Blackie, Glasgow, 222 pp.

Turner, J., Colwell, S.R., Marshall, G.J. Lachlan-Cope, T.A., Carleton, A.M., Jones, P.D., Lagun, V., Reid, P.A. and Iagovkina, S. (2005) Antarctic climate change during the last 50 years, *International Journal of Climatology* **25**, 279-294.

Vogel, M., Remmert, H. and Smith, R.I.L. (1984) Introduced reindeer and their effects on the vegetation and the epigeic invertebrate fauna of South Georgia (subantarctic), *Oecologia* **62**, 102-109.

Walton, D.W.H. (1984) The terrestrial environment, in R.M. Laws (ed), *Antarctic Ecology* Volume 1. Academic Press, London, pp 1-60.

Weller, G. (1998) Regional impacts of climate change in the Arctic and Antarctic, *Annals of Glaciology* **27**, 543-552.

Whinam, J. Copson, G. and Chapuis, J–L. (2006) Subantarctic terrestrial conservation and management, in D.M. Bergstrom, P. Convey, and A.H.L. Huiskes, (eds.) *Trends in Antarctic Terrestrial and Limnetic Ecosystems: Antarctica as a Global Indicator*, Springer, Dordrecht (this volume).

2. THE PHYSICAL SETTING OF THE ANTARCTIC

D. M. BERGSTROM
Department of Environment and Heritage
Australian Government Antarctic Division
203 Channel Highway
Kingston, Tasmania 7050, Australia
dana.bergstrom@agad.gov.au

D.A. HODGSON
British Antarctic Survey, Natural Environment Research Council
High Cross, Madingley Road
Cambridge CB3 0ET, United Kingdom
daho@bas.ac.uk

P. CONVEY
British Antarctic Survey, Natural Environment Research Council
High Cross, Madingley Road
Cambridge CB3 0ET, United Kingdom
p.convey@bas.ac.uk

Introduction

DEFINITION OF AREA COVERED BY THIS BOOK

The Antarctic terrestrial and freshwater biome examined here includes the main continental landmass, the maritime Antarctic including the Antarctic Peninsula and associated islands and archipelagos (South Shetland, South Orkney, South Sandwich Islands, Bouvetøya) and the subantarctic islands which lie on or about the Antarctic Polar Frontal Zone (PFZ), an oceanic and climate boundary where the Antarctic Circumpolar Current (ACC) meets warmer waters. These geographic regions are also meaningful biogeographical regions (see discussions in Skottsberg 1904,

D.M. Bergstrom et al. (eds.), Trends in Antarctic Terrestrial and Limnetic Ecosystems, 15–33.
© 2006 *Springer.*

Pickard and Seppelt 1984, Smith 1984, Longton 1988, Chown and Convey this volume, and see Fig. 1 in Huiskes et al. this volume), if still under refinement (Peat et al. in press) and provide useful platforms for the discussions below. The purpose of this chapter is to briefly review the environmental factors, both past and present, that are or have been evolutionary forcing variables influencing the development of the present-day biological diversity of Antarctica.

THE FORMATION OF THE ANTARCTIC CONTINENT

A sensible place to begin an examination of the influence of the physical environment on past biodiversity is some 100 million years ago (MA) when Antarctica was part of the supercontinent Gondwana and the subantarctic islands were non existent. During the early Cenozoic (>100-60MA), Gondwana began to fragment and the first major oceanic barriers to the movement of terrestrial species were formed (McLoughlin 2001). Antarctica, one of the elements of the former Gondwana supercontinent, drifted over the southern polar region. The continent's high latitude location, inevitably linked with seasonal periods of complete darkness, did not immediately lead to massive extinction of terrestrial fauna and flora, which remained typical of south–temperate rainforest regions for a long period subsequently (Feldmann and Woodburne 1988, Clarke and Crame 1989, Poole and Cantrill 2001, Francis and Poole 2002). Even after the commencement of ice sheet formation, the Antarctic experienced periods when this biota could show local expansion, until as recently as 8-10MA.

In the early Miocene, Antarctica finally became fully isolated from the Gondwana supercontinent with the full opening and deepening of the Drake Passage (28-23MA) and the separation of the Tasman Rise from Antarctica (33.5MA) (Livermore et al. 2005, Scher and Martin 2006). These tectonic processes resulted not only in the elimination of the last land-bridge connections with lower and more temperate latitudes but also in the onset of the deep-water circulation around the Antarctic continent of the Antarctic Circumpolar Current (ACC). The ACC eventually resulted in the isolation of a cool body of water (the Southern Ocean) and the establishment of the Polar Frontal Zone (PFZ). This PFZ further isolated Antarctica, its outlying archipelagos and the Southern Ocean from other continents and oceans, climatically, thermally and oceanographically and set the stage for the development of distinct Antarctic biological communities adapted for survival in the southern polar region (Barnes et al. 2006).

These movements of the Antarctic continental plate were associated with many periods of volcanism. The plate is encircled by divergent plate boundaries along roughly 95% of its perimeter and is broken internally by numerous rift-structures suggesting a plate-wide extensional tectonic regime. Within this environment, the Antarctic Plate evolved into one of the great alkaline volcanic provinces of the world (LeMasurier and Thompson 1990). Volcanism has been and continues to be involved in the formation of the many of the subantarctic and Antarctic islands. Some of these consist of volcanic rocks overlying continental basement rocks, such as the convergent plate margin volcanoes of the South Sandwich Islands. Iles Kerguelen and the Heard and McDonald Islands which, although they specifically have lacustrine bases, are part of the mostly underwater Kerguelen Plateau, which contains Gondwana fragments overlain by more recent volcanic rocks (Gladczenko and Coffin 2001). Marion and Prince Edward Islands and Iles Crozet are shield volcanoes, erupting 0.11-0.21MA and 0.2-9MA (LeMasurier and Thomson 1990).

South Georgia however, contains continental elements whilst Macquarie Island is the aerial portion of a ophiolite complex, a piece of largely intact seafloor at the junction of the Australasian and Pacific tectonic plates, emerging from the sea approximately 0.6MA (Adamson et al. 1995). The Antipodes, Auckland and Campbell Islands lie on the Campbell Plateau, underlain by continental crust that was part of Antarctica in pre-Cenozoic time (McLoughlin 2001).

The majority of subantarctic and Antarctic islands are substantially younger than the Antarctic continent with the oldest, Iles Kerguelen, only 39MA. There are at least 16 Antarctic and subantarctic volcanoes that are known to be active (including subantarctic Heard Island, most of the maritime Antarctic South Sandwich Islands, Deception Island and Bouvetøya, and Mounts Erebus, Melbourne and Rittman in continental Antarctic Victoria Land) and a further 32 suspected of Holocene activity (LeMasurier and Thompson 1990, Convey et al. 2000a, Fitch et al. 2001, Anon 2005).

The continental ice sheet

The Earth was in a state of extreme global warmth from the Cretaceous (144-65MA) to the early Eocene (c. 55MA). However, by the middle to late Eocene (42MA), there were a series of several small glaciations and one major transient glaciation of the Antarctic continent. These abrupt climate reversals were possibly associated with the first opening of the Drake Passage at 41MA (Scher and Martin 2006). Studies of the oxygen isotope composition of marine calcite suggest that the greenhouse to icehouse transition was closely coupled to the evolution of atmospheric carbon dioxide, and that negative carbon cycle feedbacks may have initially prevented the permanent establishment of large ice sheets (Tripati et al. 2005). However, by 34MA, at the Eocene-Oligocene transition, large ice sheets appeared on the Antarctic continent, evidenced by decreasing atmospheric carbon dioxide concentrations and a deepening of the calcite compensation depth in the world's oceans, coinciding with changes in seawater oxygen isotope ratios (glaciation in the Northern Hemisphere began much later, between 10 and 6MA). There are two theories why the ice sheet formed at this time. The first, involves the separation of Antarctica from the Australian and South American continents and the opening of the ocean gateways that allowed the establishment of the east to west flow of the circumpolar currents in the Southern Ocean (Kennett 1977). This led to the thermal isolation of the continent, cooling and the formation of sea ice and the continental ice sheet. The second theory puts more emphasis on the role of global atmospheric CO_2, orbital forcing and ice-climate feedbacks, with the opening of the Southern Ocean gateways playing a secondary role (DeConto and Pollard 2003). The rapid decrease in CO_2 at about 34MA brought with it a decrease in temperature sufficient for viable snow and ice to remain present throughout the year. After 15MA, a further cooling is believed to have caused the transition from an ephemeral to a permanent Antarctic ice sheet (Barret 2003). However, since its formation, this ice sheet has been far from stable and over time has exerted strong physical controls on where biological communities have survived and hence underpins many of the biogeographical patterns seen today.

During the most recent geological period, the Quaternary, that spans approximately the last 2MA, the polar ice sheets developed their characteristic cycle of slow build up to full glacial conditions, followed by rapid ice melting and deglaciation to interglacial conditions (Williams et al. 1998). These frequent changes in the configuration of the ice sheets have been driven by the cyclical changes in the Earth's orbital path around the sun (Milankovitch cycles). The most influential of these are the 41kyr (thousand years) obliquity cycle and the 100kyr eccentricity cycle (Williams et al. 1998). The continuous cycles of expansion and contraction of the east and west Antarctic ice sheets has resulted in the regular displacement of the terrestrial and freshwater environments suitable for survival or successful colonisation and establishment of biota.

What is most remarkable about the Quaternary history of Antarctica is that the periods of greatest habitat availability, the interglacials, have been relatively short-lived and unusual. The ice core record from Dome C shows that, in the period from for 430-740kyr BP when climate variability was dominated by the 41kyr obliquity cycle, the Antarctic has been c. 50% in the interglacial phase, although these were weaker interglacials than experienced at present. However, in the last 430kyr BP, when climate variability has been dominated by the 100kyr Milankovitch eccentricity cycle, the Antarctic has been c. 90% in the glacial phase (EPICA 2004) and some cold periods have been sustained for more than 60kyr (eg 140-200kyr: Jouzel et al. 1993). Thus, with only c. 10% of the late Quaternary being in full interglacial conditions, for most of this time displacement and retreat of the Antarctic biota, either to refugia or possibly to lower latitudes, appears to be the norm. However, there are some examples of parts of some Antarctic oases remaining ice free through the Last Glacial Maximum (LGM) (18-22kyr) based on evidence contained in lake sediments and other terrestrial deposits (Hodgson et al. 2001, Hodgson et al. 2005, Cromer et al. in press) and some areas such as the Prince Charles mountains have been ice-free for possibly millions of years (Fink et al. 2000). Similarly, some nunataks, which remained above the altitudinal limit of the

LGM ice sheet, contain evidence of a refuge fauna (Marshall and Pugh 1996, Marshall and Coetzee 2000, Convey and McInnes 2005) that may have subsequently been available to recolonise surrounding areas after the ice retreated and suitable habitats became available.

Despite these refugia, a result of the glaciological history of Antarctica is that the majority of the continental high-latitude habitats are likely to have formed in the present, Holocene, interglacial whilst the maritime Antarctic and some warmer subantarctic islands are likely to have had longer periods of exposure, at least in some areas. For example, the consistently low elevation of Macquarie Island, since its emergence 600 000 years ago has meant that glaciation has played a minor role in shaping the island's landscape (Selkirk *et al.* 1990, Adamson *et al.* 1995). However, along the Antarctic Peninsula and associated archipelagos, and the linked Scotia arc subantarctic island of South Georgia, there remains an apparent contradiction between ice sheet and glaciological reconstructions at LGM, which require considerable expansion in ice depth and extent (the latter to the continental shelf edge) with implicit obliteration of all low altitude terrestrial habitats, and increasing biological evidence in support of an ancient and vicariant indigenous terrestrial biota (Clapperton and Sugden 1982, 1988, Larter and Vanneste 1995, O Cofaigh *et al.* 2002, Convey this volume, Chown and Convey this volume).

A schematic model of Antarctic biodiversity

Present-day biodiversity in the Antarctic is the result of a number of factors that can be summarised in a simple schematic model (Fig. 1). The main elements of biodiversity are the continued existence of past biota, the presence or creation of habitat suitable for colonisation, the arrival and establishment of new colonisers, and the adaptation and selection of new taxa in response to environmental forcing variables associated with environmental change.

The expansion and contraction of the Antarctic ice sheet has undoubtedly led to the local extinction of biological communities on the Antarctic continent during glacial periods (Hodgson et al. 2006). Subsequent re-colonisation and the resulting present-day biodiversity is then a result of whether the species were vicariant (surviving the glacial maxima in refugia, then recolonising deglaciated areas), arrived through post-glacial dispersal from lower latitude islands and continents that remained ice free (Pugh et al. 2002), or are present through a combination of both mechanisms. Evidence can be found to support both vicariance (Marshall and Pugh 1996, Marshall and Coetzee 2000, Stevens and Hogg 2003 this volume, Allegrucci et al. 2005, Cromer et al. in press) and dispersal (Hodgson et al. 2006) for a variety of different species, and is based on the level of cosmopolitanism (dispersal model) or endemism (vicariance model) (Gibson et al. this volume), on direct palaeolimnological evidence or, most recently, on molecular phylogenetic and evolutionary studies (Skotniki and Selkirk this volume, Stevens and Hogg this volume).

On the oceanic islands, the biotas will have originally arrived via long-distance over ocean dispersal, with vicariance and dispersal playing subsequent roles in shaping the biodiversity across glacial cycles (Marshall and Convey 2004). Species on Southern Ocean islands show conventional island biogeographic relationships, with variance in indigenous species richness explained by factors including area, mean surface air temperature, and age and distance from continental land masses (Chown et al. 1998). For aquatic species, at least some groups such as the diatoms, diversity is controlled by the 'connectivity' among habitats with the more isolated regions developing greater degrees of endemism (W. Vyverman pers. comm.).

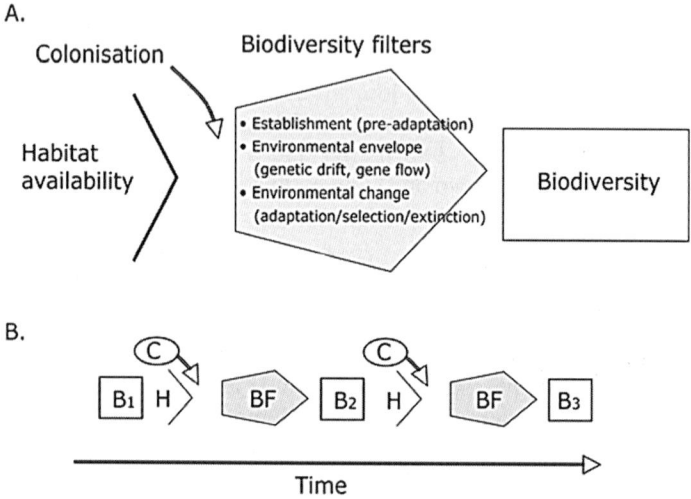

Figure 1. Schematic model illustrating factors influencing the formation of the biodiversity in a location at one time (A.), and how this changes with the passing of time and associated climate change (B).

PAST BIODIVERSITY

As described above, there is some evidence for the survival of biodiversity on the Antarctic continent and subantarctic islands through the LGM. Some taxa are recognised as Antarctic endemics. In the near future, the use of modern genomic tools will provide greater insight into the proportion of the current biota that are glacial survivors, either original species or vicariant derivatives and for what length of time these organisms have been present in these landscapes (Peck et al. 2005).

HABITAT AVAILABILITY

The first requirement of establishment is that there needs to be a substrate and habitat on and within which it is possible for colonisers to become successfully established. The main habitats for terrestrial life in Antarctica include ice-free areas of the islands and continent and inland nunataks that are surrounded by ice. Combined, these ice-free areas account for only 0.35% of the Antarctic continental area (Fox and Cooper 1994). On subantarctic islands, ice-free areas range from small coastal pockets interspersed with glaciers (South Georgia and Heard Island) to completely ice-free islands such as Macquarie and the Iles Crozet. Water bodies range from small holes in the ice (cryconites) to ponds and very large freshwater or saline lakes (Gibson et al. this volume and Quesada et al. this volume). Wet seepage areas are also present in many ice free regions. Adjacent to coasts lakes can be formed either by isolation of basins from the sea during postglacial isostatic rebound, or proglacially as glaciers and ice sheets retreat on the land (Hodgson et al. 2004, Lyons et al. this volume, Quesada et al. this volume). The majority of these habitats are Holocene in age (last 11 ka). However, some areas have survived glaciation due to their altitude or glaciological settings. Examples of at least parts of some Antarctic oases remaining ice free through the Last Glacial Maximum include the Larsemann Hills (Hodgson et al. 2001), the Schirmacher Oasis (Schwab, 1998) and possibly the Bunger Hills (Gore et al. 2001) based on evidence contained in lake sediments and other terrestrial deposits. Similarly, some nunataks have remained ice free through a number of glacial cycles. Subglacial lakes have also likely been present in some form for many millennia (Doran et al. 2004), but are outside the scope of this volume.

COLONISATION

Colonisers are carried in the air, on visiting birds and mammals (including humans) or sometimes borne on rafts (logs, kelp rafts) on the surface of the ocean (Hughes et al. this volume). The terrestrial environments of the Antarctic have been in a fairly constant if low level state of invasion and reinvasion from temperate and closer regions (Barnes et al. 2006) and are currently facing a considerable increase in invasion pressure through the direct (deliberate and accidental transport) and indirect (climate change acting to reduce the hurdles required for either successful natural long distance transport or establishment) (Frenot et al. 2005a, Convey et al. this volume b). For example, two vascular plant species on Macquarie and a further two on Heard Island have been identified as having colonised and established only in the last 200 years, with one of Heard Island species (*Leptinella plumosa*) being represented by only one plant found in 2003 (Bergstrom et al. in press, Turner et al. 2006). In recent decades, new populations of the hair grass *Deschampsia antarctica* and small cushion plant *Colobanthus quitensis* have been appearing in various locations on the Antarctic Peninsula (Fowbert and Smith 1994, Grobe et al. 1997, and review by Convey 2003).

Table 1. Present-day distribution and colonisation attributes of Antarctic biota in the subantarctic, maritime and continental Antarctic biogeographic zones (S, M, C, respectively).

Group/taxon	Distribution S, M, C	Colonization diaspora	Major dispersal methods
Mites	S,M,C	none	self-propulsion, zoochory
Insects	S,M	adults	aerial, zoochory
Nematodes	S,M,C	eggs, anhydrobiotic stages	aerial, zoochory
Collembola	S,M,C	active juveniles and adults	aerial or water surface
Crustaceans	S,M,C	adult resting stages (some groups), resistant eggs	aerial or water
Tardigrades	S,M,C	eggs, resistant adult state (tun)	aerial, zoochory
Molluscs	S	eggs	water
Testate amoeba	S,M,C	resistant stages	aerial or water
Spiders	S	adults and young	aerial
Annelids	S,M	juveniles and adults	water
Angiosperms	S,M	seeds, fruits	aerial or water, birds
Ferns	S	spores	aerial
Mosses	S,M,C	spores, plant fragments	aerial, birds
Algae	S,M,C	spores	aerial or water, birds
Lichens	S,M,C	spores and other propagules, fragments	aerial, birds
Freshwater phytoplankton	S,M,C	spores and fragments	aerial, birds
Bacteria	S,M,C	spores	aerial, birds
Cyanobacteria	S,M,C	cysts, fragments	aerial, birds
Yeasts	S,M,C	spores	aerial
Macrofungi	S,M	spores	aerial, birds
Microfungi, mycorrhizae	S,M,C	spores	aerial

Some groups of organisms are better adapted to dispersal (Table 1) and this determines the frequency with which they arrive at new habitats, but not their ability to establish there. Other organisms have limited motility but have established and survived in various refugia on account of their ability to withstand a series of selective pressures. In an examination of patterns in biogeography of Southern Ocean Islands, Greve et al. (2005) identified the level of vagility within biotic groups as an important characteristic, with more vagile taxa supporting the hypothesis of single origins of Southern Ocean biota, while those less vagile supporting a multi-regional scenarios. In a separate analysis, Muñoz et al. (2004) identified an over-riding influence of wind dispersal as an explanatory factor underlying biogeographical patterns seen across the subantarctic islands.

ESTABLISHMENT (PRE-ADAPTATION)

Once colonising organisms are present they are subjected to a variety of selection pressures or biodiversity filters either before they become established or after. Organisms that establish after colonisation in the Antarctic are, by evidence of their successful establishment event, pre-adapted to the local conditions, either through a particular character state or being phenotypically plastic. This is not to say that such organisms are operating optimally – indeed optimality is an erroneous assumption often applied even to indigenous Antarctic biota with no real evidence - and ecophysiological studies of Antarctic organisms have demonstrated for example, that optimal temperatures for particular processes or activities are often well above ambient Antarctic or subantarctic air or water temperatures (Convey 1996, Hennion et al. this volume). The continued persistence of biota operating under sub-optimal conditions is further encouraged by the generally low importance of negative biotic factors (ie competition, predation) relative to abiotic environmental variables in such habitats (Convey 1996). The survival threshold can be understood as the achievement, over time, of a net gain in carbon and biomass, allowing viability to be maintained and some form of reproduction (requisite for the long-term development of a viable population) to occur. Failure to reach an effective population size will ultimately lead to extinction and, indeed, it is likely that many initially successful establishment events do not go on to establish populations of the species concerned. Convey et al. (2000b) suggested that evidence for such a pattern of frequent dispersal and extinction could be seen in interpreting differences in biodiversity associated with geothermally-influenced ground found in two surveys of the maritime Antarctic South Sandwich Islands separated by ~ 30y.

ENVIRONMENTAL ENVELOPE (GENETIC DRIFT, GENE FLOW)

Descriptions of environmental conditions and features found in Antarctic and subantarctic environments are available from a number of sources (Walton 1984,

Selkirk et al. 1990, Kennedy 1993, Convey 1996, Bargagli 2005). The most prominent environmental drivers in terrestrial environments are photoperiod (varying with the degree of latitude south), the period temperatures are above zero, the upper and lower limits of the thermal range and the, related, availability of free water. In order to sustain ecosystems based on carbon fixation by terrestrial autotrophs, conditions must allow for summer carbon storage to be greater than year–round respiratory loss and for selection of adaptations for a resting stage or dormancy. In limnetic environments, key features again include the availability of free water in addition to the degree of salinity, depth, nutrient availability and the longevity and depth of surface ice, although the selection of mixotrophy (Gibson et al. this volume) reduces the reliance on carbon fixation by autotrophic processes. The degree of temperature fluctuation is dependant on the depth of the water body. As in other environments, stochastic events also play a major role (see below), particularly with regard to local extinction. For example, satellite imagery of volcanic McDonald Island suggests that most vegetation has been destroyed with recent substantial volcanic activity, while the geothermal plant and animal communities associated with active fumaroles described from the South Sandwich Islands and Deception Island (Convey et al. 2000a, Smith 2005) are temporally defined by the geologically ephemeral persistence of specific individual habitats.

After establishment, changes to the genetic population structure can occur through population processes such as genetic drift and gene flow. Some moss populations in continental Antarctica, have been found to have had as much genetic variation as temperate populations, while others are extremely limited (Skotknicki and Selkirk this volume). As mosses in the continental Antarctica are not known to undergo sexual reproduction to any significant extent (Smith and Convey 2002), genetic variation can be attributed to mutation, genetic drift or multiple colonisation events. Over the longer term (million year timescale) molecular phylogenetic studies of differentiation within Victoria Land springtails have been used to propose the existence of, and expansion from, refugia (Stevens and Hogg 2003, this volume), and the allopatric differentiation of populations isolated by glacial advances followed by contact being re-established (Nolan et al. in press). Over an even longer timescale of tens of millions of years, a molecular clock approach applied to species of Diptera endemic to the Antarctic Peninsula and subantarctic South Georgia has proposed separation events carrying a signal of the geological separation of the different tectonic elements linking this region with southern South America (Allegrucci et al. 2006).

ENVIRONMENTAL CHANGE (ADAPTATION/SELECTION EXTINCTION)

The key processes of adaptation, selection and extinction are inevitably linked with environmental change. The gross extinctions of Tertiary ecosystems across the Antarctic and Iles Kerguelen and Heard Island (McLoughlin 2001 and Truswell et al. 2005) are major examples of the impact of climate change on ecosystems. Extinctions associated with cooling (eg leading up to the LGM and previous events) would have been reflected both in species loss due to environmental conditions shifting below their climatic envelope in addition to the loss of habitat from ice advance.

Species that had colonised the region pre-Tertiary and survived glaciations in refugia, may have been pre-adapted with sufficiently plasticity to cope with the environmental changes to be experienced, or have evolved in *situ in* response to the selection pressures of changed environmental conditions. The break-up of Gondwana, the creation of islands and loss of habitat through ice advance isolated many environments, thus the role of genetic drift and mutations most likely played a greater role in populations, both in terrestrial and freshwater environments, than gene flow.

Although, as already mentioned, Antarctic biota are not necessarily functioning under optimal conditions, many species display both southerly and northerly limits to their distribution within the region. Fig. 2 illustrates the distribution of *Acaena magellanica*. This species, producing fruits with four barbed spines that are dispersed as an aggregate containing hundreds of fruits, is highly vagile. Thus, its distribution pattern is not limited by dispersal capacity and illustrates a degree of physiological tolerance. Balls of fruits are often seen caught on the bodies of migratory birds such as the Brown Skua and are readily transported accidentally by humans caught on clothing (Whinam et al. 2004). The southern distributional limit of *A. magellanica* is on South Georgia and the warmer (eastern) side of Heard Island, while the species has not been found in Tasmania, New Zealand or on New Zealand's southern islands. It seems highly unlikely that propagules of such a highly vagile species have not dispersed to these warmer localities where other members of the genus occur. This is an example of a species whose life history strategies have been selected for survival under cooler subantarctic environmental conditions (Convey et al. this volume a).

While climate change research is heavily focussed on the identification of large scale trends, the role of extreme and/or local events under environmental change must not be overlooked. Fig. 3 illustrates that in biological systems it is extreme events that can have the most impact in a local population, be it under a more general cooling or warming environment. Such extreme events, that exceed the physiological thresholds of a species, can lead to local extinction despite the general trend in environmental change occurring within the operating climatic envelope of the species.

The 2001/02 austral summer may provide a recent example of such an extreme event: in this season air temperatures exceeding 10°C were recorded in the Ross Sea region, flooding occurred in the Dry Valleys from melting glaciers and rain occurred at the coastal continental Dumont D'Urville station (Lyons et al. this volume, Wall in press, D. Bergstrom, E. Woehler and M. Pook unpubl. data). This was despite analyses identifying a general decadal–long cooling trend in this region (Doran et al. 2002). Increased free water associated with rain can increase opportunities for population growth and expansion, however increased temperatures without accompanied free-water can lead to increased stress from drought in terrestrial environments (Convey 2003, this volume) or increased salinity in freshwater environments.

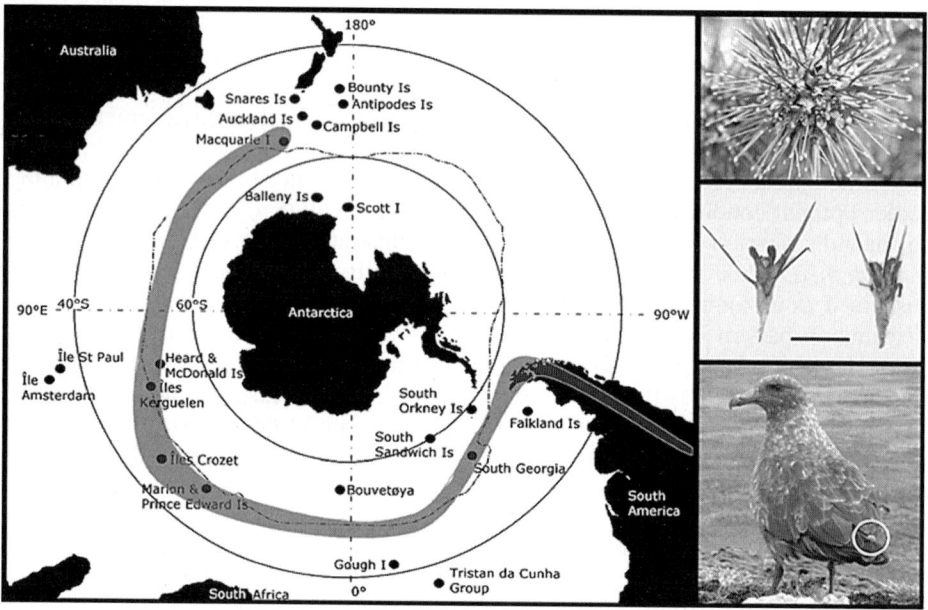

Figure 2. A map of the current distribution of the subantarctic herb Acaena magellanica. *The three figures illustrate an* Acaena *fruit head showing abundant spines (top), two fruits - the scale bar is 3mm (middle) and an* Acaena *fruit head on a subantarctic skua. Photographs by K. Kiefer.*

Biotic interactions (enhancement/inhibition/extinction)

Species richness can be exceedingly low in some Antarctic environments, but species generally do not exist alone. The simplest soil faunal communities on the planet are found in Antarctica, with those of the Victoria Land Dry Valleys being restricted to nematodes, tardigrades, rotifers and protozoan groups (Freckman and Virginia 1997), and those of inland Ellsworth Land nunataks losing even the nematodes (Convey and McInnes 2005). Even these extremely simple communities contain autotrophic, consumer and predatory trophic levels. Therefore, the species involved are subjected to a range of interactions with other species, ranging through mutually beneficial, neutral and negative impacts from competition, predation and parasitism. Bergstrom and Chown (1999), however, have also noted that on Southern Ocean islands many functional groups were missing from communities, a trend that is thought to be even more exaggerated in more southerly and less complex ecosystems. Thus the range of interactions is thought to be reduced compared with warmer ecosystems and, as already stated, biotic factors are thought to become relatively unimportant as evolutionary selective pressures.

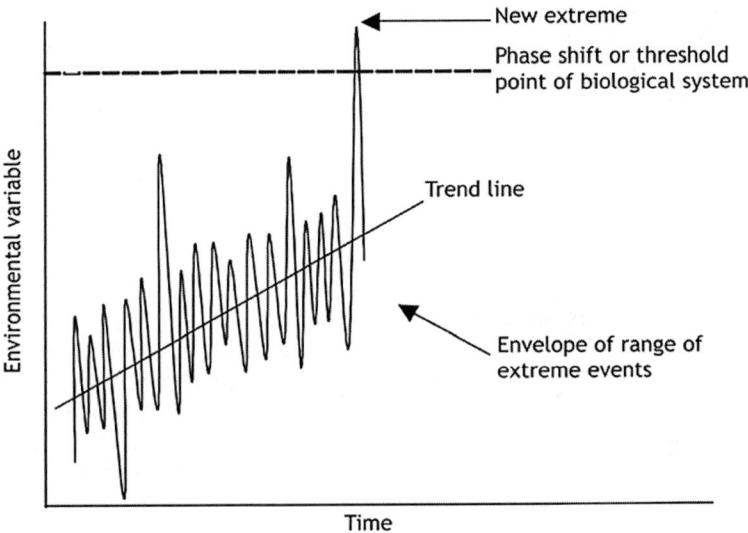

Figure 3. Schematic illustrating the potential importance of extreme climatic events even during a period of apparently ameliorating climatic regimes.

In the subantarctic, there are some indications of the impact of competition. The cushion plant species *Azorella selago* and *A. macquariensis* are limited primarily to higher altitude locations on their respective subantarctic islands, ascribed to the low capacity of these species to cope with competition from taller plants (Bergstrom and Selkirk 2000). It is often the case that introductions of non-indigenous species bring stronger competitors into communities – for instance, the introduction of a predatory carabid beetle to lowland terrestrial habitats of South Georgia and Kerguelen has had clear and considerable impacts on other indigenous invertebrate species (Ernsting et al. 1995, Chevrier et al. 1997, see reviews by Frenot et al. 2005a, Convey et al. this volume b).

Examples of positive biotic interactions include the presence of mycorrhizae in the tissues of plants, including liverworts in Antarctica and the roots of many subantarctic vascular plants (Williams et al. 1994, Laursen et al. 1997, Frenot et al. 2005b, K. Newsham and R. Upson, unpubl. data). Mycorrhizae have been identified as playing an important role in nutrient uptake by plants, and widespread observations across many environments support the idea that they can influence the colonisation capacity of plants in cold and low-nutrient environments such as subantarctic glacier forelands (Frenot et al. 2005b).

An interesting and unforseen negative biotic interaction is currently occurring on some sub- and maritime Antarctic islands. Over the last two decades, populations of fur seals have been expanding at a substantial rate, following their virtual extermination, and that of the great whales, by humans during the 19th and early 20th Centuries (Hodgson and Johnston 1997). As a result, terrestrial habitats occupied by seals during their breeding and moulting seasons are increasing. This is leading to extensive vegetation destruction and excessive manuring, of local vegetation communities, particularly in the South Shetland and South Orkney Islands and on South Georgia and Heard Islands, and to rapid eutrophication of previously oligotrophic limnetic ecosystems (Smith 1988, Butler 1999).

Summary

Present-day Antarctic biodiversity is an inventory of the species present at one time, while the factors that control biodiversity are operating more or less continuously and at a variety of different scales. Thus, determining which species survive and evolve in the Antarctic region, and for that matter in any environment, is dependent on many factors.

Antarctica and its surrounding islands have undergone significant evolution of their landscapes and climate over the last 60 MA. The same biological processes (habitat availability → colonisation → biodiversity filters → resultant biodiversity) that occur elsewhere on the planet operate in both limnetic and terrestrial ecosystems in ice–free refugia in Antarctic and peri-antarctic islands, but the impact of some processes or environmental stressors is greater than in other localities. Currently, what separates the Antarctic from other ecosystems in the world is the combination of extremes that it experiences: extreme isolation, extreme past loss of habitat from ice formation and extreme selection pressure particularly from temperature and water stress in the terrestrial environment.

What we see as a result of these factors at a continental scale is drastically reduced biodiversity at the species level. We now have a situation where many species are at their distributional limits, both northern and southern, and the environment in many places is changing rapidly and thus selection leading towards future biodiversity is most acute. Convey (this volume) and Lyons et al. (this volume) expand on evidence for recent change and discuss the future under current climate change scenarios.

With a long, complex and extreme history, Antarctica and its surrounding islands currently provide relatively simple natural laboratories where small changes impact greatly and this impact is perceivable by the human observer. Apart from its obvious but value-based attributes such as wilderness, aesthetic and existence values, this one attribute makes the region a precious asset to the human race.

References

Adamson, D.A., Selkirk, P.M., Price, D.M. and Ward, N. (1995) Uplift of the Macquarie Ridge at Macquarie Island: Pleistocene evidence from raised beaches and topography. Climate Succession and Glacial History of the Southern Hemisphere over the Past Five Million Years. *ANARE Research Notes* **94**, 3-5.

Allegrucci, G., Carchini, G., Todisco, V., Convey, P. and Sbordoni, V. (2006) A molecular phylogeny of Antarctic Chironomidae and its implications for biogeographical history, *Polar Biology*, **29**, 320-326.

Anonymous (2005) Recent volcanic activity at McDonald Island, *http://www.heardisland.aq*

Bargagli, R. (2005) Antarctic ecosystems: environmental contamination, climate change, and human impact. Ecological Studies Vol. 175, 395 pp. Springer, Berlin.

Barnes, D.K.A., Hodgson, D.A., Convey, P., Allen, C.S. and Clarke, A. (2006) Incursion and excursion of Antarctic biota: past, present and future, *Global Ecology and Biogeography*, **15**, 121-142.

Barrett, P.J. (2003) Cooling a continent. *Nature*, **421**, 221-223.

Bergstrom, D.M. and Chown, S.L (1999) Life at the front: history, ecology and change on southern ocean islands. *Trends in Ecology and Evolution*, **14**, 472-477.

Bergstrom, D.M. and Selkirk, P.M. (2000) Terrestrial vegetation and environments on Heard Island. *Papers and Proceedings of the Royal Society of Tasmania*, **133**, 33-46.

Bergstrom, D.M., Turner, P.A.M., Scott, J.J. Copson G. and Shaw J. (in press) Restricted plant species on subantarctic Macquarie and Heard Islands. *Polar Biology*, DOI 10.1007/s00300-005-0085-2.

Butler, H.G. (1999) Seasonal dynamics of the planktonic microbial community in a maritime Antarctic lake undergoing eutrophication. *Journal of Plankton Research,* **21**, 2393-2419.

Chevrier, M., Vernon, P. and Frenot, Y. (1997) Potential effects of two alien insects on a subantarctic wingless fly in the Kerguelen islands, in B. Battaglia, J. Valencia and D.W.H. Walton (eds.), *Antarctic Communities: Species, Structure and Survival*. Cambridge University Press, Cambridge, UK, pp. 424-431.

Chown, S.L. and Convey, P. (2006) Biogeography, in D.M. Bergstrom, P. Convey, and A.H.L. Huiskes, (eds.) *Trends in Antarctic Terrestrial and Limnetic Ecosystems: Antarctica as a Global Indicator*. Springer, Dordrecht (this volume)

Chown, S.L., Gremmen, N.J.M. and Gaston, K.J. (1998) Ecological biogeography of Southern Ocean islands: Species-area relationships, human impacts, and conservation, *American Naturalist* **152**, 562-575.

Clapperton, C.M. and Sugden, D.E. (1982) Late quaternary glacial history of George VI Sound area, West Antarctica, *Quaternary Research* **18**, 243-267.

Clapperton, C.M. and Sugden, D.E. (1988) Holocene glacier fluctuations in South America and Antarctica, *Quaternary Science Reviews* 7, 185-198.

Clarke. A. and Crame. A. (1989) The Origins of the Southern Ocean Marine fauna, in J.A. Crame (ed), *Origins and Evolution of the Antarctic Biota*, Geological Society Special Publication No 47, pp 253-268.

DeConto, R. M. and Pollard, D., (2003) Rapid Cenozoic glaciation of Antarctica induced by declining atmospheric CO_2, *Nature* **421**, 245-249.

Convey, P. (1996) The influence of environmental characteristics on life history attributes of Antarctic terrestrial biota, *Biological Reviews* **71**, 191-225.

Convey, P. (2003) Maritime Antarctic climate change: signals from terrestrial biology. In E. Domack, A. Burnett, A. Leventer, P. Convey, M. Kirby and R. Bindschadler (eds.), *Antarctic Peninsula Climate Variability: Historical and Palaeoenvironmental Perspectives*, Antarctic Research Series, Vol. 79, American Geophysical Union, Washington, D.C., pp. 145-158.

Convey, P. (2006) Antarctic climate change and its influences on terrestrial ecosystems, in D.M. Bergstrom, P. Convey, and A.H.L. Huiskes (eds), *Trends in Antarctic Terrestrial and Limnetic Ecosystems: Antarctica as a Global Indicator, Springer, Dordrecht (this volume)*

Convey, P. and McInnes, S.J. (2005) Exceptional tardigrade-dominated ecosystems in Ellsworth Land, Antarctica. *Ecology* **86**, 519-527.

Convey, P., Greenslade, P. and Pugh, P.J.A. (2000b) Terrestrial fauna of the South Sandwich Islands, *Journal of Natural History* **34**, 597-609.

Convey, P., Chown, S.L., Wasley, J. and Bergstrom, D.M. (2006b) Life history traits, in D.M. Bergstrom, P. Convey, and A.H.L. Huiskes (eds.) *Trends in Antarctic Terrestrial and Limnetic Ecosystems: Antarctica as a Global Indicator*, Springer, Dordrecht (this volume).

Convey, P., Frenot, Y., Gremmen, N. and Bergstrom, D.M. (2006a) Biological invasions, in D.M. Bergstrom, P. Convey, and A.H.L. Huiskes (eds.) *Trends in Antarctic Terrestrial and Limnetic Ecosystems: Antarctica as a Global Indicator*, Springer, Dordrecht (this volume).

Convey, P., Smith, R.I.L., Hodgson, D.A and Peat, H.J. (2000a) The flora of the South Sandwich Islands, with particular reference to the influence of geothermal heating, *Journal of Biogeography* **27**, 1279-1295.

Cromer, L., Gibson, J.A.E., Swadling, K.M. and Hodgson, D.A. (2006) Evidence for a lacustrine faunal refuge in the Larsemann Hills, East Antarctica, during the Last Glacial Maximum, *Journal of Biogeography* (in press).

Doran P.T., Priscu J.C., Berry Lyons W., Powell R.D., Andersen D.T., Poreda R.J. (2004) Paleolimnology of extreme cold terrestrial and extraterrestrial environments, in R. Pienitz, M.S.V. Douglas, J.P. Smol (eds.), *Developments in Palaeoenvironmental Research. Volume 8. Long-term Environmental Change in Arctic and Antarctic Lakes*, Springer, Dordrecht, p 475-507

Doran, P.T., Priscu J.C., Lyons, W.B., Walsh, J.E., Fountain, A.G., McKnight, D.M., Moorhead, D.L., Virginia, R.A., Wall, D.H., Clow, G.D., Fritsen, C. H., McKay, C. P., Parsons, A.N. (2002) Antarctic climate cooling and terrestrial ecosystem response. *Nature* **415**, 517-520.

EPICA, (2004) Eight glacial cycles from an Antarctic ice core, *Nature*, **429**, 623-628.

Ernsting, G., Block, W., MacAlister, H., and Todd, C. (1995) The invasion of the carnivorous carabid beetle *Trechisibus antarcticus* on South Georgia (sub-Antarctic) and its effect on the endemic herbivorous beetle *Hydromedion sparsutum*. Oecologia **103** 34-42.

Feldmann, R.M. and Woodburne, M.O. (eds.) (1988) Geology and paleontology of Seymour Island, Antarctic Peninsula, *Geological Society of America Memoir* **169**, 566pp.

Fink, D., McKelvey, B., Hannan, D., and Newsone, D. (2000) Cold rock, hot sands: In-situ cosmogenic applications in Australia at ANTARES, *Nuclear Instruments and Methods in Physics Research*, **B**, **172**, 838-846.

Fitch, S., Lock, A., Klok, C.J., Kiernan, K. and McConnell, A. (2001) Heard Island (Indian Ocean) Photographs of upper slopes substantiate reports of activity from two distinct vents, *Bulletin of the Global Volcanism Network*, **26**, 10-11.

Fowbert, J.A. and Smith, R.I.L. (1994) Rapid population increases in native vascular plants in the Argentine Islands, Antarctic Peninsula, *Arctic and Alpine Research*, **26**, 290-296.

Fox, A.J. and Cooper A.P.R. (1994) Measured properties of the Antarctic ice sheet derived from the SCAR Antarctic digital database, *Polar Record*, **30**, 201-206.

Francis, J.E. and Poole, I. (2002) Cretaceous and early Tertiary climates of Antarctica: evidence from fossil wood. *Palaeogeography, Palaeoclimatology, Palaeoecology*, 182, 47-64.

Freckman, D.W and Virginia, R.A. (1996) Low-diversity antarctic soil nematode communities: distribution and response to disturbance, *Ecology* 78, 363-369.

Frenot,Y., Bergstrom, D.M., Gloaguen, J.C., Tavenard, R. and Strullu D.G. (2005b) The first record of mycorrhizae on subantarctic Heard Island: a preliminary examination, *Antarctic Science*, **17**, 205-217.

Frenot, Y., Chown, S.L., Whinam, J., Selkirk, P., Convey, P., Skotnicki, M. and Bergstrom, D. (2005a) Biological invasions in the Antarctic: extent, impacts and implications, *Biological Reviews*, **80**, 45-72.

Gibson, J.A.E., Wilmotte, A., Taton, A., Van De Vijver, B., Beyens, L. and Dartnall, H.J.G. (2006) Biogeographic trends in Antarctic lake communities, in D.M. Bergstrom, P. Convey, and A.H.L. Huiskes (eds.), *Trends in Antarctic Terrestrial and Limnetic Ecosystems: Antarctica as a Global Indicator*, Springer, Dordrecht (this volume).

Gladczenko, T.P. and Coffin, M.F. (2001) Kerguelen Plateau crustal structure and basin formation from seismic and gravity data, *Journal of Geophysical Research*, **18**, 16583-16601.

Gore D.B., Rhodes E.J., Augustinus P.C., Leishman M.R., Colhoun E.A. and Rees-Jones J. (2001) Bunger Hills, East Antarctica: Ice free at the Last Glacial Maximum, *Geology*, **29**:1103-1106.

Greve, M., Gremmen, N.J.M., Gaston, K.J. and Chown, S.L. (2005) Nestedness of South Ocean island biotas: ecological perspectives on a biogeographical conundrum. *Journal of Biogeography* **32**, 155-168.

Grobe, C.W., Ruhland C.T. and Day T.A. (1997) A new population of *Colobanthus quitensis* near Arthur Harbor, Antarctica: correlating recruitment with warmer summer temperatures. *Arctic and Alpine Research* **29**, 217-221.

Hennion, F., Huiskes, A.H.L., Robinson, S. and Convey, P. (2006) Physiological traits or organisms in a changing environment, in D.M. Bergstrom, P. Convey, and A.H.L. Huiskes (eds.) *Trends in Antarctic Terrestrial and Limnetic Ecosystems: Antarctica as a Global Indicator*, Springer, Dordrecht (this volume).

Hodgson, D.A. and Johnston, N.M. (1997) Inferring seal populations from lake sediments, *Nature*, **387**, 30-31.

Hodgson D.A., Doran P.T., Roberts D. and McMinn A. (2004) Paleolimnological studies from the Antarctic and subantarctic islands, in R. Pienitz, M.S.V. Douglas, J.P. Smol (eds.), *Developments in Palaeoenvironmental Research. Volume 8. Long-term Environmental Change in Arctic and Antarctic Lakes,* Springer, Dordrecht, pp 419-474

Hodgson D.A., Verleyen E., Squier A.H., Sabbe K., Keely B.J., Saunders K.M. and Vyverman W. (2006) Interglacial environments of coastal east Antarctica: comparison of MIS 1 (Holocene) and MIS 5e (Last Interglacial) lake-sediment records. *Quaternary Science Reviews,* **25**, 179-197.

Hodgson D.A., Verleyen E., Sabbe K., Squier A.H., Keely B.J., Leng M.J., Saunders K.M. and Vyverman W. (2005) Late Quaternary climate-driven environmental change in the Larsemann Hills, east Antarctica, multi-proxy evidence from a lake sediment core, *Quaternary Research* **64**, 83-99

Hodgson D.A., Noon P.E., Vyverman W., Bryant C.L., Gore D.B., Appleby P., Gilmour M., Verleyen E., Sabbe K., Jones V.J., Ellis-Evans J.C. and Wood P.B. (2001) Were the Larsemann Hills ice-free through the Last Glacial Maximum? *Antarctic Science* **13**, 440-454.

Hughes, K.A., Ott, S., Bölter, M. and Convey, P. (2006) Colonisation processes, in D.M. Bergstrom, P. Convey, and A.H.L. Huiskes (eds.) *Trends in Antarctic Terrestrial and Limnetic Ecosystems: Antarctica as a Global Indicator*, Springer, Dordrecht (this volume).

Huiskes, A.H.L., Convey, P. and Bergstrom, D.M. (2006) Trends in Antarctic terrestrial and limnetic ecosystems: Antarctica as a global indicator, in D.M. Bergstrom, P. Convey, and A.H.L. Huiskes (eds.) *Trends in Antarctic Terrestrial and Limnetic Ecosystems: Antarctica as a Global Indicator*, Springer, Dordrecht (this volume).

Jouzel J., Barkov N.I., Barnola J.M., Bender M., Chapellaz J., Genthon C., Kotlyakov V.M., Lipenkov V., Lorius C., Petit J.R., Raynaud D., Raisbeck G., Ritz T., Sowers C., Stievenard M., Yiou F. and Yiou P. (1993) Extending the Vostok ice-core record of palaeoclimate to the penultimate glacial period, Nature **364**, 407-412

Kennedy, A.D. (1993) Water as a limiting factor in the Antarctic terrestrial environment: a biogeographical synthesis, *Arctic and Alpine Research* **25**, 308-315.

Kennett, J.P. (1977) Cenozoic evolution of Antarctic glaciation, the circum-Antarctic Ocean, and their impact on global paleoceanography, *Journal of Geophysical Research* **82**, 3843-3860.

Larter, R.D. and Vanneste, L.E. (1995) Relict subglacial deltas on the Antarctic Peninsula outer shelf, *Geology* **23**, 33-36.

Laursen, G.A., Treu, R., Seppelt, R.D. and Stephenson, S.L. (1997) Mycorrhizal assessment of vascular plants from subantarctic Macquarie Island, *Arctic and Alpine Research* **29**, 483-491.

LeMasurier, W.E. and Thomson, J.W. (1990) Volcanoes of the Antarctic Plate and Southern Oceans. *American Geophysical Union, Antarctic Research Series* **48**, 487 pp.

Livermore, R.A., Nankivell, A.P., Eagles, G. and Morris, P. (2005) Paleogene opening of Drake Passage. *Earth and Planetary Science Letters* **236**, 459-470.

Longton, R.E. (1988) *The biology of polar bryophytes and lichens,* Cambridge University Press, Cambridge, 391 pp.

Lyons, W.B., Laybourn-Parry, J., Welch, K.A. and Priscu, J.C. (2006) Antarctic lake systems and climate change, in D.M. Bergstrom, P. Convey, and A.H.L. Huiskes (eds.) *Trends in Antarctic Terrestrial and Limnetic Ecosystems: Antarctica as a Global Indicator,* Springer, Dordrecht (this volume).

Marshall, D.J. and Coetzee, L. (2000) Historical biogeography and ecology of a Continental Antarctic mite genus, *Maudheimia* (Acari, Oribatida): evidence for a Gondwanan origin and Pliocene-Pleistocene speciation, *Zoological Journal of the Linnean Society* **129**, 111-128.

Marshall, D.J. and Convey, P. (2004) Latitudinal variation in habitat specificity of ameronothroid mites. *Experimental and Applied Acarology* **34**, 21-35.

Marshall, D.J. and Pugh, P.J.A. (1996) Origin of the inland Acari of Continental Antarctica, with particular reference to Dronning Maud Land, *Zoological Journal of the Linnean Society* **118**, 101-118.

McLoughlin, S. (2001) The breakup history of Gondwana and its impact on pre-Cenozoic floristic provincialism, *Australian Journal of Botany* **49**, 271-300.

Muñoz, J., Felicísimo, A.M., Cabezas, F., Burgaz, A.R. and Martínez, I. (2004) Wind as a long-distance dispersal vehicle in the Southern Hemisphere. *Science* **304**, 1144-1147.

Nolan, L., Hogg, I.D., Stevens, M.I. and Haase, M. (In press) Molecular support for a secondary contact zone among glacial refugia for the springtail *Gomphiocephalus hodgsoni* (Collembola) in Taylor Valley, continental Antarctica, *Polar Biology.*

Ó Cofaigh, C., Pudsey, C.J., Dowdeswell, J.A. and Morris, P. (2002) Evolution of subglacial bedforms along a paleo-ice stream, Antarctic Peninsula continental shelf, *Geophysical Research Letters* **29**, 1199, doi: 10.1029/2001GLO14488.

Peat, H.J., Clarke, A. and Convey, P. (in press) Diversity and biogeography of the Antarctic flora. *Journal of Biogeography.*

Peck, L.S., Clark, M.S., Clarke, A., Cockell, C.S., Convey, P., Detrich III, H.W., Fraser, K.P.P., Johnston, I.A., Methe, B.A., Murray, A.E., Römisch, K. and Rogers, A.D. (2005) Genomics: applications to Antarctic Ecosystems, *Polar Biology* **28**, 351-365.

Pickard, J. and Seppelt, R.D. (1984) Phytogeography of Antarctica. *Journal of Biogeography* **11**, 83-102.

Poole. I. and Cantrill D.J. (2001) Fossil wood from Williams Point beds, Livingston Island, Antarctica: a late Cretaceous southern high latitude flora, *Palaeontology* **44**, 1081-1112.

Pugh, P.J.A., Dartnall, H.J.G. and Mcinnes, S.J. (2002) The non-marine Crustacea of Antarctica and the Islands of the Southern Ocean: biodiversity and biogeography, *Journal of Natural History.* **36**, 1047-1103

Quesada, A., Vincent, W.F. Kaup, E., Hobbie, J.E., Laurion, I., Pienitz, R., López-Martínez, J. and Durán, J.-J. (2006) Landscape control of high-latitude lakes in a changing climate, in D.M. Bergstrom, P. Convey, and A.H.L. Huiskes (eds.) *Trends in Antarctic Terrestrial and Limnetic Ecosystems: Antarctica as a Global Indicator,* Springer, Dordrecht (this volume).

Scher, H.D. and Martin, E.E. (2006) Timing and climatic consequences of the opening of Drake Passage. *Science,* **312**, 428-430.

Schwab, M.J. (1998) Rekonstruktion der spätquartären Klima- und Umweltgeschichte der Schirmacher Oase und des Wohlthat Massivs (Ostantarktika), *Berichte zur Polarforschung* **293**, 1-128.

Selkirk, P.M., Seppelt, R.D. and Selkirk, D.R. (1990) *Subantarctic Macquarie Island. Environment and Biology.* Studies in Polar Research Cambridge University Press, Cambridge, 285 pp.

Skotnicki, M.L. and Selkirk, P.M. (2006) Plant biodiversity in an extreme environment: genetic studies of origins, diversity and evolution in the Antarctic, in D.M. Bergstrom, P. Convey, and A.H.L. Huiskes (eds.) *Trends in Antarctic Terrestrial and Limnetic Ecosystems: Antarctica as a Global Indicator,* Springer, Dordrecht (this volume).

Skottsberg, C. (1904) On the zonal distribution of South Atlantic and Antarctic Vegetation, *Geography Journal* **24**, 655-663.

Smith, R.I.L. (1984) Terrestrial Plant Biology of the Subantarctic and the Antarctic, in R.M. Laws (ed.), *Antarctic Ecology* Volume 1, Academic Press, London, pp. 61-162.

Smith, R.I.L. (1988) Destruction of Antarctic terrestrial ecosystems by a rapidly increasing fur seal population, *Biological Conservation* **45**, 55-72.

Smith, R.I.L. (2005) The bryophyte flora of geothermal habitats on Deception Island, Antarctica, *Journal of the Hattori Botanical Laboratory* **97**, 233-248.

Smith, R.I.L. and Convey, P. (2002) Enhanced sexual reproduction in bryophytes at high latitudes in the maritime Antarctic, *Journal of Bryology* **24**, 107-117.

Stevens, M.I. and Hogg, I.D. (2003) Long-term isolation and recent range expansion from glacial refugia revealed for the endemic springtail *Gomphiocephalus hodgsoni* from Victoria Land, Antarctica. *Molecular Ecology* **12**, 2357-2369.

Stevens, M.I. and Hogg, I.D. (2006) The molecular ecology of Antarctic terrestrial and limnetic invertebrates and microbes, in D.M. Bergstrom, P. Convey, and A.H.L. Huiskes (eds.) *Trends in Antarctic Terrestrial and Limnetic Ecosystems: Antarctica as a Global Indicator*, Springer, Dordrecht (this volume).

Tripati, A., Backman, J., Elderfield, H. and Ferretti, P. (2005) Ecocene bipolar glaciation assocaited with global carbon cycle changes, *Nature* **436**, 3410346.

Truswell, E.M., Quilty P.G., Mcminn A., Macphail M.K., Wheller G.E. (2005) Late Miocene vegetation and palaeoenvironments of the Drygalski Formation, Heard Island, Indian Ocean: evidence from palynology, *Antarctic Science* **17**, 427-422 doi: 10.1017/S0954102005002865.

Turner P.A.M., Scott J.J. and Rozefelds A. (2006) Probable long distance dispersal of Leptinella plumosa Hook. f. to Heard Island: habitat, status and discussion of its arrival, *Polar Biology*, **29**, 160-168.

Wall, D.H. (In press) Above-Belowground Interactions in a Low Diversity Ecosystem: The McMurdo Dry Valleys, *Philosophical Transactions of the Royal Society.*

Walton, D.W.H. (1984) The terrestrial environment, in R.M. Laws (ed), *Antarctic Ecology* Volume 1. Academic Press, London, pp 1-60.

Whinam, J., Chilcott, N. and Bergstrom, D.M. (2004) Subantarctic hitchhikers: expeditioners as vectors for the introduction of alien organisms, *Biological Conservation* **121**, 207-219.

Williams, P.G., Roser, D.J. and Seppelt, R.D. (1994) Mycorrhizae of hepatics in continental Antarctica, *Mycological Research* **98**, 34–36.

Williams, M., Dunkerley, D., DeDeckker, P., Kershaw, P. and Chappell, J. (1998) *Quaternary Environments,* Arnold, London, 352 pp.

3. COLONISATION PROCESSES

K.A. HUGHES

British Antarctic Survey, Natural Environment Research Council
High Cross, Madingley Road
Cambridge CB3 0ET, United Kingdom
kehu@bas.ac.uk

S. OTT

Botanisches Institut
Heinrich-Heine Universität Düsseldorf
Universitätsstr. 1, D-40225 Düsseldorf
Germany
otts@uni-duesseldorf.de

M. BÖLTER

Institute for Polar Ecology
University of Kiel
Wischhofstr. 1-3, 24148 Kiel
Germany
mboelter@ipoe.uni-kiel.de

P. CONVEY

British Antarctic Survey, Natural Environment Research Council
High Cross, Madingley Road
Cambridge CB3 0ET, United Kingdom
p.convey@bas.ac.uk

Introduction

About 200 million years ago (MYA), the Antarctic continent formed, together with Australia, Africa, South America, India and New Zealand, the supercontinent

D.M. Bergstrom et al. (eds.), Trends in Antarctic Terrestrial and Limnetic Ecosystems, 35–54.

Gondwana. As the supercontinent broke up, giving the continents with which we are familiar today, Antarctica finally lost contact with Australia (c.45-50MYA) and South America (c. 30MYA). Throughout this process, the Antarctic landmass always lay at high southern latitudes, drifting southwards to reach its current position c.45 MYA. Separation from South America, which allowed the formation of oceanic and atmospheric circulation patterns isolating the continent from lower latitudes, was followed by an enormous cooling process. Nowadays, the continent of Antarctica, which is larger than Australia and comparable in area to Western Europe, is 99.6% covered by permanent ice with an average thickness of 2km and maximum of over 4km. The rocky summits of buried mountain ranges (nunataks), standing above the surrounding ice-sheet form, together with coastal landscapes, the contemporary ice-free terrestrial habitats of the Antarctic.

Today, Antarctica is the remotest continent on Earth. It is surrounded by the Southern Ocean, which at its narrowest is 1000 km wide between South America and the northern tip of the Antarctic Peninsula. Oceanic and atmospheric circulations around Antarctica form major barriers to potential colonisers (Smith 1991, Wynn-Williams 1991, Clarke et al. 2005, Barnes et al. 2006). Broadly, biodiversity decreases with increasing latitude as conditions get colder and drier and the length of the growing season decreases (Smith 1994, Convey 2001a, Kappen 2004). The greatest terrestrial biodiversity is present on the subantarctic islands, a ring of remote islands and island groups surrounding the Antarctic continent in the Southern Ocean (see Fig. 1 in Huiskes et al. this volume). In the maritime Antarctic (western Antarctic Peninsula and Scotia Arc archipelagos) although numerous moss and lichen species are found, higher plants are limited to Antarctica's only two flowering plant species (*Deschampsia antarctica* and *Colobanthus quitensis*). Higher invertebrates are limited to two native Diptera (*Belgica antarctica* and *Parochlus steinenii*) and true terrestrial vertebrates are absent. Finally, other than at a few very restricted 'oases', continental Antarctica is occupied by only the most durable species, such as lichens and microorganisms. However, some caution should be applied if this general pattern, derived from observation of visually obvious macrobiota, is to be applied to microbes. Recently, prokaryote diversity in the Antarctic has been described as surprisingly high (Tindall 2004) while, at least within specific soil habitat types, patterns of eukaryotic microbial diversity can also be opposite to or inconsistent with those of higher groups (Lawley et al. 2004).

It is axiomatic that successful colonisation by terrestrial biota can only take place in areas which become ice and snow free at some point during the year. There is a steep decrease in the size and frequency of exposed terrestrial ground suitable for colonisation with progression from the subantarctic southwards into the maritime Antarctic, along the Antarctic Peninsula and into the continent. Some parts of the maritime Antarctic, including the South Orkney and South Shetland archipelagos, and islands and coastline around the Argentine Islands and Marguerite Bay, are characterised by a relatively high diversity of species and communities, while a few coastal continental areas also include species-rich oases (eg the Windmill Islands, Vestfold Hills, Schirmacher Oasis, locations along the Victoria Land coastline). However, while diversity and extent of habitat decrease drastically within the continental interior, colonisation processes clearly continue, with the simple communities found including both relictual species and more recent colonists (Smith 2000, Stevens and Hogg 2003, Convey and McInnes 2005, Bergstrom et al. 2006).

Colonisation processes in the Antarctic can be considered in at least two different spatial and temporal contexts (Bergstrom et al. 2006). At one extreme, the origin of the contemporary Antarctic fauna and flora must be placed in an appropriate geohistorical context – following separation from the other Gondwanan fragments, isolation, cooling and extensive continental ice formation, does the contemporary biota include true relictual species, and how did other 'ancient' biota arrive on the continent? At the other extreme, what are the processes of recent and contemporary colonisation? Here, in addition to the long-distance processes bringing new biota to the continent, are also included the dispersal of organisms among suitable habitats isolated by short distances and, often, only recently exposed by ice retreat after the Last Glacial Maximum (LGM).

Colonisation success depends upon the following factors: (1) survival of propagules during transfer, (2) the physiological and biochemical capacities of propagules on deposition and (3) establishment of a reproducing population sustainable over subsequent years (Ellis-Evans and Walton 1990, Wynn-Williams 1991, Clarke 2003). The use of the term 'adaptation' as applied to success in colonisation can be understood either in a precise evolutionary sense as a pre-existing specialisation on the genetic level ('pre-adaptation') as a prerequisite for primary colonisation of terrestrial Antarctic habitats, or in an ecological context of ensuring survival and development in an established habitat.

Climate change

It is generally accepted that contemporary global climate change is primarily linked with anthropogenic activities, particularly the release of greenhouse gases (Houghton et al. 2001). In a regional context, the Antarctic Peninsula has experienced one of the most dramatic changes seen worldwide, with increases in annual average temperatures of 2°C over the last 50 years (Hansen et al. 1999, King et al. 2003, Vaughan et al. 2003). In this region, some observed changes have been linked through teleconnection with El Niño Southern Oscillation (ENSO) events, although generally links among Antarctic regional and global processes are yet to be fully understood, or replicated in Global Circulation Models of climate processes (Turner et al. 1997, King et al. 2003). Focussing on the Antarctic Peninsula and Scotia Arc island groups, current trends appear to be linked with a shift in the pattern of atmospheric depressions, resulting in a greater frequency of events moving (warmer and wetter) air from the north into the region. In addition to increasing temperatures, this link in theory could also have a direct impact on colonisation processes, through increasing the frequency of transfer of South American air masses and aerobiota southwards.

Climate change consequences are proposed to include dramatic ice shelf retreat (Vaughan and Doake 1996), deglaciation (Smith et al. 1999), reduction in snow cover (Fox and Cooper 1998) and increased precipitation over some coastal areas of the Peninsula (Turner et al. 1997). Terrestrial habitats of the subantarctic islands are also faced with trends of temperature increase combined with both increases and decreases in precipitation and water availability, depending on location (Bergstrom and Chown 1999, Convey this volume). Increased temperature and water availability can significantly increase the duration of the growing season for terrestrial Antarctica biota and, by implication, the period over which physiologically active colonising propagules may become established, but this effect will be counteracted by any increase in desiccation stress. Long-term studies have linked an increase in the abundance and local distribution of both *D. antarctica* and *C. quitensis* in the Antarctic Peninsula region over the last 25 years with climate change (Corner and Smith 1973, Fowbert and Smith 1994, Smith 1994). Rapid expansion of bryophyte ground coverage has also been observed in the maritime Antarctic (Smith 1990) and field manipulations confirm that warming may have widespread effects on terrestrial ecosystems (eg Day et al. 1999 and Convey 2001b, 2003 for summaries).

Climate amelioration does not only affect native Antarctic species (Kennedy 1995). Convey (1997) predicted that climate change may lead to:

> "an increase in rate of colonisation by species new to the Antarctic, thereby increasing diversity, biomass, trophic complexity and habitat structure, with possible loss of existing Antarctic species and communities through increased competition".

Although there are few records to date of unassisted *de novo* species colonisation, new non-native bryophyte species have become established on Signy Island (South Orkney Islands) as a result of long distance colonisation (Convey and Smith 1993). The potential synergy between anthropogenic climate change processes, which lower the environmental barriers to colonisation processes in Antarctica, and increased human activity in the region (research, tourism and industry), which increases the probability of incidental transfer of colonists, has been emphasised by Frenot et al. (2005). Indeed, these authors concluded that transfer of non-indigenous species with human assistance is likely to (already) far outweigh transfer through 'natural' processes and that while the subantarctic islands are by far the most likely to experience successful assisted colonisation events, such transfers also occur regularly into both the maritime and continental Antarctic zones.

Dispersal processes

Successful transport to and establishment in the terrestrial Antarctic environment is likely to be an infrequent event, as sites with appropriate physical and microclimatic characteristics suitable for colonisation are rare, isolated and are often surrounded by large areas of snow, ice or ocean (Block 1984, Bölter et al. 2002). Several modes of dispersion exist: organisms or their propagules can be transported long distances in air or water currents (Benninghoff and Benninghoff 1985, Marshall 1996, Coulson et al. 2002), associated with migratory birds and mammals (Schlichting et al. 1978, Bailey and James 1979, Pugh 1997, Barnes et al. 2004), or in water attached to natural or anthropogenic flotsam (Barnes 2002, Barnes and Fraser 2003). The activities and rapid movement of humans into and within the continent by ship and aircraft is a biologically recent and also extremely efficient means of dispersal (Bölter and Stonehouse 2002, Frenot et al. 2005)

PROPAGULES

The success of colonisation processes relies on the existence and transfer of viable propagules from a source location, arrival at a suitable substrate possessing appropriate exogenous factors and the maintenance of suitable microclimatic conditions during the period of arrival and initial establishment. A wide range of biotic groups appear well-adapted with respect to the first of these requirements, having developed a range of stress-resistant propagules which allow the possibility of survival during extended dispersal events (eg Jahns 1982, Longton 1988, Convey 1996a). Included amongst these groups are bacteria, cyanobacteria, fungi, mosses and the invertebrate groups of rotifers, tardigrades, nematodes and some micro-arthropods. Other groups are presented with particular problems to overcome. For instance, lichens face the challenge of maintaining their symbiotic state – an association of a fungal (mycobiont) and an algal (photobiont) component – in their new location. They have developed dispersal mechanisms depending either on vegetative diaspores (containing both bionts) or on ascospores (containing only the mycobiont). Therefore, while both are eminently suitable for long-distance aerial dispersal, the subsequent development of a new lichen thallus from an ascospore requires the re-establishment of symbiotic contact with a suitable photobiont after germination, meaning that the separate bionts must be present simultaneously at a suitable site for successful lichenisation. Furthermore, selectivity and compatibility of the symbionts are a prerequisite for the relichenisation process (Romeike et al. 2002, Schaper and Ott 2003), and there is evidence that greater diversity is possible in the algal biont of particular lichen species from Antarctica than is typical at lower latitudes (S. Ott unpubl. data). Other examples of symbioses, for instance that between specific plants and mycorrhizal fungi (Williams et al. 1994) are faced with the same challenge.

Propagule characteristics, in particular size/mass, will have a considerable influence on the likely success of transfer by aerial routes, including the relative probability of long vs short-range dispersal. This has been considered specifically in mosses, where spores >20μm in diameter are predominantly deposited within tens of metres of their source, while smaller spores have a greater (though still small) chance of being carried to higher altitudes and dispersed far greater distances (During 1979, Miles and Longton 1992). In addition to spores, mosses also produce a range of asexual propagules (Longton 1988), predominantly associated with local dispersal. Similarly in lichens, ascospores and vegetative soredia are usually associated with long-distance dispersal while isidia function more on shorter distances (Jahns 1982). In groups such as mosses and lichens, the flexibility to disperse using propagules with different characteristics can be seen as a particular advantage for successful colonisation of harsh environments.

AIRBORNE TRANSPORT

Winds are thought to be the dominant agents of biological transport among and between terrestrial sites (both into and within Antarctica) for many biological groups, as they can transport many biological particles over long distances. The typical flow of cyclones around Antarctica is generally effective in limiting direct north-south transfer of biological material into (or out of) the continent, although this 'West Wind Drift' (Mason 1971) has itself been proposed as a mechanism for dispersal of lichen propagules from Southern Hemisphere continents (Galloway 1987). In parallel, the eastward-flowing Antarctic Circumpolar Current generally limits opportunity for north-south dispersal in seawater (Clarke et al. 2005, Barnes et al. 2006), although biological debris (eg tree trunks and branches) may still reach the continent and offshore archipelagos having circumnavigated the Southern Ocean (Convey et al. 2002a).

In the most comprehensive Antarctic aerobiological study to date, Marshall (1996) showed that cyclones moving around the continent could also be associated with dramatic influxes of airborne biological material into the South Orkney Islands from South America. He estimated that suitable weather patterns for rapid transfer of large quantities of biological material (and material of geological origin, such as volcanic ash) from South America to the South Orkney Islands occurred on average every 18 months, although less dramatic influxes are likely to be more common. It is not clear to what extent his estimates can be generalised and applied to other Antarctic locations, as no comparable studies have been completed. At continental sites, katabatic winds flowing off the high continental plateau toward the coast may inhibit local aerobiological transfer of propagules towards inland sites, rather encouraging outbound transfer, although as yet little evidence exists to support this (but see Convey and McInnes 2005). However, an analogous process operating at larger scale must underlie the presence of lichen species (c. 50% of the recorded lichen flora) otherwise endemic to the Antarctic continent and Peninsula on the remote maritime Antarctic South Sandwich Islands (Convey et al. 2000), as these islands are volcanic and recent (0.5 – 3 million years) in origin. Nevertheless, other markers of long distance propagule transfer into Antarctica are present, including the occurrence of exotic pollen trapped in moss cushions (Linskens et al. 1993) and exotic plants on volcanically warmed soils in both continental and maritime Antarctica (Bargagli et al. 1996, Convey et al. 2000). Taking a different approach, in a recent analysis of patterns of regional plant biodiversity amongst groups of Southern Ocean islands, Muñoz et al. (2004) highlighted the importance of 'wind connectivity' over geographical proximity as explanatory factors for present-day distributions.

Biological material present in Antarctic air may include moss spores, pollen (eg from *D. antarctica*, *Nothofagus* spp.), fungal spores, bacteria, viruses, lichen propagules, tardigrade cysts, nematodes, arthropod fragments and particles of marine origin (Wynn Williams 1991, Marshall 1996). The majority of material is likely to be of local rather than intercontinental origin (eg moss spores, Marshall and Convey 1997). Currently, Antarctic summer temperatures are likely to be near or below minimum threshold temperatures for many physiological processes of potential colonists. However, even a minor change in physical conditions associated with climate change may allow a colonising propagule to survive transfer and/or establish successfully which, in turn, may impact on existing ecosystems. For example, members of the fungal genus *Glomus* form the normal mycorrhizal partner of *Deschampsia* grasses in temperate areas. Currently, in Antarctica the roots of *D. antarctica* (a grass distributed from southern Marguerite Bay (c.69°S) to the high Andes of Peru) are not colonised by these fungi (Demars and Boerner 1995), but regional warming may facilitate this happening, with inevitable but unknown effects on individuals and populations of these plants. Simple experiments using screens placed over Antarctic soil in order to reduce the severity of environmental conditions stimulate rapid growth from the normally dormant soil propagule bank present in the soils (Smith 1993, McGraw and Day 1997, Clarke 2003).

A recent study (Hughes et al. 2004) examined the diversity of prokaryotes in air collected over Rothera Point (Adelaide Island, western Antarctic Peninsula) using molecular biological techniques. The closest matches for many of the 16s rDNA sequences obtained were from Antarctic clones already present in databases (including Antarctic soil, marine and lichen genera) or from other cold environments, which strongly suggests that much of the aerobiota was of regional origin. This finding is in agreement with Marshall and Convey (1997), who found that spores of local mosses dominated in air samples collected on Signy Island (South Orkney Islands).

In the study by Hughes et al. (2004), air mass back trajectory data (Kottmeier and Fay 1998) showed that the sampled air may have travelled over the Antarctic Peninsula and Weddell Sea immediately before reaching Rothera Point, and therefore that a proportion of the detected biota may have had a more distant, though still Antarctic, origin. Winds may also aid the transfer of larger organisms. For example, Greenslade et al. (1999) and Convey (2005) document the transfer of moths from Australia to Macquarie Island, and South America to South Georgia, respectively, and there are many records of vagrant birds reaching particularly subantarctic islands, but also maritime Antarctic locations (Burger et al. 1980, Copson and Whinam 2001, Gauthier-Clerc et al. 2002, Frenot et al. 2005), in some cases already becoming established. In contrast, Pugh (2003) concluded that airborne transfer is very unlikely to provide a viable route of intercontinental transport for one of the main arthropod groups of Antarctica (Acari). Studies of airborne transfer into Antarctica, though few, strongly suggest that winds are a major means of transport for a wide variety of native and non-native species.

WATER-BORNE TRANSPORT

Dispersal on or near the surface of the ocean has been suggested as a possible route of entry into the Antarctic for certain terrestrial biota, although this proposal suffers from the same limitation as applied to airborne dispersal, in that the Antarctic Circumpolar Current will carry surface water and any associated biota eastwards around the continent rather than allowing for more rapid north-south transport.

Nevertheless, groups well-represented in Antarctica such as Collembola have well documented abilities to raft and survive on the water surface (Gressitt 1964), and even periods of immersion, as do some prostigmatid mites (eg Nanorchestidae), while other mites (particularly oribatids) have been shown to be capable of surviving the extended periods of immersion in sea (and also fresh) water that would be required during intercontinental dispersal (Coulson et al. 2002, Moore 2002). However, no direct evidence of trans-oceanic dispersal into the Antarctic by this route exists and it seems unlikely given the large distances required and generally poor sea conditions seen at high southern latitudes. Rather, dispersal on the water surface is likely to be an important route of transfer of these biota between terrestrial locations within Antarctica on a local scale, either via freshwater bodies (streams) or along the coast on the surface of the sea.

INCIDENTAL TRANSPORT ON OTHER ORGANISMS OR DEBRIS

Many vascular plant propagules (seeds) are able to travel trans-oceanic distances while remaining viable, although this mechanism is not yet implicated in explaining the presence of any contemporary Antarctic flora. At non-polar latitudes incidental transport on various natural and anthropogenic objects (hydrochory) is well known, involving both marine and terrestrial taxa and a range of life stages (Barnes 2002). While the Antarctic Circumpolar Current generally prevents such material crossing the Polar Frontal Zone into the Southern Ocean, this does occasionally occur (Barber et al. 1959, Convey et al. 2002a). Barnes and Fraser (2003) give the first clear report of marine taxa being dispersed within the Southern Ocean by rafting on an anthropogenic object. However, while there are several plausible examples (Barnes et al. 2006), assuming that the extended transfer distances and times will drastically reduce survival chances of non-marine species, this dispersal mechanism has been considered unlikely to be a common or important explanatory factor for terrestrial or freshwater biota (Gressitt et al. 1960).

A considerably faster long-distance transport mechanism, zoochory, is provided by association with migratory vertebrates (birds and seals). These can travel long distances between locations within the Antarctic, or from lower latitude locations, on timescales between daily and seasonal. Bird transport has been implicated in the transport of some microbiota into the Antarctic (Schlichting et al. 1978). However, other than for certain parasitic or commensal microarthropods actively associated with bird or mammal species (eg ticks, lice, feather mites), there remains little evidence for zoochory playing a significant role in explaining the long distance colonisation and contemporary distribution of free living Antarctic invertebrates (Pugh 1997). On a shorter scale of distance, this mechanism is likely to play an important role in the local dispersal among adjacent sites. For instance, the habit of Subantarctic *Catharacta lonnbergi* and South Polar *C. maccormicki* skuas of lining their nests with local vegetation (eg grass, moss and lichen) is almost inevitably associated with the direct transfer of invertebrate biota, in addition to viable plant fragments and propagules.

HUMAN-MEDIATED TRANSPORT

The general subject of import of non-indigenous species into the Antarctic region, and the over-riding importance of human activity in influencing this process, are dealt with elsewhere (Convey et al. 2006), and of a recent benchmark review (Frenot et al. 2005), and only a brief overview will be presented here. It is now clear that, in the subantarctic, human activity over only the last two centuries has underlain major changes in the structure and functioning of terrestrial ecosystems. These changes have been driven by the import of a very wide range of non-indigenous plants and animals, some introduced deliberately and others - the majority - inadvertently. While some of the most visible consequences have been associated with the introduction of vertebrates (including farm animals – sheep, mouflon, reindeer, cattle, horses – rabbits, rats, cats, freshwater fish), it is also now clear that a proportion of introduced plants and animals have become aggressive invaders, displacing (and even driving locally extinct) indigenous species and communities. The presence or impacts of introduced invertebrates are less well documented or understood but, as the known examples include species representing higher trophic levels not previously present in these communities (eg Chevrier et al. 1997, Ernsting et al. 1999), significant consequences can only be expected. To date, the presence or impacts of non-indigenous microbiota (including disease-causing organisms) are barely known and have received little study. While fewer cases of establishment are known from the maritime Antarctic, and none to date from the continental Antarctic, it is clear that many similar transfers of biota do occur (Whinam et al. 2005, Frenot et al. 2005).

This form of human-assisted colonisation already far outweighs in importance, in terms of both frequency of transfer events and numbers transferred, other natural dispersal and colonisation processes (Pugh 1997, Frenot et al. 2005). In effect, human assistance negates the effectiveness of the dispersal filter imposed by the physical isolation of Antarctic terrestrial habitats, removing the need for either physiological strategies or life history stages appropriate to survive the stresses of unassisted transfer by whatever route. It has long been known that a wide range of both plant and invertebrate species of lower latitudes possess ecophysiological and life history features that allow them to survive the more extreme conditions of Antarctic locations. Now, in the context of regional climate change trends that generally result in amelioration of existing climate regimes, it is likely that the ecophysiological barriers to the transfer and establishment of these 'assisted invaders' will be lowered, resulting in a greater frequency of successful establishment events and in a greater proportion of these becoming aggressive invaders (Frenot et al. 2005).

Survival on arrival

Dispersal is only the initial stage of the colonisation process. Assuming propagule survival during transfer, ecological, physiological and biochemical capacities are required to permit continued survival on deposition and subsequent establishment (Wynn-Williams 1991, see also Hennion et al. this volume, for a more detailed consideration of ecophysiological adaptation to environmental stress in Antarctica). This may include the ability to survive a continued period of anabiosis after arrival in Antarctica, a period which has the potential to become much extended if propagules are able to maintain viability when becoming trapped within a snow/ice column or permafrost (Nienow and Friedmann 1993, Gilichinsky 2001). Antarctica is an extreme environment. Variation in vegetation, soil or rock surface temperature during a single day can be >50°C, (eg +20 to -30°C) and annual temperature ranges may be >90°C (Smith 1988, Bölter 1992, Convey 1996b). To deal with these dramatic temperature fluctuations, many Antarctic organisms possess differences in detail of biochemistry and physiology in comparison with related temperate species (eg in membrane structure or enzyme activities) so that they can function at lower temperatures (Block 1990, Russell 2000). Consequently, many Antarctic microorganisms are psychrophilic (optimum growth rates obtained at below 15°C) or psychrotolerant (optimum growth rates at higher temperatures, eg 20-25°C, but will grow at temperatures below 0°C), both of which increase the amount of time that they can be metabolically active under polar conditions and hence their growth season and reproductive potential. Psychrophilic microbiota are particularly well represented in more environmentally stable Antarctic habitats, such as in sublithic and lake sediment communities (Bowman et al. 1999). As a broad generalisation, flexibility, or the ability to tolerate a wide range of environmental conditions in comparison with temperate biota, appears to be a key feature of the biology of many Antarctic terrestrial biota (Bölter 1990, 2004a,b, Convey 1996a).

Despite the potentially extreme temperature stress, liquid water availability is probably the major limitation to life across much of continental and maritime Antarctica (Kennedy 1993, Block 1996) as, once frozen, water is unavailable and metabolic processes cease. To cope with this, many Antarctic microorganisms, plants and animals have biochemical and physiological adaptations that allow them to withstand prolonged periods of both freezing and desiccation (Block 1990, Potts 1994, Wharton and Ferns 1995, Wharton 2002). These include the production of (1) sugars or sugar alcohols (such as trehalose and glycerol) to protect cell membranes, (2) compatible solutes to maintain internal osmotic pressures, (3) extracellular polysaccharides to trap water and keep it near the cell (and reduce the rate of drying experienced by cells, which presumably allows better intracellular preparation) and (4) the rapid triggering of metabolic pathways in the presence of water to allow maximum use of hydrated periods for growth and reproduction.

Solar ultraviolet radiation is another major stress faced by Antarctic terrestrial biota, especially during periods of ozone depletion when more biologically damaging short-wavelength UV-B radiation is transmitted to ground level (Cockell and Knowland 1999). Four main techniques are available for organisms to protect themselves from high UV doses (Quesada and Vincent 1997): (1) synthesis of UV-screening pigments such as mycosporine-like amino acids, MAAs (see also Newsham et al. 2002), (2) production of molecules (eg carotenoids) to quench the damaging free radicals generated when UV interacts with the cell, (3) rapid repair of

UV damage to cell components, or (4) have multiple copies of important biomolecules (such as DNA) giving a degree of redundancy.

When specific biochemical responses are not possible, many organisms develop behavioural responses or select niches that limit exposure to UV radiation. Antarctic fungi will generally show negative phototaxis when UV levels are high, or may produce protective pigments such as melanin (eg *Phoma herbarum* Hughes et al. 2003). The problem becomes more difficult for phototrophs that rely on light for photosynthesis, but may still suffer UV damage. Lichen photobionts are often protected by dark pigments synthesised by the mycobiont (Lud et al. 2001) and cyanobacteria such as *Nostoc commune* may generate their own UV protective pigments such as scytonemin or MAAs (Edwards et al. 2000). Small unicellular algae may be not be physically large enough to contain enough protective pigment to prevent damage to their cell components by ambient fluxes of UV, yet must still rely on visible light for their photosynthetic machinery. They may overcome this problem by selecting sites in fissures or beneath soil particles, where UV doses are too low to cause significant damage and try to absorb all available photosynthetically active radiation (PAR) using exceptionally sensitive light capturing systems (Cockell and Stokes 2004).

A particular challenge for the establishment process in maritime and continental Antarctic terrestrial habitats is the fact that initial growth and developmental processes must progress sufficiently during the very short summer season to allow preparation for and survival of the subsequent winter. Possession of flexibility in this element of the life history may provide a considerable advantage to colonising biota. In this context, prolonging of the early developmental processes after initial colonisation can be seen as a specialisation. For instance, in the northern maritime Antarctic, juvenile development may be prolonged substantially as has been seen in the fruticose lichen *Usnea antarctica* on Livingston Island, South Shetland Islands (Ott 2004). During the very early stages of development, the young thallus consists mainly of the mycobiont, with only a very limited element of the algal photobiont, which defines the overall energy budget through carbon fixation. In contrast, early development of crustose lichens may face different restrictions, as energy is only necessary for the development of a horizontal thallus. Sancho and Pintado (2004) describe the colonisation and development of the crustose lichen *Rhizocarpon geographicum* in new habitats close to a glacier in the South Shetland Islands, finding a relatively high growth rate.

The role played by these various ecophysiological, behavioural and life history mechanisms in any response to changing environmental conditions is likely to be complex and difficult to predict, either for indigenous or colonising species and much will depend on the precise detail of the environmental changes experienced. Clearly, a simple consequence of either increased temperature or water availability will be an increase in the opportunity for biological processes to occur, either in terms of rate or duration of activity. However, the integrated consequences of different climate change processes may be to increase environmental stress on biota. For instance, increased temperature may lead to earlier spring snowmelt and later autumn freeze and a longer apparent season, but may also result in exhaustion of water supplies in midsummer or increased evaporative stress (Convey et al. 2002b, 2003). Reduced precipitation, as seen in diverse locations across the subantarctic, may impose considerable stress on sensitive life stages of some organisms such as plant seedlings or developing roots (Hennion et al. this volume). Finally, while many Antarctic terrestrial biota appear well capable of avoiding or recovering from the adverse effects of UV exposure, their responses presuppose conditions suitable for biological activity. While this is generally true in hydrated habitats during the polar summer, it is not the case while organisms are either frozen or in a dehydrated (eg anhydrobiotic) state. In these latter cases, in the event of damage biota will be unable to respond until activity is resumed at a later point, a factor that may be particularly pertinent in considering the consequences of the continent-wide Antarctic ozone hole, which forms and is at its peak early in the austral spring, before many biota resume physiological activity. Many, but not all, such biota will receive sufficient protection from even a thin covering of winter snow (Cockell et al. 2002) but here, again, it becomes important to integrate the consequences of changes in unconnected climatic variables, as warming will lead to early loss of this protection.

ANTARCTIC ENDOLITHIC MICROBIAL COMMUNITIES - SURVIVAL UNDER THE MOST EXTREME CONDITIONS

As altitude or latitude increases, conditions become increasingly severe. Eventually external environmental conditions may be too severe for life and organisms are forced to adopt an endolithic existence (meaning literally 'within rock': Friedmann 1982, Fike et al. 2002). Endolithic habitats are not unique to the Antarctic, being well known from hot deserts and also from less severe environments such as the Niagara escarpment (Canada). The microbial colonisation of rock surfaces at Mars Oasis, Alexander Island (southern Antarctic Peninsula) provides an excellent model demonstrating a variety of techniques for coping with extreme environmental stresses (Hughes and Lawley 2003). At this location, large boulders are found that have developed thin (2-15 mm), partially transparent, gypsum crusts on their surfaces. Small microbial communities (2-3 mm in diameter) have colonised and developed within these crusts and are able to survive and reproduce in this unusual habitat due to two factors:

a) *Selection of a favourable micro-niche.* The ability of endolithic microorganisms to survive in extreme regions may be due to physiological adaptations, but selection of habitat is probably of greater importance. In this example, the microscopic sites within microfractures in the Mars Oasis gypsum crust reduce the physical stress to levels favourable for life. The crust is generally several degrees warmer (during summer) than the surrounding air temperature, due to solar radiative heating. This allows any water to remain in the liquid state for longer and for metabolic processes to occur more rapidly. The crust also rapidly absorbs any water or melted snow landing on its surface, which then travels through pores to the microorganisms. An important property of gypsum is that it completely filters out biologically harmful

UV radiation, while allowing transmission of low levels of light suitable for photosynthesis. This limitation results in low rates of carbon fixation, and estimates point to growth rates of as few as 4-38 cell divisions per year (Hughes and Lawley, 2003). Such slow growing organisms also illustrate a situation where biological processes approach geological timescales, as the maintenance of these communities requires local colonisation to counteract the loss of habitat through exfoliation of the mineral surface (Sun and Friedmann 1999).

b) *Community biodiversity and physiological capacity*. Culture techniques suggest that biodiversity in the Mars Oasis endolithic community is very low, as also reported from the Dry Valley sandstones. The cyanobacterium *Chloroglea* sp. is the dominant phototroph and its excess metabolites provide organic carbon for the few fungal and heterotrophic bacterial species in the community. Biochemical studies show that the heterotrophic species have broad enzymatic competence, which is a common characteristic in extreme environments with low levels of competition and biodiversity (Hughes and Lawley 2003).

Establishment and community development

Establishment involves the initiation of a long-term viable and reproducing population and differs from simple survival, as it requires the resources and habitat to permit reproduction, development and completion of the life cycle. Establishment of biota on ground newly exposed from beneath snow or ice involves particular problems. Initially, the number of species at a site will be limited and even insufficient to support community development (Naeem 1998) and the existence of the correct range of functional genes in the community as a whole may strongly influence colonisation success. Many polar soils are nitrogen-limited (Davey and Rothery 1993, Arnold et al. 2003) and colonisation by some species may be impossible until nitrogen-fixing cyanobacteria, lichens (containing nitrogen-fixing photobionts) or bacteria are present. In response to this, microbes isolated from extreme polar terrestrial environments, where biodiversity is low, often show a greater degree of physiological diversity (Siebert et al. 1996). Consequently, a relatively low number of species will be able to deliver all the biochemical processes required for colonisation and persistence, as illustrated by Antarctic endolithic communities.

Nevertheless, other factors may strongly influence colonization success. Stabilisation of pristine soil habitats by various microbiota (Davey et al. 1991, Wynn-Williams 1990, 1993), and consequential modification of local micro-environmental conditions (los Rios et al. 2003), have been found to be important precursors encouraging further colonisation by heterotrophic microbiota. However, a 'classical' view of establishment and community development as a linear successional process - initial colonisation and habitat stabilisation by algae and nitrogen-fixing cyanobacteria, facilitating successive colonisation by heterotrophic microbes and then invertebrates, and by macroscopic plants (cryptogams and phanerogams) - may not always be appropriate. Smith (1995) found that colonisation of glacial forelands by plants (lichens) was rapid due to the presence of nutrients provided by bird droppings, without which colonisation by some species may have been impossible. The grass *Deschampsia antarctica* is itself a very successful primary colonist of newly exposed glacial soils in sub- and maritime Antarctic locations (Smith 2001), apparently requiring little or no microbial 'priming' of the habitat. In certain situations, redistribution of 'dead' organic matter into newly exposed and unoccupied habitats can encourage colonisation and maintain populations of higher trophic levels before the establishment of their expected primary food source – this is the case with the colonisation of ground newly exposed by glacial retreat on High Arctic Svalbard (Hodkinson et al. 2002), although has not yet been observed in Antarctic terrestrial habitats. Establishment and community development on existing barren ground is unlikely to proceed unless some environmental variable changes (eg increase in temperature or nutrient availability, as illustrated by the many screen or greenhouse style manipulation experiments). However, the stochastic arrival of a key species with essential physiological and biochemical capabilities for establishment at a site may itself alter the environmental conditions and permit establishment of other species.

Biodiversity

Levels of biodiversity in Antarctic terrestrial environments have traditionally been considered to be low and this is certainly true for higher plants and animals (Convey 2001a). However, recent work suggests that microbial biodiversity is more than previously thought in soil and lake ecosystems in the maritime Antarctic (Pearce et al. 2003, Lawley et al. 2004, Tindall 2004). It has also been proposed that, unlike multicellular organisms, many microorganisms may be distributed globally (the 'global ubiquity hypothesis') and selectively utilise different parts of their genomes to survive in the different environments that they encounter (Finlay, 2002). Thus, there may be a fundamental difference in the dispersal and colonisation abilities of (at least some) microbiota relative to the more visible macroscopic groups. Genes encoding for biochemical systems that will allow adaptation to polar existence may already be present in the genome of colonizing organisms. For example, the green alga *Stichococcus bacillaris* is able to survive in temperate and tropical regions, but is also cultured from soils found in the La Gorce Mountains - only 400 km from the South Pole. Nevertheless, Antarctica is at one end of a steep natural gradient of environmental stresses found on Earth (Peck et al. 2006) and the most recent evidence continues to suggest that microbial biodiversity decreases with the more extreme conditions found in continental Antarctic sites (Lawley et al. 2004) though less rapidly than plant and animal biodiversity.

Conclusions

Native Antarctic organisms use a range of biochemical, physiological and behavioural adaptations to survive the extreme conditions of the Antarctic terrestrial environment. Limited evidence suggests that, for organisms producing propagules suitable for airborne transfer, distribution into Antarctica is not a limiting factor. Rather, the environmental conditions are often too severe for their survival and subsequent establishment. With recent climate change-associated temperature and precipitation changes over particularly the sub- and maritime Antarctic, environmental conditions may become less severe. If so in the near future, if not already, non-indigenous species could colonise and displace native species, with unknown implications for biodiversity and biogeochemical cycling. This process will likely be accelerated by the inadvertent and inevitable transfer of biota in association with human activity which evidence suggests may already far outweigh natural dispersal and colonisation processes in Antarctica.

References

Arnold, R. J., Convey, P., Hughes, K. A., and Wynn-Williams, D. D. (2003) Seasonal periodicity of physical factors, inorganic nutrients and microalgae in Antarctic fellfields, *Polar Biology* **26**, 396-403.
Bailey, R.H. and James, P. W. (1979) Birds and the dispersal of lichen propagules, *Lichenologist* **11**, 105.
Barber, H.N., Dadswell, H.E. and Ingle, H.D. (1959) Transport of driftwood from South America to Tasmania and Macquarie Island. *Nature* **184**, 203-204.
Bargagli, R., Broady, P.A. and Walton, D.W.H. (1996) Preliminary investigation of the thermal biosystem of Mount Rittmann fumaroles (northern Victoria Land, Antarctica), *Antarctic Science* **8**, 121-126.
Barnes, D.K.A. (2002) Invasions by marine life on plastic debris, *Nature* **416**, 808-809.
Barnes, D.K.A. and Fraser, K.P.P. (2003) Rafting by five phyla on man-made flotsam in the Southern Ocean, *Marine Ecology Progress Series* **262**, 289-291.
Barnes, D.K.A., Hodgson, D.A., Convey, P., Allen, C. and Clarke, A. (2006) Incursion and excursion of Antarctic biota: past, present and future, *Global Ecology and Biogeography* **15**, 121-142.
Barnes, D.K.A., Warren, N., Webb, K., Phalan, B. and Reid, K. (2004) Polar pedunculate barnacles piggy-back on pycnogona, penguins, pinniped seals and plastics, *Marine Ecology Progress Series*, **284**, 305-310.
Benninghoff, W.S. and Benninghoff, A.S. (1985) Wind transport of electrostatically charged particles and minute organisms in Antarctica. In W.R. Siegfried, P.R. Condy and R.M. Laws (eds.) *Antarctic nutrient cycles and food webs*, Springer-Verlag, Berlin Germany, pp 592-596.
Bergstrom, D.M. and Chown, S.L. (1999) Life at the front: history, ecology and change on southern ocean islands, Trends in Ecology and Evolution **14**, 472-477.
Bergstrom, D.M., Hodgson, D.A. and Convey, P. (2006) The physical setting of the Antarctic, in D.M. Bergstrom, P. Convey, and A.H.L. Huiskes (eds.), *Trends in Antarctic Terrestrial and Limnetic Ecosystems: Antarctica as a Global Indicator*, Springer, Dordrecht (this volume).
Block, W. (1984) Terrestrial Microbiology, Invertebrates and Ecosystems, in R.M. Laws (ed.), *Antarctic Ecology Volume 1*. Academic Press, London, pp. 163-236.
Block, W. (1990) Cold tolerance of insects and other arthropods, *Philosophical Transactions of the Royal Society of London Series B* **326**, 613-633.
Block, W. (1996) Cold or drought – the lesser of two evils for terrestrial arthropods? *European Journal of Entomology* **93**, 325-339.
Bölter, M. (1990) Microbial ecology of soils from Wilkes Land, Antarctica. II. Patterns of microbial activity and related organic and inorganic matter, *Proceedings of the NIPR Symposia on Polar Biology* **3**, 120-132.
Bölter, M. (1992) Environmental conditions and microbiological properties from soils and lichens from Antarctica (Casey Station, Wilkes Land), *Polar Biology* **11**, 591-599.
Bölter, M. (2004a) Soil: an extreme habitat for microorganisms? *Pedosphere* **14**, 137-144.

Bölter, M. (2004b) Ecophysiology of psychrophilic and psychrotolerant microorganisms. In S. Shivaji (ed.) *Microbes from cold habitats: biodiversity, biotechnology and cold adaptation. Cellular and Molecular Biology Special Issue* **50**, 563-573.

Bölter, M. and Stonehouse, B. (2002) Uses, preservation, and protection of Antarctic coastal regions. In: Beyer, L. and Bölter, M. (eds.) *Geoecology of Antarctic ice-free coastal landscapes.* Springer-Verlag, Berlin, Germany, *Ecological Studies* **154**, 393-407.

Bölter, M., Beyer, L. and Stonehouse, B. (2002) Antarctic coastal landscapes: characteristics, ecology and research. In L. Beyer and M. Bölter (eds) *Geoecology of Antarctic ice-free coastal landscapes.* Springer-Verlag, Berlin, Germany, *Ecological Studies* **154**, 5-22.

Bowman, J.P., Rea, S.M., Brown, M.V., McCammon, S.A., Smith, M.C. and McMeekin, T.A. (1999) Community structure and psychrophily in Antarctic microbial ecosystems. In C.R. Bell, M. Brylinsky and P. Johnson-Green (eds.) *Microbial biosystems: new frontiers.* Proceedings of the 8th International Symposium on Microbial Ecology, Atlantic Canada Society for Microbial Ecology, Halifax, Canada.

Burger, A.E., Williams, A.J. and Sinclair, J.C. (1980) Vagrants and the paucity of land bird species at the Prince Edward Islands, *Journal of Biogeography* **7**, 305-310.

Chevrier, M., Vernon, P. and Frenot, Y. (1997) Potential effects of two alien insects on a subantarctic wingless fly in the Kerguelen Islands. In B. Battaglia, J. Valencia and D.W.H. Walton (eds.), *Antarctic Communities: Species, Structure and Survival*, Cambridge University Press, Cambridge, pp. 424-431.

Clarke, A. (2003) Evolution, adaptation and diversity: global ecology in an Antarctic context. In: A.H.L. Huiskes, W.W.C. Gieskes, J. Rozema, R.M.L. Schorno, S.M. van der Vries and W.J. Wolff (eds.), *Antarctic Biology in a Global Context*, Backhuys Publishers, Leiden, The Netherlands, pp. 3-17.

Clarke, A., Barnes, D.K.A. and Hodgson, D.A. (2005) How isolated is Antarctica? *Trends in Ecology and Evolution* **20**, 1-3.

Cockell, C.S. and Knowland, J. (1999) Ultraviolet radiation screening compounds, *Biological Reviews* **74**, 311-345.

Cockell, C.S., Rettberg, P., Horneck, G., Wynn-Williams, D.D., Scherer, K. and Gugg-Helminger, A. (2002) Influence of ice and snow covers on the UV exposure of terrestrial microbial communities: dosimetric studies, *Journal of Photochemistry and Photobiology B: Biology* **68**, 23-32.

Cockell, C.S. and Stokes, M.D. (2004) Widespread colonization by polar hypoliths, *Nature* **431**, 414.

Convey, P. (1996a) The influence of environmental characteristics on life history attributes of Antarctic terrestrial biota, *Biological Reviews* **71**, 191-225.

Convey, P. (1996b) Overwintering strategies of terrestrial invertebrates from Antarctica - the significance of flexibility in extremely seasonal environments, *European Journal of Entomology* **93**, 489-505.

Convey, P. (1997) Environmental change: possible consequences for the life histories of Antarctic terrestrial biota, *Korean Journal of Polar Research* **8**, 127-144.

Convey, P. (2001a) Antarctic Ecosystems. In S. Levin (ed.), *Encyclopaedia of Biodiversity*, vol 1., Academic Press, San Diego, USA, pp. 171-184.

Convey, P. (2001b) Terrestrial ecosystem response to climate changes in the Antarctic. In G.-R. Walther, C.A. Burga and P.J. Edwards (eds.), *"Fingerprints" of climate change - adapted behaviour and shifting species ranges*, Kluwer, New York, pp 17-42.

Convey, P. (2003) Maritime Antarctic climate change: signals from terrestrial biology. In E. Domack, A. Burnett, A. Leventer, P. Convey, M. Kirby and R. Bindschadler (eds.), *Antarctic Peninsula Climate Variability: Historical and Palaeoenvironmental Perspectives*, Antarctic Research Series, Vol. 79, American Geophysical Union, Washington, D.C., pp. 145-158.

Convey, P. (2005) Recent lepidopteran records from sub-Antarctic South Georgia, *Polar Biology* **28**, 108-110.

Convey, P. (2006) Antarctic climate change and its influences on terrestrial ecosystems, in D.M. Bergstrom, P. Convey, and A.H.L. Huiskes (eds.), *Trends in Antarctic Terrestrial and Limnetic Ecosystems: Antarctica as a Global Indicator.* Springer, Dordrecht (this volume).

Convey, P. and McInnes, S.J. (2005) Exceptional, tardigrade dominated, ecosystems from Ellsworth Land, Antarctica, *Ecology* **86**, 519-527.

Convey, P. and Smith, R.I.L. (1993) Investment in sexual reproduction by Antarctic mosses, *Oikos* **68**, 293-302.

Convey, P., Barnes, D.K.A. and Morton, A. (2002a) Artefact accumulation on Antarctic oceanic island shores, *Polar Biology* **25**, 612-617.

Convey, P., Block, W. and Peat, H.J. (2003) Soil arthropods as indicators of water stress in Antarctic terrestrial habitats? *Global Change Biology* **9**, 1718-1730.

Convey, P., Smith, R.I.L., Hodgson, D.A. and Peat, H.J. (2000) The flora of the South Sandwich Islands, with particular reference to the influence of geothermal heating, *Journal of Biogeography* **27**, 1279-1295.

Convey, P., Pugh, P. J. A., Jackson, C., Murray, A. W., Ruhland, C. T., Xiong, F. S. and Day, T. A. (2002b) Response of Antarctic terrestrial arthropods to multifactorial climate manipulation over a four year period. *Ecology* **83**, 3130-3140.

Copson, G. and Whinam, J. (2001) Review of ecological restoration programme on subantarctic Macquarie Island: pest management progress and future directions, *Ecological Management and Restoration* **2**, 129-138.

Corner, R.W.M. and Smith, R.I.L. (1973) Botanical evidence of ice recession in the Argentine Islands, *British Antarctic Survey Bulletin* **35**, 83-86.

Coulson, S.J., Hodkinson, I.D., Webb, N.R. and Harrison, J.A. (2002) Survival of terrestrial soil-dwelling arthropods on and in seawater: implications for trans-oceanic dispersal, *Functional Ecology* **16**, 353-356.

Davey, M.C., Davidson, H.P.B., Richard, K.J. and Wynn-Williams, D.D. (1991) Attachment and growth of Antarctic soil cyanobacteria and algae on natural and artificial substrata, *Soil Biology and Biochemistry* **23**, 185-191.

Davey, M.C. and Rothery, P. (1993) Primary colonization by microalgae in relation to spatial variation in edaphic factors on antarctic fellfield soils, *Journal of Ecology* **8**, 335-343.

Day, T.A., Ruhland, C.T., Grobe, C.W. and Xiong, F. (1999) Growth and reproduction of Antarctic vascular plants in response to warming and UV radiation reductions in the field, *Oecologia* **119**, 24-35.

Demars, B.G. and Boerner, R.E.J. (1995) Mycorrhizal status of *Deschampsia antarctica* in the Palmer Station area, Antarctica, *Mycologia* **87**, 451-453.

During, H.J. (1979) Life strategies of bryophytes: a preliminary review, *Lindbergia* **5**, 2-18.

Edwards, H.G.M., Garcia-Pichel, F., Newton, E.M. and Wynn-Williams, D.D. (2000) Vibrational Raman spectroscopic study of scytonemin, the UV-protective cyanobacterial pigment, *Spectrochimica Acta Part A - Molecular and Biomolecular Spectroscopy* **56**, 193-200.

Ellis-Evans, J.C. and Walton, D. (1990) The process of colonization in Antarctic terrestrial and freshwater ecosystems, *Proceedings of the NIPR Symposium on Polar Biology* **3**, 151-163.

Ernsting, G., Brandjes, G.J., Block, W. and Isaaks, J.A. (1999) Life-history consequences of predation for a subantarctic beetle: evaluating the contribution of direct and indirect effects, *Journal of Animal Ecology* **68**, 741-752.

Fike, D.A., Cockell, C., Pearce, D. and Lee, P. (2002) Heterotrophic microbial colonization of the interior of impact-shocked rocks from Haughton impact structure, Devon Island, Nunavut, Canadian High Arctic, *International Journal of Astrobiology* **1**, 311-323.

Finlay, B.J. (2002) Global dispersal of free-living microbial eukaryote species, *Science* **296**, 1061-1063.

Frenot, Y., Chown, S.L., Whinam, J., Selkirk, P., Convey, P., Skotnicki, M. and Bergstrom, D. (2005) Biological invasions in the Antarctic: extent, impacts and implications, *Biological Reviews*, **80**, 45-72.

Friedmann, E. I. (1982) Endolithic microorganisms in the Antarctic cold desert *Science* **215**, 1045-1053.

Fowbert, J.A. and Smith, R.I.L. (1994) Rapid population increases in native vascular plants in the Argentine Islands, Antarctic Peninsula, *Arctic and Alpine Research* **26**, 290-296.

Fox, A.J. and Cooper, A.P.R. (1998) Climate-change indicators from archival aerial photography of the Antarctic Peninsula, *Annals of Glaciology* **27**, 636-642.

Galloway, D. (1987) Austral lichen genera: some biographical problems, *Progress and Problems in Lichenology in the Eighties. Bibl. Lichenol.* **25**, 385-399. J. Cramer, Berlin-Stuttgart, Germany.

Gauthier-Clerc, M., Jiguet, F. and Lambert, N. (2002) Vagrant birds at Possession Island, Crozet Islands and Kerguelen Island from December 1995 to December 1997, *Marine Ornithology* **30**, 38-39.

Gilichinsky, D.A. (2001) Permafrost a model of extraterrestrial habitat. In G. Horneck and C. Baumstark-Khan (eds.) *Astrobiology. The quest for conditions of life.* Springer-Verlag, Berlin, Germany, pp. 125-142.

Greenslade, P., Farrow, R.A. and Smith, J.M.B. (1999) Long distance migration of insects to a subantarctic island, *Journal of Biogeography* **26**, 1161-1167.

Gressitt, J.L. (1964) Insects of Campbell Island, *Pacific Insects Monograph* **7**, 1-663.

Gressitt, J.L., Larch, R.E. and O'Brien, C.W. 1960. Trapping air-borne insects in the Antarctic area. *Pacific Insects* **2**, 245-250.

Hansen, J., Ruedy, R., Glasgoe, J. and Sato, M. (1999) GISS analysis of surface temperature change, *Journal of Geophysical Research* **104**, 30997-31022.

Hennion, F., Huiskes, A.H.L., Robinson, S. and Convey, P. (2006) Physiological traits or organisms in a changing environment, in D.M. Bergstrom, P. Convey, and A.H.L. Huiskes (eds.), *Trends in Antarctic Terrestrial and Limnetic Ecosystems: Antarctica as a Global Indicator,* Springer, Dordrecht (this volume).

Hodkinson, I.D., Webb, N.R. and Coulson, S.J. (2002) Primary community assembly on land – the missing stages: why are the heterotrophic organisms always there first? *Journal of Ecology* **90**, 569-577.

Houghton, J.T., Ding, Y., Griggs, D.J., Noquer, M., van der Linden, P.J., Dai, X., Maskell, K. and Johnson, C.A. (eds) (2001) *Climate change 2001: The scientific basis: contribution of working group I to the third assessment report of the intergovernmental panel on climate change*, Cambridge University Press, Cambridge, UK.

Hughes, K. A. and Lawley, B. (2003) A novel Antarctic microbial endolithic community within gypsum crust *Environmental Microbiology* **5**, 555-565.

Hughes, K. A., Lawley, B. and Newsham, K. K. (2003) Solar UV-B radiation inhibits the growth of Antarctic terrestrial fungi *Applied and Environmental Microbiology* **69**, 1488-1491.

Hughes, K. A., McCartney, H. A., Lachlan-Cope, T. A. and Pearce D. A. (2004) A preliminary study of airborne microbial biodiversity over peninsular Antarctica, *Cellular and Molecular Biology* **50**, 537-542.

Huiskes, A.H.L., Convey, P. and Bergstrom, D.M. (2006) Trends in Antarctic terrestrial and limnetic ecosystems: Antarctica as a global indicator In: Bergstrom, D.M., Convey, P. and Huiskes, A.H.L. (eds) *Trends in Antarctic Terrestrial and Limnetic Ecosystems: Antarctica as a Global Indicator.* Springer, Dordrecht (this volume).

Jahns, H.M. (1982) The cyclic development of mosses and the lichen *Baeomyces rufus* in an ecosystem, *Lichenologist* **14**, 261-265.

Kappen, L. (2004) The diversity of lichens in Antarctica, a review and comments. *Contributions to Lichenology. Festschrift in Honour of Hannes Hertel. P. Döbbeler and G. Rambold (eds.) Bibl. Lichenol.* **88**, 331-343. J. Cramer, Berlin-Stuttgart.

Kennedy, A.D. (1993) Water as a limiting factor in the Antarctic terrestrial environment: a biogeographical synthesis *Arctic and Alpine Research* **25**, 308-315.

Kennedy, A. (1995) Antarctic terrestrial ecosystem response to global environmental change, *Annual Review of Ecology and Systematics* **26**, 683-704.

King, J.C., Turner, J., Marshall, G.J., Connolley, W.M. and Lachlan-Cope, T.A. (2003) Antarctic Peninsula climate variability and its causes as revealed by analysis of instrumental records, *Antarctic Research Series* **79**, 17-30.

Kottmeier, C. and Fay, B. (1998) Trajectories in the Antarctic lower troposphere, *Journal of Geophysical Research* **103**, 10947-10959.

Lawley, B., Ripley, S., Bridge, P. and Convey, P. (2004) Molecular analysis of geographic patterns in eukaryotic diversity of Antarctic soils, *Applied and Environmental Microbiology* **70**, 5963-5972.

Linskens, H.F., Bargagli, R., Cresti, M. and Focardi, S. (1993) Entrapment of long-distance transported pollen grains by various moss species in coastal Victoria Land, Antarctica, *Polar Biology* **13**, 81-87.

Longton, R.E. (1988) *The biology of polar bryophytes and lichens,* Cambridge University Press, Cambridge, 391 pp.

Lud, D., Huiskes, A.H.L., Moerdijk, T.C.W. and Rozema, J. (2001) The effects of altered levels of UV-B radiation on an Antarctic grass and lichen, *Plant Ecology* **154**, 87-99.

Marshall, W.A. (1996) Biological particles over Antarctica, *Nature,* **383**, 680.

Marshall, W.A. and Convey, P. (1997) Dispersal of moss propagules on Signy Island, maritime Antarctic, *Polar Biology* **18**, 376-383.

Mason, B. J. (1971) Global atmospheric research programme, *Nature* **233**, 382-388.

McGraw J.B. and Day, T.A. (1997) Size and characteristics of a natural seed bank in Antarctica, *Arctic and Alpine Research* **29**, 213-216.

Miles, C.J. and Longton, R.E. (1992) Deposition of moss spores in relation to distance from parent gametophytes, *Journal of Bryology* **17**, 355-368.

Moore, P.D. (2002) Springboards for springtails, *Nature* **418**, 381.

Muñoz, J., Felicísimo, A.M., Cabezas, F., Burgaz, A.R. and Martínez, I. (2004) Wind as a long-distance dispersal vehicle in the Southern Hemisphere, *Science* **304**, 1144-1147.

Naeem, S. (1998) Species redundancy and ecosystem reliability, *Conservation Biology* **12**, 39-45.

Newsham, K.K., Hodgson, D.A., Murray, A.W.A., Peat, H.J. and Smith, R.I.L. (2002) Response of two Antarctic bryophytes to stratospheric ozone depletion, *Global Change Biology* **8**, 972-983.

Nienow, J.A. and Friedmann, E.I. (1993) Terrestrial lithophytic (rock) communities. In E.I. Friedmann (ed.) *Antarctic microbiology*. Wiley-Liss, New York, USA, pp. 343-412.

Ott, S. (2004) Early stages of development in *Usnea antarctica* Du Rietz in the South Shetland Islands, northern maritime Antarctic, *The Lichenologist*, **36**, 413-423.

Pearce, D.A., van der Gast, C.J., Lawley, B. and Ellis-Evans, J.C. (2003) Bacterioplankton community diversity in a maritime Antarctic lake, determined by culture-dependent and culture independent techniques, *FEMS Microbial Ecology* **45**, 59-70.

Peck, L.S., Convey, P. and Barnes, D.K.A. (2006) Environmental constraints on life histories in Antarctic ecosystems: tempos, timings and predictability, *Biological Reviews*.

Potts, M. (1994) Desiccation tolerance of prokaryotes, *Microbiological Reviews* **58**, 755-805.

Pugh, P.J.A. (1997) Acarine colonization of Antarctica and the islands of the Southern Ocean: the role of zoohoria, *Polar Record* **33**, 113-122.

Pugh, P.J.A. (2003) Have mites (Acarina: Arachnida) colonised Antarctica and the islands of the Southern Ocean via air currents? *Polar Record* **39**, 239-244.

Quesada, A. and Vincent, W.F. (1997) Strategies of adaptation by Antarctic cyanobacteria to ultraviolet radiation, *European Journal of Phycology* **32**, 335-342.

los Rios, A., Wierzchos, J., Sancho, L. and Ascaso, C. (2003) Acid microenvironments in microbial biofilms of Antarctic endolithic microecosystems, *Environmental Microbiology* **5**, 231-237.

Romeike, J., Friedl, T., Helms, G. and Ott, S. (2002) Genetic diversity of algal and fungal partners in four species of *Umbilicaria* (lichenized Ascomycetes) along a transect of the Antarctic Peninsula, *Molecular Evolution and Biology* **19**, 1209-1217.

Russell, N.J. (2000) Toward a molecular understanding of cold activity of enzymes from psychrophiles, *Extremophiles* **4**, 83-90.

Sancho, L.G. and Pintado, A. (2004) Evidence of high annual growth rate for lichens in the maritime Antarctic, *Polar Biology* **27**, 312-319.

Schaper, T. and Ott, S. (2003) Photobiont selectivity and interspecific interactions in lichen communities. I. Culture experiments with the mycobiont *Fulgensia bracteata*, Plant Biology **5**, 441-450.

Schlichting, H.E., Speziale, B.J. and Zink, R.M. (1978) Dispersal of algae and protozoa by Antarctic flying birds, *Antarctic Journal of the USA* **13**, 147-149.

Siebert, J., Hirsh, P., Hoffman, B., Gliesche, C.G., Peisse, K. and Jendrach, M. (1996) Cryptoendolithic microorganisms from Antarctic sandstones of Linnaeus Terrace (Asgard Range): diversity, properties and interactions, *Biodiversity and Conservation* **5**, 1337-1363.

Smith, A.M., Vaughan, D.G., Doake, C.S.M. and Johnson, A.C. (1999) Surface lowering of the ice ramp at Rothera Point, Antarctic Peninsula, in response to regional climate change, *Annals of Glaciology* **27**, 113-118.

Smith, R.I.L. (1988) Recording bryophyte microclimate in remote and severe environments, In J.M. Glime (ed.), *Methods in Bryology*, Hattori Botanical Laboratory, Nichinan, Japan, pp. 275-284.

Smith, R.I.L. (1990) Signy Island as a paradigm of biological and environmental change in Antarctic terrestrial ecosystems, in K.R. Kerry and G. Hempel (eds), *Antarctic Ecosystems, Ecological Change and Conservation*, Springer, Berlin, Germany, pp32-50.

Smith, R.I.L. (1991) Exotic sporomorpha as indicators of potential immigrant colonists in Antarctica, *Grana* **30**, 313-324.

Smith, R.I.L. (1993) The role of bryophyte propagule banks in primary succession: case-study of an Antarctic fellfield soil, in: J. Miles and D.W.H. Walton, (eds), *Primary Succession on Land*, British Ecological Society Special Publication, Blackwell Scientific Publications, Oxford, UK, pp 55-78.

Smith, R.I.L. (1994) Vascular plants as indicators of regional warming in Antarctica, *Oecologia* **99**, 322-328.

Smith, R.I.L. (1995) Colonization by lichens and the development of lichen-dominated communities in the maritime Antarctic, *Lichenologist* **27**, 473-483.

Smith, R.I.L. (2000) Plants of extreme habitats in Antarctica, in B. Schroeter, M. Schlensog and T.G.A. Green (eds.), *New Aspects in Cryptogamic Research. Contributions in Honour of Ludger Kappen. Bibl. Lichenol. 75: 405-419.* J. Cramer, Berlin-Stuttgart, Germany.

Smith R.I.L. (2001) Plant colonization response to climate change in the Antarctic, *Folia Fac. Sci. Nat. Univ. Masarykianae Brunensis, Geographia* **25**, 19-33.

Stevens, M.I., and Hogg, I.D. (2003) Long-term isolation and recent range expansion from glacial refugia revealed for the endemic springtail *Gomphiocephalus hodgsoni* from Victoria Land, Antarctica, *Molecular Ecology* **12**, 2357-2369.

Sun, H.J. and Friedmann, E.I. (1999) Growth on geological time scales in the Antarctic cryptoendolithic microbial community, *Geomicrobiological Journal* **16**, 193-202.

Tindall, B.L. (2004) Prokaryotic diversity in the Antarctic: the tip of the iceberg, *Microbial Ecology* **47**, 271-283.

Turner, J., Colwell, S.R. and Harangozo, S. (1997) Variability of precipitation over the coastal western Antarctic Peninsula from synoptic observations, *Journal of Geophysical Research* **102**, 13999-14007.

Vaughan, D.G. and Doake, C.S.M. (1996) Recent atmospheric warming and retreat of ice shelves on the Antarctic Peninsula, *Nature* **379**, 328-331.

Vaughan, D.G., Marshall, G.J., Connolley, W.M., Parkinson, C., Mulvaney, R., Hodgson, D.A., King, J.C., Pudsey, C.J. and Turner, J. (2003) Recent rapid regional climate warming on the Antarctic Peninsula, *Climate Change* **60**, 243-274.

Wharton, D.A. (2002) *Life at the limits: organisms in extreme environments*, Cambridge University Press, Cambridge, UK, 307 pp.

Wharton, D.A. and Ferns, D.J. (1995) Survival of intracellular freezing by the Antarctic nematode *Panagrolaimus davidi*, *Journal of Experimental Biology* **198**, 1381-1387.

Whinam, J., Chilcott, N. and Bergstrom, D.M. (2005) Subantarctic hitchhikers: expeditioners as vectors for the introduction of alien organisms, *Biological Conservation* **121**, 207-219.

Williams, P.G., Roser, D.J. and Seppelt, R.D. (1994) Mycorrhizas of hepatics in continental Antarctica, *Mycological Research* **98**, 34-36.

Wynn-Williams, D.D. (1990) Microbial colonization processes in Antarctic fellfield soils – an experimental overview, *Proceedings of the NIPR Symposia on Polar Biology* **3**, 164-178.

Wynn-Williams, D.D. (1991) Aerobiology and colonization over Antarctica – the BIOTAS programme, *Grana* **30**, 380-393.

Wynn-Williams, D.D. (1993) Microbial processes and initial stabilization of fellfield soil. In J. Miles and D.W.H. Walton (eds.) *Primary succession on land*, Blackwell, Oxford, UK, pp. 17-32.

4. BIOGEOGRAPHY

S.L. CHOWN
Centre for Invasion Biology
Department of Botany and Zoology
Stellenbosch University
Private Bag X1
Matieland 7602
South Africa
slchown@sun.ac.za

P. CONVEY
British Antarctic Survey, Natural Environment Research Council
High Cross, Madingley Road
Cambridge CB3 0ET, United Kingdom
p.convey@bas.ac.uk

Introduction

The biogeography of Antarctica and its surrounding islands is both complex and contentious. Why this should be the case is readily apparent when considering the region and its biotas. The Antarctic continent alone has a surface area of c.45 million km^2, originated following accretion of several very different terrains, and now has less than 0.32% of this surface exposed above its extensive ice cover (British Antarctic Survey 2004, Peck et al. 2006). Its surrounding islands, which are distributed sparsely within the vast Southern Ocean between approximately 45° - 60°S, likewise have varied geological origins, ranging from Upper Jurassic/Lower Cretaceous continental crust in the case of South Georgia to recent (0.5 million years) basaltic volcanism in the case of the Prince Edward Islands (Le Masurier and Thompson 1990, Peck et al. 2006), and their glacial histories differ markedly from archipelago to archipelago, and even among islands within a group (Hall 2002). Making use of this geological chessboard are a group of pieces whose origins,

55

D.M. Bergstrom et al. (eds.), Trends in Antarctic Terrestrial and Limnetic Ecosystems, 55–69.
© 2006 *Springer.*

identities, positions, moves and roles are often poorly understood. Many ice-free areas have yet to be systematically explored and investigations of several areas are surprisingly recent (Broady and Weinstein 1998, Convey et al. 2000a,b, Marshall and Chown 2002, Stevens and Hogg 2002, Bargagli et al. 2004, Convey and McInnes 2005, H.J. Peat et al. unpubl. data). Moreover, no comprehensive database of the distributions of Antarctic and subantarctic species yet exists (see Griffiths et al. 2003 for a marine example), although several non-digital compilations are now becoming available (eg Pugh 1993, Bednarek-Ochyra et al. 2000, Øvstedal and Lewis Smith 2001, Pugh et al. 2002, Pugh and Scott 2002, Ochyra et al. in press) and the RiSCC biodiversity database has accumulated 80 000 terrestrial and freshwater records. Systematic information is also absent for many taxa, the number of systematists working on the biota of the region is surprisingly low (ie a substantial taxonomic impediment exists, Samways 1994) and comprehensive phylogenies based on modern molecular methods are rare (for an exception see Allegrucci et al. 2006). In consequence, biogeographic assessments have to rely on incomplete data and methods that often make use of species as if they were wholly independent of each other and shared no phylogenetic history, when other, more powerful phylogeographic or similar approaches are available (van Veller and Brooks 2001). It is little wonder then that Antarctic biogeography remains pre-occupied with many of the same questions that puzzled the region's early biogeographers, although in some cases this pre-occupation is waning given recent solutions to these problems.

Perhaps best known among the questions facing Antarctic terrestrial biogeography are:

- The role that dispersal and vicariance have played in determining the current distribution of organisms across the region and the significance of the order of break-up of Gondwana for both these taxa and their relatives on the surrounding Southern Hemisphere continents (Brundin 1966, Darlington 1970, Craig 2003, Bergstrom et al. this volume, Gibson et al. this volume).

- The age of the Kerguelen Plateau and Iles Crozet, and the origin of the biotas both of these islands and others in this region, that is often known as the Kerguelen Biogeographic province or the South Indian Ocean Province (Udvardy 1987, Kuschel and Chown 1995, Craig et al. 2003).

- Whether most Antarctic organisms are recent, post-glacial colonists, or whether, at least in some taxa, palaeoendemism is the norm (Chown and Convey 2006).

In this chapter, we provide a review of the historical and ecological biogeography of terrestrial systems in the Antarctic, realizing that the distinction between the two approaches to biogeography is largely artificial. It represents little more than a convenient way to discuss the same patterns of biodiversity from the perspective of different scales (Ricklefs 2004). In doing so, we also recognize that both Antarctica and at least some of its surrounding islands have not always been occupied by their current biotas. The continent was once home to a diverse flora (eg Quilty 1990) and a fauna that included dinosaurs (Hammer and Hickerson 1994), the earliest representatives of the globe's modern avifauna (Clarke et al. 2005) and presumably a wide variety of insects (Ashworth and Kuschel 2003). Likewise, extensive fossil floras characterized the oldest of the subantarctic islands on the Kerguelen Plateau (Chastain 1958, Quilty and Wheller 2000). Over shorter time scales, palynological evidence has demonstrated compositional change in the floras of many sites (eg Scott 1985) and lake sediment cores have also revealed substantial variation in the abundances of terrestrial and freshwater invertebrates such as mites and crustaceans (Hodgson and Convey 2005, Cromer et al. in press). However, synthetic review of temporal trends and processes in the Antarctic biota is largely beyond the scope of this chapter and we refer to such processes only in the context of explaining current spatial distributions.

Biogeographic regions

Early biogeographic work divided the Antarctic region into the continental and maritime Antarctic and subantarctic biogeographic zones (reviewed in Lewis Smith 1984, Peck et al. 2006, see Fig. 1 Huiskes et al. this volume). The continental Antarctic is by far the largest of the zones, including most of the continental area, the east side and southern regions of the Antarctic Peninsula and the Balleny Islands. The maritime Antarctic is typically described as including the western coastal regions of the Antarctic Peninsula southwards to Alexander Island (approximately 72°S) (although this is not equivalent to the West Antarctic geological region, which includes the entire Antarctic Peninsula), along with additional Scotia Arc island groups including the South Shetland, South Orkney and South Sandwich archipelagos and the isolated Bouvetøya and Peter I Øy (Peck et al. 2006). Recent work has identified a much clearer distinction between the Antarctic Peninsula and the remainder of continental Antarctica than previously recognised. Indeed, the lack of overlap at species level among the representatives of several higher taxonomic groups is striking. Thus, the most recent taxonomic treatments indicate that there is no overlap at species level among these regions for nematodes (Andrássy 1998) and free-living mites (Pugh 1993) and that only a single springtail species is shared (Greenslade 1995). Other groups, such as tardigrades, show an intermediate level of species overlap (c. 50%, Convey and McInnes 2005). Although bryophytes appear to show a completely different pattern of biodiversity, with very few (< 5 species) or possibly even no endemism across the entire continent, the strength of the separation between the Peninsula and continental Antarctica is so robust that Chown and Convey (2006) argued that it should be recognized by a formal biogeographic boundary (Fig. 1). They called this the 'Gressitt Line' in honour of the work done by J.L. Gressitt in the region.

The third region, the subantarctic, has been the subject of some controversy. Typically when eco-climatic criteria (temperature and the presence/absence of trees or woody shrubs) are used, it has been argued that the New Zealand subantarctic islands (Auckland, Campbell, Antipodes, Snares) and other islands, such as the Falklands, St Paul, Gough and others, should be excluded from this zone or region because they are 'cool temperate' islands. Rather, the only islands that should be termed subantarctic are South Georgia, Prince Edward Islands, Iles Crozet, Iles Kerguelen, Heard and McDonald Islands and Macquarie Island (see discussion in Lewis Smith 1984, Chown et al. 1998). However, other authors have suggested that the New Zealand subantarctic islands and others should be included in the subantarctic region (see Dingwall 1995), largely because they share many species or genera with the more typically recognized subantarctic islands. To some extent, the problem is tied up with the question of whether the biotas can be considered part of a larger Insulantarctica biogeographic region, or whether they have such divergent origins and histories that they should be dealt with as part of the continental biotas from which they are presumably derived (reviewed in Greve et al. 2005 and see also below). Recent work has shown that this depends very much on the taxa under consideration (Greve et al. 2005) and the questions that are being investigated. Moreover, it seems likely that this perspective will continue to hold even when phylogeographic work has been done for a variety of taxa. Therefore, we think that the philosopher John Dewey's advice, to get over the question, rather than to answer it, is best heeded in this instance.

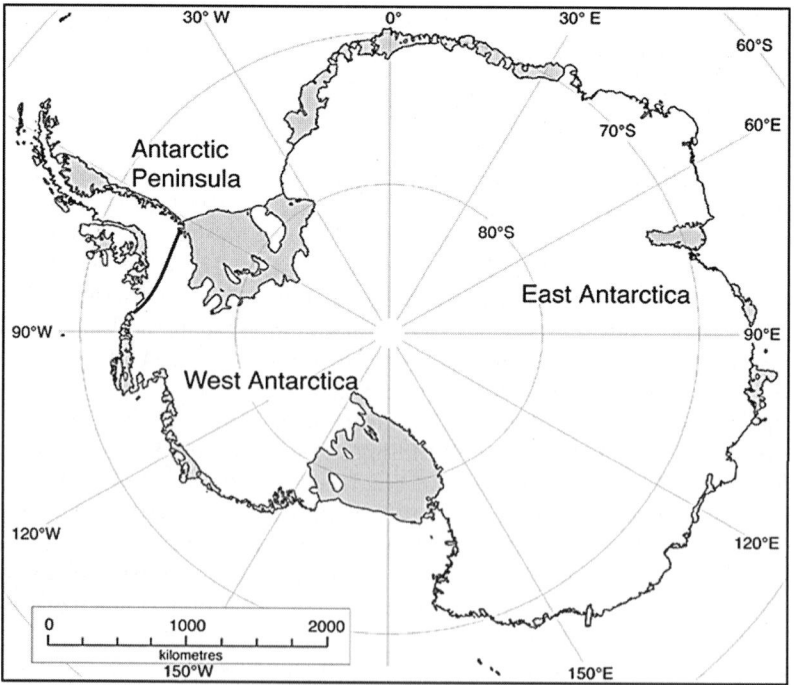

Figure 1. Map of the 'Gressitt Line' (shown in black across the base of the Antarctic Peninsula), a strong biogeographic region of separation between the biotas of the Antarctic Peninsula and continental Antarctica (Chown and Convey 2006).

Historical biogeography

Owing to its large area and complex geological origins (biologically relevant reviews in Clarke 2003, Peck et al. 2006), the historical biogeography of the terrestrial Antarctic must by necessity be relatively complicated. Compounding this situation is the fact that by the mid-Miocene continental ice-sheets were substantial and most of the biota had probably been driven extinct either by then or soon thereafter. Thus, the early history of the region has largely been lost (especially since so little ice-free land that might contain fossils is available). Moreover, very low species endemism in some groups, particularly the mosses (Bednarek-Ochyra et al. 2000, Ochyra et al. in press, H.J. Peat et al. unpubl. data), combined with substantially more extensive glaciation of the Antarctic during the last glacial maximum than at present (Peck et al. 2006), has encouraged the view that the majority of Antarctic species are relatively recent arrivals, with perhaps a few microbial or protozoan taxa being substantially older.

However, recent work has altered this perspective. It now seems that mosses may have an atypical pattern of endemism compared with most other major groups of Antarctic flora and fauna and that the generalisation of an assumption of recent origin may be little more than dogma. In the continental Antarctic, it is clear that several areas (eg parts of the Victoria Land Dry Valleys, Transantarctic Mountains, some inland nunatak groups) have remained ice free since at least the end of the Miocene (Boyer 1979, Prentice et al. 1993). Careful re-assessment of some continental Antarctic mite distributions has also indicated that the majority of species are probably pre-Pleistocene endemics, and that speciation in some groups, such as the endemic oribatid genus *Maudheimia*, is in keeping with models of the development of the East Antarctic ice sheet (Marshall and Pugh 1996, Marshall and Coetzee 2000). The west Antarctic remains problematic, because the large majority of biota are present in coastal and low altitude locations, which glacial and ice sheet reconstructions indicate would have been obliterated at glacial maximum by ice sheets extending out to the point of continental shelf drop-off. Thus, the existence and location of potential refuge regions remain hypothetical (Convey 2003). Nevertheless, evidence is increasing for the presence of an ancient and vicariant biota, such as species of endemic midges (Diptera, Chironomidae) whose age of evolutionary separation has been estimated using a "molecular clock" approach at 20-40 million years, in keeping with the opening of the Drake Passage (Allegrucci et al. 2006). It is also clear that many Antarctic nematodes are endemic to either the continental or maritime Antarctic, strongly suggesting that they are glacial survivors rather than post-glacial colonists (Andrássy 1998, Maslen and Convey, in press).

Precisely how the biotas of different refuges and ice-free areas in Antarctica are related and how these relationships might be concatenated to indicate the history of the major regions of the continent is, nonetheless, still some way off. As we have noted previously in this chapter, many ice-free areas remain unsurveyed, and although molecular work at the population level is being done (Stevens and Hogg this volume, Chown and Convey 2006), appropriate phylogeographic work is rare (although now common for many marine groups: Bargelloni et al. 2000, Page and Linse 2002).

The origins and relationships of the subantarctic biota also remain contentious, for two major reasons. First, the geological history of areas such as the Kerguelen Plateau has, until recently (Wallace et al. 2002) been contentious, while for others, such as the Iles Crozet it remains so (Craig 2003). Second, almost no molecular systematic work has been undertaken on the biotas of these islands. In consequence, it is still not clear what are the closest relatives of many groups (such as the *Ectemnorhinus*-group weevils and *Pringleophaga* spp. tineid moths), despite the fact such relationships have been mooted on the basis of conventional, morphological systematics (Jeannel 1964, Kuschel and Chown 1995). The absence of such studies for a wide variety of taxa has also compounded the question of whether many of the Southern Ocean islands should be considered a single biogeographic region or simply part of the regions to which they are closest. On the one hand, it has been suggested that all of the islands share enough history and a sufficient proportion of their biotas to be considered a single biogeographic province (Holdgate 1960, Skottsberg 1960, Van de Vijver and Beyens 1999), often known as Insulantarctica (Udvardy 1987). The other, and perhaps more widely accepted, multi-regional view is that these similarities are not large enough to warrant inclusion of the biotas within a single biogeographic province. Proponents of the multi-regional hypothesis argue that across the Southern Ocean, island biotas differ substantially in their origins, histories, source areas, and endemicity (Gressitt 1970, Lewis Smith 1984, Morrone 1998, Cox 2001, Broughton and McAdam 2002), and as a consequence there are considerable regional differences in the structure and composition of the biotas.

Despite several recent analyses, the question of the origin and relationships of the Southern Ocean Island biotas appears to have done little but shift back and forth among these competing hypotheses, providing no convincing argument why either of the ideas should enjoy primacy. Whilst the obvious solution to these problems might seem to be a concerted, multitaxon phylogeographic study of the region, too few marine and terrestrial taxa have been examined in sufficient detail for this to be done at present (though see Bargelloni et al. 2000). Moreover, the results of such an analysis might also be subject to domination by the most speciose taxa. Recently, however, nestedness analyses of a subset of terrestrial taxa have provided considerable insight into both this conundrum and the ways in which it might be solved (Greve et al. 2005).

A perfectly nested distribution, based on species incidences, is one in which species occurring at the site of interest are always present in a more species-rich site, whereas species absent from the site of interest never occur in a more depauperate one. In the Insulantarctica scenario, significant nestedness might be expected, at least at a generic level, owing to similarities in the source pool of colonists across the region, or to vicariance of a once larger landmass. Alternatively, in the multi-regional scenario, differences in endemicity, source pools and biogeographic history are likely to mean little in the way of significant nestedness across the Southern Ocean islands at either the generic or species levels. However, within a particular archipelago, significant nestedness should not only occur, but should also be very much greater than that found across the region (Greve et al. 2005). Because one of the mechanisms underlying nestedness is differential colonization ability, the extent of nestedness should vary considerably with the dispersal ability of the taxon concerned: organisms with poor dispersal abilities are apt to show much less nestedness than taxa with well-developed dispersal abilities (Greve et al. 2005). In the case of the Southern Ocean island taxa for which comprehensive information on distributions is available (Chown et al. 1998), nestedness might be expected to increase in likelihood as follows: insects < vascular plants < land birds < seabirds, inter-island benthos and fish < plankton, whilst acknowledging that within each of these groups dispersal ability will vary considerably.

In the Southern Ocean island taxa investigated by Greve et al. (2005) species, considerable nestedness was found at the species and generic levels across the region, but there are also substantial differences between nestedness at this scale and within archipelagos (specifically the New Zealand subantarctic islands) (Greve et al. 2005). Thus, both the Insulantarctic and multi-regional scenarios appeared to enjoy support, which is perhaps not surprising given the complexity of the region (Gressitt 1970). However, much of this complexity is a consequence of differences of vagility of the groups (insects, vascular plants, land birds, seabirds). Highly mobile taxa, such as seabirds, that, at least in some instances, have demonstrably panmictic populations in the region, show considerable nestedness. By contrast, less mobile taxa, such as the insects, are hardly nested at all when compared to both the seabirds and to assemblages from elsewhere (Wright et al. 1998). In consequence, distribution patterns shown by groups with poor dispersal ability, such as weevils (Morrone 1998), are dominated by the influence of regional source pools (ie continental areas and large islands located close to particular archipelagos), thus providing support for a multi-regional hypothesis, whilst the distribution patterns of more mobile taxa are less subject to this constraint (Barrat and Mougin 1974). This is true also of several marine taxa, with considerable dispersal abilities, investigated by Greve et al. (2005), and of more vagile taxa such as testate amoebae, which are highly nested and marginally more so than the pelagic seabirds (Greve et al. 2005). Curiously, the bryophyte presence-absence matrix used by Muñoz et al. (2004), which they claimed supports an hypothesis of dispersal by wind across the southern continents and islands, is only slightly less nested than that of the vascular plant matrix examined by Greve et al. (2005). Clearly, bryophyte dispersal across the region might not be as efficient as might at first be expected for species that can spread by means of spores and gemmae.

Although the nestedness approach has not resolved the conundrum regarding the origins of the subantarctic biotas (only a phylogeographic approach will be capable of doing so), it has demonstrated that dispersal plays a major role in determining biogeographic patterns in the region, and the way that they are interpreted (see also McDowall 2002, Craig et al. 2003, Muñoz et al. 2004).

Ecological biogeography

Essentially, ecological biogeography concerns variation in biodiversity over shorter time scales, recognizing that historical events often set the rules of the board and the identities of the pieces. What determines species incidences and abundances in Antarctica are questions that have long occupied ecologists working in the region (Janetschek 1967, 1970, Lewis Smith 1984). The determinants of local scale occupancy and abundance have been extensively investigated across a range of sites. Water availability, temperature (which also influences water availability), protection from wind, the availability of nutrients (often nitrogen, but also carbon – many continental Antarctic systems are carbon-poor), the extent of lateral water movement, and the extent of soil movement and ice formation all have a pronounced effect on the suitability of sites for colonization, growth and reproduction (eg Janetschek 1970, Lewis Smith 1984, Kennedy 1993, Ryan and Watkins 1989, Convey 1996, Freckman and Virginia 1997, Convey et al. 2000a, Sinclair 2001, Smith et al. 2001). Of these, water availability (and the elevated temperatures that drive it) is thought to be most significant on the Antarctic continent and Peninsula, whilst nutrient availability, soil water movement and temperature are most significant in the subantarctic (eg Gabriel et al. 2001, Smith et al. 2001). Most authors have concluded that, at least on the continent, extreme abiotic conditions preclude life in many ice-free areas and that, unlike the situation across most of the planet, abiotic rather than biotic stressors exert a controlling influence on life histories (Janetschek 1970, Convey 1996, Wall and Virginia 1999).

Translated to the assemblage level, these processes mean that many ice-free areas seem totally devoid of life, in others life is restricted to depauperate endolithic communities in sandstone rocks or gypsum crusts and that some sites are characterized by less than a dozen invertebrate and plant species (Broady and Weinstein 1998, Wall and Virginia 1999, Hughes and Lawley 2003, Cockell and Stokes 2004, Convey and McInnes 2005). Nonetheless, substantial spatial complexity in the richness and identity of species is found across Antarctica. At a large scale, species richness increases with a decrease in latitude (Block 1984, Broady 1996, Smith 1996), but the pattern is spatially complex rather than monotonic (Clarke 2003). For example, in the maritime Antarctic, soil eukaryote microbial diversity is as high at c. 71-72°S as it is at c. 60-67°S and only decreases steeply beyond 74°S (Lawley et al. 2004). Likewise, in West Antarctic Alexander Island and East Antarctic Dronning Maud Land, the metazoan microfauna shows evidence of a complex pattern of richness across ice-free areas (Sohlenius et al. 1996, Sohlenius and Boström 2005, Maslen and Convey in press). Similar patterns of complexity are emerging for mosses and lichens (Clarke 2003, H.J. Peat et al. unpubl. data), as taxonomic resolution and sampling coverage improves. Complex spatial variation in richness is not unexpected, given that it is characteristic of life elsewhere on the planet (Gaston 2000). However, owing to relatively poor sampling across the ice-free areas of the Antarctic continent it is not possible to discern whether any systematic trends characterize this large-scale richness variation.

By comparison with the continental and maritime Antarctic, the Southern Ocean islands are species rich and well surveyed, at least for taxa such as insects and vascular plants (eg Vernon and Voisin 1990, Dreux and Voisin 1993, Chown et al. 1998, Gremmen and Smith 1999, Frenot et al. 2001, Jones et al. 2003, Chown et al. 2006). Moreover, patterns in species richness and the mechanisms underlying these patterns have been comprehensively investigated (Chown et al. 1998, 2001, 2005). Across the islands, species richness of indigenous and of alien vascular plants covaries significantly with available energy and the same is true of indigenous and alien insect species richness. The richness of alien insects and vascular plants also covaries with human visitor frequency to the islands. The latter correlates suggest that alien species richness is high in high energy areas for two reasons: the ecological processes that enable high numbers of species to coexist (see Evans et al. 2005 for review) and the historical processes that have meant enhanced propagule pressure as a consequence of high visitor frequency (eg Lonsdale 1999). However, visitor frequency also covaries positively with both area and energy availability (Chown et al. 2005). Thus, the effects of energy availability and human visitor frequency on the richness of exotic species on these islands are confounded. In consequence, distinguishing the relative importance of the ecological (energy) and historical (human occupancy) effects requires that the latter be held relatively constant.

Comparing patterns of invasion on Heard and Gough Islands, which have had virtually identical numbers of visits, but which differ substantially in climate and indigenous species richness, allows this to be done. Gough Island is comparatively warm, with a mean annual temperature of 11.5°C that has remained little changed since the 1950s (Jones et al. 2003). It has a well-developed indigenous terrestrial vascular flora (70 species) and is home to 28 indigenous insect species. Heard Island is cold, with a mean annual temperature of 1.7°C and extensive, though rapidly receding glaciation. It is species poor, with just 12 indigenous vascular plant and nine indigenous insect species. Humans first landed on Gough Island in 1675. However, until 1802 only one other recorded landing took place. Thereafter, the numbers of visits increased – first for sealing purposes and later (after 1955) for science (Fig. 2), resulting in a total of 239 landings.

Heard Island was discovered in 1853, experienced a rapid increase in visits owing to sealing until this resource was exhausted and is once again experiencing an increase in visits for science and tourism (Fig. 2). In total, there have been 232 landings on the island. There are two non-indigenous plant and insect species known from Heard Island, of which the grass *Poa annua* is thought to have arrived before 1955 and the thrip *Apterothrips apteris* in the early 1990s. Alien species have thus established on average once every 72 years or every 116 visits. Gough Island is highly invaded, with 74 non-indigenous insects and plants generally widespread across the island. These species have established at a rate of on average one every four to five years, or with every third visit (Chown et al. 2005). These results provide support for the hypothesis that on these islands energy availability is a major correlate of exotic species richness, and presumably a determinant thereof via the ecological processes that result in an association among energy availability, high numbers of individuals and elevated diversity in all species. Positive relationships between energy availability and species richness have been documented in many systems and at a wide variety of spatial scales (Waide et al. 1999, Hawkins et al. 2003) and are considered one of only a few ecological laws (Rosenzweig 1995). They also suggest that as climates warm in the Antarctic region, so sites will become more susceptible to invasion (see also Walther et al. 2002, Frenot et al. 2005).

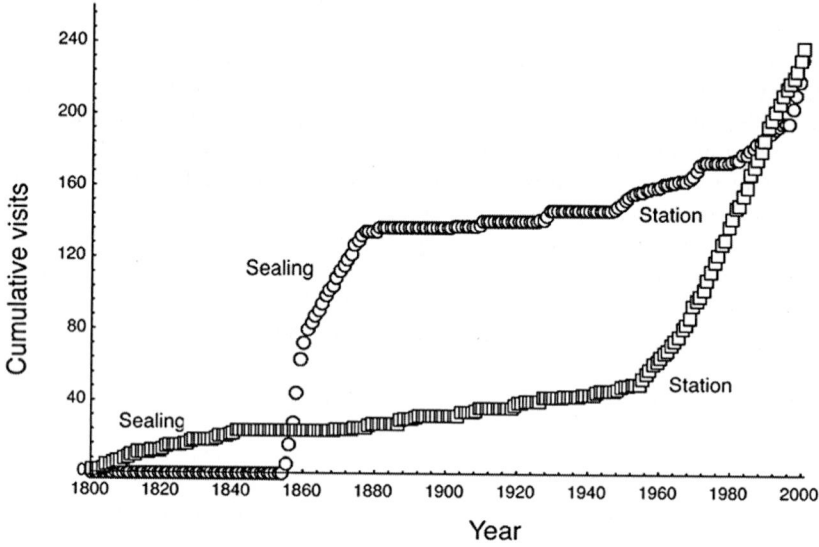

Figure 2. Cumulative number of landings at Heard Island (circles) and Gough Island (squares) since 1800. Before this, only two landings were made at Gough Island and none at Heard Island. The sealing boom years for each island are indicated, as are the dates at which permanent scientific stations were established (1947-1954 for Heard Island, 1956 onwards for Gough Island). Redrawn from Chown et al. 2005.

Conclusions

What land there is in the Antarctic region presently exists as a set of islands surrounded by vast areas of either ice or water. Provision of a comprehensive understanding of the relationships among these islands, the extent to which their biotas have been shaped by dispersal or vicariance, and the mechanisms underlying current variation in diversity will require that humans minimally alter current patterns in the abundance and distribution of the Antarctic biota (at least directly – little can be done about anthropogenic climate change in the absence of global recognition of the problem). Although reasonably recent natural dispersal among areas (Skotnicki et al. 2000, Turner et al. 2006) has been documented, it is typically infrequent (see also Fanciulli et al. 2001, Frati et al. 2001). However, anthropogenic dispersal events are much more common (Frenot et al. 2005). How much these have contributed to homogenization of the Antarctic terrestrial biota through transfer of 'indigenous' organisms, and dilution of the biogeographic signal, is not yet clear (but see Stevens and Hogg 2003). Given that global climate change is likely to make colonization of at least some areas (such as the Peninsula and subantarctic islands) more straightforward (Walther et al. 2002), and that human traffic to the region is increasing, so increasing propagule pressure (Whinam et al. 2005), it might be expected that in the future, separation of biogeographic signal from noise might become more problematic. This is likely to be true also if biogeographers themselves do not take adequate quarantine precautions when visiting different sites. Therefore, a substantial responsibility rests on biogeographers and systematists in the region not only to survey a range of sites and produce molecular phylogenies for the biota of the Antarctic, but also to set the quarantine standard. This would also allow temporal trends associated with broader scale environmental changes to be distinguished from local, anthropogenic propagule transfers.

Acknowledgements

SLC is supported by grants from the Centre for Invasion Biology and a South African National Antarctic Programme grant to B. Jansen van Vuuren. PC is supported by the British Antarctic Survey BIOPEARL project.

References

Allegrucci, G., Carchini, G., Todisco, V., Convey, P. and Sbordoni, V. (2006) A molecular phylogeny of Antarctic Chironomidae and its implications for biogeographical history, *Polar Biology,* **29**, 320-326.

Andrássy, I. (1998) Nematodes in the Sixth Continent, *Journal of Nematode Morphology and Systematics,* **1**, 107-186.

Ashworth, A.C. and Kuschel, G. (2003) Fossil weevils (Coleoptera: Curculionidae) from latitude 85°S Antarctica, *Palaeogeography, Palaeoclimatology, Palaeoecology* **191**, 191-202.

Bargagli, R., Skotnicki, M.L., Marri, L., Pepi, M., Mackenzie, A. and Agnorelli, C. (2004) New record of moss and thermophilic bacteria species and physico-chemical properties of geothermal soils on the northwest slope of Mt. Melbourne (Antarctica), *Polar Biology* **27**, 423-431.

Bargelloni, L., Zane, L., Derome, N., Lecointre, G. and Patarnello, T. (2000) Molecular zoogeography of Antarctic euphausiids and notothenioids: from species phylogenies to intraspecific patterns of genetic variation, *Antarctic Science* **12**, 259-268.

Barrat, A. and Mougin, J.L. (1974) Donneés numériques sur la zoogéographie de l'avifaune Antarctique et Subantarctique, *Comité Nationale Français des Recherches Antarctiques* **33**, 1-18.

Bednarek-Ochyra, H., Väna, J., Ochyra, R. and Lewis Smith, R.I. (2000) *The liverwort flora of Antarctica,* Polish Academy of Sciences, Institute of Botany, Cracow

Bergstrom, D.M., Hodgson, D.A. and Convey, P. (2006) The physical setting of the Antarctic, in D.M. Bergstrom, P. Convey, and A.H.L. Huiskes (eds.), *Trends in Antarctic Terrestrial and Limnetic Ecosystems: Antarctica as a Global Indicator,* Springer, Dordrecht (this volume).

Block, W. (1984) Terrestrial Microbiology, Invertebrates and Ecosystems, in R.M. Laws (ed.), *Antarctic Ecology Volume 1.* Academic Press, London, pp. 163-236.

Boyer, S.J. (1979) Glacial geologic observations in the Dufek Massif and Forrestal Range, 1978-79. *Antarctic Journal of the United States of America* **14**, 46-48.

British Antarctic Survey (2004) *Antarctica 1:10,000,000 map.* BAS (misc) 11. Cambridge, British Antarctic Survey.

Broady, P.A. (1996) Diversity, distribution and dispersal of Antarctic terrestrial algae, *Biodiversity and Conservation* **5**, 1307-1335.

Broady, P.A. and Weinstein, R.N. (1998) Algae, lichens and fungi in la Gorce Mountains, Antarctica, *Antarctic Science* **10**, 376-385.

Broughton, D.A. and McAdam, J.H. (2002) *The Vascular Flora of the Falkland Islands: An Annotated Checklist and Atlas,* Falklands Conservation, Stanley.

Brundin, L. (1966) Transantarctic relationships and their significance as evidenced by chironomid midges, *Svenska Vteneskapsakademie Handlunge,* **11**, 1-472.

Chastain, A. (1958) La flore et la vegetation des Îles de Kerguelen. Polymorphisme des espèces australes, *Memoires Museum National d'Histoire Naturelle B* **11**, 1-136.

Chown, S.L. and Convey, P. (2006) Spatial and temporal variability across life's hierarchies in the terrestrial Antarctic. *Philosophical Transactions of the Royal Society of London B*, in press.

Chown, S.L., Greenslade, P. and Marshall, D.J. (2006) Terrestrial invertebrates of Heard Island. In *Heard Island: Southern Ocean Sentinel*, K. Green and E.J. Woehler (eds.), pp. 91-104, Surrey-Beatty & Sons, Chipping Norton, Sydney.

Chown, S.L., Gremmen, N.J.M. and Gaston, K.J. (1998) Ecological biogeography of southern ocean islands: Species-area relationships, human impacts, and conservation, *American Naturalist* **152**, 562-575.

Chown, S.L., Hull, B. and Gaston, K.J. (2005) Human impacts, energy availability, and invasion across Southern Ocean Islands, *Global Ecology and Biogeography* **14**, 521-528.

Chown, S.L., Rodrigues, A.S.L., Gremmen, N.J.M. and Gaston, K.J. (2001) World Heritage status and the conservation of Southern Ocean islands. *Conservation Biology* **15**, 550-557.

Clarke, A. (2003) Evolution, adaptation and diversity: global ecology in an Antarctic context, in: A.H.L. Huiskes, W.W.C. Gieskes, J. Rozema, R.M.L. Schorno, S.M. van der Vries and W.J. Wolff (eds.), *Antarctic Biology in a Global Context*, Backhuys Publishers, Leiden, The Netherlands, pp. 3-17.

Clarke, J.A., Tambussi, C.P., Noriega, J.I., Erickson, G.M. and Ketcham, R.A. (2005) Definitive fossil evidence for the extant avian radiation in the Cretaceous, *Nature* **433**, 305-308.

Cockell, C.S. and Stokes, M.D. (2004) Widespread colonization by polar hypoliths, *Nature* **431**, 414-414.

Convey, P. (1996) The influence of environmental characteristics on life history attributes of Antarctic terrestrial biota, *Biological Reviews* **71**, 191-225.

Convey, P. (2003) Maritime Antarctic climate change: signals from terrestrial biology, in: E. Domack, A. Burnett, A. Leventer, P. Convey, M. Kirby and R. Bindschadler (eds.), *Antarctic Peninsula Climate Variability: Historical and Palaeoenvironmental Perspectives*, Antarctic Research Series, Vol. 79, American Geophysical Union, Washington, D.C., pp. 145-158.

Convey, P. and McInnes, S.J. (2005) Exceptional tardigrade-dominated ecosystems in Ellsworth Land, Antarctica, *Ecology* **86**, 519-527.

Convey, P., Greenslade, P. and Pugh, P.J.A. (2000b) Terrestrial fauna of the South Sandwich Islands, *Journal of Natural History* **34**, 597-609.

Convey, P., Lewis Smith, R.I., Hodgson, D.A. and Peat, H.J. (2000a) The flora of the South Sandwich Islands, with particular preference to the influence of geothermal heating, *Journal of Biogeography* **27**, 1279-1296.

Cox, C.B. (2001) The biogeographic regions reconsidered, *Journal of Biogeography* **28**, 511-523.

Craig, D.A. (2003) Deconstructing Gondwana - words of warning from the Crozet Island Simuliidae (Diptera). *Cimbebasia* **19**, 157-164.

Craig, D.A., Currie, D.C. and Vernon, P. (2003) *Crozetia* Davies (Diptera: Simuliidae): redescription of *Cr. crozetensis, Cr. sequyi*, number of larval instars, phylogenetic relationships and historical biogeography, *Zootaxa* **259**, 1-39.

Cromer, L., Gibson, J.A.E., Swadling, K.M. and Hodgson, D.A. (In press) Evidence for a lacustrine faunal refuge in the Larsemann Hills, East Antarctica, during the Last Glacial Maximum, *Journal of Biogeography*.

Darlington, P.J. (1970) A practical criticism of Hennig-Brundin "phylogenetic systematics" and Antarctic biogeography, *Systematic Zoology* **19**, 1-18.

Dingwall, P.R. (Ed) (1995) *Progress in conservation of the Subantarctic Islands. Proceedings of the SCAR/IUCN Workshop on Protection, Research and Management of Subantarctic Islands*, Paimpont, France, 27-29 April, 1992. World Conservation Union, Gland.

Dreux, P. and Voisin, J. (1993) Faune entomologique de l'ile des Pingouins (archipel Crozet), *Bulletin de la Societe Entomologique de France* **97**, 453-464.

Evans, K.L. and Gaston, K.J. and Warren, P.H. (2005) Species-energy relationships at the macroecological scale: a review of the mechanisms, *Biological Reviews* **80**, 1-25.

Fanciulli, P.F., Summa, D., Dallai, R. and Frati, F. (2001) High levels of genetic variability and population differentiation in *Gressitacantha terranova* (Collembola, Hexapoda) from Victoria Land, Antarctica, *Antarctic Science* **13**, 246-254.

Frati, F., Spinsanti, G. and Dallai, R. (2001) Genetic variation of mtCOII gene sequences in the collembolan *Isotoma klovstadi* from Victoria Land, Antarctica: evidence for population differentiation, *Polar Biology* **24**, 934-940.

Freckman, D.W. and Virginia, R.A. (1997) Low-diversity Antarctic soil nematode communities: distribution and response to disturbance, *Ecology* **78**, 363-369.

Frenot, Y., Gloaguen, J.C., Masse, L. and Lebouvier, M. (2001) Human activities, ecosystem disturbance and plant invasions in subantarctic Crozet, Kerguelen and Amsterdam Islands, *Biological Conservation* **101**, 33-50.

Frenot, Y., Chown, S.L., Whinam, J., Selkirk, P.M., Convey, P., Skotnicki, M. and Bergstrom, D.M. (2005) Biological invasions in the Antarctic: extent, impacts and implications, *Biological Reviews* **80**, 45-72.

Gabriel, A.G.A., Chown, S.L., Barendse, J., Marshall, D.J., Mercer, R.D., Pugh, P.J.A. and Smith, V.R. (2001) Biological invasions on Southern Ocean islands: the Collembola of Marion Island as a test of generalities, *Ecography* **24**, 421-430.

Gaston, K.J. (2000) Global patterns in biodiversity, *Nature* **405**, 220-227.

Gibson, J.A.E., Wilmotte, A., Taton, A., Van De Vijver, B., Beyens, L. and Dartnall, H.J.G. (2006) Biogeographic trends in Antarctic lake communities, in D.M. Bergstrom, P. Convey, and A.H.L. Huiskes (eds.), *Trends in Antarctic Terrestrial and Limnetic Ecosystems: Antarctica as a Global Indicator.* Springer, Dordrecht (this volume).

Greenslade, P. (1995) Collembola from the Scotia Arc and Antarctic Peninsula including descriptions of two new species and notes on biogeography, *Polskie Pismo Entomologiczne* **64**, 305-319.

Gremmen, N.J.M. and Smith, V.R. (1999) New records of alien vascular plants from Marion and Prince Edward Islands, sub-Antarctic, *Polar Biology* **21**, 401-409.

Gressitt, J.L. (1970) Subantarctic entomology and biogeography. *Pacific Insects Monograph* **23**, 295-374.

Greve, M., Gremmen, N.J.M., Gaston, K.J. and Chown, S.L. (2005) Nestedness of South Ocean island biotas: ecological perspectives on a biogeographical conundrum, *Journal of Biogeography* **32**, 155-168.

Griffiths, H.J., Linse, K. and Crame, A. (2003) SOMBASE – Southern Ocean Molluscan Database: A tool for biogeographic analysis in diversity and ecology, *Organisms, Diversity and Evolution* **3**, 207-213.

Hall, K. (2002) Review of Present and Quaternary periglacial processes and landforms of the maritime and sub-Antarctic region, *South African Journal of Science* **98**, 71-81.

Hammer, W.R. and Hickerson, W.J. (1994) A crested theropod dinosaur from Antarctica, *Science* **264**, 828-830.

Hawkins, B.A., Field, R., Cornell, H.V., Currie, D.J., Guegan, J.F., Kaufman, D.M., Kerr, J.T., Mittelbach, G.G., Oberdorff, T., O'Brien, E.M., Porter, E.E. and Turner, J.R.G. (2003) Energy, water, and broad-scale geographic patterns of species richness, *Ecology* **84**, 3105-3117.

Hodgson, D.A. and Convey, P. (2005) A 7000-year record of oribatid mite communities on a Maritime-Antarctic island: Responses to climate change, *Arctic, Antarctic and Alpine Research* **37**, 239-245.

Holdgate, M.W. (1960) The fauna of the mid-Atlantic islands, *Proceedings of the Royal Society of London B* **152**, 550-567.

Hughes, K.A. and Lawley, B. (2003) A novel Antarctic microbial endolithic community within gypsum crust, *Environmental Microbiology* **5**, 555-565.

Huiskes, A.H.L., Convey, P. and Bergstrom, D.M. (2006) Trends in Antarctic terrestrial and limnetic ecosystems: Antarctica as a global indicator, in: Bergstrom, D.M., Convey, P. and Huiskes, A.H.L. (eds) *Trends in Antarctic Terrestrial and Limnetic Ecosystems: Antarctica as a Global Indicator.* Springer, Dordrecht (this volume).

Janetschek, H. (1967) Arthropod ecology of South Victoria Land. In *Entomology of Antarctica* (ed. J.L. Gressitt), pp. 205-293. Washington: American Geophysical Union.

Janetschek, H. (1970) Environments and ecology of terrestrial arthropods in the high Arctic. In *Antarctic ecology vol. 2*, M.W. Holdgate (ed.), pp. 871-885. London: Academic Press.

Jeannel, R. (1964) Biogéographie des terres Australes de l'Océan Indien, *Revue Francaise d'Entomologie* **31**, 319-417.

Jones, A.G., Chown, S.L., Webb, T.J. and Gaston, K.J. (2003) The free-living pterygote insects of Gough Island, South Atlantic Ocean, *Systematics and Biodiversity* **1**, 213-273.

Kennedy, A.D. (1993) Water as a limiting factor in the Antarctic terrestrial environment: A biogeographical synthesis, *Arctic and Alpine Research* **25**, 308-315.

Kuschel, G. and Chown, S.L. (1995) Phylogeny and systematics of the *Ectemnorhinus*-group of genera (Insecta: Coleoptera), *Invertebrate Taxonomy* **9**, 841-863.

Lawley, B., Ripley, S., Bridge, P. and Convey, P. (2004) Molecular analysis of geographic patterns of eukaryotic diversity in antarctic soils, *Applied and Environmental Microbiology* **70**, 5963-5972.

Le Masurier, W.E. and Thomson, J.W. (1990) *Volcanoes of the Antarctic Plate and Southern Oceans*, American Geophysical Union, Washington DC.

Lewis Smith, R.I. (1984) Terrestrial plant biology of the sub-Antarctic and Antarctic, in R.M. Laws (ed.), *Antarctic Ecology,* Academic Press, London, pp 61-162.

Lonsdale, W.M. (1999) Global patterns of plant invasions and the concept of invisibility, *Ecology* **80**, 1522-1536.

Marshall, D.J. and Chown, S.L. (2002) The acarine fauna of Heard Island, *Polar Biology* **25**, 688-695.

Marshall, D.J. and Coetzee, L. (2000) Historical biogeography and ecology of a Continental Antarctic mite genus, *Maudheimia* (Acari, Oribatida): evidence for a Gondwanan origin and Pliocene-Pleistocene speciation, *Zoological Journal of the Linnean Society* **129**, 111-128.

Marshall, D.J. and Pugh, P.J.A. (1996) Origin of the inland Acari of Continental Antarctica, with particular reference to Dronning Maud Land, *Zoological Journal of the Linnean Society* **118**, 101-118.

Maslen, N.R. and Convey, P. (In press) Nematode diversity and distribution in the southern maritime Antarctic – clues to history? *Soil Biology and Biochemistry.*

McDowall, R.M. (2002) Accumulating evidence for a dispersal biogeography of southern cool temperate freshwater fishes, *Journal of Biogeography* **29**, 207-219.

Morrone, J.J. (1998) On Udvardy's Insulantarctica province: a test from the weevils (Coleoptera: Curculionoidea), *Journal of Biogeography* **25**, 947-955.

Muñoz, J., Felicísimo, A.M., Cabezas, F., Burgaz, A.R. and Martínez, I. (2004) Wind as a long-distance dispersal vehicle in the Southern Hemisphere, *Science* **304**, 1144-1147.

Ochyra, R., Bednarek-Ochyra, H. and Lewis Smith, R.I. (In press) *The Moss Flora of Antarctica.* Cambridge University Press, Cambridge.

Øvstedal, D.O. and Lewis Smith, R.I. (2001) *Lichens of Antarctica and South Georgia A Guide to their Identification and Ecology,* Cambridge University Press, Cambridge.

Page, T.J. and Linse, K. (2002) More evidence of speciation and dispersal across the Antarctic Polar Front through molecular systematics of Southern Ocean *Limatula* (Bivalvia: Limidae), *Polar Biology* **25**, 818-826.

Peck, L. Convey, P. and Barnes, D.K.A. (2006) Environmental constraints on life histories in Antarctic ecosystems: tempos, timings and predictability. *Biological Reviews* **81**, 75-109.

Prentice, M.L., Bockheim, J.G., Wilson, S.C., Burckle, L.H., Hodell, D.A., Schlüchter, C. and Kellogg, D.E. (1993) Late Neogene Antarctic glacial history: evidence from central Wright Valley, in: *The Antarctic Paleoenvironment: A Perspective on Global Change*, Antarctic Research Series **60**, 207-250.

Pugh, P.J.A. (1993) A synonymic catalogue of the Acari from Antarctica, the sub-Antarctic Islands and the Southern Ocean, *Journal of Natural History* **27**, 323-421.

Pugh, P.J.A., Dartnall, H.J.G. and McInnes, S.J. (2002) The non-marine Crustacea of Antarctica and the islands of the Southern Ocean: biodiversity and biogeography, *Journal of Natural History* **36**, 1047-1103.

Pugh, P.J.A. and Scott, B. (2002) Biodiversity and biogeography of non-marine Mollusca on the islands of the Southern Ocean, *Journal of Natural History* **36**, 927-952.

Quilty, P.G. (1990) Significance of evidence for changes in the Antarctic marine environment over the last 5 million years, in: *Antarctic Ecosystems. Ecological Change and Conservation*, K.R. Kerry and G. Hempel (eds.), pp. 3-8, Springer, Berlin.

Quilty, P.G. and Wheller, G. (2000) Heard Island and the McDonald Islands: A window into the Kerguelen Plateau, *Papers and Proceedings of the Royal Society of Tasmania* **133**, 1-12.

Ricklefs, R.E. (2004) A comprehensive framework for global patterns in biodiversity, *Ecology Letters* **7**, 1-15.

Rosenzweig, M.L. (1995) *Species Diversity in Space and Time*, Cambridge University Press, Cambridge.

Ryan, P.G. and Watkins, B.P. (1989) The influence of physical factors and ornithogenic products on plant and arthropod abundance at an inland nunatak group in Antarctica, *Polar Biology* **10**, 151-160.

Samways, M.J. (1994) *Insect conservation biology*, Chapman and Hall, London.

Scott, L. (1985) Palynological indications of the Quaternary vegetation history of Marion Island (sub-Antarctic), *Journal of Biogeography* **12**, 413-431.

Sinclair, B.J. (2001) On the distribution of terrestrial invertebrates at Cape Bird, Ross Island, Antarctica. *Polar Biology* **24**, 394-400.

Skotnicki, M.L., Ninham, J.A. and Selkirk, P.M. (2000) Genetic diversity, mutagenesis and dispersal of Antarctic mosses – a review of progress with molecular studies, *Antarctic Science* **12**, 363-373.

Skottsberg, C. (1960) Remarks on the plant geography of the southern cold temperate zone, *Proceedings of the Royal Society of London B* **152**, 447-457.

Smith, H.G. (1996) Diversity of Antarctic terrestrial protozoa, *Biodiversity and Conservation* **5**, 1379–1394.

Smith, V.R., Steenkamp, M. and Gremmen, N.J.M. (2001) Terrestrial habitats on sub-Antarctic Marion Island: their vegetation, edaphic attributes, distribution and response to climate change, *South African Journal of Botany* **67**, 641-654.

Sohlenius, B. and Boström, S. (2005) The geographic distribution of metazoan microfauna on East Antarctic nunataks, *Polar Biology* **28**, 439-448.

Sohlenius, B. and Boström, S. and Hirschfelder, A. (1996) Distribution patterns of microfauna (nematodes, rotifers and tardigrades) on nunataks in Dronning Maud land, East Antarctica, *Polar Biology* **16**, 191-200.

Stevens, M.I. and Hogg, I.D. (2002) Expanded distributional records of Collembola and Acari in southern Victoria Land, Antarctica, *Pedobiologia* **46**, 485-495.

Stevens, M.I. and Hogg, I.D. (2003) Long-term isolation and recent range expansion from glacial refugia revealed for the endemic springtail *Gomphiocephalus hodgsoni* from Victoria Land, Antarctica, *Molecular Ecology* **12**, 2357-2369.

Stevens, M.I. and Hogg, I.D. (2006) The molecular ecology of Antarctic terrestrial and limnetic invertebrates and microbes, in: D.M. Bergstrom, P. Convey, and A.H.L. Huiskes (eds.), *Trends in Antarctic Terrestrial and Limnetic Ecosystems: Antarctica as a Global Indicator.* Springer, Dordrecht (this volume).

Turner, P.A.M., Scott J.J. and Rozefelds, A. (2006) Probable long distance dispersal of *Leptinella plumosa* Hook. f. to Heard Island: habitat, status and discussion of its arrival, *Polar Biology*, **29**, 160-168.

Udvardy, M.D.F. (1987) The biogeographical realm Antarctica: a proposal, *Journal of the Royal Society of New Zealand* **17**, 187-194.

van de Vijver, B. and Beyens, L. (1999) Biogeography and ecology of freshwater diatoms in Subantarctica: a review, *Journal of Biogeography* **26**, 993-1000.

van Veller, M.G.P. and Brooks, D.R. (2001) When simplicity is not parsimonious: *a priori* and *a posteriori* methods in historical biogeography, *Journal of Biogeography* **28**, 1-11.

Vernon, P. and Voisin, J.-F. (1990) Faune entomologique de la Grande Ile des Apôtres (Archipel Crozet, Océan Indien Austral), *Bulletin de la Societe Entomologique de France* **95**, 263-268.

Waide, R.B., Willig, M.R., Steiner, C.F., Mittelbach, G., Gough, L., Dodson, S.I., Juday, G.P. and Parmenter, R. (1999) The relationship between productivity and species richness, *Annual Review of Ecology and Systematics* **30**, 257-300.

Wall, D.H. and Virginia, R.A. (1999) Controls on soil biodiversity: insights from extreme environments, *Applied Soil Ecology* **13**, 137-150.

Wallace, P.J., Frey, F.A., Weis, D. and Coffin, M.F. (2002) Origin and evolution of the Kerguelen Plateau, Broken Ridge, and Kerguelen Archipelago: Editorial, *Journal of Petrology* **43**, 1105-1108.

Walther, G.-R., Post, E., Convey, P., Menzel, A., Parmesan, C., Beebee, T.J.C., Fromentin, J.-M., Hoegh-Guldberg, O. and Bairlein, F. (2002) Ecological responses to recent climate change, *Nature* **416**, 389-395.

Whinam J., Chilcott, N., Bergstrom, D.M. (2005) Subantarctic hitchhikers: expeditioners as vectors for the introduction of alien organisms, *Biological Conservation* **121**, 207-219.

Wright, D.H., Patterson, B.D., Mikkelson, G.M., Cutler, A. and Atmar, W. (1998) A comparative analysis of nested subset patterns of species composition, *Oecologia* **113**, 1-20.

5. BIOGEOGRAPHIC TRENDS IN ANTARCTIC LAKE COMMUNITIES

J.A.E. GIBSON
Institute of Antarctic and Southern Ocean Studies, University of Tasmania, Private Bag 77, Hobart, Tasmania 7001, Australia
John.Gibson@utas.edu.au

A. WILMOTTE
Centre d'ingénierie des protéines, Institut de Chimie B6, Université de Liège, B-4000 Liège, Belgium
awilmotte@ulg.ac.be

A. TATON
Centre d'ingénierie des protéines, Institut de Chimie B6, Université de Liège, B-4000 Liège, Belgium
A.Taton@student.ulg.ac.be

B. VAN DE VIJVER
Universiteit Antwerpen (CMI), Departement Biologie/PLP, Groenenborgerlaan 171, B-2020 Antwerp, Belgium
vandevijver@br.fgov.be

L. BEYENS
Universiteit Antwerpen (CMI), Departement Biologie/PLP, Groenenborgerlaan 171, B-2020 Antwerp, Belgium
louis.beyens@ua.ac.be

H.J.G. DARTNALL
Department of Biological Sciences, Macquarie University, Sydney, NSW 2109, Australia

D.M. Bergstrom et al. (eds.), Trends in Antarctic Terrestrial and Limnetic Ecosystems, 71–99.

Introduction

Water bodies that contain liquid water for at least part of the year are a surprisingly common feature of the Antarctic landscape. While these water bodies are generically termed lakes, there is a remarkable variety of types present both on the continent and on subantarctic islands, some of which do not have counterparts in more temperate regions. They range widely in size from cryoconite holes less than a metre across to large lakes with areas over 100 km², in depth from a few centimetres to over 300 m, in salinity from some of the freshest lakes in the world to others that have salinity approaching that of the Dead Sea and in age from a single summer season to more than 300 000 years. Most of these upper extremes are exceeded by the sub-glacial lakes located deep beneath the Antarctic ice cap.

All lakes that have been investigated in Antarctica have been found to contain organisms of some sort. Only in the most saline case – Don Juan Pond in the McMurdo Dry Valleys (see Fig. 1 for a map showing the locations of places mentioned in the text) – is evidence for active metabolism equivocal (Vincent 1988). While some lacustrine plants and animals typically found in more temperate regions are missing from Antarctic lakes and are largely absent from subantarctic islands,

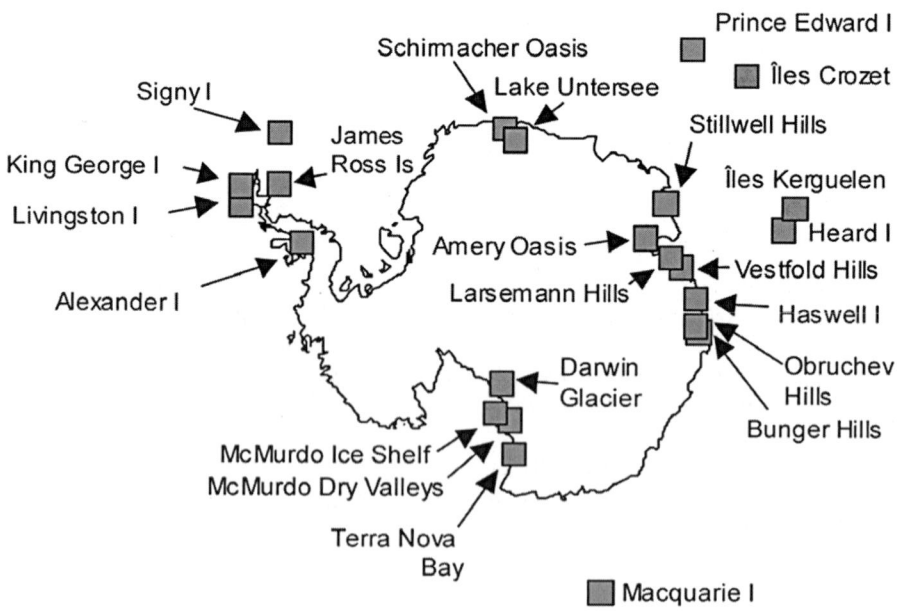

Figure 1. Map of Antarctica, showing the locations mentioned in the text.

notably fish and emergent vegetation, many taxonomic groups, ranging in complexity from bacteria to crustacea, are present (examples of the flora and fauna of Antarctic lakes are presented in Fig. 2).

The origins of the Antarctic lake fauna and flora and their position in global biogeography have been of interest since the heroic age of Antarctic exploration (eg Murray 1910, Rühe 1914). Throughout the next 60 years, scattered reports of the biota of lakes in various areas appeared (eg Schmitt 1945, Korotkevich 1958) and slowly a broad picture of the biogeographical patterns emerged (Goldman 1970). This material had wider appeal: Crawford (1974) used the distribution of the cladoceran genus *Daphniopsis*, which occurs in Antarctica, South America, Australia and Tibet, as evidence for inclusion of all these areas in a greater Gondwanaland. Since 1975 there has been a greater emphasis on lake studies in national programs, which has either directly or indirectly added to our knowledge of lacustrine biogeography. Various reviews have appeared (eg Heywood 1984, Wright and Burton 1981, Ellis-Evans 1996) in addition to a valuable bibliography containing details of many hard-to-get references (Block 1992).

In this chapter, we discuss the types and origins of Antarctic lakes and the factors that control biodiversity in these lakes. We review present-day understanding of biodiversity for a series of important taxonomic groups and discuss these data in terms of previous views of Antarctic biogeography.

Types and habitats of Antarctic lakes

A range of different lake types is present in Antarctica (Fig. 3). Some of these lakes types are well-known, but other are more cryptic or isolated and little studied. The most studied lakes are rock-bound lakes that occur in ice-free oases such as the McMurdo Dry Valleys, the Vestfold Hills and the Antarctic Peninsula. They are also widespread on subantarctic islands. Deeper lakes contain liquid water throughout the year, with either seasonal or perennial ice insulating the water against heat loss and complete freezing. Shallow lakes that freeze to the base during winter are often referred to as ponds.

Epiglacial lakes, which are bounded by both rock and ice, are located at the feet of glaciers both on the continent and on subantarctic islands (eg Lakes Fryxell and Hoare, McMurdo Dry Valleys), along active and passive margins of ice free areas (eg in the Vestfold and Bunger Hills: Pickard and Adamson 1983, Gibson et al. 2002; Lake Untersee, Dronning Maud Land: Wand et al. 1997, and Lake Wilson, Darwin Glacier, Southern Victoria Land: Webster et al. 1996). These lakes can be deep (>100 m) and are in nearly all cases permanently ice-covered. Epiglacial lakes are probably far more common and widespread than currently recognised, as they will occur in many places where nunataks occur.

Epishelf lakes contain fresh or brackish water that is dammed by a floating ice shelf or glacier and has a hydraulic connection to the marine ecosystem. In some

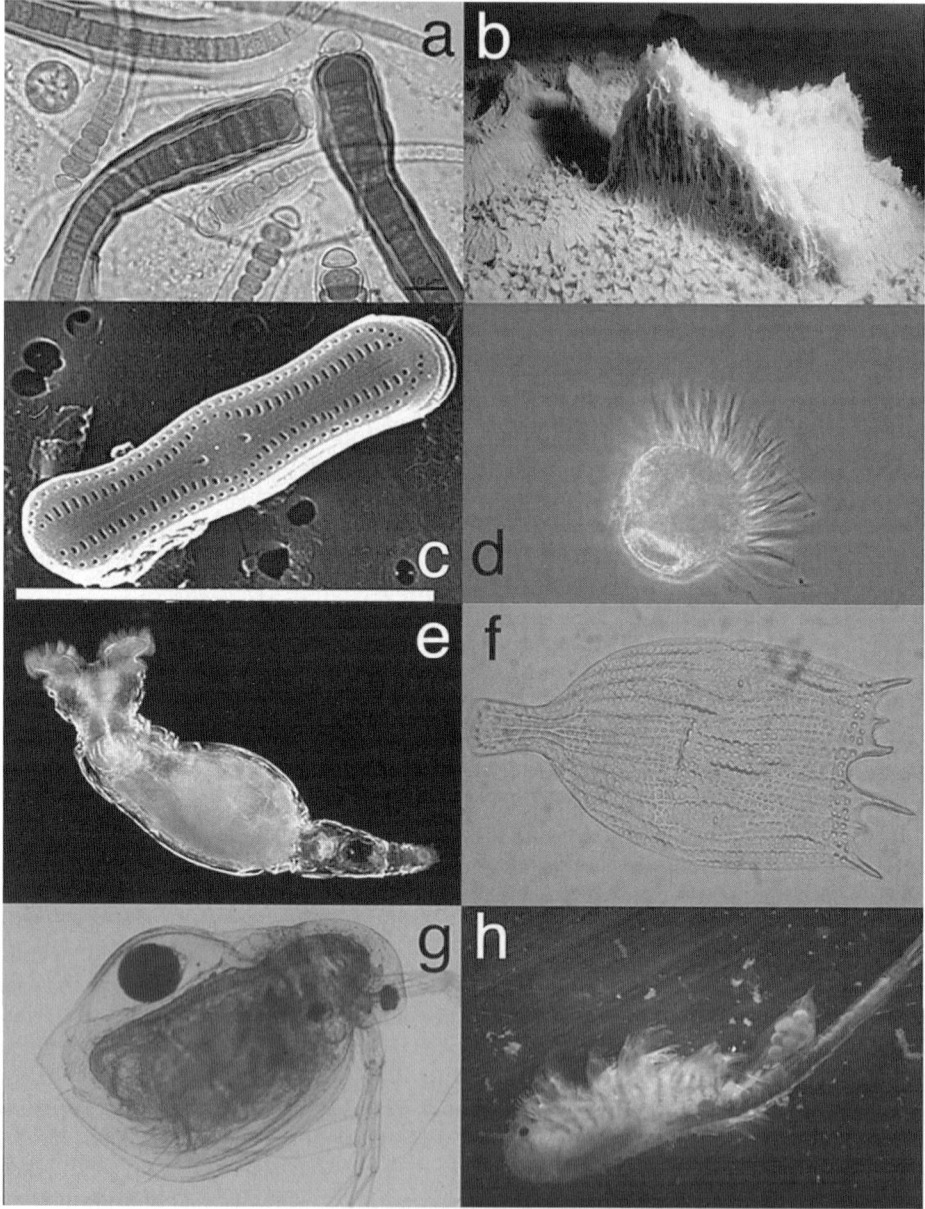

*Figure 2. Examples of organisms found in Antarctic lakes: (a) trichomes
of the filamentous cyanobacterium* Calothrix *sp. from Firelight Lake, Bølingen Islands,
Larsemann Hills (typical trichome width 10-15μm) (Photo: A. Taton),
(b) microbial mat dominated by cyanobacteria, Lake Vanda, McMurdo Dry Valleys.
Triangular part ca 10 cm across. (Photo: D. Andersen),*

(c) freshwater diatom Diadesmis ingeae *Van de Vijver from Ile de la Possession,
Iles Crozet. Scale bar 10 μm (Photo: B. Van de Vijver),
(d) ciliate* Askenasia sp. *from Lake Fryxell, McMurdo Dry Valleys (diameter 40 μm)
(Photo: S. Spaulding), (e) Antarctic endemic rotifer* Philodina gregaria *Murray from Signy
Island (length ca. 500 μm) (Photo: J. Walsh, Micrographia.com) (f) sub-fossil lorica of the
rotifer* Notholca walterkostei *De Paggi from Lake Boeckella, Northern Antarctic Peninsula
(length 150 μm) (Photo: J. Gibson), (f) cladoceran* Daphniopsis studeri *Rühe from Crooked
Lake in the Vestfold Hills (length ca 2 mm) (Photo: J. Laybourn-Parry) and (g) anostracan*
Branchinecta gaini *Daday from a pool on the Antarctic Peninsula (length ca 2 cm) (Photo:
British Antarctic Survey).*

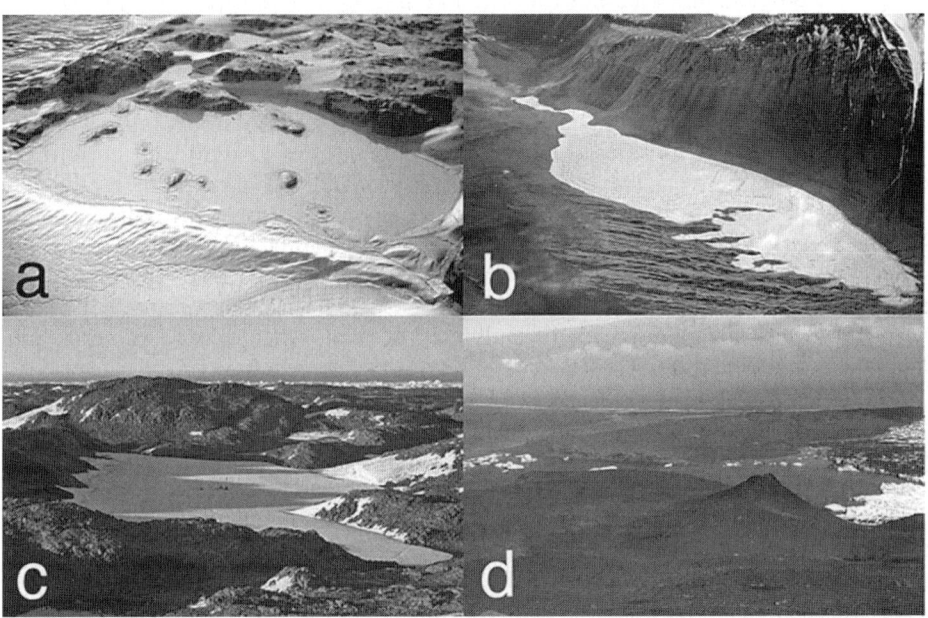

*Figure 3. Examples of Antarctic lakes: (a) epishelf Lake Pol'anskogo, Bunger Hills (note the
tide-cracks around the lake margin) (Photo: D. Andersen), (b) Lake Vanda, McMurdo Dry
Valleys (Photo: D. Andersen), (c) marine-derived Abraxas Lake, Vestfold Hills (Photo: L.
Cromer) and (d) epiglacial Stephenson Lagoon, Heard Island, formed after the retreat of the
Stephenson Glacier (right) (Photo: K. Kiefer).*

examples the fresh water floats directly on the denser sea water, while in other cases
the connection is less direct (Gibson and Andersen 2002). Epishelf lakes are known
only from Antarctica (Alexander Island (southern Antarctic Peninsula), Schirmacher
Oasis, Amery Oasis, Bunger Hills), if the ice shelf breaks out, catastrophic draining
of the freshwater basin and loss of any freshwater biota occurs, as happened recently
for the only known Northern Hemisphere epishelf lake, which now no longer exists
(Mueller et al. 2003).

Many, probably short-lived, supraglacial lakes occur on the surface of large glaciers and the ice sheet and little is known about these environments. Better-studied supraglacial lakes occur on ice shelves, particularly near ice-free areas. The surface of the McMurdo Ice Shelf near Ross Island is characterised by the presence of numerous irregular, productive and biodiverse lakes and ponds that have developed due to the presence of dark, heat-absorbing moraine material (Suren, 1990, James et al. 1995, Vincent and James 1996). A similar environment that has received considerable attention is cryoconite holes, which are small, cylindrical holes (typically 0.5 m across and 0.5 m deep) in the ice that collect mineral sediment, that contain liquid water during summer and that can develop relatively complex biological communities (Wharton et al. 1985). The final lake type is sub-glacial lakes, which are located many thousands of metres beneath the ice surface. These lakes are not discussed in more detail here.

Most of the biota of Antarctic lakes lives in the water column (plankton), or associated with the sediment surface (benthos). Many Antarctic lakes contain luxuriant benthic cyanobacterial mats that are the sites of a significant proportion of primary productivity (Moorhead et al. 2005) and are also more biodiverse than the planktonic communities. The permanent ice-cover of lakes in the McMurdo Dry Valleys has also been identified as a habitat for cyanobacteria and other organisms (Priscu et al. 1998). Numerous meromictic (permanently-stratified) lakes occur in Antarctica (Gibson 1999) and these provide further, anoxic habitats for photosynthetic sulphur bacteria, other anoxic bacteria and methanogenic Archaea.

Origins of Antarctic lakes

The majority of rock-bound Antarctic lakes are thought to have been formed during the last 12 000 yrs, as the wasting of the ice cap and retreat of major outlet glaciers after the last glacial maximum (LGM) resulted in the formation of ice free land (eg Zale and Karlén 1989, Roberts and McMinn 1999, Verkulich et al. 2002). As glacially-formed basins in the exposed rock were exposed they filled with fresh water, biota arrived and organic sedimentation began. Alternatively, the reduced ice loading on the continental margin resulted in isostatic uplift that isolated marine basins, forming saline lakes (Zwartz et al. 1998). The biota in these lakes were largely derived from that present in the basin at the time of isolation.

Exceptions to this general rule do occur. Some of the lakes of the McMurdo Dry Valleys have existed in one form or another for over 300 000 years (Hendy 2000) and pre-LGM sediments have also been recovered from lakes in the Larsemann Hills (Hodgson et al. 2001) and the Schirmacher Oasis (Schwab 1998). It is possible that other coastal oases in East Antarctica were at least partly ice-free at that time (Gore et al. 2001) and that lakes existed.

Lakes on peri-antarctic islands appear to be of similar age to the majority of the continental lakes: radiocarbon dates on basal sediment from lakes on Signy Island

and Macquarie Island have not exceeded ca. 12 500 ^{14}C yr BP (Selkirk et al. 1988, Jones et al. 2000), implying that these environments are in general of similar age to the majority of the lakes studied in continental Antarctica. However, older sediments (>20 000 yr) have been recorded in drained palaeo-lakes on Macquarie Island (Selkirk and Saffigna 1999).

Little reliable information is available regarding the ages of epiglacial and epishelf lakes. Epiglacial Lake Untersee appears to have been present at the LGM (Schwab 1998) and on very limited data Bardin (1990) suggested an age of 20 000 – 90 000 yrs for epishelf Beaver Lake. Coring in Transkriptsii Gulf, another epishelf lake in the Bunger Hills, recovered apparently pre-Holocene sediments dated at 24 140 ^{14}C yr BP (Kulbe 1997). However, this may have been the result of reworking of older sediments. From these limited data, it appears that these lake types could be significantly older than many rock-bound lakes.

General trends in biogeography of Antarctic lakes

The current understanding of Antarctic biogeography is by necessity based on incomplete data, as in many regions lakes have been poorly studied, or in many cases not at all. For example, nothing is known of the biota of rock-bound lakes in the numerous ice free areas of Enderby and Kemp Lands (eg Stillwell Hills) or the vast majority of epiglacial lakes. The biodiversity present in lakes close to long-term scientific research stations, such as the McMurdo Dry Valleys, the Vestfold Hills, Signy Island and the subantarctic islands is generally well known, but even in these regions surprises occur. It has long been thought that crustacea were absent from the lakes of the McMurdo Dry Valleys due to the extreme environment, but the recent report of an as yet unidentified copepod from Lake Joyce (Roberts et al. 2004) means that this conclusion has to be revisited.

A further problem is the difficulty in assignation of individuals to particular species. Many lower organisms can be difficult to identify and furthermore, there has been a tendency to assign Antarctic populations incorrectly to temperate species (Sabbe et al. 2003). As discussed further below, molecular genetic techniques are beginning to unravel these problems.

Antarctica and the peri-antarctic islands of the Southern Ocean can be broken up into three climatic zones (Holdgate 1977):

1. subantarctic islands, which include the Prince Edward Islands, Iles Crozet, Iles Kerguelen, Heard Island, Macquarie Island and South Georgia.

2. maritime Antarctica, which includes the western side of the Antarctic Peninsula to approximately 70°S, the South Shetland Islands, the island groups of the Scotia Arc and Bouvetøya.

3. continental Antarctica, which includes the remainder of the continent and nearby offshore islands.

It appears that the major biogeographic provinces are closely aligned with these zones (see below). An important question to be answered is whether the different climate of these zones has led to this distribution of animals, or whether the similarities in the biogeographical and climate zones reflect colonisation processes? For example, most of the fauna and flora that occurs on the Antarctic Peninsula apparently had origins in South America. The lack of ability of these species to spread more widely may be attributable to the physical boundaries of the polar ice sheet. The occurrence of species from the maritime Antarctic in other regions, eg the copepod *Boeckella poppei* (Mrázek) (Bayly et al. 2003), suggests that it is not purely climate that defines the boundaries of the biogeographic zones.

Factors controlling the biogeography and diversity of the biota of Antarctic lakes

The present day biodiversity in Antarctic lakes is the result of the interplay of characteristics such as location, salinity, temperature, light, nutrient status and food availability and the nature of individual species or populations that (potentially) inhabit them.

Antarctic lakes are isolated habitats that are located in small groups on islands, ice-free areas of the continent, or on the ice itself. There are often limited physical connections between lakes and therefore few chances for organisms to invade new habitats directly by transfer in water. For example, the biota of individual cryoconite holes on the Canada Glacier (southern Victoria Land) have been shown to develop independently of each other (Mueller and Pollard 2004) and drainage systems linking lakes are rare, at least on the continent. In many cases regions containing newly-formed lakes would have been located many hundreds, if not thousands, of kilometres from similar environments and therefore the efficiency of transport of organisms from one lake to another and of colonisation once they had arrived would become important factors in determining the development of the biota.

Exceptions to this view of a biota possibly limited by the efficiency of colonisation are marine-derived lakes that occur in many of the coastal oases on the continental margin. These lakes were formed when local sea level fell, resulting in the isolation of pockets of seawater. The biota of these lakes is largely derived from that present in the lake when final isolation occurred and has developed since isolation by species loss as the environmental tolerances of particular species have been exceeded (Cromer et al. 2005). However, these lakes can still be colonised by new organisms: the algae in marine-derived but now brackish Highway Lake in the Vestfold Hills are derived from the initial marine input, but the sole crustacean in the lake, the cladoceran *Daphniopsis studeri* Rühe, has a non-marine origin (Laybourn-Parry et al. 2002)

The colonisation of Antarctic or subantarctic lakes will progress by short distance transport of propagules from both local habitats, or longer distance transport from other continental oases or subantarctic islands, or from South America, Australia, Africa, New Zealand and even further afield. There is growing evidence of a significant degree of endemism in the biota of Antarctic lakes, suggesting that these organisms have not colonised Antarctica from other continents during the Holocene. There is the further implication that individual lakes existed throughout the LGM providing habitat for these species. Alternatively, a species may have passed through a series of shorter-lived lakes that have overlapped temporally. Such a relictual origin was recently suggested for the copepod *Gladioferens antarcticus* Bayly, which is only found in epishelf lakes in the Bunger Hills (Bayly et al. 2003). No extra-continental populations of this species have been recorded and as it is the most primitive of its subgenus, it cannot be derived from other species extant in Australia or New Zealand. It (or its ancestors) appears to have survived in Antarctica since the separation of Australia and Antarctica. Intra-continental dispersal of endemic species may be limited, as their survival mechanism does not rely on dispersal to new habitat. However such dispersal could occur and it is also possible that truly endemic Antarctic species could disperse to more temperate regions.

Species that have reached Antarctica or the peri-antarctic islands since the onset of glaciation must be adapted to long-distant transport. The likelihood of a propagule reaching a lake will depend both on the nature of the propagule and the efficiency by which it can be transported. Some organisms, for example some cyanobacteria and other protists, can be transported by wind and therefore there is a relatively high chance that propagules of these species will reach most new habitats over a relatively short time (Ellis-Evans and Walton 1990). Furthermore, the continual arrival of propagules will serve to reinforce the occurrence of a species in a lake and could also lead to significant genetic variation. There is a constant traffic of cyanobacterial mat pieces between lakes and their terrestrial surroundings, even through the ice covers (Cowan and Tow 2004) and strong winds may then disperse these pieces. Frozen and dried mats contain viable cyanobacteria and algae, revealing a very resilient character that should facilitate their dispersal (Hawes et al. 1992). For Metazoa, the problems of dispersal are greater and it is likely that colonisation over larger distances has occurred as a result of exceptional transport events, with local dispersion more efficient after that. It is possible that a species could bypass good habitat if no propagule transport to that area had occurred, resulting in patchy distributions controlled less by the nature of the habitat than by stochastic colonisation events. Due to the more difficult dispersion of the Metazoa, it is probable that there is reduced genetic diversity in isolated populations.

A further consideration for a potential colonising species is its ability to be able to survive and reproduce in its new environment. Each species has particular thermal, chemical and physical requirements and if these are not met in a new environment, viable populations will not develop. The major considerations in Antarctic lakes include temperature, salinity, light, nutrients and food.

Antarctica is rightfully considered a cold continent, but the presence of liquid water in lakes provides a thermal buffer against the extremes of the external climate.

In freshwater lakes, the minimum temperature is 0°C, even if the air temperature is -50°C. Therefore, the 'harshness' of the lake environment is decoupled from that of the external environment and for species adapted to existence at near freezing temperatures, survival further south should not provide any greater difficulty. It is also interesting to consider that many marine organisms that live in the Southern Ocean complete their entire life cycles at temperatures below 0°C.

For ponds that freeze completely in winter the situation is a little different, as the temperature of the ice is not limited to 0°C. Any organism surviving in these lakes must be able to produce a resting stage that can survive these conditions (eg the cysts of the anostracan *Branchinecta gaini* Daday (Peck 2004)) or able to enter a cryptobiotic state.

Salinity is a significant controller of biodiversity only in closed saline lakes, where changes in water balance result in increases or decreases in salinity. The biodiversity of the saline lakes of the Vestfold Hills has been controlled in part by the salinity tolerances of the originally marine inhabitants (Eslake et al. 1991, Roberts and McMinn 1999).

Light for photosynthesis in Antarctic lakes is both highly seasonal and attenuated by the presence of ice, though in many cases the clarity of the ice is high (Vincent et al. 1998). Species adapted to both low and seasonal light will be more likely to establish populations. Two adaptations assist survival: photoautotrophs can be highly efficient at harvesting light for photosynthesis (Seaburg et al. 1983, Lizotte and Priscu 1992, Neale and Priscu 1995, Hawes and Schwarz 2001) and the ability to switch nutritional modes between autotrophy and heterotrophy (termed mixotrophy) is a common trait in Antarctic lakes that allows species to survive low-light periods (Laybourn-Parry 2002).

Many Antarctic lakes are nutrient poor, with nutrient regeneration occurring largely in the surface sediments rather than within the water column. Benthic microbial mats consisting largely of cyanobacteria are adapted to taking advantage of this nutrient source and biomass in most Antarctic lakes is dominated by such mats. If nutrient use and regeneration in the mats is tightly coupled, little nutrient is available for productivity in the water column and therefore planktonic species will struggle. The short growing season and the extended periods of ice cover in continental lakes also contribute to a limited planktonic role. This may be an important factor behind the occurrence of small picocyanobacteria in the plankton of continental Antarctic lakes (but less so in maritime and subantarctic lakes), as the high surface area:volume ratio of these organisms will enhance nutrient transport.

For herbivores to be successful in Antarctic lakes they must be adapted to feeding on the cyanobacterial mats, as they constitute both the bulk of the organic carbon available and provides an annual food source, or to the scarce, seasonal plankton. Most Metazoa found in Antarctic lakes have benthic or nekto-benthic habits (eg Almada et al. 2004), indicating that they are adapted to using the more abundant food source.

The physical and chemical conditions in any lake are not likely to be optimal for a particular organism. This has been best documented in Antarctic lakes for bacteria, including cyanobacteria (Tang et al. 1997), which are rarely true psychrophiles. Continued survival of a species will depend on how well it is able to interact, both in terms of competition and cooperation, with other species present in the lake. Sub-optimal habitat does not necessarily limit species distribution. However, the ability to grow and compete over a wide range of fluctuating environmental conditions – phenotypic flexibility – will be an important feature of successful colonising species.

Climate change continues today, and the lake communities will continue to develop in part due to habitat changes resulting from global warming. Significant chemical and physical changes in the lakes of the Antarctic Peninsula region magnified relative to that of the surrounding catchment have been recorded (Quayle et al. 2002), and these changes will undoubtedly affect the biota, possibly in unpredictable ways (Winder and Schindler 2004). Some species may not be able to adapt to the warmer conditions and may become locally extinct, while new species may now be able to invade the lakes from more temperate zones and establish permanent populations.

Palaeobiogeography of Antarctic lakes

There is little information available as to how biodiversity in Antarctic lakes has changed with time. In particular, there is no definitive information as to whether the species currently present are recent arrivals that have dispersed from more temperate zones, or relict species that have survived harsher glacial conditions in refugia across their current distributions. Records of change are preserved in the sediments of lakes and study of these has provided a few interesting insights. For example, the anostracan *Branchinecta gaini*, which is currently restricted to the maritime Antarctic, occurred for a significant period a few thousand years ago in pools on James Ross Island on the eastern side of the Antarctic Peninsula (Björck et al. 1996). This record could be viewed as an indicator that the climate in this area was milder at the time, or purely as a colonisation event outside the current distribution that ultimately failed, as survival of a species in a lake may not be due purely to external climatic conditions. To completely understand Antarctic biogeography, such an historical view is imperative.

There is more information available on how the lakes themselves and therefore the environments in which the biota lived, have changed during the Holocene. For example, the lakes of the McMurdo Dry Valleys were reduced to concentrated brine pools approximately 3000 years ago (Lyons et al. 1998). Salinity changes associated with water levels will have a dramatic influence on the biodiversity in these lakes: the species present today either had to be able to survive these changes either through euryhalinity or the ability to recolonise the lakes after return to wetter conditions. These climate-driven changes in the lakes have resulted in the development of meromixis and therefore greater habitat diversity.

Biogeography of selected groups in Antarctic lakes

The following sections discuss aspects of the currently known biogeography of some important groups of organisms found in Antarctic lakes. They were chosen in part to illustrate points made above. Further groups not covered here, including *inter alia* bacteria (reviewed in Tindall 2004), flagellate algae, fungi (Göttlich et al. 2003), amoebae, heliozoa, nematodes (reviewed in Andrássy 1998), platyhelminths, gastrotrichs, annelid worms and tardigrades (reviewed in McInnes and Pugh 1998), have been recorded in Antarctic lakes. In some cases these organisms are known to be widespread and abundant, but others have been very poorly studied so far.

CYANOBACTERIA

Cyanobacteria are the dominant photoautotrophic organisms in freshwater ecosystems in Antarctica, where they accumulate conspicuous biomasses (Vincent 2000). They are present in both planktonic and benthic microbial communities in lakes, where they are important primary producers and have been recorded from as far south as the Darwin Glacier area (80°S) (Vincent and Howard-Williams 1994). About 500 species have been identified in Antarctica on morphological grounds.

Early studies on the diversity and biogeographical distribution of cyanobacteria were based on the identification of the organisms using morphological criteria. Unfortunately, cyanobacteria have often quite simple morphologies and some of these characters exhibit plasticity so that their taxonomic usefulness can be limited. The unsatisfactory state of the cyanobacterial taxonomy based on morphology is reflected in current revisions (eg Komárek and Anagnostidis 1989). In addition, a number of identifications of Antarctic cyanobacteria were made with floral guides written for temperate species without taking into account their ecology (Komárek 1999), which could give the impression that mostly cosmopolitan taxa were found on the continent. Out of 68 species found in various microbiotopes of ice-free areas of King George Island, Komárek (1999) determined that about 60% were probably endemic to Antarctica. The problems encountered in identifying cyanobacteria on the basis of morphological criteria have prompted the use of molecular tools for diversity studies. A view has been promoted that retrieval of 16S rRNA sequences directly from the environment could teach us more about diversity than the study of strains isolated from that same sample (Hugenholtz and Pace 1996). However, a polyphasic approach to diversity studies (combining phenotypic and genotypic characterisations) seems the best solution, as experience has also shown that the molecular tools have some significant weaknesses (Wintzingerode et al. 1997, Speksnijder et al. 2001).

Use of these molecular tools has allowed a new view of biodiversity and endemism in Antarctic cyanobacteria. We present here a synthesis of the biodiversity and distribution of Antarctic cyanobacteria based on the rRNA gene sequences recorded in the web-based databases, to which we have added data from microbial mats in four lakes from Eastern Antarctica (A. Taton and A. Wilmotte unpubl. data). We have included all sequences of cultured strains (Rudi et al. 1997, Smith et al. 2000, Vincent et al. 2000, Nadeau et al. 2001, D. Casamatta and M. Vis unpubl. data) and for uncultivated cyanobacteria (Priscu et al. 1998, Bowman et al. 2000, Smith et al. 2000, De la Torre et al. 2003, Christner et al. 2003, Taton et al. 2003, Coolen et al. unpubl. data), one representative sequence per operational taxonomic unit (an OTU is a group of sequences that exhibited more than 97.5% similarity) per clone library or study. Each OTU might correspond to more than one species, following bacteriological standards, but is surely distinct from other OTUs at the specific level (Stackebrandt and Goebel 1994). The determination of the molecular diversity of cyanobacteria is limited by the number of samples that are available, which at present are limited to Prydz Bay (PB) coastal oases (including Vestfold Hills, Larsemann Hills and Rauer Islands), Southern Victoria Land (SVL: McMurdo Dry Valleys and Bratina Island, McMurdo Ice Shelf) and Dronning Maud Land (DML).

A total of 53 OTUs were identified (Fig. 4), of which 38 have not been recorded for non-Antarctic sites (or at least were not present amongst the circa 3000 cyanobacterial sequences now present in the databases). This suggests a high degree of endemism. Of the endemic Antarctic OTU, 28 were recorded in only one sample and they may be limited to the areas in which they were found. A further five OTUs were found in several samples, but restricted to one region. The remaining five Antarctic OTUs were found in at least two regions, as were nine non-polar OTUs. Thus, 60% of the non-polar OTUs were widely distributed, whereas this was the case of only 13% of the Antarctic ones. This would fit with the idea that non-polar OTUs had to be well adapted to transport and dissemination and thus were quite successful in spreading to new Antarctic habitats in different regions. Alternatively, it would be interesting to test whether the Antarctic genotypes are more specialised and less widely distributed.

It is noteworthy that in each sample studied several new OTUs were found. Furthermore, most of the genotypic data came from samples from the PB and the SVL regions and therefore, other regions should be studied to have a more complete view of the geographical distribution of OTU. These results show that the bulk of the cyanobacterial genomic diversity in Antarctic lakes still remains to be discovered and that it would be very interesting to test the hypothesis of endemism by studying similar microbial mats along geographical gradients (maritime Antarctica, subantarctic islands, Southern Hemisphere continents), in addition to in mountain and Arctic lakes.

FRESHWATER DIATOMS

Freshwater diatoms comprise one of the most abundant algal groups in (sub-) Antarctic lakes (Jones 1996, Van de Vijver and Beyens 1999). They are present in the entire region from the highly diverse subantarctic lakes (eg Le Cohu and Maillard 1986, Oppenheim 1990, Van de Vijver et al. 2002) to the more

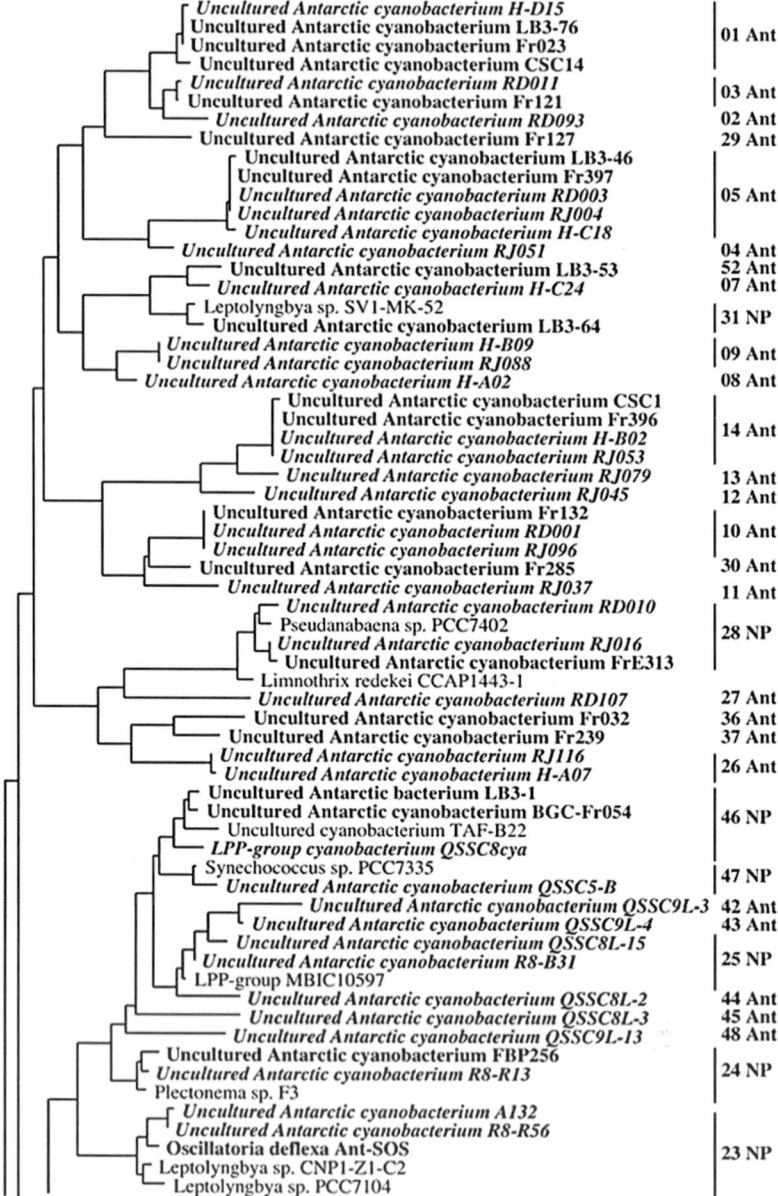

Figure 4. Distance tree based on partial 16S rRNA gene sequences corresponding to
E. coli positions 405 to 780. The tree was constructed by the neighbor-joining method (Saitou
and Nei 1987) with the software package ARB [http://www.arb-home.de/]. The dissimilarity
values corrected with the equation of Jukes and Cantor (1969) were used to calculate the
distance matrix. Indels and ambiguous nucleotides were not taken into account.

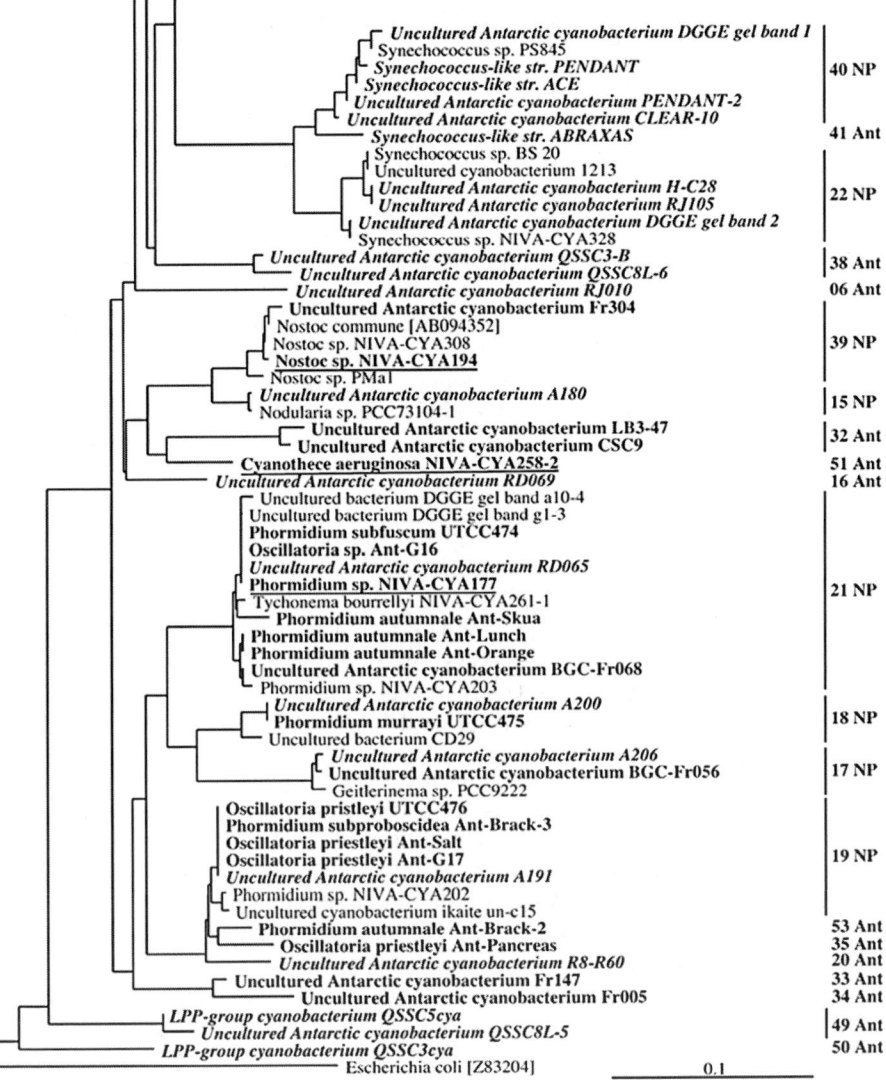

The E. coli *sequence is used as the outgroup. Antarctic sequences from
the Prydz Bay region are in boldface italic type, those from southern Victoria Land are in
boldface roman type and these from Dronning Maud Land are in boldface roman type and
underlined. Non-Antarctic nearest neighbors indicated by BLAST were included when they
exhibited more than 97.5% sequence similarity with the query. Only the first hit was retained
and if the hit was from uncultured clones, we added the sequence of the closest cultured
strain. The evolutionary distance between two sequences is obtained by adding the lengths of
the horizontal branches connecting them and using the scale bar (0.1 mutation per position).*

species-poor lakes of continental Antarctica (eg Pankow et al. 1991, Sabbe et al. 2003, Cremer et al. 2004).

A striking feature of Antarctic freshwater diatoms is the degree of endemism, contrary to what has been reported in the past. Schmidt et al. (1990), for instance, gave an estimate of 63% of cosmopolitan species in their palaeoecological study of lakes on King George Island (South Shetlands). On Ile de la Possession (Iles Crozet), almost 50% of all reported species are only found in the subantarctic region (Van de Vijver et al. 2002). One of the best examples of this problem is from the genus *Stauroneis* Ehrenberg. Until recently, only typical cosmopolitan species have been reported from Antarctic locations (Fukushima 1965, Oppenheim 1990, Jones 1993). Van de Vijver et al. (2004) reinvestigated the Antarctic material and 17 new taxa were described. All of them had a very limited distributional range within the southern zone, whereas only 5 cosmopolitan taxa were observed. Due to taxonomic deficiencies such as the use of a (too) broad species concept, the true number of endemic species in continental Antarctic lakes is still unknown (Sabbe et al. 2003, Cremer et al. 2004).

A decrease in species richness is observed with an increase in latitude. The freshwater diatom flora of the islands of the subantarctic zone comprises far more species than the other zones. On Iles Kerguelen and Iles Crozet, between 200 and 250 diatom species have been recorded (Le Cohu and Maillard 1986, Van de Vijver et al. 2002). Moving southward, the number decreases below 150 for South Georgia (Hirano 1965, Van de Vijver and Beyens 1996), to fewer than 100 for the maritime Antarctic zone (eg Oppenheim and Greenwood 1990, Jones 1993, Van de Vijver and Beyens 1997) and to 40 for continental Antarctica (Fukushima 1962, Sabbe et al. 2003, Cremer et al. 2004).

Biogeographical relationships among the different parts of the Antarctic region are difficult to make due to the taxonomic confusion that has prevailed in past studies on Antarctic diatoms (Sabbe et al. 2003). Still, based on the most recent data (after 2002), several tendencies can be observed. There are marked differences in species composition resulting in the delimitation of three main entities.

The islands south of the Indian Ocean (Kerguelen Province) form a separate entity, presenting a high degree of intra-similarity. They are well separated from the other parts of the region (eg islands south of the Atlantic Ocean). Studies on diatoms from lakes and pools on the various islands (Iles Crozet, Iles Kerguelen, Heard and Marion/Prince Edward Islands) have shown that more than 40% of all species on the islands are shared (Van de Vijver et al. unpubl. data) but absent in the other parts of the Antarctic Region. Roughly speaking, these islands coincide with the subantarctic region with the exception of Macquarie Island. The position of the latter is at the moment rather unclear due to the lack of appropriate data. The older records point to a relatively high similarity with the Kerguelen Province islands but more recent (so far unpublished) results contradict this hypothesis.

Several species show clear latitudinal preferences, with *Kobayasiella subantarctica* Van de Vijver and Vanhoutte and *Psammothidium abundans* (Manguin) Bukhtiyarova and Round, decreasing in relative abundance from Marion to Heard Islands while *Luticola muticopsis* (Van Heurck) Mann and *Chamaepinnularia australomediocris* (Lange-Bertalot and Schmidt) Van de Vijver increase in importance moving southwards.

A second group is formed by the islands in the maritime Antarctic. The species number is lower than in the subantarctic region. Still, a number of species are shared by both regions but this similarity is more based on cosmopolitan species and less on Antarctic endemics.

The Antarctic continent presents an entity on its own and can be easily separated from the other subdivisions. It has a very typical but impoverished diatom flora with a very high number of truly endemic species such as *Luticola murrayi* (West and West) Mann and *Navicula shackletoni* West and West (Sabbe et al. 2003). Subdividing the continent in western and eastern zones is rather difficult due to the paucity of data.

CILIATES

Ciliates are a diverse group of unicellular eukaryotes that play an important role in planktonic and benthic foodwebs of aquatic ecosystems. However, relatively few investigations of ciliates have been made for Antarctic lakes. The earliest data are from Murray (1910) and there have been only scattered reports since then (see references in Ellis-Evans 1996). Furthermore, in most studies the entire lake system (water column and benthos) was not sampled, rather a particular subset of the diversity. From these patchy and incomplete data, Ellis-Evans (1996) concluded that there was no obvious division in diversity for ciliates among continental and maritime sites.

A recent and thorough comparative study of ciliates in lakes from various limnetic habitats (eg cyanobacterial mats, filamentous green algal strands, aquatic moss, epilithion, sediments) of pools, lakes and streams on Byers Peninsula, Livingston Island and Deception Island (South Shetland Islands, 63°S) and Terra Nova Bay (Victoria Land, 75°S) indicated a significant gradient in diversity (Petz, 2003). One hundred and twenty species were recorded in the maritime Antarctic sites and 59 in coastal continental Antarctica, which indicated a distinct decrease in diversity towards higher latitudes. Approximately 20% of the species were new to science and therefore may be endemic. Highest ciliate diversity was usually found associated with cyanobacterial mats, reflecting the importance of microbial mats in Antarctic lakes, with somewhat fewer species associated with sediments and in the plankton. While Litostomatea (predominantly haptorids) and Stichotrichia (mostly oxytrichid hypotrichs) dominated the aquatic community in both areas (each between 19-29% of the species), the proportion of cyanobacterial feeders and scuticociliate taxa increased by three to fourfold in Victoria Land. Peritrichs, suctorians and some other groups decreased towards the south while colpodid ciliates were entirely absent at Terra Nova Bay.

Further comparisons have been made between the planktonic ciliates of lakes in the Vestfold Hills and the McMurdo Dry Valleys. In this case the trend discussed above is reversed: ciliate diversity was greater in the more southerly Dry Valley lakes (15 species) than in the Vestfold Hills (< 10 species) (Laybourn-Parry 1997). The reasons for this are not immediately clear, but possibly reflect the far longer period that the McMurdo Dry Valleys lakes have existed and therefore the period available for colonisation and development of more mature ecosystems. Furthermore, the complex history of the McMurdo Dry Valleys lakes would result in numerous environmental niches for colonisation. In contrast, Beaver Lake was species poor (Laybourn-Parry et al. 2001), though the species present were similar to those recorded at the other sites. Beaver Lake is thought to be older than the Vestfold Hills lakes, so there are apparently other factors, including the lack of environmental niches in the water column of the lake and less complex history, that may influence ciliate diversity. The biodiversity of ciliates in the microbial mats of these lakes, however, has not been investigated and may yield a different distributional pattern.

The origins of the fauna are uncertain. Kepner et al. (1999) proposed that marine ciliates colonized the lakes of the McMurdo Dry Valleys from the nearby Ross Sea. However, the occurrence of numerous cosmopolitan species in the study of Petz (2003) suggested that colonisation from more northerly regions has been an efficient process. The moderate level of endemism found in this study may reflect the occurrence of a true Antarctic fauna, but could also be due to limited studies in areas such as Patagonia from which the fauna may have been sourced.

ROTIFERS

Rotifers are common inhabitants of Antarctic lakes across a wide range of environmental conditions and geographical areas. Most species are found associated with benthic vegetation (Armitage and House 1962, Opalinski 1972, Dartnall 1983), with the corollary that planktonic species are uncommon. Their appearance in cryoconite holes (Christner et al. 2003) attests to their ability to colonise new aquatic habitats.

More than 150 species of rotifer have been reported from the Antarctic region. This total could be an overestimate as the number of taxa almost certainly represents a lower number of species when mistaken identities and some sensible attributions are invoked. Nevertheless some patterns do emerge. Adjacent locations share more species than widely separated ones. For example, South Georgia and the South Orkney Islands share 26 species, the Larsemann and Vestfold Hills share 14, Heard Island and Iles Kerguelen share 11, and the South Orkney Islands and the South Shetland Islands 10. There is also a reduction in the number of species present with increasing latitude, as demonstrated by diversity on islands in the Scotia Arc (59 species occur at South Georgia, 38 on the South Orkney Islands, 19 on the South Shetland Islands and nine on the Antarctic Peninsula), while the subantarctic islands - South Georgia, Iles Kerguelen, Heard Island and Macquarie Island - are all richer than continental sites to their south.

The rotifers found at any location are a mixture of cosmopolitan, widespread, localized and endemic species. Examples of cosmopolitan species include

Collotheca ornata cornuta (Dobie), *Cephalodella catellina* (O. F. Müller), *Epiphanes senta* (O. F. Müller), *Lepadella patella* (O. F. Müller), *Resticula gelida* (Harring and Myres) and the bdelloid rotifers *Adineta barbata* Janson, *A. gracilis* Janson and *Habrotrocha constricta* (Dujardin).

Of particular interest are localized and endemic species, which include members of the genera *Notholca*, *Adineta*, *Philodina* and *Keratella*. Species of *Notholca* are found at most Antarctic and subantarctic locations. *Notholca walterkostei* De Paggi (Fig. 2) is present at the southern tip of South America (De Paggi and Koste 1995), on the Falkland Islands (H. Dartnall unpubl. data), at South Georgia and on the South Orkney Islands (Dartnall and Hollowday 1985), on the South Shetland Islands (De Paggi 1982) and down the Antarctic Peninsula (Heywood 1977). *Notholca hollowdayi* Dartnall is present on Iles Kerguelen (de Smet 2001) and Heard Island (Dartnall 1995a) and *Notholca salina* Foche, *N. squamula* (O. F. Müller) and other common species of the '*squamula* complex' are present in the South Shetlands and South Orkney Island (De Paggi 1982, Dartnall and Hollowday 1985). *Notholca verae* Kutikova was described from the Bunger Hills (Kutikova 1958a, b), but more recent collections have failed to find animals consistent with the original description (H. Dartnall unpubl. data). The species present in the Larsemann and Vestfold Hills has been assigned to *Notholca verae* (Everitt 1981), but is more likely to be another species in the '*squamula* complex' (Dartnall 1995b, 2000). There are two forms in the Vestfold Hills, one of which is found in freshwater and the other in brackish to saline lakes. Thus, *Notholca* appears to be an ancient Antarctic genus that has evolved new species in different areas.

Keratella sancta Russell was originally described by Russell (1944) from South Island, New Zealand and has subsequently been reported from Iles Kerguelen (Russell 1959, de Smet 2001) and Macquarie Island (Dartnall 1993). The specimens from these three locations - New Zealand, Macquarie Island and Iles Kerguelen - show differences in some measurements, the length and thickness of the spines and the degree of pustulation of the dorsal plate so that they are instantly recognizable one from another. This indicates considerable speciation and possibly indicates that the species has been on the islands as separate populations for a considerable time.

By far the most interesting species are the Antarctic continental endemics: the bdelloid rotifers *Adineta grandis* Murray, *Philodina alata* Murray, *P. antarctica* Murray and *P. gregaria* Murray (Fig. 2). *Philodina gregaria* and *Adineta grandis* are invariably found together in shallow pools alongside seal and penguin colonies. These are found in the Bunger, Vestfold and Larsemann Hills (Korotkevich 1958, Kutikova 1958b, Dartnall 1995b, 2000) on the Antarctic Peninsula (Schmitt 1945, Heywood 1977) and up to the South Orkney Islands (Dartnall and Hollowday 1985, Dartnall 1992), but are not present on any of the subantarctic islands. *Philodina antarctica* has been reported from both McMurdo Sound and the Antarctic Peninsula (Heywood 1977), it probably occurs elsewhere but may be confused with *P. gregaria*. *Philodina alata* is restricted to east Antarctica, notably southern Victoria Land and the Obruchev and Bunger Hills

(Korotkevich 1958, Kutikova 1958b). These species may have survived the LGM in Antarctica in refugia, but further investigations are needed to confirm the occurrence of relict populations.

CRUSTACEA

Crustacea – cladocera, copepods, anostraca and ostracods – are widespread in lakes on the subantarctic islands, but more poorly represented species-wise on the continent itself. It is often stated that the crustacea are rare or absent in continental lakes (eg Laybourn-Parry 1997). However, a critical review of recent distributional and functional data suggests otherwise (Hansson and Tranvik 1997, Swadling and Gibson 2000, Almada et al. 2004, Peck 2004). Crustacea occur in lakes throughout the maritime Antarctic zone including the Antarctic Peninsula, where they are often abundant (eg Peck 2004). In continental Antarctica copepods occur in most if not all fresh and slightly brackish lakes in many of the major coastal oases, including the Vestfold, Bunger and Larsemann Hills and the Amery Oasis. Other areas, notably the Schirmacher Oasis and most of the lakes of the McMurdo Dry Valleys, lack crustacea.

Pugh et al. (2002) recently reviewed the distribution of Antarctic lacustrine crustacea and from their data a general picture of distributional patterns emerges (Fig. 5). Firstly, the fauna of South Georgia, at the northern end of the maritime Antarctic zone, has strong similarities to the South American fauna, while the fauna of the Antarctic Peninsula is a subset of that of South Georgia, with all the species present also found in South America. These observations argue strongly for a South

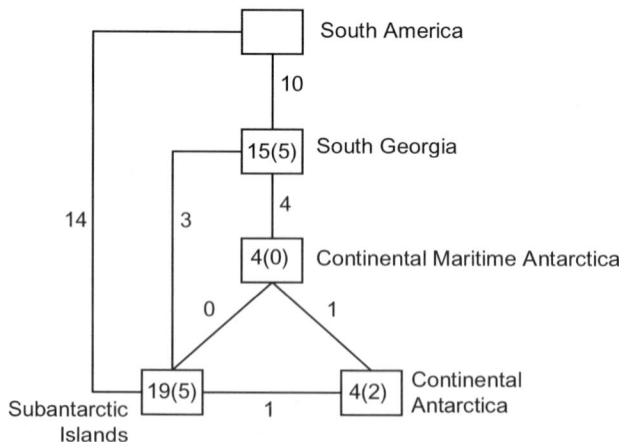

Figure 5. Relationships among the occurrences of non-marine-derived crustacean species in Antarctica. The numbers in the boxes are the numbers of species found in each location (data from Pugh et al. 2002) and those on the linking lines the number shared between regions. The number of endemic species is shown in parentheses.

American source for this fauna, though there are some apparently endemic species on South Georgia.

Freshwater lakes in continental Antarctica are species poor, though further marine-derived species occur in saline lakes of the Vestfold Hills. One freshwater species is shared with maritime Antarctica and another with the subantarctic islands. Thus there are two endemic species (*Gladioferens antarcticus* Bayly and *Acanthocyclops mirnyi* Borutsky and Vinogradov) and there is evidence that a third, the copepod *Boeckella poppei* (Mrázek), which occurs both on the Antarctic Peninsula and in epishelf Beaver Lake and other lakes of the Amery Oasis (Fig. 1), has inhabited continental Antarctica for an extended period of time (Bayly et al. 2003). Centropagid copepods such as *Boeckella poppei* have very poor dispersal abilities, consistent with a relictual origin for the continental populations. The southernmost record of this species from the Antarctic Peninsula is from epishelf Ablation Lake and nearby ponds on Alexander Island, which is near the southern boundary of the maritime Antarctic zone (Heywood 1977). The possibility that this is a further relict population of this species cannot be discounted.

The remaining continental species, the cladoceran *Daphniopsis studeri* Rühe, occurs in the Vestfold and Larsemann Hills and at least one lake in Enderby Land and also on all the islands of the subantarctic zone with the exception of Bouvetøya and Macquarie Island (Gibson et al. 1998). Molecular genetic studies of this species indicate little variation and that it passed through a population bottleneck at some stage (Wilson et al. 2002). This species may be an Antarctic endemic that might survive glaciations by sequential colonisation of the subantarctic islands. However, this would imply the occurrence of lakes on the subantarctic islands through the LGM, for which at present there is no evidence.

Most subantarctic islands share few species with South Georgia, but considerably more with South America, though many of the species present are cosmopolitan (Pugh et al. 2002). It is probable that these species reached the islands by transport in the westerly winds and easterly currents that occur in this region, though transport from South Africa, Australia or New Zealand is also possible. Furthermore, the fauna of Macquarie Island differs from the other subantarctic islands, some species shared by all the other islands are absent and there are others that are endemic to this island.

Conclusions

The basic biogeographic zones proposed many years ago – the subantarctic islands, maritime Antarctica and continental Antarctica – continue to hold up, though they cannot be seen as absolute dividers of biodiversity. For example, subantarctic Macquarie Island appears to be biogeographically separate from the islands of the Kerguelen Province and on the continent there are species that are present in lakes of more than one zone. Furthermore, there are numerous lake environments that have yet to be investigated and it is probable that some of these lakes could turn up surprises that will bring into question these basic divisions.

An important question to be answered is whether these biogeographic zones reflect climate attributes, or whether they were moulded long ago by barriers to dispersal. Again, our imperfect knowledge of Antarctic lacustrine biogeography means that this question cannot at present be answered. However, as discussed elsewhere (Chown and Convey this volume), there are indications of a strong biogeographical boundary for terrestrial species between the maritime and continental Antarctic zones.

A palaeolimnological approach will assist in answering this question: understanding how Antarctic biogeography has developed through time will provide necessary insights into current distributions. A prime example is the occurrence of the copepod *Boeckella poppei* in Beaver Lake. Pugh et al. (2002) initially concluded that this species was an anthropogenic introduction, then Bayly et al. (2003) provided morphological evidence for long habitation in the area of Beaver Lake. Recent palaeolimnological work has shown that the species has been present in nearby Lake Terrasovoje for at least 9000 yrs (Bissett et al. 2005). Even though this lake has only existed in the Holocene, cosmogenic exposure dates in the same area of exposed rock can exceed 10^6 years (Fink et al. 2006). From these observations it can be concluded that *Boeckella poppei* has been associated with the Beaver Lake area for at least the entire Holocene and probably well back into the Pleistocene and that its occurrence outside its 'preferred' biogeographical zone (maritime Antarctica) is not a reflection of current climate, rather of history.

The majority of our knowledge regarding Antarctic lacustrine biodiversity and biogeography has come from classic taxonomic studies, where the morphology (or biochemistry for bacteria) has been of greatest importance. In many cases this has led to questionable identification: correct identification of species is paramount if the true biodiversity and biogeography of Antarctica is to be deduced. It is only in the last few years that the more objective approach of molecular genetics has been applied to Antarctic lacustrine organisms and then only for more cryptic groups, such as bacteria and cyanobacteria. As more samples and organisms are studied by these methods it is likely that new relationships among species distributions will be found. Due to the limited number of species in Antarctica (compared to more temperate zones), it may be possible in the future to record the make-up of selected genes of most, if not all, of the biota, which will allow more precise analysis.

There is increasing evidence for endemism amongst the inhabitants of lakes both on the Antarctic continent and the subantarctic islands, from bacteria to crustacea. Use of molecular genetic techniques to identify more cryptic species will most likely add to the list of putative endemics. It is clear, however, that recent colonisation and current climate also play important roles in the distribution of the biota, as most of the lakes in Antarctica are of relatively recent (Holocene) origin. Colonising species have to be adapted to transport from source areas, which can either involve inter- or intra-continental movement, in addition to survival on arrival at potential habitat. Flexibility in nutritional and habitat requirements is an important factor in determining whether a species will be a successful coloniser. The buffering to environmental extremes provided by the liquid water habitat means that conditions further south will not be as harsh as those experienced by their terrestrial counterparts.

As the climate changes in the future it will be interesting to note the effects of these changes on the lacustrine biota. Will new species colonise the Antarctic Peninsula where temperatures are increasing? In the longer term, the biogeography of Antarctic lakes will continue to be dynamic. New species will arrive, others will become extinct. The biogeographic zones long-proposed may continue to hold, though more precise knowledge of current distributions and responses to climate change may refine our view.

Acknowledgements

The authors thank Clive Howard-Williams and Cynan Ellis-Evans for comments on earlier drafts of this manuscript. John Gibson received funding through an Australian Research Council Discovery Grant (DP0342815). The cyanobacterial study reported here was funded by the European Union through the MICROMAT project (BIO4-CT98-0040) and the Federal Science Policy of Belgium through the project LAQUAN (EV/02/1B). Annick Wilmotte is a research associate of the National Fund for Scientific Research (Belgium). Arnaud Taton had a fellowship from the Funds for Research Formation in Industry and Agriculture (Belgium). They thank Dominic Hodgson (British Antarctic Survey, UK), Johanna Laybourn-Parry (University of Nottingham, UK) and collaborators and the Australian Antarctic Division, under whose auspices some material discussed in the chapter was collected.

References

Almada, P., Allende, L., Tell, G. and Izaguirre, I. (2004) Experimental evidence of the grazing impact of *Boeckella poppei* on phytoplankton in a Maritime Antarctic lake, *Polar Biology* **28**, 39-46.

Andrássy, I. (1998) Nematodes in the sixth continent, *Journal of Nematode Morphology and Systematics* **1**, 107-186.

Armitage, K.B. and House, H.B. (1962) A limnological reconnaissance in the area of McMurdo Sound, Antarctica, *Limnology and Oceanography* **7**, 36-41.

Bardin, V.I. (1990) Antarctic lakes as a source of paleo-glacial information, in A. Louro, M.A. Van Zele, and I. Velasco (eds.), *Actas de la Primera Conferencia Latinoamericana sobre Geofísica, Geodesia e Investigación Espacial Antárticas*, Centro Latinoamericano de Física, Buenos Aires, Argentina.

Bayly, I.A.E., Gibson, J.A.E., Wagner, B. and Swadling, K.M. (2003) Taxonomy, ecology and zoogeography of two Antarctic freshwater calanoid copepod species: *Boeckella poppei* and *Gladioferens antarcticus*, *Antarctic Science* **15**, 439-448.

Bissett, A., Gibson, J.A.E., Jarman, S.N., Swadling, K.M. and Cromer L. (2005) Isolation, amplification and identification of ancient copepod DNA from lake sediments, *Limnology and Oceanography: Methods* **3**, 533-542.

Björck, S., Olsson, S., Ellis-Evans, C., Håkansson, H., Humlum, O. and de Lirio, J.M. (1996) Late Holocene palaeoclimatic records from lake sediments on James Ross Island, Antarctica, *Palaeogeography, Palaeoclimatology, Palaeoecology* **121**, 195-220.

Block, W. (1992) *Annotated bibliography of Antarctic invertebrates (terrestrial and freshwater)*, British Antarctic Survey, Cambridge.

Bowman, J.P., Rea, S.M., McCammon, S.A. and McMeekin TA (2000) Diversity and community structure within anoxic sediment from marine salinity meromictic lakes and a coastal meromictic marine basin, Vestfold Hills, Eastern Antarctica, *Environmental Microbiology* **2**, 227-237.

Christner, B.C., Kvitko, B.H. and Reeve, J.N. (2003) Molecular identification of Bacteria and Eukarya inhabiting an Antarctic cryoconite hole, *Extremophiles* **7**, 177-183.

Chown, S.L. and Convey, P. (2006) Biogeography, in D.M. Bergstrom, P. Convey, and A.H.L. Huiskes (eds.), *Trends in Antarctic Terrestrial and Limnetic Ecosystems: Antarctica as a Global Indicator*. Springer, Dordrecht (this volume).

Cowan, D. and Tow, L.A. (2004) Endangered Antarctic Environments, *Annual Reviews of Microbiology* **58**, 649-690.

Crawford, A.R. (1974) Greater Gondwanaland, *Science* **184**, 1179-1181.

Cremer, H., Gore, D., Hultzsch, N., Melles, M. and Wagner, B. (2004) The diatom flora and limnology of lakes in the Amery Oasis, East Antarctica, *Polar Biology* **27**, 513-531.

Cromer, L., Gibson, J.A.E., Swadling, K.M. and Ritz, D.A. (2005) Faunal microfossils: Indicators of ecological evolution in a saline Antarctic lake, *Palaeogeography, Palaeoclimatology, Palaeoecology* **221**, 83-97.

Dartnall, H.J.G. (1983) Rotifers of the Antarctic and Subantarctic, *Hydrobiologia* **104**, 57-60.

Dartnall, H.J.G. (1992) The reproductive strategies of two Antarctic rotifers, *Journal of Zoology, London* **227**, 145-162.

Dartnall, H.J.G. (1993) The Rotifers of Macquarie Island, *ANARE Research Notes* **89**, 1-41.

Dartnall, H.J.G. (1995a) The Rotifers of Heard Island: preliminary survey, with notes on other freshwater groups, *Papers and Proceedings of the Royal Society of Tasmania* **129**, 7-15.

Dartnall, H.J.G. (1995b) Rotifers, and other aquatic invertebrates from the Larsemann Hills, Antarctica, *Papers and Proceedings of the Royal Society of Tasmania* **129**, 17-23.

Dartnall, H.J.G. (2000) A limnological reconnaissance of the Vestfold Hills, *ANARE Reports* **141**, 1-55.

Dartnall, H.J.G. and Hollowday, E.D. (1985) Antarctic Rotifers, *British Antarctic Survey Scientific Reports* **100**, 1-46.

de la Torre, J.R., Goebel, B.M., Friedmann, E.I. and Pace, N.R. (2003) Microbial diversity of cryptoendolithic communities from the McMurdo Dry Valleys, Antarctica, *Applied and Environmental Microbiology* **69**, 3858-3867.

De Paggi, S.B.J. (1982) *Notholca walterkostei* sp. nov. y ostros Rotiferos dulceacuicolas de la Peninsula Potter. Isla 25 de Mayo (Shetland del Sur, Antartida), *Revista Asociacion de Ciencias Naturales del Litoral* **13**, 81-95.

De Paggi, S.B.J. and Koste, W. (1995) Additions to the checklist of rotifers of the superorders monogononta recorded from neotropis, *International Revue der Gesamte Hydrobiologie* **80**, 133-140.

De Smet, W.H. (2001) Freshwater Rotifera from plankton of the Kerguelen Islands (Subantarctica), *Hydrobiologia* **446/447**, 261-272.

Ellis-Evans, J.C. (1996) Microbial diversity and function in Antarctic freshwater ecosystems, *Biodiversity and Conservation* **5**, 1395-1431.

Ellis-Evans, J.C. and Walton, D. (1990) The process of colonization in Antarctic terrestrial and freshwater ecosystems, *Proceedings of the NIPR Symposium on Polar Biology* **3**, 151-163.

Eslake, D., Kirkwood, R., Burton, H. and Zipan, W. (1991) Temporal changes in zooplankton composition in a hypersaline, Antarctic lake subject to periodic seawater incursions, *Hydrobiologia* **210**, 93-99.

Everitt, D.A. (1981) An ecological study of an Antarctic freshwater pool with particular reference to tardigrada and rotifera, *Hydrobiologia*, **83**, 225-237.

Fink, D.A., McKelvey, B., Hambrey, M.J., Fabel, D. and Brown, R. (2006) Pleistocene deglaciation chronology of the Amery Oasis and Radok Lake, northern Prince Charles Mountains, Antarctica, *Earth and Planetary Science Letters* **243**, 229-243.

Fukushima, H. (1962) Notes on the diatom vegetation of the Kasumi Rock ice-free area, Prince Olav Coast, Antarctica, *Antarctic Record* **15**, 39-52.

Fukushima, H. (1965) Preliminary report on diatoms from South Georgia, *JARE Scientific Report Series E Biology* **24**, 18-30.

Gibson, J.A.E. (1999) The meromictic lakes and stratified marine basins of the Vestfold Hills, East Antarctica, *Antarctic Science* **11**, 175-192.

Gibson, J.A.E. and Andersen, D.T. (2002) Physical structure of epishelf lakes of the southern Bunger Hills, East Antarctica, *Antarctic Science* **14**, 253-262.

Gibson, J.A.E., Dartnall, H.J.G. and Swadling, K.M. (1998) On the occurrence of males and production of ephippial eggs in populations of *Daphniopsis studeri* (Cladocera) in lakes of the Vestfold and Larsemann Hills, East Antarctica, *Polar Biology* **19**, 148-150.

Gibson, J.A.E., Gore, D.M. and Kaup, E. (2002) Algae River: An extensive drainage system in the Bunger Hills, East Antarctica, *Polar Record*, **38**, 141-152.

Goldman, C.R. (1970) Antarctic freshwater ecosystems, in R.M. Holdgate (ed.), *Antarctic Ecology*, Vol. 2, Academic Press, pp 609-627.

Gore, D.B., Rhodes, E.J., Augustinus, P.C., Leishman, M.R., Colhoun, E.A. and Rees-Jones, J. 2001. Bunger Hills, East Antarctica: Ice free at the Last Glacial Maximum, *Geology* **29**, 1103-1106.

Göttlich, E., de Hoog, G.S., Genilloud, O., Jones, B.E. and Marinelli, F. (2003) MICROMAT: Culturable fungal diversity in microbial mats of Antarctic lakes, in A.H.L. Huiskes, W.W.C. Gieskes, R. Rozema, R.M.L. Schorno, S.M. van der Vies and W.J. Wolff (eds) *Antarctic Biology in a Global Context*, Backhuys Publishers, pp 251-254.

Hansson, L.-A. and Tranvik, L.J. (1997) Algal species composition and phosphorus recycling at contrasting grazing pressure: an experimental study in subantarctic lakes with two trophic levels, *Freshwater Biology* **37**, 45-53.

Hawes, I. and Schwarz, A.M.J. (2001) Absorption and utilization of irradiance by cyanobacterial mats in two ice-covered Antarctic lakes with contrasting light climates, *Journal of Phycology* **37**, 5-15.

Hawes, I., Howard-Williams, C. and Vincent, W.F (1992) Desiccation and recovery of antarctic cyanobacterial mats, *Polar Biology* **12**, 587-594.

Hendy, C.H. (2000) Late Quaternary lakes in the McMurdo Sound region of Antarctica, *Geografiska Annaler* **82A**, 411-432.

Heywood, R.B. (1977) A limnological survey of the Ablation Point area, Alexander Island, Antarctica, *Philosophical Transactions of the Royal Society, Series B* **279**, 39-54.

Heywood, R.B. (1984) Antarctic inland waters, in R.M. Laws (ed.) *Antarctic Ecology*, Academic Press, pp 279-334.

Hirano, M. (1965) Freshwater algae in the Antarctic regions, in: J. Van Mieghem, P. Van Oye and J. Schell (eds.), *Biogeography and Ecology in Antarctica*, W. Junk, The Hague, pp. 127-193.

Hodgson, D.A., Noon, P.E., Vyverman, W., Bryant, C.L., Gore, D.B., Appleby, P., Gilmour, M., Verleyen, E., Sabbe, K., Jones, V.J., Ellis-Evans, J.C. and Wood, P.B. (2001) Were the Larsemann Hills ice-free through the Last Glacial Maximum? *Antarctic Science* **13**, 40-454.

Holdgate, M.W. (1977) Terrestrial ecosystems in Antarctica, *Philosophical Transactions of the Royal Society, Series B,* **279**, 5-25.

Hugenholtz, P. and Pace, N.R. (1996) Identifying microbial diversity in the natural environment: a molecular phylogenetic approach, *Trends in Biotechnology* **14**,190-197.

James, M.R., Pridmore, R.D. and Cummings, V.J. (1995) Planktonic communities of melt ponds on the McMurdo Ice Shelf, *Journal of Plankton Research* **15**, 555-567.

Jones, V.J. (1993) The use of diatoms in lake sediments to investigate environmental history in the Maritime Antarctic: an example from Sombre Lake, Signy Island, in R.B. Heywood (ed.) University Research in Antarctica 1989-1992: Proceedings of the British Antarctic Survey Antarctic Special Topic Award Scheme Symposium, British Antarctic Survey, Cambridge, pp. 91-95.

Jones, V.J. (1996) The diversity, distribution and ecology of diatoms from Antarctic inland waters, *Biodiversity and Conservation* **5**, 1433-1449.

Jones, V.J., Hodgson, D.A. and Chepstow-Lusty, A. (2000) Palaeolimnological evidence for marked Holocene environmental changes on Signy Island, Antarctica, *The Holocene* **10**, 43-60.

Jukes, T.H. and Cantor, C.R. (1969) Evolution of protein molecules, in H.N. Munro (ed.) *Mammalian protein metabolism*, Vol. 3., Academic Press, pp. 21-132.

Kepner, R.L., Wharton, R.A. and Coats, D.W. (1999) Ciliated protozoa of two Antarctic lakes: analysis by quantitative protargol staining and examination of artificial substrates, *Polar Biology* **21**, 285-294.

Komárek, J. (1999) Diversity of cyanoprokaryotes (cyanobacteria) of King George Island, maritime Antarctica-a survey, *Archiv für Hydrobiologie* **94**, 181-193.

Komárek, J. and Anagnostidis, K. (1989) Modern approach to the classification system of cyanophytes. 4. Nostocales, *Archiv für Hydrobiologie* **56 (Suppl. 82/3)**, 247-345.

Korotkevich, V.S. (1958) Nasalinie vodoemov oazisov v Vostochnoy Antarktide, *Informatsionnyi Byulleten sovietskoi antarkticheskoi Ekspeditsii* **3**, 91-98.

Kulbe, T. (1997) Die spätquartäre Klima- und Umweltgeschichte der Bunger-Oase, Ostantarktis, *Berichte zur Polarforschung* **254**, 1-130.

Kutikova, L.A. (1958a) O novoy kolovratke v Antarktide, *Informatsionnyi Byulleten sovietskoi antarkticheskoi Ekspeditsii* **2**, 45-46.

Kutikova, L.A. (1958b) K fauna kolovratok s poberezh'ya antarkticheskoi, *Informatsionnyi Byulleten sovietskoi antarkticheskoi Ekspeditsii* **3**, 99.

Laybourn-Parry, J. (1997) The microbial loop in Antarctic lakes, in W.B. Lyons, C. Howard-Williams and I. Hawes (eds.), *Ecosystem processes in Antarctic ice-free landscapes,* A.A. Balkema, Rotterdam, pp. 231-240.

Laybourn-Parry, J. (2002) Survival mechanisms in Antarctic lakes, *Philosophical Transactions, Royal Society of London. Biological Sciences* **357**, 863-869.

Laybourn-Parry, J., Quayle, W. and Henshaw, T. (2002) The biology and evolution of Antarctic saline lakes in relation to salinity and trophy, *Polar Biology* **25**, 542-552.

Laybourn-Parry, J., Quayle, W.C., Henshaw, T. Ruddell, A. and Marchant, H.J. (2001) Life on the edge: the plankton and chemistry of Beaver Lake, an ultra-oligotrophic epishelf lake, Antarctica, *Freshwater Biology* **46**, 1205-1217.

Le Cohu, R. and Maillard, R. (1986) Diatomées d'eau douce des îles Kerguelen (à l'exclusion des Monoraphidées), *Annales de Limnologie* **22**, 99-118.

Lizotte, M.P. and Priscu, J.C. (1992) Photosynthesis irradiance relationships in phytoplankton from the physically stable water column of a perennially ice-covered lake (Lake Bonney, Antarctica), *Journal of Phycology* **28**, 179-185.

Lyons, W.B., Tyler, S.W., Wharton, R.A., McKnight, D.M. and Vaughn, B.H. (1998) A Late Holocene desiccation of Lake Hoare and Lake Fryxell, McMurdo Dry Valleys, Antarctica, *Antarctic Science* **10**, 247-256.

McInnes, S.J. and Pugh, P.J.A. (1998) Biogeography of limno-terrestrial Tardigrada, with particular reference to the Antarctic fauna, *Journal of Biogeography* **25**, 31-36.

Moorhead, D., Schmeling, J. and Hawes, I. (2005) Modelling the contribution of benthic microbial mats to net primary production in Lake Hoare, McMurdo Dry Valleys, *Antarctic Science* **17**, 33-45.

Mueller D.R. and Pollard W.H. (2004) Gradient analysis of cryoconite ecosystems from two polar glaciers, *Polar Biology* **27**, 66-74.

Mueller, D.R., Vincent, W.F. and Jeffries, M.O. (2003) Break-up of the largest Arctic ice shelf and associated loss of an epishelf lake, *Geophysical Research Letters* **30**, Article no. 2031.

Murray, J. (1910) Antarctic Rotifera, in J. Murray (ed.) *British Antarctic Expedition, 1907-09. Reports on the Scientific Investigation, Biology*, Vol. 1, William Heineman, pp. 41-65.

Nadeau, T.L., Milbrandt, E.C. and Castenholz, R.W. (2001) Evolutionary relationships of cultivated Antarctic Oscillatoriaceans (cyanobacteria), *Journal of Phycology* **37**, 650-654.

Neale, P.J. and Priscu, J.C. (1995) The photosynthetic apparatus of phytoplankton from a perennially ice-covered Antarctic lake - acclimation to an extreme shade environment, *Plant and Cell Physiology* **36**, 253-263.

Opalinski, K.W. (1972) Freshwater fauna and flora in Haswell Island (Queen Mary Land, eastern Antarctica), *Polskie Archiwum Hydrobiologii* **19**, 377-381.

Oppenheim, D.R. (1990) A preliminary study of the benthic diatoms in contrasting lake environments, in K.R. Kerry and G. Hempel (eds.) *Antarctic Ecosystems. Ecological change and conservation.* Springer-Verlag, Berlin, pp. 91-99.

Oppenheim, D.R. and Greenwood, R. (1990) Epiphytic diatoms in two freshwater maritime lakes, *Freshwater Biology* **24**, 303-314.

Pankow, H., Haendel, D. and Richter W. (1991) Die Algenflora der Schirmacheroase (Ostantarktika), *Nova Hedwigia* **103**, 1-197.

Peck, L.S. (2004) Physiological flexibility: the key to success and survival for Antarctic fairy shrimps in highly fluctuating extreme environments, *Freshwater Biology,* **49**, 1195-1205.

Petz, W. (2003) Ciliate biodiversity in Antarctic and Arctic freshwater habitats – a bipolar comparison, *European Journal of Protistology*, **39**, 491-494.

Pickard, J. and Adamson, D.A. (1983) Perennially frozen lakes at glacier/rock margins, East Antarctica, in R.L. Oliver, P.R. James and J.B. Jago (eds.), *Antarctic Earth Science*, Australian Academy of Science, Canberra, pp. 470-472.

Priscu, J.C., Fritsen, C.H., Adams, E.E., Giovannoni, S.J., Paerl, H.W., McKay, C.P., Doran, P.T., Gordon, D.A., Lanoil, B.D. and Pinckney, J.L. (1998) Perennial Antarctic lake ice: an oasis for life in a polar desert, *Science* **280**, 2095-2098.

Pugh, P.G.H., Dartnall, H.J.G. and McInnes, S.J. (2002) The non-marine crustacea of Antarctica and the islands of the Southern Ocean: biodiversity and biogeography, *Journal of Natural History* **36**, 1047-1103.

Quayle, W.C., Peck, L.S., Peat, H., Ellis-Evans, J.C. and Harrigan, P.R. (2002) Extreme responses to climate change in lakes, *Science* **295**, 645.

Roberts, D. and McMinn, A. (1999) A diatom-based palaeosalinity history of Ace Lake, Vestfold Hills, Antarctica, *The Holocene* **9**, 401-408.

Roberts, E.C., Priscu, J.C., Wolf, C., Lyons, W.B. and Laybourn-Parry, J. (2004) The distribution of microplankton in the McMurdo Dry Valley Lakes, Antarctica: response to ecosystem legacy or present-day climate controls? *Polar Biology* **27**, 238-249.

Rudi, K., Skulberg, O.M., Larsen, F. and Jakobsen, K.S. (1997) Strain characterization and classification of oxyphotobacteria in clone cultures on the basis of 16S rRNA sequences from the region V6, V7, and V8, *Applied and Environmental Microbiology* **63**, 2593-2599

Rühe, E. (1914) Die süsswassercrustaceen der Deutschen Südpolar-Expedition 1901-1903 mit Ausschluss des ostracoda, *Deutschen Südpolar Expedition 1901-1903, 16, Zoology* **8**, 5-66.

Russell, C.R. (1944) A New Rotifer from New Zealand, *Journal of the Royal Microscopical Society* **64**, 121-123.

Russell, C.R. (1959) Rotifera, *B.A.N.Z. Antarctic Research Expedition, 1929-31, reports, Series B* **8**, 81-88.

Sabbe, K., Verleyen, E., Hodgson, D.A., Vanhoutte, K. and Vyverman, W. (2003) Benthic diatom flora of freshwater and saline lakes in the Larsemann Hills and Rauer Islands (East-Antarctica), *Antarctic Science* **15**, 227-248.

Saitou, N. and Nei, M. (1987) The neighbor-joining method: a new method for reconstructing phylogenetic trees, *Molecular Biology and Evolution* **4**, 406-425.

Schmidt, R., Mäusbacher, R. and Müller, J. (1990) Holocene diatom flora and stratigraphy from sediment cores of two Antarctic lakes (King George Island), *Journal of Paleolimnology* **3**, 55-74.

Schmitt, W.L. (1945) Miscellaneous zoological material collected by the United States Antarctic Service Expedition, 1939-1941, *Proceedings of the American Philosophical Society* **89**, 297.

Schwab, M.J. (1998) Rekonstruktion der spätquartären Klima- und Umweltgeschichte der Schirmacher Oase und des Wohlthat Massivs (Ostantarktika), *Berichte zur Polarforschung* **293**, 1-128.

Seaburg, K.G., Kaspar, M. and Parker, B.C. (1983) Photosynthetic quantum efficiencies of phytoplankton from perennially ice covered Antarctic lakes, *Journal of Phycology* **19**, 446-452.

Selkirk, J.M. and Saffigna, L.J. (1999) Wind and water erosion of a peat and sand area on subantarctic Macquarie Island, *Arctic, Antarctic and Alpine Research* **31**, 412-420.

Selkirk, D.R., Selkirk, P.M., Bergstrom, D.M. and Adamson, D.A. (1988) Ridge top peats and paleolake deposits on Macquarie Island, *Papers and Proceedings of the Royal Society of Tasmania* **122**, 83-90.

Smith, M.C., Bowman, J.P., Scott, F.J. and Line, M.A. (2000) Sublithic bacteria associated with Antarctic quartz stones, *Antarctic Science* **12**, 177-184.

Speksnijder A. G. C. L., Kowalchuk G. A., De Jong S., Kline E., Stephen J. R. and Laanbroek H. J. (2001) Microvariation artifacts introduced by PCR and cloning of closely related 16SrRNA gene sequences, *Applied and Environmental Microbiology* **67**, 469-472.

Stackebrandt, E. and Goebel, B.M. (1994) Taxonomic note: a place for DNA-DNA reassociation and 16S rRNA sequence analysis in the present species definition in bacteriology, *International Journal of Systematic Bacteriology* **44**, 846-849.

Suren, A. (1990) Microfauna associated with algal mats in melt ponds of the Ross Ice Shelf, *Polar Biology* **10**, 329-335.

Swadling, K.M. and Gibson, J.A.E. (2000) Grazing rates of a calanoid copepod (*Paralabidocera antarctica*) in a continental Antarctic lake, *Polar Biology* **23**, 301-308.

Tang, E.P.Y., Tremblay, R. and Vincent, W.F. (1997) Cyanobacterial dominance of polar freshwater ecosystems: are high-latitude mat-formers adapted to low temperature? *Journal of Phycology* **33**, 171-181.

Taton, A., Grubisic, S., Brambilla, E., De Wit, R. and Wilmotte, A. (2003) Cyanobacterial diversity in natural and artificial microbial mats of Lake Fryxell (McMurdo Dry Valleys, Antarctica): a morphological and molecular approach, *Applied and Environmental Microbiology* **69**, 5157-5169.

Tindall, B.J. (2004) Prokaryotic diversity in the Antarctic: The tip of the iceberg, *Microbial Ecology* **47**, 271-283.

Van de Vijver, B. and Beyens, L. (1996) Freshwater diatom communities of the Stromness Bay area, South Georgia, *Antarctic Science* **8**, 359-368.

Van de Vijver, B. and Beyens, L. (1997) A preliminary study of freshwater diatoms of small islands in the Maritime Antarctic Region, *Antarctic Science* **9**, 418-425.

Van de Vijver, B. and Beyens, L. (1999) Biogeography and ecology of freshwater diatoms in Subantarctica: a review, *Journal of Biogeography* **26**, 993-1000.

Van de Vijver, B., Beyens, L. and Lange-Bertalot, H. (2004) *The genus* Stauroneis *in Arctic and Antarctic locations*, Bibliotheca Diatomologica 50.

Van de Vijver, B., Frenot, Y. and Beyens, L. (2002) *Freshwater diatoms from Ile de la Possession (Crozet Archipelago, Subantarctica)*, Bibliotheca Diatomologica 46.

Verkulich, S.R., Melles, M., Hubberten, H.-W. and Pushina, Z.V. (2002) Holocene environmental changes and development of Figurnoye Lake in the southern Bunger Hills, East Antarctica, *Journal of Paleolimnology* **28**, 253-267.

Vincent, W.F. (1988) *Microbial Ecosystems of Antarctica*. Cambridge University Press, Cambridge.

Vincent, W.F. (2000) Cyanobacterial dominance in the polar regions, in B.A. Whitton and M. Potts (eds.) *The ecology of cyanobacteria*, Vol. 12. Kluwer Academic Publishers, Dordrecht, pp. 321-340.

Vincent, W.F. and Howard-Williams, C. (1994) Nitrate-rich inland waters of the Ross Ice Shelf region, Antarctica, *Antarctic Science* **6**, 339-346.

Vincent, W.F. and James, M.R. (1996) Biodiversity in extreme aquatic environments: Lakes, ponds and streams of the Ross Sea Sector, Antarctica, *Biodiversity and Conservation* **5**, 1451-1471.

Vincent, W.F., Rae, R., Laurion, I., Howard-Williams, C. and Priscu, J. (1998) Transparency of Antarctic ice-covered lakes to solar UV radiation, *Limnology and Oceanography* **43**, 618-624.

Vincent, W.F., Bowman, J.P., Rankin, L.M. and McMeekin, T.A. (2000) Phylogenetic diversity of picocyanobacteria in Arctic and Antarctic ecosystems, in C. Bel, M. Brylinsky, M. and Johnson-Green (eds.) *8th International Symposium on Microbial Ecology*, Atlantic Canada Society for Microbial Ecology, Halifax, Canada, pp. 317-322.

Wand, U., Schwarz, G., Bruggemann, E. and Brauer, K. (1997) Evidence for physical and chemical stratification in Lake Untersee (central Dronning Maud Land, East Antarctica), *Antarctic Science* **9**, 43-45.

Webster, J., Hawes, I., Downes, M., Timperley, M. and Howard-Williams, C. (1996) Evidence for regional climate change in the recent evolution of a high latitude pro-glacial lake, *Antarctic Science* **8**, 49-59.

Wharton, R.A., McKay, C.P., Simmons, G.M. and Parker, B.C. (1985) Cryoconite holes on glaciers, *BioScience*, **35**, 499-503.

Wilson, C., Swadling, K. and Gibson, J. (2002) Understanding the origins of Antarctic freshwater crustaceans, *Newsletter for the Canadian Antarctic Research Network* **14**, 4-5.

Winder, M. and Schindler, D.E. (2004) Climate change uncouples trophic interactions in an aquatic ecosystem, *Ecology* **85**, 2100-2106.

Wintzingerode, F. von, Göbel, U.B. and Stackebrandt, E. (1997) Determination of microbial diversity in environmental samples: pitfalls of PCR-based rRNA analysis, *FEMS Microbiology Reviews* **21**, 213-229.

Wright, S.W. and Burton, H.R. (1981) The biology of Antarctic saline lakes, *Hydrobiologia* **82**, 319-338.

Zale, R. and Karlén, W. (1989) Lake sediment cores from the Antarctic Peninsula and surrounding islands, *Geografiska Annaler* **71A**, 211-220.

Zwartz, D., Bird, M., Stone, J. and Lambeck, K. (1998) Holocene sea-level change and ice-sheet history in the Vestfold Hills, East Antarctica, *Earth and Planetary Science Letters* **155**, 131-145.

6. LIFE HISTORY TRAITS

P. CONVEY
British Antarctic Survey, Natural Environment Research Council
High Cross, Madingley Road
Cambridge CB3 0ET, United Kingdom
p.convey@bas.ac.uk

S.L. CHOWN
Centre for Invasion Biology
Department of Botany and Zoology
Stellenbosch University
Private Bag X1
Matieland 7602
South Africa
slchown@sun.ac.za

J. WASLEY
Department of Environment and Heritage
Australian Government Antarctic Division
203 Channel Highway
Kingston, Tasmania 7050, Australia
jane.wasley@agad.gov.au

D.M. BERGSTROM
Department of Environment and Heritage
Australian Government Antarctic Division
203 Channel Highway
Kingston, Tasmania 7050, Australia
dana.bergstrom@agad.gov.au

D.M. Bergstrom et al. (eds.), Trends in Antarctic Terrestrial and Limnetic Ecosystems, 101–127.
© 2006 *Springer.*

Introduction

From our anthropocentric point of view, Antarctic terrestrial habitats are potentially stressful in many respects, facing a range of 'extreme' conditions, both in terms of chronic exposure and extreme or acute events. It would be more accurate to say that the conditions experienced in the Antarctic, with respect to many variables, lie at the extreme end of the range or continuum of conditions found worldwide (Peck et al. 2006, Chown and Convey 2006a). In an evolutionary context it can, therefore, be argued that selection imposed on organisms by the Antarctic environment may be expected to be strong, leading to more clearly identifiable consequences with fewer confounding factors than elsewhere. This chapter provides an overview of the evolutionary consequences of environmental stress as seen in the life history strategies present, and assesses the likely consequences and vulnerabilities of current trends of environmental change for biota with this suite of contemporary strategies.

Over the year, both Antarctic and Arctic terrestrial habitats experience similar extreme low temperatures during the winter months. However, the Antarctic also experiences chronically low summer temperatures, even in comparison with those of the Arctic (Convey 1996a, Danks 1999). This means that low thermal energy input is a constraint faced by most Antarctic terrestrial biota. Clearly, the temperature (and hence thermal energy) available to an organism is most accurately described by its microenvironment, rather than the longer term averages of air temperature that are the basis of these generalisations but, in the absence of many robust and extended microclimatic datasets, the latter provide at least a reasonable baseline. It is not only the long term temperature average, but also the scales and patterns of temperature variation that are of significance in terms of potential impacts on biology. Here, such variables as the upper and lower extremes experienced, diurnal and annual ranges, short-term means and rates of change, and the predictability of changes, will also be influential (for general discussion see Gaines and Denny 1993, Kingsolver and Huey 1998, Sinclair 2001, Vasseur and Yodzis 2004).

The chronically low temperatures of the Antarctic are likely to be near minimum threshold temperatures for many physiological processes, even during the short summer season of the continent or Antarctic Peninsula, and will spend long periods of the year below these thresholds. Even in the subantarctic, where the normal seasonal pattern of temperature variation tends to be broken down by the over-riding influence of thermal damping from the surrounding ocean, the resulting temperatures are chronically low (Chown and Crafford 1992, Convey 1996a, Smith 2002). This leads to a particular prediction in the context of climate change - a small temperature increment experienced by an organism in an environment chronically near to its operational threshold will have a relatively greater biological impact than the same increment experienced in a less extreme environment (Convey 2001), or one that is more variable.

Despite the more obvious factor of chronically low temperature, the significance of the lack of liquid water in many habitats and periods of the year is regarded as being at least as important in understanding the biology of Antarctic terrestrial biota (Kennedy 1993a, Sømme 1995, Block 1996). Water availability in most terrestrial habitats across the globe is governed by precipitation patterns (ie rainfall). However, other than in the subantarctic and, increasingly, in the maritime Antarctic during summer, precipitation as rain is either unusual or unknown in the Antarctic. Therefore, water availability in these terrestrial habitats is more accurately governed by patterns of thaw, both directly from thawing of snowfall, and indirectly through the melt of long term ice and glaciers. Thus it is normal for free water availability to be separated temporally from the timing of precipitation and it is often the case that it is separated spatially. Associated with desiccation stress is the impact of wind, which acts both as a stressor and a disturbance factor (Bergstrom and Selkirk 2000).

The final major environmental variable to be considered in the context of environmental selection pressures is that of solar radiation. Here, two elements may be significant in the context of climate change. First, the pattern of direct insolation experienced is affected by variables including cloud and snow cover, both of which are expected to or already seen to change as part of the overall climate change processes seen in the Antarctic (Convey 2006). Changes in irradiance received obviously have implications for autotrophic primary production, the basis of ecosystem processes. Second, much attention has been given to the potentially deleterious consequences of changes in exposure to shorter wavelength ultra-violet (UV-B) radiation that may be experienced as a consequence of the formation of the spring ozone hole over Antarctica. This is a recent anthropogenic phenomenon (Farman et al. 1985). Although it is separate from the processes that have led to climate warming and associated changes in most regions, some interactions have been suggested. For example, recent work has shown that increases in temperature and decreases in precipitation at Marion Island, and possibly at other South Indian Ocean Province islands, are related to phase changes in the semi-annual oscillation, which in turn are linked to Antarctic stratospheric ozone depletion (Rouault et al. 2005). In the context of selection on the biota, these three major environmental variables are often unlikely to act in isolation. Rather, their interactions and predictability will be important (Convey 1996b).

Key life history features of Antarctic biota

Southwood (1977) provided a key impetus in developing evolutionary models of the route by which environmental characteristics could drive the evolution of biological life history strategies, and such models have provided a considerable contribution to the general ecological literature, and as a source of continuing debate. In the context of Antarctic terrestrial biology, the life history strategies seen are generally 'adversity' or 'stress' selected (Convey 1996b) (*sensu* Southwood 1977, 1988, Greenslade 1983, Grime 1988). Features of Antarctic life history strategies documented include considerable investment and development of characters relating to stress tolerance, lack of competitive ability or investment in dispersal strategies, reduced reproductive investment and output, and extended lifespans and life cycles (Cannon and Block 1988, Block 1990, Sømme 1995, Convey 1996b, 1997, 2000). As we will note, these features have been principally identified in invertebrates (the dominant terrestrial fauna of much of the Antarctic); indeed, in comparison with the Antarctic terrestrial fauna, relatively limited attention has been paid to studies of the life history strategies adopted by Antarctic plants or the microbial flora. However the possession of considerable flexibility in many life history and physiological characteristics, rather than any specific single adaptation has been highlighted as a key element in the response of many Antarctic terrestrial biota, both animal and plant, to environmental variation (Convey 1996b). This is easy to understand in the context of the typically rapid and unpredictable variation that is often experienced at microclimate level in Antarctic terrestrial habitats. In this chapter, we will examine key life history strategies in some terrestrial biota but with the caveat that our current knowledge is skewed in the direction of macro- rather than micro-organisms and the main focus of this type of work has been with invertebrates. We will not discuss microbial groups as research questions on these life forms have generally not been focused on life history traits.

The general life history models mentioned above have an evolutionary basis. Features that have been extensively researched, such as the considerable levels of investment in ecophysiological strategies relating to cold and desiccation tolerance, indicate that the evolutionary process has further enhanced the effectiveness of capabilities that were present in the ancestors of the contemporary biota (Cannon and Block 1988, Longton 1988, Block 1990, Sømme 1995, Convey 1996b, 1997, 2000). However, in some cases it may be inappropriate to invoke an evolutionary explanation for life history features, even when they are apparently consistent with adversity or stress selection. For instance, the often–used example of life cycle extension as a response to the extreme environmental conditions of the maritime and continental Antarctic has a simpler proximate explanation in terms of straightforward thermodynamic limitations - the low levels of energy available in summer and long periods of enforced winter inactivity directly constrain the development rates that can be achieved, carrying the corollary of extended development times and the necessity of repeated overwintering before maturity can be reached. This is perhaps best seen as a selective filter (or biodiversity filter, *sensu* Bergstrom et al. 2006), in other words that only biota whose life cycles do not contain specific and limited over–wintering life stages are likely to be able to establish in the Antarctic terrestrial environment.

Growth and the life cycle

In the subantarctic, indigenous arthropods tend to have long life cycles (1-5 years), low reproductive rates, and metabolic rate–temperature (R-T) relationships with slopes that are shallower than those found for similar, north temperate species (Crafford 1984, 1990, Davies 1987, Chown and Scholtz 1989, Chown 1997, Arnold and Convey 1998, Ernsting et al. 1999, Barendse and Chown 2000, Addo-Bediako et al. 2002). Diapause is typically also not common, although the South Georgian diving beetle *Lancetes angusticollis* is an exception (Arnold and Convey 1998). Long life cycles, low reproductive rates and shallow R-T relationships are largely thought to be a consequence of low temperatures and little seasonal variability in temperature (Crafford et al. 1986, Convey 1996b). Nonetheless, seasonal variation in abundance is seen, although it may range through a wide variety of forms, from showing little variation over the annual cycle to peaking in virtually any season (Brown 1964, Watt 1970, West 1982, Bellido and Cancela da Fonseca 1988, Chown and Scholtz 1989, Barendse and Chown 2000, 2001, Nondula et al. 2004). Whether this variety is indicative of flexibility or of some form of programmed strategy is a matter of ongoing debate (Convey 1996a,b, Danks 1999, Barendse and Chown 2001).

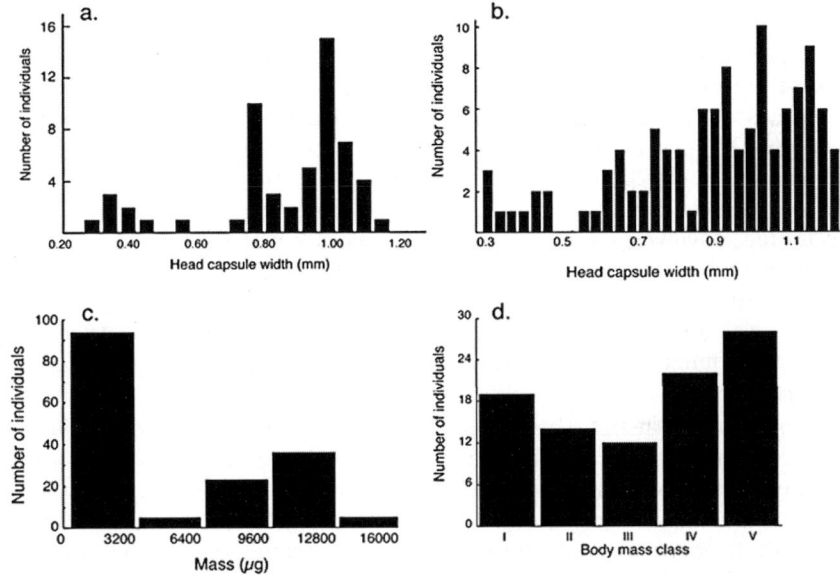

Figure 1. Head capsule size distributions for Palirhoeus eatoni *in a. January (Heard Island) and b. April (Marion Island) and body size distributions in mid summer (January and February) for* Embryonopsis halticella *on c. Heard and d. Marion Islands (from Chown et al. 2006).*

There is evidence that seasonal patterns of temperature can lead to synchronization of life cycles. The weevil *Palirhoeus eatoni* and the moth *Embryonopsis halticella* occur on both Marion and Heard Islands. The former island, which lies to the north of the Antarctic Polar Frontal Zone, experiences less climatic seasonality than the latter, which lies to the south of the frontal zone (Chown and Klok 2003). Thus, on Marion Island, overlapping generations are found in both species, whilst on Heard Island the generations are far more discrete (Fig. 1) (Chown and Klok 2003, Chown et al. 2006). In the more extreme and obviously seasonal habitats of the maritime Antarctic, where invertebrates typically have free–running life cycles (Convey 1996b) lasting multiple years, seasonality can lead to an analogous synchronisation of important life history stages or events, such as moulting (Convey 1994a).

Extreme temperature seasonality dominates the maritime and continental Antarctic environment and as a result plant communities show strongly seasonal patterns in growth and productivity (Davey and Rothery 1996, Schlensog et al. 2004), in contrast to subantarctic environments which are characterised by long growing seasons, equable annual temperature regimes and in some habitats, resulting high productivity (Jenkin and Ashton 1970, Smith and Steenkamp 1990). Bryophyte and lichen communities, which dominate the continental terrestrial vegetation, remain dormant during the sub–zero winter months (Schlensog et al. 2004). Freeze–induced dormancy is possible for these plants as they contain cryoprotective compounds, such as trehalose, which facilitate membrane function during freeze–thaw events (Roser et al. 1992, Montiel 2000). Lichens in particular are especially well adapted to cold dry conditions, as demonstrated by the ability of some species to photosynthesise under snow (Kappen 1993) and at sub-zero temperatures (Kappen et al. 1995, Schroeter and Scheidegger 1995, Kappen et al. 1996, Hennion et al. 2006). Some species also have the ability to uptake water vapour from snow (Pannewitz et al. 2003). The insulating properties of snow can slow warming, however, and keep lichens and mosses under snow inactive at sub-zero temperatures for a prolonged time. In general, lichens appear to be most active at the final disappearance of snow cover and for up to two weeks following (Pannewitz et al. 2003). After initial snow melt, lichen photosynthetic activity across the Antarctic summer is reliant entirely on precipitation for water supply (Hovenden et al. 1994).

The continental Antarctic seasonality in water availability is also inherently linked to annual temperature fluctuations. For most of the year, free water is biologically unavailable in the forms of snow and ice and only occurs during the few months (or even weeks or days, depending on location) of summer melt. Similar to the response to extreme cold, during periods of severe aridity, the continental Antarctic flora has the ability to shut down (Davey 1997, Robinson et al. 2000, Wasley et al. 2006). Metabolism is quickly established when water becomes available, with some species or groups having greater recovery potential than others (Schlensog et al. 2004, Wasley et al. 2006). Lichens are particularly well adapted to dry conditions and some species are able to utilise water vapour from snow when liquid water is not available (Kappen et al. 1995).

Perhaps as a consequence of these plants adopting an 'on and off' strategy, longevity in the continental flora is relatively high. Direct plant age estimates are difficult to obtain but data demonstrating low growth rates (eg < 0.6mm.yr^{-1} for the moss species *Grimmia antarctici*, Melick and Seppelt 1997), combined with relatively large plant size (eg gametophyte length >4cm, Wasley et al. 2006), suggests individuals may exceed 60 years of age. Estimates for continental Antarctic lichen longevity are even greater, their exceedingly slow growth rates of perhaps 0.01mm.yr^{-1} suggests continental Antarctic lichen communities may represent centuries of growth (Green 1985). Further contributing to this longevity is the lack of major attack from pathogens and herbivores.

In the subantarctic, the biodiversity filters of long distance dispersal and persistent low temperatures have, in general, selected against woody growth forms and selected for herbs, graminoids and cushions, sometimes, with renascent or semi–deciduous behaviour (Smith 1984, Selkirk et al. 1990). In the context of key life history strategies megaherbs are deserving of special note. These are an important element, in terms of biomass and habitat structure, of most subantarctic island floras and of the biogeographically related cool temperate islands south of New Zealand (Smith 1984, Muerk et al. 1994, Mitchell et al. 1999, Shaw 2005). The megaherb growth form is also well represented in some other extreme habitats worldwide, with those of tropical high altitude mountains perhaps being best known. A number of unrelated families have evolved regional, endemic megaherb genera (eg Apiaceae - *Stilbocarpa*, Brassicaceae - *Pringlea*, Asteraceae – *Pleurophyllum*). In addition, three tall tussock grass species, *Poa foliosa* and *P. littorosa* on Macquarie Island and *Parodiochloa flabelatta* on South Georgia also present large structural elements within subantarctic ecosystems with leaf lengths in excess of 1 m and may develop tussocks 1.5-2.0 m tall. Meurk et al. (1984a) and Mitchell et al. (1999) proposed that megaherb development reflects selection under a regime devoid of natural herbivores. The megaherb habit has been identified as having potentially several adaptive benefits for survival in subantarctic environments, with large leaves successful in harvesting low light while focusing radiation towards apical reproductive structures and buds (Wardle 1991). Furthermore, large leaves might intercept marine derived aerosols (Meurk et al. 1994b). Erskine et al. (1988) recorded volatised nitrogen above penguin colonies and traced this nitrogen into plant communities.

Megaherbs present a mixture of morphological traits that are generally described elsewhere as being mutually exclusive because of resource trade-offs, including substantial allocation to leafy biomass and underground storage tissue, high seed mass and abundant seed output and rain (eg seed rain *Stilbocarpa polaris* >10 000 seeds.m^{-2} over 35 days, *Pleurophyllum hookeri* >13 300.m^{-2} over 42 days: Shaw (2005), single season's seed output 150 000 *Pringlea antiscorbutica* seeds.m^{-2}, D. Bergstrom and K. Kiefer unpubl. data). The first two reflect responses to long growing seasons and high nutrient input and the last combination strongly supports the proposition of selection devoid of herbivores. With the introduction into the subantarctic of alien vertebrate herbivores over the last 200 years (Frenot et al. 2005, Convey et al. 2006), megaherbs and tall tussock grasses have all been subjected to grazing, in some cases leading to significant reductions or changes in local distributions (Leader-Williams et al. 1987, Leader-Williams 1988, Pye 1993, Copson and Whinam 1998, Frenot et al. 2005). Shaw et al. (2005) also reported rats destroying up to 80% of racemes of *Pleurophyllum hookeri* on Macquarie Island and impacting upon initial recruitment and seedling survival.

Although solar radiation, water availability and temperature are generally described as the main environmental drivers in the Antarctic region, the importance of exposure to wind must also be recognised, both as a component of water stress and also through disturbance and abrasion (Berjak 1979). The subantarctic islands lie in the realms of the 'roaring forties' and 'furious fifties' and Briggs et al. (in press) has reported the first evidence of contractile stems in angiosperms in *P. hookeri*, arguing that this is an adaptation to wind stress, with the mechanism enhancing anchorage of the plant by its rhizome. Similarly, contractile roots have been reported in the subantarctic endemic fellfield cushion plant *Azorella maquariensis* (Jenkin and Ashton 1979). Avoidance of wind (in addition to desiccation) as a life trait is found at its extreme expression in endolithic microbial communities in continental and maritime Antarctica (Friedmann 1982, Green 1985, Hughes and Lawley 2003). Exposure to wind, and hence the risk of dislocation from suitable habitat, is also as a factor proposed to contribute to the preponderance of more cryptic invertebrates (with behavioural strategies reducing the likelihood of exposure), and in the reduction of the use of flight in insects, amongst the faunas typical of Antarctic terrestrial habitats (Convey 1996b but see also Crafford et al. 1986).

Reproduction

Predictably, subantarctic plants are more cued into the highly seasonal light regime. Flowering onset in most species takes place in spring, with reports mainly of variation of a perennial life pattern, including species with very high and/or highly germinable seed production, and long-lived perennial stayers which often do not complete a sexual reproductive cycle in one year, with both preformation of flower buds the season previous and post winter ripening (Dorne 1977, Walton 1982, Bergstrom et al 1987, Hennion and Walton 1997, Shaw 2005). Late summer floral initiation also takes place in the two vascular plants that occur in the maritime Antarctic (Edwards 1974). Only a few, small, species exhibit an annual life history pattern, with this pattern more common in alien species (Frenot et al. 2005), but even here, the most common alien vascular species, *Poa annua* has also been observed to exhibit perennial behaviour (Taylor 1955).

The feature that links the 55 native angiosperms (from 14 families) found in the subantarctic is that sexual reproduction, as a strategy, appears to be important (Tallowin and Smith 1977, Jenkin and Ashton 1979, Lawrence and McClintock 1989, Bergstrom et al. 1997, Shaw 2005, D. Bergstrom and K. Kiefer unpubl. data). In a comparative study of reproductive output and investment in temperate grasses, Wilson and Thompson (1989) noted in general that perennials showed much lower levels of investment of resources than annuals. Estimates of investment in the subantarctic (South Georgia) and maritime Antarctic perennial vascular plants *Deschampsia antarctica* and *Colobanthus quitensis* (Convey 1996c), found high levels comparable to temperate annuals rather than perennials, a feature suggested to be more consistent with an 'r-selected' rather than a stress-selected life history. However, data of a comparable form are not available for most other subantarctic plant species and this area is worthy of future research. Sexual reproduction is also a widespread feature in lichens of the Antarctic and other extreme environments (Seymour et al. 2005), with the additional feature that the fruiting bodies involved may themselves be perennial structures of the organism. Again, estimates of the level of investment in sexual or asexual reproduction in this group are not available.

Species do not live in isolation but interact. One of the life history features found in insects on subantarctic islands, reduction in the use of flight (reflecting an adaptation to low energy and extremely windy environments) may have had a major flow–on effect in selection for pollination strategies in subantarctic flowering plants. In an area still needing much more attention, a number of modes of pollination have been observed. Many species, particularly the grasses and sedges have typically wind-pollinated flowers, ideal for encouraging potential cross-fertilization and thus it is assumed that wind pollination occurs, and this must be the case in the dioecious grass species, *P. foliosa* (Du Puy et al. 1993). For other flowers, there is limited size or colour variation, with the majority being small, white or green (Walton 1982), which is standard for attracting generalist pollinators such as flies. However, amongst the indigenous fauna and flora of most, if not all, subantarctic islands it is thought that there are no obligate invertebrate pollinators or plants requiring pollination (except for the two orchid species on Macquarie Island whose pollinator is suspected to be the island's native fungus gnat, *Bradysia watsonii*). This is one area in which the introduction of non-indigenous flowering plants and/or invertebrates has the potential to have a major impact on subantarctic biology.

If the generalisation of reduced flight in pollinators is a real feature, the potential for cross-pollination is reduced, and hence a greater tendency toward geitonogamy (pollination by different individuals on the same flower) is expected. Furthermore, the observations of cleistogamy (Werth 1911, Walton 1982, Bergstrom et al. 1997), and the numerically high seed output of many species indicates forms of selfing and self-compatibly, which is a common feature of island ecosystems (Barrett 1996) and cold climate ecosystems (Billings 1974) and conforms to Baker's Rule (Baker 1955) that self-compatible rather than self-incompatible plants will be favoured in establishment following long-distance dispersal. Barrett (1996) highlights that, although self-fertilization is obviously an advantageous life strategy in island ecosystems, 'escape from homozygosity' is essential for subsequent radiation and diversification in island groups

In experimental and observational studies of reproduction of mosses on subantarctic Macquarie Island, Bergstrom and Selkirk (1987) noted that of 43% (36 species) of the flora documented at the time had been observed with sporophytes, with 11 of these species being dioecious. Of the remaining species, 57% (47 species, 19 dioecious) had never been observed to fruit. Three reproductive strategies were identified: species with sexual reproduction but with low spore germination rates (although note that very few specific studies have attempted to identify optimum germination conditions for any Antarctic bryophytes), which included mainly species from sparse, harsh fellfield environments with low competitive pressure; species that relied on asexual reproduction only, particularly the copious production of specialised asexual structures (gemmae); and species which displayed both sexual and asexual modes of reproduction. This included the dioecious species *Dicranella cardotii* that displayed five different types of asexual reproduction, including specialized structures and regeneration from deciduous leaf tips and leaf fragments. Ochyra (1998) reported a similar ratio of sexual to asexual species (29:32 species) on maritime King George Island, although only 18 species had been observed to have produced mature capsules, with these having a bias to monoecious species where both gametes are formed by the same plant (see also discussion of bryophyte reproductive strategies in Longton 1988). In the maritime Antarctic, sporophyte production can be relatively high in micro-oases and is reported to be higher (in terms of the total number of species recorded with sporophytes rather than at the individual cushion or shoot level) at high latitudes than previously expected (Smith and Convey 2002). Sexual reproductive effort in maritime Antarctic bryophytes is greater in annual short-lived species than most perennial species (Convey and Smith 1993), although the sexual behaviour of some longer–lived species examined by Convey and Smith (1993) shows some surprisingly large levels of investment in sexual reproduction. In continental Antarctica, sexual reproduction in mosses is extremely rare (Filson and Willis 1975, Seppelt et al. 1992, Skotnicki and Selkirk 2006).

Finally, the investment of energy into reproductive reservoirs such as seeds, spores or asexual propagule banks allow for survival across environmental variability in addition to environmental change. Calculations of seed bank sizes on Macquarie Island across a range of vegetation types ranged from 12 650 to 96 333 seeds.m^{-2} (W. Misak and D. Bergstrom unpubl. data). These estimates contrasted sharply with the range at several Arctic sites (0 to 3367 seeds.m^{-2}, reviewed by McGraw and Vavrek 1989). Furthermore, Bergstrom and Selkirk (1999) found that the viable moss spore bank at relative high altitude on Macquarie Island included species found at lower altitude, thus there existed potential for establishment of new species at the sites under changed or warmer conditions (cf also Smith 1990, 2001, Convey 2006, Hughes et al. 2006).

Physiological responses to low temperature

One of the most widely explored suites of traits in sub- and maritime Antarctic species is cold hardiness, or the ways in which these species respond to decreases in temperature below the freezing point of water (Hennion et al. 2006). Many species of 'higher' insect, such as most weevils, and the caterpillars of the tineid moth, *Pringleophaga marioni*, are moderately freeze-tolerant (Klok and Chown 1997, van der Merwe et al. 1997, Sinclair and Chown 2005a). Thus, they survive freezing, but only to relatively high sub-zero temperatures. At least some species in this group also appear to be incapable of rapid responses to low temperatures of the kind typically associated with rapid cold hardening (Lee et al. 1987). This is believed to be a consequence of moderate freeze tolerance already conferring a strategy that allows rapid responses to changing low temperatures (Sinclair and Chown 2003). However one species, the maritime Antarctic midge, *Belgica antarctica*, provides an exception, as rapid cold hardening improves freeze tolerance in larvae, but not adults, of this species (Lee et al. 2006).

A second group of subantarctic species is freeze-intolerant. These species lower their body freezing temperatures (by, *inter alia*, biochemical changes and cessation of feeding) to well below the freezing point of water (Klok and Chown 1998, Worland 2005, J. Deere et al. unpubl. data). Amongst these species there is also little indication of a rapid cold hardening response (Slabber 2005, E. Marais et al. unpubl. data), which is surprising given the preponderance of rapid cold hardening in other species both on the continent (Worland and Convey 2001, Sinclair et al. 2003) and elsewhere (Chown and Nicolson 2004). At present, no explanation has been proffered for the absence of rapid cold hardening in these subantarctic species, but it should be noted that the cue(s) for changes in supercooling points may not be solely temperature related.

In the last group of species, individuals can vary between freeze-tolerant and freeze-intolerant. In the rove beetle, *Halmaeusa atriceps*, adults and larvae both appear to be moderately freeze-tolerant in winter and chill susceptible in summer (Slabber and Chown 2005). Such a seasonal change in response is more in line with 'typical' patterns in cold hardiness found amongst many other more temperate insects (Worland and Convey 2001, Chown and Nicolson 2004). In larvae of the subantarctic perimylopid beetle, *Hydromedion sparsutum*, however, the situation is quite different. Following repeated exposure to low temperatures in early winter, the population differentiates into individuals that retain a relatively high freezing point, are moderately freeze-tolerant, and capable of surviving repeated freezing events, and those that substantially depress their freezing point for prolonged periods, but are unable to survive freezing should it take place (Bale et al. 2001). This seems to be a 'bet-hedging' strategy in this species. Ontogenetic differences in cold hardiness strategy have also been recorded in two subantarctic fly species (Vernon and Vannier 1996, Klok and Chown 2001). The degree of 'bet-hedging' or variation displayed in life history strategies is a topic for much needed research for all biota, as it is the flexibility that is the key survival of environmental change.

Although lethal temperatures and insect responses to them have formed much of the focus for insect environmental physiology, sub–lethal effects, such as chilling injury, are also of considerable significance (Chen et al. 1987, McDonald et al. 1997, 2000). What the effects of repeated sub-lethal stresses are for insects has not been well investigated generally (but see Brown et al. 2004). However, this has been done for caterpillars of *Pringleophaga marioni* on Marion Island. Repeated exposure to freezing events (once a day for five days) had little effect on caterpillars and did not alter survival. However, repeated exposure to sub–lethal cold resulted in a decrease in mass, gut mass and feeding activity (Sinclair and Chown 2005a). The consequences of the decrease in feeding activity lasted well beyond the week–long period of repeated cold exposure. Treated caterpillars took approximately one month to regain mass to that of pre-treatment controls, over which period control caterpillars had grown substantially. Further investigations of caterpillar distributions in the field have indicated the significance of these findings. It has long been known that these caterpillars are most readily found and are abundant in recently abandoned Wandering Albatross *Diomedea exulans* nests, and this has been assumed to be the result of elevated nutrients in and around the nests (Joly et al. 1987). However, on Marion Island, nests either do not differ from or have lower nutrient contents than the surrounding environment, despite higher biomass and less variability of caterpillar biomass in them. By contrast, adult birds and chicks maintain nest temperatures approximately 5°C higher than the surrounding environment. Higher temperatures reduce the risk of repeated cold exposure and maintain the environment close to the optimum temperature for feeding in the caterpillars (Sinclair and Chown 2005b), thus providing a proximate reason for higher biomass in the nests. Nonetheless, just how the higher biomass arises (preferential oviposition by females, higher growth rates, caterpillar dispersal to nests) remains to be identified.

Cold tolerance in plants has equally received significant attention and physiological aspects of survival are discussed in detail by Hennion et al. (2006). Patterns in cold tolerance include ice-nucleation outside of cells (Schroeter and Scheidegger 1995) and dehydration to a level of effective suspension of most metabolic activity, followed by rapid repair and resumption of metabolic activity. This strategy is common in mosses and lichens. Many species have high cell lipid content and concentrations of other 'anti–freeze compounds' such as trehalose and proline (Jackson and Seppelt 1997, Montiel 2000). Antarctic, and some subantarctic, flowering plants have been observed to contain a range of secondary metabolites, such as proline and various polyamines, known to confer cold hardiness (Hennion et al. 2006).

Recovery from freezing appears to depend on the absolute depth of low temperature experienced (Kennedy 1993b). More disruptive to metabolic processes than periods of constant freezing are repeated freeze–thaw cycles (Kennedy 1993b). The maritime and continental Antarctic floras have the ability to withstand freeze–thaw cycles, but suffer photoinhibition (Lovelock et al. 1995a,b) and loss of soluble carbohydrates (Tearle 1987, Melick and Seppelt 1992) during such events.

Competition and predation

Life history responses to predation, parasitism and competition have enjoyed less attention, largely because these interactions are relatively rare, or perceived to be rare, in the subantarctic and Antarctic. However, they are not absent, and the lack of attention paid to them is unfortunate, especially given that many alien species are predators or competitors with substantial impacts on indigenous species (Frenot et al. 2005, Convey et al. 2006). Among the indigenous species, Davies (1987) suggested that interspecific competition between the subantarctic carabids *Amblystogenium pacificum* and *A. minimum* has resulted in non–overlapping adult size distributions where they occur in sympatry in fellfield habitats on Ile de la Possession, but that in mire habitats where *A. pacificum* also occurs separately, adults are smaller and overlap with the size distribution of fellfield *A. minimum*. Analyses of body size ratios of weevils on the Prince Edward Islands also seem to provide evidence of interspecific competition in the older epilithic biotope because resident species have size ratios larger than those expected from random (Chown 1992). However, this was not the case in the younger vegetated biotope and on other South Indian Ocean Province Islands, suggesting that local and regional processes interact to produce different outcomes on different islands. Altitudinal variation in size (Chown and Klok 2003) does not seem to be a cause of the patterns found on the Prince Edward Islands (Chown 1992), but one of the proposed mechanisms - competition for refuge from dry conditions - was also not supported (Chown 1993). Although experimental evidence for interspecific competition among epilithic weevils on the Prince Edward Islands has yet to be provided, theoretical analyses suggest that such a process is not unexpected in moderately harsh environments (Menge and Sutherland 1987). Thus, competition cannot yet be discounted as an unimportant process in all Antarctic habitats.

A number of studies on plants has indicated wide ecological amplitude and/or a lack of competitive ability in indigenous subantarctic or Antarctic species (Bergstrom and Selkirk 2000, Smith et al. 2002). A pertinent example is the cushion plant genus *Azorella* on subantarctic islands. On Heard Island, *A. selago* is the dominant and most widespread species (Bergstrom and Selkirk 2000) occurring in most vegetation types and from the coast up to the upper altitudinal limits of vegetation. Although not dominant except in fellfield environments, *A. macquariensis* on Macquarie Island also occurs from the coast to the upper limit of vegetation. Both species are slow growing, show slow sexual reproductive development and are out-competed by taller, faster growing or more aggressive species (Bergstrom et al. 1997, Bergstrom and Selkirk 2000, le Roux and McGeogh 2004, le Roux et al. 2005). In a short-term simulation experiment, le Roux et al. (2005) demonstrated that *A. selago* lost cushion integrity in response to shading. In another example, Pammenter et al. (1986) suggested that reduced competitive ability of a native subantarctic species, *Agrostis magellanica* compared with the alien, congeneric *A. stolonifera* may be due to a greater resource allocation into structural material rather than total leaf area (therefore photosynthetic area). This structural material allows the species to occupy a range of microhabitats including more wind–exposed sites, but make it less able to compete in sheltered areas. Since its arrival in the region in the 1950s, *A. stolonifera* has invaded numerous habitats and is most successful in protected wet slopes and riverbanks, where it has reduced native species richness by up to 50% (Gremmen et al. 1998). This example identifies a pertinent point: although there are limited studies identifying the lack of competitive ability of native Antarctic organisms, the fact that alien taxa are out–competing them (Frenot et al. 2005, Convey et al. 2006) demonstrates this feature.

Likely responses to climate change

Unlike most other areas of the planet, the scale of geographical isolation of the Antarctic and its terrestrial habitats allows the separate consideration of two fundamental elements of biotic responses to climate change (Convey 2001). This isolation means that there is currently little or no influence on the biology of indigenous species of the continental and maritime Antarctic from colonising alien competitors and, thus the response of this biota directly to climate change parameters can be studied in the absence of the confounding variable of competition from alien invasives. However, it is also widely accepted that colonisation by alien species, and consequential changes in patterns and importance of competition (including predation), will itself be one of the major outcomes of climate change for Antarctic terrestrial ecosystems (Kennedy 1995c, Convey 2003a, Frenot et al. 2005, Convey et al. 2006, Hull and Bergstrom 2006).

The subantarctic islands already provide an exceptional opportunity for the study of this latter process, combining the advantage of relatively simple indigenous ecosystems with documentary evidence of colonisation and establishment by a wide range of taxa (Frenot et al. 2005, Chown and Convey 2006b, Convey et al. 2006). One particularly apposite example is given by the impacts seen over recent decades after the introduction of species of predatory carabid beetle, probably from the Falkland Islands, to two subantarctic islands. In both cases there are no comparable indigenous invertebrate predators in the native terrestrial ecosystems. *Oopterus soledadinus* was probably introduced to Iles Kerguelen with agricultural supplies in the early 20th Century (Chevrier et al. 1997), while both *O. soledadinus* and its closely-similar relative *Trechisibus antarcticus* appear to have been introduced to South Georgia much more recently (early 1980s), probably in association with tourist or military shipping activity (Ernsting 1993). Both have had substantial impacts on population sizes of their indigenous prey, leading to considerable local reduction and even local extinction (Ernsting et al. 1995, 1999, Chevrier et al. 1997, Lebouvier and Frenot 2005). On South Georgia, one prey species has been proposed to have responded to this new and intense selective pressure by accelerating early development such that juveniles spend a shorter period at smaller and more vulnerable sizes, while it has also been impacted (reduced body size) by an enforced change of diet to a less nutritious grass species as a separate consequence of grazing pressure from alien reindeer (Ernsting et al. 1995, 1999, Chown and Block 1997).

Life history flexibility

Within the indigenous biota, life history flexibility is likely to play an integral role in any response to climate change. One response in life cycles that are determined simply by energy availability, with few or no fixed elements such as obligate diapause, would be to allow increases in development rate which would result in shortening of the life cycle. Such a response has been proposed for the predatory diving beetle, *Langustis angusticollis*, in lakes on South Georgia (Arnold and Convey 1998). While this species does possess an environmentally (thermally) cued diapause (Nicolai and Droste 1984), this operates only in the final larval instar. At present, this beetle has a two–year life cycle in its field habitats on South Georgia, as the limited thermal energy available does not allow larvae to develop past the final instar before late summer conditions attain the point of cueing the diapause state. However, calculations indicate that an increase of only 1°C in lake temperatures would be sufficient to permit development past the stage at which diapause is cued, allowing an annual life cycle to be completed (Arnold and Convey 1998). This order of temperature increase appears very realistic, as it has already been seen in records of air temperatures, while separate studies of lakes on maritime Antarctic Signy Island have reported that these systems can actually magnify the response seen in air temperatures (Quayle et al. 2002, 2003). Should such a change in life cycle duration occur, it is likely to generate major though unknown consequences within the trophic web, as *L. angusticollis* is the top predator in these lake ecosystems.

The most widely quoted examples of a rapid response to recent climate warming in part of the Antarctic relates to the recent increases in populations of the two native Antarctic higher or flowering plants (*Deschampsia antarctica* and *Colobanthus quitensis*) (Fowbert and Smith 1994, Smith 1994, Grobe et al. 1997, Convey 2006). These include increases amounting to one to two orders of magnitude, and to the colonisation, at least on a local scale, of new areas of ice-free ground (although there is no overall increase in the species' ranges). These population increases are most likely underlain by a change in an important element of the life history strategy - the successful production of mature seeds, which can now occur much more frequently than previously (Convey 1996c cf Edwards 1974) and, possibly, the reactivation of seeds that have remained dormant in soil propagule banks (McGraw and Day 1997). In species such as *D. antarctica* and *C. quitensis* that can survive the normal stresses of the Antarctic terrestrial environment, such population increases may not be surprising.

Many other groups have biological attributes that would be expected to allow similar responses to any relaxation of environmental stresses, although comparable data do not exist. For instance, sexual reproduction in many mosses is determined by temperature (Longton 1988). Changes in the successful utilisation of this mode of reproduction would thus be expected and could provide a sensitive barometer of change (Frahm and Klaus 2001). Sexual reproduction in mosses also leads to the production of highly dispersible spores, possibly providing opportunities not available to the range of asexual propagules also produced, which will alter the pattern of long and short distance colonization achieved. Although data on spore production in Antarctic mosses are available (Convey and Smith 1993, Convey 1994b, Smith and Convey 2002), specific comparative data are not available over time, while comparisons within species across latitude are currently equivocal. Another important group of the Antarctic flora, the lichens, show a similar range of asexual and sexual propagules, with the added complication of requiring both elements of the symbiotic relationship to be present in order to permit successful establishment (Hughes et al. 2006). Again, changes in the balance of utilisation of the different strategies with climate change would be expected, but no data are currently available against which to test this expectation.

In the context of studies of the impact of an entirely different stress, short wavelength ultraviolet (UV-B) radiation, the importance of response flexibility in resource allocation to protective pigments has recently also received attention, with patterns of variation observed in some Antarctic bryophytes being best correlated with changes in UV-B exposure caused through variations in the depth of the ozone hole over a specific location (Newsham et al. 2002, 2005, Newsham 2003). These findings concur with more general overviews of the biochemical impact of exposure to UV-B (Rozema 1999, Paul 2001) including changes in the activity of reaction pathways involved in specific pigment production. Although the magnitude of investment of resources or energy in these responses is currently unknown, there may very clearly be important ecological implications in any response that requires significant alterations in resource allocation strategies within the organism. Wasley et al. (2006) identified morphological and physiological traits of the endemic moss

Grimmia antarctici that, collectively, may reduce this species' competitive ability compared with more widely distributed and physiologically flexible co-occurring moss species. If this is the case, it provides a salient example of how past selection pressures might reduce the capacity within a species to cope with current environmental changes.

Interactions among environmental variables

A further caveat is the likelihood that the consequences even of direct environmental changes might not always be straightforward to ascertain. For example, many subantarctic islands are showing increases in mean annual temperature (Bergstrom and Chown 1999). To date, there has been no suggestion that, even at the microclimate level, the increases are likely to exceed the upper lethal limits of most arthropods. However, in some areas, such as Marion Island, it is not only mean temperature that is predicted to change in line with current trends. Rather, the frequency of freeze–thaw events and occurrence of minimum temperatures are also predicted to increase because of a greater frequency of cloud-free skies and a lower frequency of snow (which is a thermal insulator) (Smith and Steenkamp 1990, Smith 2002). An increase in the frequency and intensity of freeze–thaw events could very readily exceed the tolerance limits of many arthropods, as recent work both on Marion Island and other south temperate locations has shown (Sinclair 2001, Sinclair and Chown 2005b, Slabber 2005). Other biota, such as continental bryophytes and lichens, may also be pushed beyond their tolerance limits if freeze– thaw frequency increases, especially given the physiological effects of this stress such as soluble carbohydrate loss (Tearle 1987, Melick and Seppelt 1992). Thus, one of the major consequences of climate change might paradoxically not be an increase in upper lethal stress, but rather an increase in stress at the other end of the temperature spectrum. How organisms are likely to respond to this kind of challenge has not been well investigated, though it is clear that lower lethal temperatures show substantial capacity for both phenotypic plasticity and evolutionary change (Chown 2001).

In ice–dominated continental Antarctica, changes to temperature are intimately linked to fluctuations in water availability. Changes to this latter variable will arguably have a greater effect on vegetation and faunal dynamics than that of temperature alone (Convey 2006). Future regional patterns of water availability are unclear, but increasing aridity is likely on the continent in the long-term (Robinson et al. 2003). Plant species which show high tolerance of desiccation, such as *Ceratodon purpureus,* or species such as *Bryum pseudotriquetrum,* which have a high degree of physiological flexibility with respect to tolerance of desiccation, are more likely to persist under increased aridity than the relatively desiccation-sensitive and physiologically inflexible *Grimmia antarctici* (Wasley et al. 2006). Changes to water availability that cause an increased frequency of desiccation events are likely to negatively impact hydric maritime bryophytes more than mesic or xeric species (Davey 1997).

For species occurring naturally across a range of latitudes, the environmental gradient thus presented can be used as a proxy or model for at least some of the projections of climate change models. This is based on a hypothesis that systematic patterns of climate variation across environmental gradients will have resulted in consistent pattens of variation in investment of resources within an organism. This model was the basis of the RiSCC International program. A correlational study of reproductive investment in an Antarctic oribatid mite (*Alaskozetes antarcticus*) has been based on this hypothesis, using material of the species collected between subantarctic South Georgia and Marguerite Bay in the southern maritime Antarctic (a latitudinal range of 53 – 69°S) (Convey 1998). This study gave some support to there being a variable resource allocation strategy within the species, with significantly greater investment in eggs in samples obtained from the milder subantarctic. As great flexibility in ecophysiological and life history strategies is a feature of Antarctic invertebrates (Convey 1996b), such a response might be expected to apply widely amongst this group.

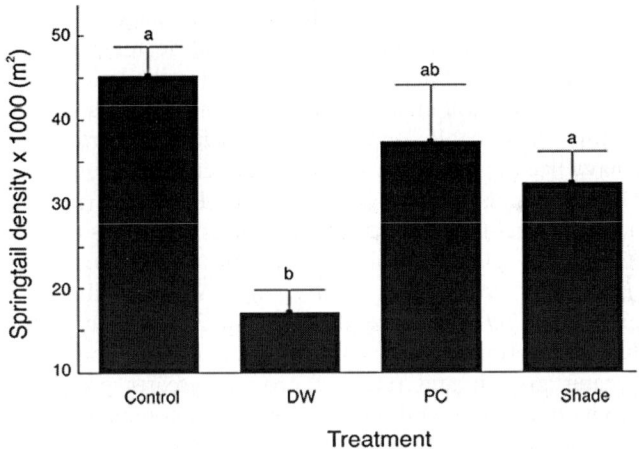

Figure 2. Springtail density following a year-long manipulation of climate of Azorella selago *cushions on Marion Island using rain-out shelters to mimic the ongoing trend of warming and drying. Controls were not manipulated, dry-warm treatments (DW) were covered by polycarbonate rain-out shelters, procedural controls (PC) were covered by perforated polycarbonate, and shade treatments were covered by shade cloth which reduced light transmission by 80%. Different letters indicate significant differences in springtail density based on generalized linear models (McGeoch* et al. *2006).*

It is also the case that many of the pigments involved in UV-B stress protection belong to classes (eg polyphenolics) known to have an influence on digestibility or palatability to herbivores, while they are also often refractory and may further affect the accessibility of dead material to detritivores (Newsham et al. 1999, Paul et al. 1999, Rozema 1999, Rozema et al. 1999). At first sight, such responses may appear to be very subtle, when seen in the context of the suite of biochemical pathways taking place within an organism. However, it is important to realise that individually insignificant changes may accumulate to have greater consequences, especially when integrated through several steps in a food web (Day 2001, Johnson et al. 2002, Convey 2003a). In the Antarctic and, undeniably, worldwide, few studies of climate change impacts have been attempted across trophic levels, either through field observation or manipulation. The only such study completed at an Antarctic location - near Anvers Island off the west coast of the Antarctic Peninsula - involved a multivariate manipulation of vegetation and the component arthropod communities using screens over a period of several years, and postulated that the negative impacts found of UV–B radiation on arthropod numbers could be related to altered food quality (Convey et al. 2002).

Studies of biological responses to climate change are necessarily but unfortunately limited by the constraints of logistics and funding processes. While, in a global context, rates of warming in parts of the Antarctic are very rapid, scales of interannual variability remain large and it remains optimistic to be able to expect to identify let alone understand responses seen within a single or a few field seasons. Thus, biologists have relied heavily on the use of a range of field manipulation techniques in order to try and identify the general features of responses to be expected under natural conditions. While such manipulations themselves can introduce a range of methodological artefacts and uncertainties (Kennedy 1995a,b), they are the only practicable approach to such long-term studies. Various manipulations have been applied in a range of studies at sites on subantarctic South Georgia and Marion Island, and along the Antarctic Peninsula (maritime Antarctic), leading to rapid and sometimes spectacular responses in studies of microbes (Wynn-Williams 1993, 1996), plants (Smith 1990, 2001, le Roux et al. 2005) and invertebrates (Kennedy 1994, 1996, Convey and Wynn-Williams 2002, Convey 2003b, McGeoch et al. 2006, but see also Sinclair 2002). The responses observed generally include activation of dormant propagule banks, considerably increased populations, greater growth and rates of growth, greater ground coverage, and changes in reproductive output and mode. However, a combination of warming and drying may have the opposite effect in the subantarctic and the maritime Antarctic (Convey et al. 2002), although many of the responses are species-specific and difficult to predict (le Roux et al. 2004, McGeoch et al. 2006). Nonetheless, a few studies have sampled sequentially over time or, as noted above, attempted to integrate responses across trophic levels. Thus, as found in a manipulation study of soils at Mars Oasis (southern Alexander Island, 72°S) at the southern limit of the maritime Antarctic, parts of the soil faunal community (microbivorous nematodes of the genus *Plectus*) may show extremely rapid population increases of several orders of magnitude within a few years, only for other omnivorous elements of the nematode community to lag behind initially, with a more 'natural' community structure then recovering towards the original balance of relative abundance over subsequent years (Convey and Wynn–Williams 2002, Convey 2003b). In all such studies, the fundamental caveat remains that there has been virtually no effort devoted to making detailed autecological studies of the biology of most Antarctic terrestrial biota in any taxonomic or functional group.

Conclusions

The strongest evidence available to date on the potential life history responses of Antarctic biota to climate change processes has been generated through the use of field manipulation methodologies, with more limited evidence, correlational in nature, obtained from direct field observations. From these studies and observations, it is clear that various consequences of life history flexibility generate potential for massive and rapid species and community responses. Possession of long–surviving propagules can also be an important life history feature, as existing soil propagule banks can be activated and exert an important influence on developing community structure.

The influence of patterns of variability in key environmental variables (ie temperature, water, solar radiation and wind) on life history traits provides some insight into the likely responses of these traits to climate change. Across the Antarctic biome, from the relatively mild, aseasonal subantarctic to the extremely cold, dry and seasonally distinct regions of continental Antarctica patterns of variability in life history traits are evident. The growth and life cycle patterns of many invertebrates and plants are fundamentally dependent on regional temperature regimes and their linkage with patterns of water availability. Distinct patterns in sexual reproduction are evident across the Antarctic flora and are most likely a function of temperature variation. In addition, phenology of flowering plants is cued to seasonality in the light regime.

In regions supporting angiosperms, wind is assumed to play a major role in pollination ecology of grasses and sedges resulting in cross-pollination. The lack of specialist pollinators in the native fauna, combined with high reproductive outputs in non-wind pollinated species implies a high reliance on selfing.

The Antarctic biota shows high development of ecophysiological adaptations relating to cold and desiccation tolerance, and displays an array of traits to facilitate survival of these conditions. While patterns in absolute low temperatures are clearly influential in determining survival, perhaps more influential is the pattern of the freeze-thaw regime, with repeated freeze-thaw events being more damaging than a sustained freeze event. How these patterns change in the future will be an area of major importance.

The final suite of life history traits discussed in the present chapter are those relating to competition and predation. The lack of attention to these traits to date is unfortunate, particularly with respect to the understanding of alien species' impacts.

It is already well known that Antarctic terrestrial biota possess very effective stress tolerance strategies, in addition to considerable response flexibility. The exceptionally wide degree of environmental variability experienced in many Antarctic terrestrial habitats, on a range of timescales between hours and years, means that predicted levels of change in environmental variables (particularly temperature and water availability) are often small relative to the range already experienced. Given the absence of colonisation by more effective competitors, predicted and observed levels of climate change may be expected to generate positive responses from resident biota of the maritime and continental Antarctic. The picture is likely to be far more complex on the different subantarctic islands, and many already host (different) alien invasive taxa, some of which already have considerable impacts on native biota.

Many responses are likely to be subtle and multifactorial, and will be underlain by changes in resource allocation strategies. However, it is fundamental to recognise that integrating these subtle responses can result in considerable and unexpected consequential impacts for communities and ecosystems. Given the importance of understanding biological responses to climate change and that, fundamentally, all such responses lie within the definition of life history strategy, it is clear that an understanding of the entire life history is a prerequisite not only in an Antarctic but also in a global context. Worryingly, research into this field is currently receives a low profile across the Antarctic scientific community.

References

Addo-Bediako, A., Chown, S.L. and Gaston, K.J. (2002) Metabolic cold adaptation in insects: a large-scale perspective, *Functional Ecology* **16**, 332-338.

Arnold, R.J. and Convey, P. (1998) The life history of the diving beetle, *Lancetes angusticollis* (Curtis) (Coleoptera: Dytiscidae), on sub-Antarctic South Georgia, *Polar Biology* **20**, 153-160.

Baker, H.G. (1955) Self –compatibility and establishment after "long–distance" dispersal, *Evolution* **9**, 347-349.

Bale, J.S., Worland, M. R. Block, W. (2001) Effects of summer frost exposures on the cold tolerance strategy of a sub-Antarctic beetle, *Journal of Insect Physiology* **47**, 1161-1167.

Barendse, J. and Chown, S.L. (2000) The biology of *Bothrometopus elongatus* (Coleoptera, Curculionidae) in a mid-altitude fellfield on sub-Antarctic Marion Island, *Polar Biology* **23**, 346-351.

Barendse, J. and Chown, S.L. (2001) Abundance and seasonality of mid-altitude fellfield arthropods from Marion Island, *Polar Biology* **24**, 73-82.

Barrett, S.C.H. (1996) The reproductive biology and genetics of island plants, *Philosophical Transactions of the Royal Society, London, Series B* **351**, 725-733.

Bellido, A. and Cancela da Fonseca, J.P. (1988) Spatio-temporal organization of the oribatid mite community in a littoral turf of the Kerguelen archipelago, *Pedobiologia* **31**, 239-246.

Bergstrom, D.M. and Chown, S.L. (1999) Life at the front: history, ecology and change on southern ocean islands, *Trends in Ecology and Evolution* **14**, 472-477.

Bergstrom, D.M. and Selkirk, P.M. (1987) Reproduction and dispersal of mosses on Macquarie Island, *Symposia Biologica Hungarica* **35**, 247-257.

Bergstrom, D.M. and Selkirk, P.M (1999) Bryophyte propagule banks in feldmarks on subantarctic Macquarie Island, *Arctic, Antarctic and Alpine Research* **31**, 202-208.

Bergstrom, D.M. and Selkirk, P.M. (2000) Terrestrial vegetation and environments on Heard Island, *Papers and Proceedings of the Royal Society of Tasmania* **133**, 33-46.

Bergstrom, D.M., Hodgson, D.A. and Convey, P. (2006) The physical setting of the Antarctic, in D.M. Bergstrom, P. Convey, and A.H.L. Huiskes (eds.), *Trends in Antarctic Terrestrial and Limnetic Ecosystems: Antarctica as a Global Indicator*, Springer, Dordrecht (this volume).

Bergstrom, D.M., Selkirk, P.M., Keenan, H.M. and Wilson, M.E. (1997) Reproductive behaviour of ten flowering plant species on subantarctic Macquarie Island, *Opera Botanica* **132**, 109-132.

Berjak P. (1979) The Marion Island flora – leaf structure in *Poa cookii* (F. Hook), *Proceedings of the Electron Microscopy Society of South Africa* **9**, 67–68.

Billings, W.D. (1974) Arctic and alpine vegetation: plant adaptations to cold summer climates, in J.D. Ives and R.G. Barry (eds) *Arctic and Alpine environments*, Methuen, London, pp. 403–443.

Block, W. (1990) Cold tolerance of insects and other arthropods, *Philosophical Transactions of the Royal Society of London, Series B* **326**, 613-633.

Block, W. (1996) Cold or drought - the lesser of two evils for terrestrial arthropods? *European Journal of Entomology* **93**, 325-339.

Briggs, C.L., Selkirk, P.M and Bergstrom, D.M. (2006) Facing the furious fifties: the contractile stem of the subantarctic megaherb *Pleurophyllum hookeri*, *New Zealand Journal of Botany* **44**, 187-197.

Brown, C.L., Bale, J.S. and Walters, K. F. A. (2004) Freezing induces a loss of freeze tolerance in an overwintering insect, *Proceedings of the Royal Society of London B* **271**, 1507-1511.

Brown, K.G. (1964) The insects of Heard Island, *A. N. A. R. E. Report* **1**, 1-39.

Cannon, R.J.C. and Block, W. (1988) Cold tolerance of microarthropods, *Biological Reviews* **63**, 23-77.

Chen, C–P., Denlinger, D.L. and Lee, R.E. (1987) Cold-shock injury and rapid cold hardening in the flesh fly *Sarcophaga crassipalpis*, *Physiological Zoology* **60**, 297-304.

Chevrier, M., Vernon, P. and Frenot, Y. (1997) Potential effects of two alien insects on a sub-Antarctic wingless fly in the Kerguelen islands, in B. Battaglia, J. Valencia, and D.W.H. Walton (eds) *Antarctic communities: Species, structure and survival,* Cambridge: Cambridge University Press, pp. 424-431.

Chown, S.L. (1992) A preliminary analysis of weevil assemblages in the Sub-Antarctic: local and regional patterns, *Journal of Biogeography* **19**, 87-98.

Chown, S.L. (1993) Desiccation resistance in six sub-Antarctic weevils (Coleoptera: Curculionidae): humidity as an abiotic factor influencing assemblage structure, *Functional Ecology* **7**, 318-325.

Chown, S.L. (1997) Thermal sensitivity of oxygen uptake of Diptera from sub-Antarctic South Georgia and Marion Island, *Polar Biology* **17**, 81-86.

Chown, S. L. (2001) Physiological variation in insects: hierarchical levels and implications, *Journal of Insect Physiology* **47**, 649-660.

Chown, S.L. and Block, W. (1997) Comparative nutritional ecology of grass-feeding in a sub-Antarctic beetle: the impact of introduced species on *Hydromedion sparsutum* from South Georgia, *Oecologia* **111**, 216-224.

Chown, S.L. and Convey, P. (2006a) Spatial and temporal variability across life's hierarchies in the terrestrial Antarctic, *Philosophical Transactions of the Royal Society of London B*, in review.

Chown, S.L. and Convey, P. (2006b) Biogeography, in D.M. Bergstrom, P. Convey, and A.H.L. Huiskes (eds.), *Trends in Antarctic Terrestrial and Limnetic Ecosystems: Antarctica as a Global Indicator,* Springer, Dordrecht (this volume).

Chown, S.L. and Crafford, J.E. (1992) Microhabitat temperatures at Marion Island (46°54'S 37°45'E), *South African Journal of Antarctic Research* **22**, 51-58.

Chown, S.L. and Klok, C.J. (2003) Altitudinal body size clines: latitudinal effects associated with changing seasonality, *Ecography* **26**, 445-455.

Chown, S.L. and Nicolson, S.W. (2004) *Insect Physiological Ecology. Mechanisms and Patterns.* Oxford University Press, Oxford, pp. 256.

Chown, S.L. and Scholtz, C.H. (1989) Biology and ecology of the *Dusmoecetes* Jeannel (Col. Curculionidae) species complex on Marion Island, *Oecologia* **80**, 93-99.

Chown, S.L., Greenslade, P. and Marshall, D.J. (2006) Terrestrial invertebrates of Heard Island, in K. Green and E.J. Woehler (eds.), *Heard Island: Southern Ocean Sentinel.* Chipping Norton, Surrey and Beatty, pp. 91-104.

Convey, P. (1994a) Growth and survival strategy of the Antarctic mite *Alaskozetes antarcticus*, *Ecography* **17**, 97-107.

Convey, P. (1994b) Modelling reproductive effort in sub- and maritime Antarctic mosses, *Oecologia* **100**, 45-53.

Convey, P. (1996a) Overwintering strategies of terrestrial invertebrates in Antarctica - the significance of flexibility in extremely seasonal environments, *European Journal of Entomology* **93**, 489-505.

Convey, P. (1996b) The influence of environmental characteristics on life history attributes of Antarctic terrestrial biota, *Biological Reviews* **71**, 191-225.

Convey, P. (1996c) Reproduction of Antarctic flowering plants, *Antarctic Science* **8**, 127-134.

Convey, P. (1997) How are the life history strategies of Antarctic terrestrial invertebrates influenced by extreme environmental conditions? *Journal of Thermal Biology* **22**, 429-440.

Convey, P. (1998) Latitudinal variation in allocation to reproduction by the Antarctic oribatid mite, *Alaskozetes antarcticus*, *Applied Soil Ecology* **9**, 93-99.

Convey, P. (2000) How does cold constrain life cycles of terrestrial plants and animals? *Cryo-Letters* **21**, 73-82.

Convey, P. (2001) Terrestrial ecosystem response to climate changes in the Antarctic, in G.-R. Walther, C.A. Burga and P.J. Edwards (eds.), *"Fingerprints" of climate change - adapted behaviour and shifting species ranges*, Kluwer, New York, pp 17–42.

Convey, P. (2003a) Maritime Antarctic climate change: signals from terrestrial biology, in E. Domack, A. Burnett, A. Leventer, P. Convey, M. Kirby and R. Bindschadler (eds.), *Antarctic Peninsula Climate Variability: Historical and Palaeoenvironmental Perspectives*, Antarctic Research Series, Vol. 79, American Geophysical Union, pp. 145-158.

Convey, P. (2003b) Soil faunal community response to environmental manipulation on Alexander Island, southern maritime Antarctic, in A.H.L. Huiskes, W.W.C. Gieskes, J. Rozema, R.M.L. Schorno, S. van der Vies and W.J. Wolff (eds.), *VIII SCAR International Biology Symposium: Antarctic Biology in a Global Context*, Backhuys, Leiden, pp. 74-78.

Convey, P. (2006) Antarctic climate change and its influences on terrestrial ecosystems, in D.M. Bergstrom, P. Convey, and A.H.L. Huiskes (eds.), *Trends in Antarctic Terrestrial and Limnetic Ecosystems: Antarctica as a Global Indicator*, Springer, Dordrecht (this volume).

Convey, P. and Smith R. I. L. (1993) Investment in sexual reproduction by Antarctic mosses, *Oikos* **68**, 293-302.

Convey, P. and Wynn-Williams, D.D. (2002) Antarctic soil nematode response to artificial environmental manipulation, *European Journal of Soil Biology* **38**, 255-259.

Convey, P., Frenot, Y., Gremmen, N. and Bergstrom, D.M. (2006) Biological invasions, in D.M. Bergstrom, P. Convey, and A.H.L. Huiskes (eds.), *Trends in Antarctic Terrestrial and Limnetic Ecosystems: Antarctica as a Global Indicator*, Springer, Dordrecht (this volume).

Convey, P., Pugh, P. J. A., Jackson, C., Murray, A. W., Ruhland, C. T., Xiong, F. S. and Day, T. A. (2002) Response of Antarctic terrestrial arthropods to multifactorial climate manipulation over a four year period. *Ecology* **83**, 3130-3140.

Copson, G. and Whinam, J. (1998) Response of vegetation on subantarctic Macquarie Island to reduced rabbit grazing, *Australian Journal of Botany* **46**, 15-24.

Crafford, J.E. (1984) Life cycle and kelp consumption of *Paractora dreuxi mirabilis* (Diptera: Helcomyzidae): A primary decomposer of stranded kelp on Marion Island, *South African Journal of Antarctic Research* **14**, 18-22.

Crafford, J.E. (1990) The role of feral house mice in ecosystem functioning on Marion Island, in K.R. Kerry and G. Hempel (eds.), *Antarctic Ecosystems. Ecological Change and Conservation*, Springer, Berlin, pp. 359-364.

Crafford, J.E., Scholtz, C.H. and Chown, S.L. (1986) The insects of sub-Antarctic Marion and Prince Edward Islands; with a bibliography of entomology of the Kerguelen Biogeographical Province, *South African Journal of Antarctic Research* **16**, 42-84.

Danks, H.V. (1999) Life cycles in polar arthropods - flexible or programmed? *European Journal of Entomology* **96**, 83-102.

Davey, M.C. (1997) Effects of continuous and repeated dehydration on carbon fixation by bryophytes from the maritime Antarctic, *Oecologia* **110**, 25-31.

Davey, M.C. and Rothery, P. (1996) Seasonal variation in respiratory and photosynthetic parameters in three mosses from the maritime Antarctic, *Annals of Botany* **78**, 719-728.

Davies, L. (1987) Long adult life, low reproduction and competition in two sub-Antarctic carabid beetles, *Ecological Entomology* **12**, 149-162.

Day, T.A. (2001) Multiple trophic levels in UV-B assessments - completing the ecosystem, *New Phytologist* **152**, 183-186.

Dorne A.J. (1977) Analysis of the germination under laboratory and field conditions of seeds collected in the Kerguelen Archipelago, in G.A. Llano (ed) *Adaptations within the Antarctic Ecosystem, 3rd SCAR Biology Symposium*, Smithsonian Institute. Washington DC, pp. 1003–1013.

Du Puy, D.J., Telford, I.R.H., Edgar, E. (1993) Poaceae, *Flora of Australia* **50**, 456-511.

Edwards, J.A. (1974) Studies in *Colobanthus quitensis* (Kunth) Bartl. and *Deschampsia antarctica* Desv.: VI.: Reproductive performance on Signy Island, *British Antarctic Survey Bulletin* **39**, 67-86.

Ernsting, G. (1993) Observations on life cycle and feeding ecology of two recently introduced beetle species at South Georgia, sub-Antarctic, *Polar Biology* **13**, 423-428.

Ernsting, G., Block, W., MacAlister, H.E. and Todd, C.M. (1995) The invasion of the carnivorous carabid beetle *Trechisibus antarcticus* on South Georgia (sub-Antarctic) and its effect on the endemic herbivorous beetle *Hydromedion sparsutum*, *Oecologia* **103**, 34-42.

Ernsting, G., Brandjes, G.J., Block, W. and Isaaks, J.A. (1999) Life-history consequences of predation for a subantarctic beetle: evaluating the contribution of direct and indirect effects, *Journal of Animal Ecology* **68**, 741-752.

Erskine, P.D., Bergstrom, D.M., Schmidt, S., Stewart, G.R., Tweedie, C.E., and Shaw, J.D. (1998) Subantarctic Macquarie Island - a model ecosystem for studying animal-derived nitrogen sources using ^{15}N natural abundance, *Oecologia* **117**, 187-193.

Farman, J.C., Gardiner, B.G. and Shanklin, J.D. (1985) Large losses of total ozone in Antarctica reveal seasonal ClO_x/NO_x interaction, *Nature* **315**, 207-210.

Filson, R.B. and J.H. Willis (1975) A fruiting occurrence of *Bryum algens* Card. in East Antarctica, *Muelleria* 3, 112-116.

Fowbert, J.A. and Smith, R.I.L. (1994) Rapid population increases in native vascular plants in the Argentine Islands, Antarctic Peninsula, *Arctic and Alpine Research* **26**, 290-296.

Frahm, J.P. and Klaus, D. (2001) Bryophytes as indicators of recent climate fluctuations in Central Europe, *Lindbergia* **26**, 97-104.

Frenot, Y., Chown, S.L., Whinam, J., Selkirk, P., Convey, P., Skotnicki, M. and Bergstrom, D. (2005) Biological invasions in the Antarctic: extent, impacts and implications, *Biological Reviews* **80**, 45-72.

Friedmann, E.I. (1982) Endolithic microorganisms in the Antarctic cold desert, *Science* **215**, 1045-1053.

Gaines, S.D. and Denny, M.W. (1993) The largest, smallest, highest, lowest, longest, and shortest: extremes in ecology, *Ecology* **74**, 1677-1692.

Green, T.G.A. (1985) Dry Valley terrestrial plant communities. The problem of production and growth, *New Zealand Antarctic Record* **6**, 40-44.

Greenslade, P.J.M. (1983) Adversity selection and the habitat templet, *American Naturalist* **122**, 352-365.

Gremmen, N.J.M., Chown, S.L., Marshall D.J. (1998) Impact of the introduced grass *Agrostis stolonifera* on vegetation and soil fauna communities at Marion Island, sub-Antarctic, *Biological Conservation* **85**, 223-231.

Grime, J.P. (1988) The C-S-R model of primary plant strategies - origins, implications and tests, in R.M. Anderson, B.D. Turner and L.R. Taylor (eds.), *Population Dynamics*, Blackwell, Oxford, pp. 123-139.

Grobe, C.W., Ruhland C.T. and Day T.A. (1997) A new population of *Colobanthus quitensis* near Arthur Harbor, Antarctica: correlating recruitment with warmer summer temperatures, *Arctic and Alpine Research* **29**, 217-221.

Hennion, F. and Walton, D.W.H (1997) Seed germination of endemic species from Kerguelen phytogeographic zone, *Polar Biology* **17**, 180–187.

Hennion, F., Huiskes, A.H.L., Robinson, S. and Convey, P. (2006) Physiological traits or organisms in a changing environment, in D.M. Bergstrom, P. Convey, and A.H.L. Huiskes (eds.), *Trends in Antarctic Terrestrial and Limnetic Ecosystems: Antarctica as a Global Indicator*, Springer, Dordrecht (this volume).

Hovenden, M.J., Jackson, A.E. and Seppelt, R.D. (1994) Field photosynthetic activity of lichens in the Windmill Islands oasis, Wilkes Land, continental Antarctica, *Physiologia Plantarum* **90**, 567-576.

Hughes K.A. and Lawley, B. (2003) A novel Antarctic microbial endolithic community within gypsum crusts, *Environmental Microbiology* **5**, 555-565.

Hughes, K.A., Ott, S., Bölter, M. and Convey, P. (2006) Colonisation processes, in D.M. Bergstrom, P. Convey, and A.H.L. Huiskes (eds.), *Trends in Antarctic Terrestrial and Limnetic Ecosystems: Antarctica as a Global Indicator*, Springer, Dordrecht (this volume).

Hull, B.B. and Bergstrom, D.M. (2006) Antarctic terrestrial and limnetic conservation and management, in D.M. Bergstrom, P. Convey, and A.H.L. Huiskes (eds.), *Trends in Antarctic Terrestrial and Limnetic Ecosystems: Antarctica as a Global Indicator*, Springer, Dordrecht (this volume).

Jackson, A.E. and R.D. Seppelt (1997) Physiological adaptations to freezing and UV radiation exposure in *Prasiola crispa,* an Antarctic terrestrial alga, in B. Battaglia, J. Valencia and D.W.H. Walton (eds.), *Antarctic Communities: Species, Structure and Survival.* Cambridge University Press, Cambridge, pp. 226-233.

Jenkin J.F. and Ashton D.H. (1970) Productivity studies on Macquarie Island vegetation, in M.W. Holdgate, *Antarctic Ecology Volume 2,* Academic Press, London, pp851–863.

Jenkin J.F. and Ashton D.H. (1979) Pattern in *Pleurophyllum* herbfields on Macquarie Island (subantarctic), *Australian Journal of Botany* **4**, 47-66.

Johnson, D., Campbell, C.D., Lee, J.A., Callaghan, T.V. and Gwynne-Jones, D. (2002) Arctic microorganisms respond more to elevated UV-B radiation than CO_2, *Nature* **416**, 82-83.

Joly, Y., Frenot, Y. and Vernon, P. (1987) Environmental modifications of a subantarctic peat-bog by the Wandering Albatross (*Diomedea exulans*): a preliminary study, *Polar Biology* **8**, 61-72.

Kappen, L. (1993) Plant activity under snow and ice, with particular reference to lichens, *Arctic* **46**, 297-302.

Kappen, L., Sommerkorn, M. and Schroeter, B. (1995) Carbon acquisition and water relations of lichens in polar regions - Potentials and limitations, *Lichenologist* **27**, 531-545.

Kappen, L., Schroeter, B., Scheidegger, C., Sommerkorn, M., Hestmark, G. (1996) Cold resistance and metabolic activity of lichens below 0 degrees C. *Life Sciences: Space and Mars Recent Results*, Vol. 18, Pergamon Press, Oxford, pp. 119-128.

Kennedy, A.D. (1993a) Water as a limiting factor in the Antarctic terrestrial environment: a biogeographical synthesis, *Arctic and Alpine Research* **25**, 308-315.

Kennedy, A.D. (1993b) Photosynthetic response of the Antarctic moss *Polytrichum alpestre* Hoppe to low temperatures and freeze-thaw stress, *Polar Biology* **13**, 271-279.

Kennedy, A.D. (1994) Simulated climate change: a field manipulation study of polar microarthropod community response to global warming, *Ecography* **17**, 131-140.

Kennedy, A.D. (1995a) Temperature effects of passive greenhouse apparatus in high-latitude climate change experiments, *Functional Ecology* **9**, 340-350.

Kennedy, A.D. (1995b) Simulated climate change: are passive greenhouses a valid microcosm for testing the biological effects of environmental perturbations? *Global Change Biology* **1**, 29-42.

Kennedy, A.D. (1995c) Antarctic terrestrial ecosystem response to global environmental change, *Annual Review of Ecology and Systematics* **26**, 683-704.

Kennedy, A.D. (1996) Antarctic fellfield response to climate change: a tripartite synthesis of experimental data, *Oecologia* **107**, 141-150.

Kingsolver, J.G. and Huey, R.B. (1998) Evolutionary analyses of morphological and physiological plasticity in thermally variable environments, *American Zoologist* **38**, 545-560.

Klok, C.J. and Chown, S.L. (1997) Critical thermal limits, temperature tolerance and water balance of a sub-Antarctic caterpillar, *Pringleophaga marioni* Viette (Lepidoptera: Tineidae), *Journal of Insect Physiology* **43**, 685-694.

Klok, C.J. and Chown, S.L. (1998) Interactions between desiccation resistance, host-plant contact and the thermal biology of a leaf-dwelling sub-antarctic caterpillar, *Embryonopsis halticella* (Lepidoptera: Yponomeutidae), *Journal of Insect Physiology* **44**, 615-628.

Klok, C.J. and Chown, S.L. (2001) Critical thermal limits, temperature tolerance and water balance of a sub-Antarctic kelp fly, *Paractora dreuxi* (Diptera: Helcomyzidae), *Journal of Insect Physiology* **47**, 95-109.

Lawrence J. and McClintock J.B. (1989) Biomass plasticity of the leaves and inflorescences of *Acaena magellanica* (Lam.) Vahl. (Roseaceae) on subantarctic Îles Kerguelen, *Polar Biology* **9**, 409-413.

Leader-Williams, N. (1988) *Reindeer on South Georgia: The Ecology of an Introduced Population*, Cambridge University Press, Cambridge, pp. 319.

Leader-Williams, N., Smith, R.I.L. and Rothery, P. (1987) Influence of introduced reindeer on the vegetation of South Georgia: results from a long-term exclusion experiment, *Journal of Applied Ecology* **24**, 801-822.

Lebouvier, M. and Frenot, Y. (2005) Impact of climate change on vegetation and soil erosion at Kerguelen Islands, in Anon (ed), *Evolution and biodiversity in Antarctica*, Abstracts of the 9th SCAR International Biology Symposium, Curitiba, Brazil 2005, p 102.

Lee, R.E., Chen, C–P. and Denlinger, D.L. (1987) A rapid cold-hardening process in insects, *Science* **238**, 1415-1417.

Lee, R.E., Elnitsky, M.A., Rinehart, J.P., Hayward, S.A.L., Sandro, L.H. and Denlinger, D.L. (2006) Rapid cold-hardening increases the freezing tolerance of the Antarctic midge *Belgica Antarctica*, *Journal of Experimental Biology* **209**, 399-406.

le Roux, P.C. and McGeoch, M.A. (2004) The use of size as an estimator of age in the subantarctic cushion plant, *Azorella selago* (Apiaceae), *Arctic, Antarctic and Alpine Research*, **36**, 509–517.

le Roux, P.C., McGeoch, M.A., Nyakatya, M.J., Chown, S.L. (2005) Effects of a short-term climate change experiment on a keystone plant species in the sub-Antarctic, *Global Change Biology* **11**, 1628-1639.

Longton, R.E. (1988) *Biology of Polar Bryophytes and Lichens*, Cambridge University Press, Cambridge.

Lovelock, C.E., Osmond, C.B. and Seppelt, R.D. (1995a) Photoinhibition in the Antarctic moss *Grimmia antarctica* Card. when exposed to cycles of freezing and thawing, *Plant, Cell and Environment* **18**, 1395-1402.

Lovelock, C.E., Jackson, A.E., Melick, D.R. and Seppelt, R.D. (1995b) Reversible photoinhibition in Antarctic moss during freezing and thawing, *Plant Physiology* **109**, 955-961.

McDonald, J.R., Bale, J.S. and Walters, K.F.A. (1997) Rapid cold hardening in the western flower thrips *Frankliniella occidentalis, Journal of Insect Physiology* **43**, 759-766.

McDonald, J.R., Head, J., Bale, J.S. and Walters, K.F.A. (2000) Cold tolerance, overwintering and establishment potential of *Thrips palmi, Physiological Entomology* **25**, 159-166.

McGeoch, M.A., le Roux, P.C., Hugo, E.A. and Chown, S.L. (2006) Species and community responses to short-term climate manipulation: microarthropods in the sub–Antarctic, *Austral Ecology*, in press.

McGraw, J.B. and Day, T.A. (1997) Size and characteristics of a natural seed bank in Antarctica, *Arctic and Alpine Research* **29**, 213-216.

McGraw J.B. and Vavrek M.C. (1989) The role of buried viable seeds in arctic and alpine plant communities, in M.A. Leck, V.T. Parker and Simpson, R.L. (eds), *Ecology of Soil Seed Banks*, Academic Press, San Diego.

Melick, D.R. and R.D. Seppelt (1992) Loss of soluble carbohydrates and changes in freezing point of Antarctic bryophytes after leaching and repeated freeze-thaw cycles, *Antarctic Science* **4**, 399-404.

Melick, D.R. and R.D. Seppelt (1997) Vegetation patterns in relation to climatic and endogenous changes in Wilkes Land, continental Antarctica, *Journal of Ecology* **85**, 43-56.

Menge, B.A. and Sutherland, J.P. (1987) Community regulation: variation in disturbance, competition and predation in relation to environmental stress and recruitment, *American Naturalist* **130**, 730-757.

Meurk, C.D., Foggo, M.N. and Wilson, J.B. (1994a) The vegetation of subantarctic Campbell Island, *New Zealand Journal of Ecology* **18**, 123–168.

Meurk, C.D., Foggo, M.N., Thompson, B.M., Bathurst, E.T.J., and Crompton, M.B. (1994b) Ion–rich precipitation and vegetation patterns on subantarctic Campbell Island, *Arctic and Alpine Research* **26**, 281–289.

Mitchell A.D., Meurk, C.D. and Wagstaff, S.J. (1999) Evolution of *Stilbocarpa,* a megaherb from New Zealand's sub–antarctic islands, *New Zealand Journal of Botany* **37**, 205-211.

Montiel, P.O. (2000) Soluble carbohydrates (trehalose in particular) and cryoprotection in polar biota, *CryoLetters* **21**, 83-90.

Newsham, K.K. (2003) UV-B radiation arising from stratospheric ozone depletion influences the pigmentation of the Antarctic moss *Andreaea regularis, Oecologia* **135**, 327-331.

Newsham, K.K., Greenslade, P.D., Kennedy, V.H. and McLeod, A.R. (1999) Elevated UV-B radiation incident on *Quercus robur* leaf canopies enhances decomposition of resulting leaf litter in soil, *Global Change Biology* **5**, 403-409.

Newsham, K.K., Geissler, P.A., Nicolson, M.J., Peat, H.J., Smith R.I.L (2005) Sequential reduction of UV-B radiation in the field alters the pigmentation of an Antarctic leafy liverwort, *Environmental and Experimental Botany* **54**, 22-32.

Newsham, K.K., Hodgson, D.A., Murray, A.W.A., Peat, H.J. and Smith, R.I.L. (2002) Response of two Antarctic bryophytes to stratospheric ozone depletion, *Global Change Biology* **8**, 972-983.

Nicolai, V., and Droste, M. (1984) The ecology of *Lancetes claussi* (Müller) (Coleoptera, Dytiscidae), the Subantarctic water beetle of South Georgia. *Polar Biology* **3**, 39-44.

Nondula, N., Marshall, D.J., Baxter, R., Sinclair, B.J. and Chown, S.L. (2004) Life history and osmoregulatory ability of *Telmatogeton amphibius* (Diptera, Chironomidae) at Marion Island, *Polar Biology* **27**, 629-635.

Ochyra, R. (1988) *The moss flora of King George Island, Antarctica*, Polish Academy of Sciences, Cracow, Poland, pp. 278.

Pammenter, N.W., Drennan, P.M., Smith, V.R. (1986) Physiological and anatomical aspects of photosynthesis of two *Agrostis* species at a sub-Antarctic island, *New Phytologist* **102**, 143-160.

Pannewitz, S., Schlensog, M., Green, T.G.A., Sancho, L.G. and Schroeter, B. (2003) Are lichens active under snow in continental Antarctica? *Oecologia* **135**, 30-38.

Paul, N. (2001) Plant responses to UV-B: time to look beyond stratospheric ozone depletion? *New Phytologist* **150**, 5-8.

Paul, N., Callaghan, T., Moody, S., Gwynne-Jones, D., Johanson, U. and Gehrke, C. (1999) UV-B impacts on decomposition and biogeochemical cycling, in J. Rozema (ed.), *Stratospheric ozone depletion, the effects of enhanced UV-B radiation on terrestrial ecosystems*, Backhuys, Leiden, pp. 117-133.

Peck, L.S., Convey, P. and Barnes, D.K.A. (2006) Environmental constraints on life histories in Antarctic ecosystems: tempos, timings and predictability, *Biological Reviews* **81**, 75-109.

Pye, T. (1993) Reproductive biology of the feral house mouse (*Mus musculus*) on Subantarctic Macquarie Island, *Wildlife Research* **20**, 745-758.

Quayle, W.C., Convey, P., Peck, L.S., Ellis-Evans, J.C., Butler, H.G. and Peat, H.J. (2003) Ecological responses of maritime Antarctic lakes to regional climate change, in E. Domack, A. Burnett, A. Leventer, P. Convey, M. Kirby and R. Bindschadler (eds.), *Antarctic Peninsula climate variability: Historical and palaeoenvironmental perspectives, Antarctic Research Series* vol. 79, American Geophysical Union, pp. 159-170.

Quayle, W.C., Peck, L.S., Peat, H., Ellis-Evans, J.C. and Harrigan, P.R. (2002) Extreme responses to climate change in Antarctic lakes, *Science* **295**, 645.

Robinson, S.A., Wasley, J. and Tobin, A.K. (2003) Living on the edge - plants and global change in continental and maritime Antarctica, *Global Change Biology* **9**, 1681-1717.

Robinson, S.A., Wasley, J., Popp, M., Lovelock, C.E. (2000) Desiccation tolerance of three moss species from continental Antarctica, *Australian Journal of Plant Physiology* **27**, 379-388.

Roser, D.J., Melick, D.R., Ling, H.U., Seppelt, R.D. (1992) Polyol and sugar content of terrestrial plants from continental Antarctica, *Antarctic Science* **4**, 413-420.

Rouault, M., Mélice, J–L., Reason, C.J. and Lutjeharms, J.R.E. (2005) Climate variability at Marion Island, Southern Ocean, since 1960, *Journal of Geophysical Research* **110**, C05007.

Rozema, J. (ed.) (1999) *Stratospheric ozone depletion, the effects of enhanced UV-B radiation on terrestrial ecosystems*. Backhuys, Leiden.

Rozema J., Kooi, B., Broekman, R. and Kuijper, L. (1999) Modelling direct (photodegradation) and indirect (litter quality) effects of enhanced UV-B on litter decomposition, in J. Rozema (ed.), *Stratospheric ozone depletion, the effects of enhanced UV-B radiation on terrestrial ecosystems* Backhuys, Leiden, pp. 135-156.

Schlensog, M., Pannewitz, S., Green, T.G.A. and Schroeter, B. (2004) Metabolic recovery of continental antarctic cryptogams after winter, *Polar Biology* **27**, 399-408.

Schroeter, B. and C. Scheidegger (1995) Water Relations in Lichens at Subzero Temperatures - Structural-Changes and Carbon-Dioxide Exchange in the Lichen *Umbilicaria aprina* from Continental Antarctica, *New Phytologist* **131**, 273-285.

Seppelt, R.D, Green, T.G.A., Schwarz, A–M.J. and Frost, A. (1992) Extreme southern locations for moss sporophytes in Antarctica, *Antarctic Science* **4**, 37-39.

Seymour, F.A., Crittenden, P.D. and Dyer, P.S. (2005) Sex in the extremes: lichen-forming fungi, *Mycologist* **19**, 51-58.

Shaw J.D (2005) Reproductive Ecology of Vascular Plants on Subantarctic Macquarie Island, PhD Thesis, University of Tasmania, Hobart.

Shaw J.D, Bergstrom, D.M and Hovenden, M. (2005) The impact of feral rats (*Rattus rattus*) on populations of subantarctic megaherb (*Pleurophyllum hookeri*), *Austral Ecology* **30**, 118-125.

Sinclair, B.J. (2001) Field ecology of freeze tolerance: interannual variation in cooling rates, freeze-thaw and thermal stress in the microhabitat of the alpine cockroach *Celatoblatta quinquemaculat*, *Oikos* **93**, 286-293.

Sinclair, B.J. (2002) Effects of increased temperatures simulating climate change on terrestrial invertebrates on Ross Island, Antarctica, *Pedobiologia* **46**, 150-160.

Sinclair, B.J. and Chown, S.L. (2003) Rapid responses to high temperature and desiccation but not to low temperature in the freeze tolerant sub-Antarctic caterpillar *Pringleophaga marioni* (Lepidoptera, Tineidae), *Journal of Insect Physiology* **49**, 45-52.

Sinclair, B.J. and Chown, S. L. (2005a) Deleterious effects of repeated cold exposure in a freeze-tolerant sub-Antarctic caterpillar, *Journal of Experimental Biology* **208**, 869-879.

Sinclair, B.J. and Chown, S.L. (2005b) Caterpillars benefit from thermal ecosystem engineering by Wandering Albatrosses on sub-Antarctic Marion Island, *Biology Letters*, doi: 10.1098/rsbl.(2005).0384.

Sinclair, B.J., Klok, C.J., Scott, M.B., Terblanche, J.S. and Chown, S.L. (2003) Diurnal variation in supercooling points of three species of Collembola from Cape Hallett, Antarctica, *Journal of Insect Physiology* **49**, 1049-1061.

Skotnicki, M.L. and Selkirk, P.M. (2006) Plant biodiversity in an extreme environment: genetic studies of origins, diversity and evolution in the Antarctic, in D.M. Bergstrom, P. Convey, and A.H.L. Huiskes (eds.), *Trends in Antarctic Terrestrial and Limnetic Ecosystems: Antarctica as a Global Indicator*, Springer, Dordrecht (this volume).

Slabber, S. (2005) *Physiological plasticity in arthropods from Marion Island: indigenous and alien species*. PhD Thesis, Stellenbosch University, South Africa.

Slabber, S. and Chown, S.L. (2005) Differential responses of thermal tolerance to acclimation in the sub-Antarctic rove beetle *Halmaeusa atriceps*, *Physiological Entomology* **30**, 195-204.

Smith, R.I.L. (1990) Signy Island as a paradigm of biological and environmental change in Antarctic terrestrial ecosystems, in K.R. Kerry and G. Hempel (eds.), *Antarctic Ecosystems, Ecological Change and Conservation*, Springer-Verlag, Berlin, pp. 32-50.

Smith, R.I.L. (1994) Vascular plants as indicators of regional warming in Antarctica, *Oecologia* **99**, 322-328.

Smith, R.I.L. (2001) Plant colonization response to climate change in the Antarctic, *Folia Facultatis Scientiarum Naturalium Universitatis Masarykiana Brunensis, Geographia*, **25**, 19-33.

Smith, R.I.L. and Convey, P. (2002) Enhanced sexual reproduction in bryophytes at high latitudes in the maritime Antarctic, *Journal of Bryology* **24**, 107-117.

Smith, V.R. (2002) Climate change in the sub-Antarctic: An illustration from Marion Island, *Climatic Change* **52**, 345-357.

Smith, V.R. and Steenkamp, M. (1990) Climatic change and its ecological implications at a subantarctic island, *Oecologia* **85**, 14-24.

Smith, V.R., Avenant, N.L. and Chown, S.L. (2002) The diet of house mice on a sub-Antarctic island, *Polar Biology* **25**, 703-715.

Sømme, L. (1995) *Invertebrates in Hot and Cold Arid Environments*, Springer, Berlin.

Southwood, T.R.E. (1977) Habitat, the templet for ecological strategies, *Journal of Animal Ecology* **46**, 337-365.

Southwood, T.R.E. (1988) Tactics, strategies and templets, *Oikos* **52**, 3-18.

Strathdee, A.T., Bale, J.S., Block, W.C., Coulson, S.J., Hodkinson, I.D. and Webb, N.R. (1993) Effects of temperature elevation on a field population of *Acyrthosiphon svalbardicum* (Hemiptera: Aphididae) on Spitsbergen, *Oecologia* **96**, 457-465.

Tallowin J.R.B. and Smith R.I.L. (1977) Studies in the reproductive biology of *Festuca contracta* T. Kirk on South Georgia: I. The Reproductive Cycles, *British Antarctic Survey Bulletin* **45**, 63-76.

Taylor, B.W. (1955) *The flora, vegetation and soils of Macquarie Island*, Australian Antarctic Division, Melbourne, pp. 192.

Tearle, P.V. (1987) Cryptogamic carbohydrate release and microbial response during spring freeze-thaw cycles in Antarctic fellfield fines, *Soil Biology and Biochemistry* **19**, 381-390.

van der Merwe, M., Chown, S.L. and Smith, V.R. (1997) Thermal tolerance limits in six weevil species from sub-Antarctic Marion Island, *Polar Biology* **18**, 331-336.

Vasseur, D.A. and Yodzis, P. (2004) The color of environmental noise, *Ecology* **85**, 1146-1152.

Vernon, P. and Vannier, G. (1996) Developmental patterns of supercooling capacity in a subantarctic wingless fly, *Experientia* **52**, 155-158.

Walton D.W.H (1982) Floral phenology in the South Georgian vascular flora, *British Antarctic Survey Bulletin* **55**, 11–25.

Wardle, P. (1991) *Vegetation of New Zealand*, Cambridge University Press, Cambridge, UK, 672 pp.

Wasley, J., Robinson, S.A., Lovelock, C.E. and Popp, M. (2006) Some like it wet – an endemic Antarctic bryophyte likely to be threatened under climate change induced drying, *Functional Plant Biology* **33**, 443–455.

Watt, J.C. (1970) Coleoptera: Perimylopidae of South Georgia, *Pacific Insects Monograph* **23**, 243-253.

Werth, E. (1911) Die Vegetation de subantarktischen Inseln Kerguelen, Possession und Heard–Eiland. *2 Deutsche Sudpolar–Expedition 1901 -1903* **8** Botanic 2, 223–371.

West, C. (1982) Life histories of three species of sub-Antarctic oribatid mite, *Pedobiologia* **23**, 59-67.

Wilson, A.M. and Thompson, K. (1989) A comparative study of reproductive allocation in 40 British grasses, *Functional Ecology* **3**, 297-302.

Worland, M.R. (2005) Factors that influence freezing in the sub-Antarctic springtail *Tullbergia Antarctica*, *Journal of Insect Physiology* **51**, 881-894.

Worland, M.R. and Convey, P. (2001) Rapid cold hardening in Antarctic microarthropods, *Functional Ecology* **15**, 515-525.

Wynn-Williams, D.D. (1993) Microbial processes and the initial stabilisation of fellfield soil, in J. Miles and D.W.H. Walton (eds.), *Primary Succession on Land*, Blackwell, Oxford, pp. 17-32.

Wynn-Williams, D.D. (1996) Response of pioneer soil microalgal colonists to environmental change in Antarctica, *Microbial Ecology* **31**, 177-188.

7. PHYSIOLOGICAL TRAITS OF ORGANISMS IN A CHANGING ENVIRONMENT

F. HENNION
*Impact des Changements Climatiques, UMR 6553, Centre National de
la Recherche Scientifique - Université de Rennes 1
Campus de Beaulieu
F-35042 Rennes cedex, France
Francoise.Hennion@univ-rennes1.fr*

A.H.L. HUISKES
*Unit for Polar Ecology,
Netherlands Institute of Ecology (NIOO-KNAW)
P.O. Box 140, 4400 AC Yerseke, The Netherlands
A.Huiskes@nioo.knaw.nl*

S. ROBINSON
*Institute for Conservation Biology,
University of Wollongong
Northfields Avenue
Wollongong, NSW 2522, Australia
sharonr@uow.edu.au*

P. CONVEY
*British Antarctic Survey, Natural Environment Research Council
High Cross, Madingley Road
Cambridge CB3 0ET, United Kingdom
p.convey@bas.ac.uk*

D.M. Bergstrom et al. (eds.), Trends in Antarctic Terrestrial and Limnetic Ecosystems, 129–159.
© 2006 *Springer.*

Introduction

Antarctic ecosystems represent one extreme of the continuum of environmental conditions across the planet. To our eyes, the environment appears harsh but, even though terrestrial biological diversity is restricted, a wide range of life is present and, locally, thrives. In the Antarctic, unusually, environments exist in which physical characteristics are dominant and overcome biological considerations. These are at the extreme ends of the ranges of many characteristics (temperature, snow, ice and solar radiation) found across environments globally. However, the Antarctic is also a large continent, comparable in area to continental Europe, and further surrounded by the cold Southern Ocean, within which lie a ring of subantarctic islands. Together, these islands and the continent give a natural environmental gradient with which to study the biological impacts of climate variables.

Antarctica is also a focus for studies of responses to regional and global change (eg Bergstrom and Chown 1999, Convey 2001, 2003, Robinson et al. 2003). Some of the fastest changing regions on earth (air temperatures along the western Antarctic Peninsula and Scotia Arc) are found here (King and Haranzogo 1998, Skvarca et al. 1998, Smith 2002, Quayle et al. 2002, 2003). Evaluations of change in this area are expected to provide a vital 'early warning system' for change consequences worldwide (Convey et al. 2003a, b). This chapter addresses an area central to our ability to understand and evaluate biotic responses to climate change predictions – that of organism physiology.

Antarctic climate change

Features of climate change as seen in the Antarctic are described in Convey this volume and are summarised briefly here. Two non-connected aspects of Antarctic climate change have received most attention – the rapid temperature increases that have been well documented along the Antarctic Peninsula and Scotia Arc and are observed to a lesser extent elsewhere and the seasonal formation in the austral spring of the Antarctic 'ozone hole'. At a continental scale some areas, particularly of inland continental Antarctica and parts of Victoria Land, are also thought to have experienced cooling over the same period, although data are sparse. In addition to any direct biological consequences of changes in temperature, indirect consequences may also be significant – for instance, temperature increases in either winter or summer may lead to a shortening of the winter season – while temperature changes will clearly also be linked with the processes controlling water availability in terrestrial habitats.

Water availability can be more important even than temperature in controlling biological activity in Antarctic terrestrial habitats (Kennedy 1993, Block 1996). In the subantarctic in particular, recent changes in precipitation patterns potentially have greater biological significance than the concurrent changes in temperature. Clearly, changes in precipitation patterns will impact water availability, although this is likely to be seen most directly in the subantarctic and parts of the maritime Antarctic (during summer), when the majority of precipitation falls as rain rather than snow and is immediately available to terrestrial biota. Summer thawing of seasonal snow banks and runoff from permanent glaciers also provide liquid water to terrestrial habitats, hence changes in the timing and magnitude of winter precipitation events, and the timing of thaws, will also be important. Consequences may also include a decrease in water availability, in instances where precipitation reduces, or finite resources of snow or ice are exhausted. Trends of both increasing and decreasing precipitation, identified in datasets covering up to the last six decades, are being seen at a variety of maritime and subantarctic locations.

The potential biological significance of seasonal ozone depletion is linked with the associated increase in shorter wavelength UV-B radiation reaching the Earth's surface. However, it is the timing of this increase rather than the absolute magnitude that is important. During periods of maximum ozone loss (typically October and November, during the austral spring) the intensity of UV-B radiation at ground level is similar to that normally experienced in mid-summer. However, early in the spring, exposed biota may be unable to respond, as they are yet to resume normal physiological activity after winter.

ORGANISMS UNDER INCREASED RADIATION

The potential effects of UV-B radiation on phototrophic organisms may be grouped into three areas: (a) changes in photosynthesis and growth (eg through trade-offs with reproductive capacity and biomass) (Teramura and Ziska 1996) (b) increased investment in UV-B absorbing or screening compounds (Karentz et al. 1991a,b) and (c) DNA damage, repair and photoreactivation (Lud et al. 2001a). Mobile organisms, such as cyanobacteria, can move to deeper layers in the soil to avoid radiative stress (Wynn-Williams 1994), while sessile organisms are particularly exposed to UV-B radiation. However, despite the clear effects seen in laboratory manipulations, those field studies that find effects generally report these to be much more minor, while others often report no detectable consequences (Jackson and Seppelt 1997, Montiel et al. 1999, Day et al. 1999, Huiskes et al. 1999, 2001, Lud et al. 2001a, Rozema et al. 2001).

Sessile phototrophic organisms such as lichens can not avoid incident solar radiation completely and they must make a trade-off between receiving sufficient levels of PAR and appreciable doses of UV-B (Day 2001). Microniche selection also plays a part, in that many utilise partially protected crevices and fissures, or locations that are protected by snow cover during the critical period (Cockell et al. 2002). To some extent they can also utilise a passive avoidance strategy: being poikilohydric, the periods when they are moist and active generally occur in the mornings or in the evenings, when UV radiation levels are low. However, it is also important to quantify the potential for uncontrolled damage during inactive periods, when repair is not possible.

CHANGES IN PHOTOSYNTHESIS AND GROWTH

Day and co-workers (Day et al. 1999, 2001, Xiong and Day 2001, Ruhland and Day 2000, 2001) have completed an extensive field manipulation study near Palmer Station (Anvers Island, western Antarctic Peninsula), separating the influence of UV-B and other environmental variables on the growth and ecophysiology of the two Antarctic flowering plants, *Colobanthus quitensis* (Kunth) Bartl. and *Deschampsia antarctica* Desv. Relative to ambient controls, exposure to solar UV-B from spring to mid-summer led to 11 - 22% less biomass and a 24 - 31% decrease in leaf area. Rates of photosynthesis were reduced when expressed relative to chlorophyll content or dry mass, but not relative to leaf area, through the development of thicker leaves containing more photosynthetic and screening pigments. Exposure to UV-B also led to reductions in quantum yield of photosystem II, based on fluorescence measurements of adaxial leaf surfaces and impaired photosynthesis in the upper mesophyll layer. The latter was suggested to be associated with light-independent enzymatic limitations. In *C. quitensis*, exposure to solar UV-B led to reductions in leaf longevity, branch production, cushion diameter growth, above-ground biomass and thickness of the non-green cushion base and litter layer. Exposure to UV-B also influenced patterns of reproductive investment, accelerating the development of reproductive structures and increasing the number of panicles (*D. antarctica*) and capsules (*C. quitensis*). Seed viability appeared to remain unchanged. In a similar field manipulation study in the Windmill Islands, East Antarctica, gametophytes of the moss *Grimmia antarctici* growing under near ambient UV radiation had a higher density of leaves than those growing under reduced UV radiation and there was evidence of morphological changes in moss exposed to ambient UV radiation (Robinson et al. 2005) (Fig. 1). However, no evidence for changes to photosynthetic parameters were detected in this species and rates of growth were too slow for any difference to be apparent over a 14 month period. Also other studies have failed to detect differences in photosynthetic parameters related to UV-B exposure. Rozema et al. (2001) reported no difference in net photosynthesis of *Deschampsia antarctica* treated with different levels of UV-B radiation in a growth chamber. They also showed a reduction in the length of shoots of *D. antarctica* when grown under experimental conditions but, under field conditions, no difference in shoot length or reproductive biomass was detected. Lud et al. (2001b) found no changes in photosynthetic activity of lichens and mosses either in field measurements or under controlled conditions. This was equally true for net photosynthesis, maximum PSII quantum efficiency and effective PSII quantum efficiency. Likewise, Montiel et al. (1999) found no changes in *D. antarctica* and *C. quitensis* and George et al. (2001) reported that some cyanobacteria were unaffected by UV-B radiation. Some other studies however reported evidence for a negative effect of UV-B radiation on photosynthesis of certain Antarctic species, including some cyanobacteria (George et al. 2001), mosses (Montiel et al. 1999) and a terrestrial alga (Post and Larkum 1993).

Figure 1. Comparison of gametophytes of Grimmia antarctici *showing the normal leaf morphology (c and d) and atypical leaf morphology (a and b), enlarged (e). Atypical leaves were characterized by short length and blunted leaf tips and were twice as common in plants grown under near ambient rather than reduced UV-B radiation (from Robinson* et al. *2005).*

INCREASES IN UV-B ABSORBING COMPOUNDS

A variety of UV-B absorbing compounds are found in Antarctic phototrophic organisms (eg Karentz et al. 1991a,b, Adamson and Adamson 1992). Phenylpropanoids produced by the shikimate pathway are thought to be predominant, often in combination with the acetate-malonate pathway (Day 2001). These pathways are responsible for the synthesis of various phenolic compounds (hydroxycinnamic acids, flavonoids, lignins and tannins). Key enzymes in these pathways are induced by light, both visible and UV (Beggs and Wellmann 1994). Flavonoids are present in *D. antarctica* (Webby and Markham 1994) and some moss species (Webby et al. 1996). They often accumulate in the epidermis, where, in higher plants, over 90% of incident UV- radiation is attenuated (Robberecht and Caldwell 1978). Recent research (Nybakken et al. 2004a,b) suggests that epidermal screening may be a constitutive feature, with little or no evidence for any dynamic response in effectiveness in Arctic and alpine plants. The only direct evidence available for plants having a dynamic ability to respond to incident levels of UV-B radiation over ecologically realistic timescales is from studies of three Antarctic mosses (Newsham et al. 2002, Newsham 2003) exposed to natural variations under the Antarctic ozone hole. These studies found that screening pigment concentrations were best correlated with levels of UV-B radiation experienced during the preceding 24 h. Species variation has also been shown for co-occurring Antarctic mosses, with the endemic species *Grimmia antarctici* possessing low levels of UV-B screening pigments whilst the two cosmopolitan species *Ceratodon purpureus* and *Bryum pseudotriquetrum* have intermediate and higher concentrations respectively (Dunn 2000, Lovelock and Robinson 2002, Robinson et al. 2005). In only one of these species, *B. pseudotriquetrum*, was evidence found that UV screening pigments correlate with UV radiation doses (Dunn 2000, Robinson et al. 2005). Experimental manipulations demonstrate that flavonoid synthesis may be enhanced by both UV-B radiation (Meykamp et al. 2001) and UV-A and visible wavelengths (Bornman and Sundby-Emanuelsson 1995). Concentrations of UV-B absorbing compounds were found to be higher in seedlings of *Colobanthus quitensis* exposed to natural UV-B levels in the field compared to those under UV-B absorbing screens (Ruhland and Day 2001).

Other secondary UV-screening compounds such as mycosporine-like amino acids (MAAs) may also be involved in UV-B responses. MAAs consist of a substituted cyclohexenone linked with an amino acid or an amino alcohol, and are found in many different organisms including fungi, eukaryotic algae, corals and starfish, Antarctic microalgae and cyanobacteria (Karentz et al. 1991b, Garcia-Pichel and Castenholz 1993, Ishikura et al. 1997, George et al. 2001). *Prasiola crispa*, a common Antarctic terrestrial green alga, contains low concentrations of UV-B absorbing compounds (including an unknown MAA) (Post and Larkum 1993, Hoyer et al. 2001, Lud et al. 2001a). However, chlorophyll concentration and photosynthesis in this alga are depressed when exposed to UV-B radiation (Post and Larkum 1993, Jackson and Seppelt 1997).

Lichens also contain a high proportion of secondary products, some not found in other groups (Huneck et al. 1984). Many are synthesised by the mycobiont and are often deposited as crystals on the surfaces of hyphae and phycobiont cells (Honegger 1986). Their role (if any) in photoprotection is unclear. For instance, although Solhaug and Gauslaa (1996) confirmed the photoprotective role of parietin in *Xanthoria parietina*, this was not the case in two other lichens examined. Swanson and Fahselt (1997) and Swanson et al. (1996) found a negative correlation between exposure to UV-B and thallus contents of lichen phenolics.

PHOTOSYNTHETIC AND PHOTOPROTECTIVE PIGMENTS IN ANTARCTIC PLANTS

In Antarctica, even with relatively low solar zenith angles, irradiance levels can be both very high and variable, being strongly influenced by meteorological conditions and the high surface albedo of surrounding snow and ice. Several Antarctic plants have photosynthetic pigment characteristics more common to high light plants. For example in three Antarctic moss species the photoprotective xanthophyll cycle and ß carotene pigments comprised between 32-42% of the total carotenoid pool, which is comparable to species grown under high light conditions (Demmig-Adams and Adams 1992, Dunn 2000, Robinson et al. 2001, Lovelock and Robinson 2002). In addition, 15-80% of the xanthophyll cycle pigments remain in their photoprotective forms even after periods under low light conditions. This suggests the use of an 'insurance' strategy, ie that plants experience frequent periods of high light under temperature conditions that preclude use of this energy for photosynthesis, hence requiring the pigments to absorb excess energy (Schlensog et al. 1997, Verhoeven et al. 1998, Dunn 2000, Lovelock and Robinson 2002).

Photoprotective pigments, such as zeaxanthin, are important in protecting certain Antarctic mosses from the negative effects of freeze-thaw events (Lovelock et al. 1995a,b). It has also been suggested that these antioxidant pigments play a role in protection from UV-B exposure, since some Antarctic bryophytes and higher plants produce increased levels of carotenoids in response to high, ambient UV-B radiation (Dunn 2000, Xiong and Day 2001, Newsham et al. 2002, Newsham 2003, Robinson et al. 2003). Some studies of Antarctic mosses further suggest that plants with low levels of UV-B pigments have correspondingly higher levels of photoprotective pigments (Dunn 2000, Robinson et al. 2001, 2003).

Chlorophyll a:b ratios of Antarctic mosses are high (>3.2) and more similar to those found in higher plants growing in high light environments, than those found in temperate and subarctic mosses which exhibit shade characteristics (Balo 1967, Martin 1980, Martin and Churchill 1982, Kershaw and Webber 1986, McCall and Martin 1991, Barsig et al. 1998, Gehrke 1998). However, evidence that chlorophyll bleaching occurs in exposed environments has been found in some species. *Grimmia antarctici* turf growing on exposed ridges and under ambient UV radiation had significantly less chlorophyll than moss in sheltered valleys or under reduced UV-radiation (Robinson et al. 2005). In addition, despite having higher levels of UV-B screening pigments, chlorophyll b content was lowered in *Colobanthus quitensis* seedlings exposed to natural UV-B levels in the field (Ruhland and Day 2001).

DNA-DAMAGE AND PHOTOREACTIVATION

DNA is a vulnerable component of the living cell and is more sensitive to UV-B damage than the photosynthetic process (Lud et al. 2001a,b). Cyclobutane pyrimidine dimer formation occurs when DNA is damaged by UV-B radiation (Buma et al. 1995), leading to the activation of repair mechanisms, including photoreactivation, excision of dimers, recombinational filling of gaps and resynthesis of DNA. Repair processes can be inhibited by low temperatures (Pakker et al. 2000), which might imply that higher damage levels would be expected in Antarctic biota exposed to UV-B radiation at low temperature. The natural growth form of some species, such as the multi-layered thallus of the alga *Prasiola crispa* and the accumulation of dead cells on the thallus surface reported for some lichen species (Crittenden 1998) could be a form of protection against UV-B (both in the context of DNA damage and other biochemical processes) as suggested previously for mat-forming cyanobacteria (Margulis et al. 1976).

There have been very few studies of DNA damage in Antarctic terrestrial biota. Damage levels encountered in the moss *Sanionia uncinata*, alga *Prasiola crispa* and experimental microbial dosimeters are substantially lower than those found in marine organisms (Buma et al. 2001, George et al. 2002). In *P. crispa* damage does not increase linearly with dose as found for unicellular marine organisms (Buma et al. 1995), indicating repair of DNA damage during the day under ambient UV-A and PAR (Lud et al. 2001a) (Fig. 2). Even after exposure to high levels of UV-B, repair overnight is sufficient to prevent accumulation of dimers and low levels of DNA damage do not appear to affect growth of *P. crispa*.

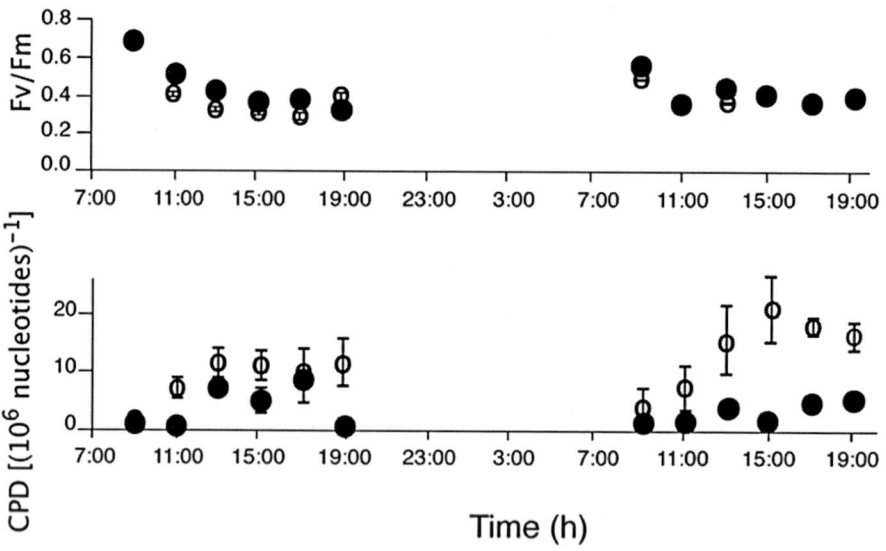

Figure 2. Values of maximum PSII quantum efficiency (Fv/Fm) (a) and cyclobutyl pyrimidine dimer (CPD) frequencies, indicating DNA damage (b) of Prasiola crispa under UV-Mini-lamps with mylar filters (blocking UV-B transmission), respectively, Cellulose Acetate (CA) filters (transparent to UV-B). Filled circles represent samples exposed to ambient UV-B (mylar lamp), open circles represent samples exposed to enhanced UV-B (CA lamp). The symbols show mean and standard deviation. Note the overnight decrease in CPD frequencies, indicating the presence of repair mechanisms (Adapted from Lud et al. 2001b).

CONSEQUENTIAL EFFECTS ON FAUNA AND THROUGH FOOD WEBS

The direct effects of UV-B radiation on plants already described above may have much wider consequential impacts. Changes in the concentrations and/or types of protective pigments can affect the quality of plant or microbial material as a food resource, with implications throughout the food web. In a multi-trophic level and multiple stress field manipulation study completed at Palmer Station, Anvers Island, Day et al. (1999 2001) identified a range of biochemical and morphological changes resulting particularly from exposure to UV-B. Convey et al. (2002) further demonstrated consequential negative impacts of UV-B exposure on the associated arthropod populations, proposing to link these with altered diet quality. Perhaps surprisingly, these studies also found that warming, while increasing growth of plants, had a negative influence on some invertebrate populations, although this may be linked with increased drying stresses at higher temperatures.

Field manipulation experiments focussing on temperature and UV radiation are known to lead to changes in leaf or shoot production, leaf length, or foliar cover, all of which may affect vegetative growth form (Smith 1990, 2001, Day et al. 1999, 2001, Sullivan and Rozema 1999, Ruhland and Day 2000). With the simple, cryptogamic, vegetation characteristic of maritime and continental Antarctic habitats, even subtle changes in growth form will alter habitat structure and microclimate characteristics. While the consequences of such changes have not been studied in the Antarctic, the fundamental importance of addressing climate change consequences across multiple trophic levels is recognised, as it is clear that subtle effects at one level may combine or interact to produce much greater consequences elsewhere in the food web (Day 2001, Searles et al. 2001).

Antarctic plants under changing temperature

PHOTOSYNTHESIS

Antarctic and subantarctic photosynthetic autotrophs maintain photosynthesis at low temperatures, with some being active well below 0°C. For instance, the subantarctic crucifer *Pringlea antiscorbutica*, the grass *Poa cookii* and the Rosaceae *Acaena magellanica* and *A. tenera* sustain photosynthesis throughout the year, even during the coldest days when the temperature remains close to 0°C (Bate and Smith 1983, Smith 1984, Aubert et al. 1999a). Photosynthetic activity has been reported in continental Antarctic lichens below -10°C (Schroeter et al. 1994). However, during summer photosynthetic organs often achieve temperatures well above zero and optimum temperatures for photosynthesis are correspondingly higher, ranging from 10-12°C in *Deschampsia antarctica* from Signy Island and the Antarctic Peninsula (Edwards and Smith 1988, Xiong et al. 1999), 12°C in *Poa cookii* from Marion

Island (Bate and Smith 1983), 15°C in *Pringlea antiscorbutica* at Iles Kerguelen (Aubert et al. 1999a), 19°C in *Colobanthus quitensis* from Signy Island (Edwards and Smith 1988) and up to 20-25°C in a variety of maritime moss species (Rastorfer 1972, Green et al. 1999, 2000, Schroeter et al. 1997, Pannewitz et al. 2003). Some species additionally show plasticity in their optimal temperature. When grown at temperatures above those experienced in the field, the optimal temperature for *Drepanocladus uncinatus* (now known as *Sanionia uncinata*) remained at 15 °C, whilst that for *Polytrichum alpestre* increased from 5-10 °C to 15 °C (Collins 1977).

Prevailing low temperatures throughout the Antarctic biome are generally considered to limit net photosynthesis (P_n) for most of the growing season (Xiong et al. 1999). However photosynthetic organs can reach relatively high temperatures and show much greater fluctuations than seen in diurnal air temperature, eg +20°C above ambient on the continent (Longton 1974, Melick and Seppelt 1994b). In contrast, in the subantarctic plant *Pringlea antiscorbutica* at Iles Kerguelen, leaf temperature is maintained far lower than maximum air temperature (23°C) on warm days, possibly through the intense leaf transpiration observed (Aubert et al. 1999a). Increased P_n under elevated temperature has been demonstrated in three continental moss species (Smith 1999) and the two maritime vascular species (Xiong et al. 2000) whilst in contrast, decreases in P_n were observed in other mosses and vascular plants (Vining et al. 1997, Xiong et al. 1999). In many species, increasing temperatures can reduce carbon gain by increasing respiratory loss (Nakatsubo 2002). Decreases in P_n under increasing temperatures might also be due to consequential increase in photoinhibition as has been shown in some continental Antarctic mosses (Kappen et al. 1989).

The relative importance of temperature and irradiance to photosynthetic rates varies. In the two maritime Antarctic vascular plants net photosynthetic rates are negligible at canopy air temperatures greater than 20 °C, with temperature, rather than high irradiance responsible for this photosynthetic depression (Xiong et al. 1999). However, it has also been demonstrated for these species that increasing vegetative growth outweighs decreases in photosynthetic rates under 20 °C daytime temperatures (Xiong et al. 2000). In subantarctic species such as the crucifer *Pringlea antiscorbutica* and the grass *Poa cookii*, net photosynthesis is very little affected by temperature within a range of naturally occurring values (2.5 – 25°C) and is much more responsive to irradiance (Bate and Smith 1983, Aubert et al. 1999a).

DEVELOPMENTAL AND METABOLIC RESPONSES

While Antarctic plants are, generally speaking, at their biological limits for sexual reproduction, diverse patterns are found. The two antarctic flowering plants, *Colobanthus quitensis* (Kunth) Bartl. and *Deschampsia antarctica* Desv., although flowering every year down to the southern limits of their distribution, do not always achieve mature seed and their reproductive performance appears to be linked with habitat favourability (Edwards 1974). In contrast, several autochthonous subantarctic plants, including endemic species from the Kerguelen phytogeographical province, show a regular and relatively high production of viable seeds and share rapid development and early seed ripening (Dorne 1977, Hennion and Walton 1997a,b, Chapuis et al. 2000). Other autochthonous, non-endemic species encounter limitations in seed fertility due to embryo immaturity or dormancy, which are also reported in their more northerly locations but are more severe in subantarctic sites and reproductive capacity of these species may be influenced by site favourability (Walton 1976, Dorne 1977, Hennion and Walton 1997a,b). In Macquarie Island, with a milder climate, diverse patterns of phenology and seed fertility are found (Bergstrom et al. 1997). Finally, while the majority of species alien to the subantarctic do not achieve successful sexual reproduction, a minority show high production of viable seeds and have become invasive (Frenot et al. 2005).

Autochthonous subantarctic plants and the two antarctic phanerogams have high temperature optima for germination (c. 20°C) as is also observed in arctic and alpine species, a feature possibly related to high soil surface temperatures during the summer (Holtom and Greene 1967, Callaghan and Lewis 1971, Dorne 1977, Hennion and Walton 1997a,b). Meanwhile, rates of germination at low temperatures are greatly improved by cold pretreatment (Holtom and Greene 1967, Dorne 1977, Walton 1977, Frenot and Gloaguen 1994, Hennion and Walton 1997a), with long-term cold storage of soil cores from the west Antarctic Peninsula promoting increased germination rates from seed banks (Ruhland and Day 2001). This suggests that enhanced temperature under climate change may, firstly, increase the germination rate of autochthonous subantarctic and Antarctic species. However, changes in temperature and water availability are also expected to affect seed output and fertility and seedling survival, making the combined consequences of climate change on the colonisation success of individual species hard to predict. Along the Antarctic Peninsula, increases in both the size and number of populations of *D. antarctica* and *C. quitensis* over the last 20 years and the frequency of successful seed maturation suggest that longer, warmer and wetter growing seasons have a positive effect on their reproductive capacity (Convey 1996a, Day et al. 1999). Thus, evaluating the impacts of climate variables on the reproductive success of autochthonous and alien phanerogams should be one of the key targets for plant ecophysiologists in the subantarctic and Antarctic.

Amongst autochthonous species, the crucifer *Pringlea antiscorbutica*, a highly successful plant of the Iles Kerguelen and Iles Crozet, Marion and Heard Islands, displays several features indicating specific adaptation to subantarctic climate. On Iles Kerguelen, *P. antiscorbutica* is found from the shore to the highest vegetated areas, and is tolerant to chilling, frequent freeze-thaw cycles and some salt exposure (Hennion and Bouchereau 1998). In contrast, all attempts to cultivate the Kerguelen cabbage under temperate conditions have failed, resulting in leaf necrosis, accelerated abscission and death after a few weeks (Dorne 1981, Hennion and Martin-Tanguy 2000). The sensitivity of seedling, especially root, development to high temperature appears as soon as germination is completed (Dufeu et al. 2003, Hummel et al. 2004a) (Fig. 3). Polyamines, growth regulators involved in stress response and reported at high levels in some cold tolerant plants, are abundant in *Pringlea antiscorbutica* and also in *Lyallia kerguelensis*, a species endemic to Iles Kerguelen (Hennion and Martin-Tanguy 1999, 2000). The level of agmatine is positively correlated with the root growth rate of *P. antiscorbutica* seedlings at low temperature (Hummel et al. 2002). The balance of polyamine metabolism is dramatically affected by heat stress in both seedlings and mature plants, with for instance accumulations of putrescine or spermidine possibly due to depressed catabolism (Hennion and Martin-Tanguy 2000, Dufeu et al. 2003, Hummel et al. 2004a). Additionally, the aromatic amines dopamine and tyramine are present at high levels in this species and are strongly responsive to heat stress (Hennion and Martin-Tanguy 2000, Dufeu et al. 2003). These developmental and metabolic characteristics indicate a particular sensitivity of *P. antiscorbutica* to enhanced temperatures at all stages and suggest that this and other autochthonous subantarctic species may not have the physiological flexibility to cope with large increases in temperature. Little is known about the metabolic responses of antarctic plants to increased temperature.

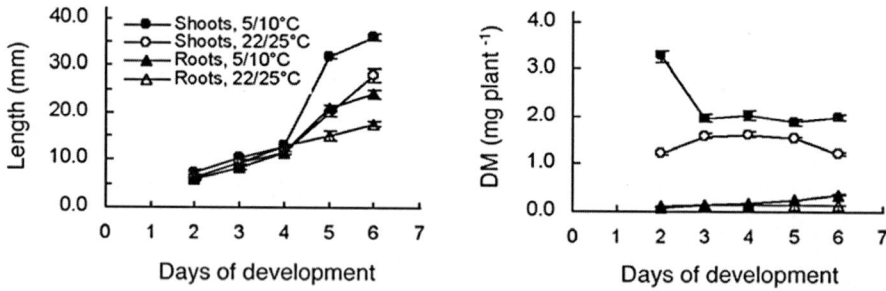

Figure 3. Changes in growth parameters (length and dry mass) during seedling development of Kerguelen cabbage, Pringlea antiscorbutica *R. Br., under a 5°C/10°C night/day (closed symbols) or a 22°C/25°C night/day (open symbols) regime. Day 0 is the time of radicle emergence, post-germination lasts until day 2 (Adapted from Dufeu et al. 2003).*

Some other biochemical features of Antarctic plants are also significant. The subantarctic *P. antiscorbutica* and maritime Antarctic *Deschampsia antarctica* display high levels of soluble carbohydrates, likely to be important in cryoprotection in addition to tolerance of desiccation and freeze-thaw events (Hennion and Bouchereau 1998, Aubert et al. 1999a, Zuñiga et al. 1996). The two maritime Antarctic vascular plants show different cold tolerance strategies. *Deschampsia antarctica* is freezing tolerant, with a low LT_{50} temperature, whereas *Colobanthus quitensis* is freeze avoiding (Alberdi et al. 2002). Reyes et al. (2003) suggested that heat shock protein (HSP70) accumulation may protect *D. antarctica* against temperatures lower or higher than its optimal photosynthetic temperature, a role reported for these proteins in several other species. Neither *D. antarctica* nor *P. antiscorbutica* has a lipid composition characteristic of cold resistant plants either in the content of polar lipids or the degree of unsaturation of fatty acids, a finding possibly related to the temperatures of leaves often being higher than air temperature in the sub- and maritime Antarctic (Dorne et al. 1987, Zuñiga et al. 1994).

Antarctic plants under changing water availability

WATER

On the Antarctic continent most water is permanently locked up as ice and snow, while large areas are accurately described as frigid desert, receiving very low or no direct precipitation and experiencing chronically low relative humidity. Organisms living here must therefore be able to survive long periods of freezing and consequent drought, year round. In the summer, water may become available from snowmelt and around melt lakes, however, this water supply is transient and repeated freeze-thaw events still occur. In the maritime and subantarctic, water is less limiting, although the maritime region experiences extended periods of freezing during winter and rainfall plays a more important role. Throughout the region water availability thus varies spatially and temporally, with large variation occurring from year to year. At a broad scale, levels of tolerance of desiccation therefore vary across the Antarctic biome and among species, with the maritime region supporting some desiccation-sensitive species, particularly in hydric habitats (Davey 1997a,b,c, Robinson et al. 2000, Lange and Kappen 1972).

On the continent, all water comes from melted snow. Uptake of water by lichens is largely from snow deposited on their surfaces which, even at subzero temperatures, can be adequate for rehydration (Kappen and Breuer 1991, Kappen 1993, 2000, Schroeter et al. 1994, 1997, Schroeter and Scheidegger 1995, Pannewitz et al. 2003). These lichens show extraordinarily high levels of tolerance of desiccation and are capable of reactivating photosynthetic activity very rapidly via uptake of water vapour (Lange and Kappen 1972, Hovenden et al. 1994, Hovenden and Seppelt 1995). Although lichens can receive sufficient light to photosynthesise under several centimetres of snow (Lange and Kappen 1972, Walton 1982), the insulating effects of snow cover may normally maintain temperatures below the minimum for photosynthesis (Pannewitz et al. 2003).

Continental Antarctic bryophytes, whilst less tolerant of increasing aridity than lichens, also have the ability to survive desiccation. Species-specific differences in tolerance of desiccation have been detected for three moss species from the

Windmill Islands, East Antarctica (Robinson et al. 2000, Wasley et al. 2006b), with two cosmopolitan species (*Bryum pseudotriquetrum, Ceratodon purpureus*) able to metabolise at lower turf water content than the endemic *Grimmia antarctici* ≈ *Shistidium antarctici*. *Bryum pseudotriquetrum* also shows greater plasticity than the other species, with plants from drier sites showing greater tolerance of desiccation that those from wetter sites, in addition to seasonal changes in desiccation tolerance (Robinson et al. 2000, Wasley et al. 2006b). The ability to survive repeated desiccation and freezing events is probably related to the high concentrations of soluble carbohydrates found in these species (Melick and Seppelt 1994b) and in particular the presence of compounds such as stachyose and trehalose in *B. pseudotriquetrum* (Robinson et al. 2000, Wasley et al. 2006b). In contrast *C. purpureus* has a much higher proportion of fatty acids/soluble carbohydrates. All three species have a high proportion of their fatty acids in polyunsaturated forms (> 67%), a feature characteristic of cold tolerant higher plants (Zuñiga et al. 1996, Wasley et al. 2006b). Quantifying the relative importance of lipids versus soluble carbohydrates in these freeze tolerant plants stands out as an interesting target for further study, it may be that lipids are a safer storage compound, since soluble carbohydrates are known to leak from bryophytes during desiccation-rehydration and freeze thaw cycles (Melick and Seppelt 1992).

In general, the evidence suggests that net photosynthesis and growth are currently limited by water availability (Fowbert 1996, Davey 1997a,b,c, Schlensog and Schroeter 2000). However this may not be the case for lichens, where high water content can cause a depression of net photosynthesis (Kappen and Breuer 1991, Hovenden et al. 1994, Pannewitz et al. 2003). Antarctic bryophytes sampled from wetter sites have higher water contents at full hydration (Robinson et al. 2000), chlorophyll concentrations (Kappen et al. 1989, Melick and Seppelt 1994a), concentrations of soluble carbohydrates (Melick and Seppelt 1994a, Robinson et al. 2000), nitrogen and potassium (Fabiszewski and Wojtun 2000), turf CO_2 concentrations (Tarnawski et al. 1992), rates of nitrogen fixation (Davey 1982, Davey and Marchant 1983) and production rates, in addition to a wider temperature range for maximal net photosynthesis (Kappen et al. 1989). On the negative side, photosynthetic efficiency decreases at higher tissue water contents (Robinson et al. 2000) and tissues freeze at higher temperatures (Melick and Seppelt 1994a) in samples collected from wet sites compared to those from dry sites. For lichens, increased water availability may decrease photosynthesis and appears to have a destabilizing effect on the symbiotic relationship, with the free-living algal and intermediate forms becoming more dominant in wet habitats (Huiskes et al. 1997).

Water is less likely to be limiting in the relatively moist maritime Antarctic. On Signy Island, whilst some xeric species are occasionally water-limited (Davey 1997c), more generally photosynthesis is not water-limited (Collins 1977). When the photosynthetic rates of a range of xeric and hydric species from this island were compared under laboratory conditions, no difference among habitats was detected

(Convey 1994). Maritime moss species from a variety of habitats (hydric, mesic, xeric) also experience increased penetration of light into the turf as drying occurs, counteracting at least in the short term the loss of productivity during periods of desiccation (Davey and Rothery 1996).

Subantarctic climates are generally described as having a strong maritime influence, with high levels of precipitation. Over recent decades, however, climate change data are becoming available from this region that, in some locations, indicate a dramatic reduction in precipitation (Bergstrom and Chown 1999). Newly experienced drought periods visibly affect autochthonous subantarctic species. Among these, *Pringlea antiscorbutica* displays several features indicative of sensitivity to water stress, such as permanently high leaf relative water content despite a low leaf diffusion resistance and stomatal closure only under severe water deprivation, all facts in accordance with this species growing exclusively in places with water-saturated soils (Dorne and Bligny 1993). Mature *P. antiscorbutica* plants show some salt tolerance with characteristics such as high levels of proline in all organs and proline accumulation in the cell cytoplasm in response to salt concentration in the surrounding medium (Hennion and Bouchereau 1998, Aubert et al. 1999a,b). In contrast, in field microenvironments as under laboratory experiments, water and salt stresses were shown to result in a drastic reduction of root growth and high mortality of seedlings, suggesting that seedling establishment of *P. antiscorbutica* may be restricted by climate change (Hummel et al. 2004b).

Water availability has been shown to influence turf and gametophyte morphology in a range of continental and maritime Antarctic mosses and this, in turn, can affect water relations (Nakanishi 1979). In general, gametophyte shoots are shorter and turf denser in drier sites (Gimingham and Smith 1971, Wilson 1990, Wasley 2004, Wasley et al. 2006b). Indeed, the changes in plant morphology and growth patterns that are reported as the norm in many long-term environmental manipulation experiments, often implicitly assumed to relate primarily to temperature increase, may equally well be explained by changes in microclimate humidity and soil moisture. Some subantarctic higher plants, such as the magellanic *Ranunculus biternatus* and *R. pseudotrullifolius* and the Kerguelen endemic *R. moseleyi*, three closely related species (Hennion and Couderc 1992, 1993), show high variability of leaf morphology in relation with environmental water availability (Hennion et al. 1994). *R. pseudotrullifolius* and *R. moseleyi* were characterized by developmental flexibility which increased the degree of heterophylly in fluctuating conditions, which could be considered adaptive to the wide variations of water level in their aquatic habitats. In contrast, *R. biternatus* had a much wider variability in leaf shape, part of which could be genetically fixed, and which could be related with the more varied, terrestrial habitats of this species (Hennion et al. 1994). As a whole, data suggest that *R. biternatus* may be better fitted to cope with present climate drying at Iles Kerguelen than the more biogeographically and ecologically restricted *R. pseudotrullifolius* and *R. moseleyi*.

In the subantarctic, altitudinal factors strongly influence the frequency of freeze-thaw events experienced by biota. For example, *P. antiscorbutica* grows at a range of altitudes and can experience frequent freeze-thaw cycles (Hennion and Bouchereau 1998). High levels of glucose present in the leaves of this species may help in protection against osmotic stresses (Hennion and Bouchereau 1998), such as that experienced in association with ice formation. In general, continental Antarctic plants can survive repeated freeze-thaw events (Melick and Seppelt 1992), whilst maritime species appear to be less tolerant (Davey 1997b). The pattern of exposure to freezing is also important - repeated freeze-thaw cycles cause a greater reduction in gross photosynthesis than constant freezing over the same time period (Kennedy 1993). Tolerance of freeze-thaw events involves interactions with other environmental parameters, in particular that of water availability. For example, desiccation before freezing reduces damage to the photosynthetic apparatus, while protection from freeze-thaw events can be provided by snow cover acting as an insulator (Lovelock et al. 1995a,b). Ice nucleation activity, measured on 19 species of lichens, mosses and flowering plants, ranked in the order: lichens > mosses > flowering plants (Worland et al. 1996). In the search for genes associated with freezing tolerance, cold-acclimation related transcripts were isolated from *Deschampsia antarctica* and appeared also responsive to water stress (Gidekel et al. 2003).

NUTRIENTS

Nutrient cycling in the Antarctic is relatively slow, due to the constraints imposed on biological activity by low temperatures and extreme aridity. Nutrient availability in Antarctic terrestrial ecosystems is patchy, with high concentration of nutrients in the vicinity of bird and seal colonies, whilst elsewhere nutrients are limited to those deposited in precipitation and through aerosol transfer from the sea (Allen et al. 1967, Greenfield 1992a,b). Nutrient requirements for Antarctic vegetation have been reported to be so low that nitrogen levels in precipitation were assumed to be non-limiting for growth of cryptogams, particularly lichens (Greenfield 1992a, Crittenden 1998). However, nutrient availability does play a role in determining patterns of species distributions in Antarctica. Two studies, in particular, have demonstrated positive correlations between vegetation patterns and nutrient availability associated with nutrient inputs from birds (Gremmen et al. 1994, Leishman and Wild 2001). A recent field manipulation study has shown that, in both moss and lichen communities, electron transport rate and chlorophyll content respond positively to nutrient inputs (Wasley et al. 2006a), suggesting that nutrient limitations may be more important than previously thought (Beyer et al. 2000). This finding is particularly relevant in the context of future climate scenarios, since plants may be unable to respond to changes in thermal climate or hydration if nutrients become limiting.

In the continental and maritime Antarctic, release of nutrients from organic matter is primarily mediated through microbial processes and is relatively slow (Smith and Steenkamp 1992). In contrast, in the subantarctic, rates of inorganic nutrient release from plant litter are enhanced by macro-invertebrate communities (Smith and Steenkamp 1992) which are absent on the continent. Smith (2005) used a modelling approach to examine decomposition-related phenomena in the subantarctic across a broad range of soils from Marion Island. Respiration rate was shown to increase with soil moisture content and the degree of this response strongly increased with temperature, especially above 10°C. The model predicted that the currently experienced drying of soils, due to climate drying, on Marion would result in a decrease in soil respiration rate, despite a slight positive effect of warming. In addition, the consequences of nutrient input from vertebrates were shown to include improved primary production, larger, more active and more diverse soil microbial populations and larger populations of microbivores (Smith 2005).

Increased nutrient availability is not automatically an advantage to Antarctic biota. In the extreme, this is obvious in the context of such events as excessive manuring by vertebrates such as Antarctic fur seals causing the death of moss banks (Smith 1988), an effect that is separate to but generally acts in concert with physical trampling. Some Antarctic plants, such as *Pringlea antiscorbutica* seedlings, display strikingly better growth under nutrient deprivation than addition (Hummel et al. 2004a), a feature which may apply more generally than currently thought in species characteristic of oligotrophic habitats. Recent studies have demonstrated mycorrhizae to be associated with several subantarctic plant species, with variable plant mycorrhizal status among sites, some data suggested that mycorrhizae might be involved in plant colonization in oligotrophic sites (Strullu et al. 1999). The sensitivity of this symbiotic association to climate change processes is unknown. Finally, as climate change is also likely to alter patterns of rock weathering and release of elements, impacts of changes in nutrient supply are to be expected on the physiological performances of plants.

Invertebrates under changes in temperature and water patterns

A range of ecophysiological studies support there being a very close relationship, both over short-term and evolutionary timescales, between tolerance of desiccation and cold (Ring and Danks 1994, Block 1996, Worland 1996, Worland et al. 1998). Ecophysiological and biochemical features associated with cold tolerance have received much attention (see Zachariassen 1985, Block 1990, Duman et al. 1991, Lee and Denlinger 1991, Wharton 1995, Danks 1996, Sinclair et al. 2003a,b, while Cannon and Block 1988, Danks et al. 1994 and Convey 1996b apply to polar studies), with cold tolerance strategies now being grouped into ecologically-useful classifications (Bale 1993, Sinclair 1999).

Freezing tolerant organisms (those that can survive the formation of ice within their body, normally in extracellular spaces) can 'seed' ice formation at relatively high sub-zero temperatures by using ice nucleators such as specific proteins or bacteria. Control of the process of ice formation, which is more easily achievable at higher sub-zero temperatures, appears to be key to organism survival here. As temperature decreases further after initial ice formation there is much variation among taxa in the degree of protection provided, with some only surviving a few degrees below the freezing point while others can then tolerate biologically unrealistic extreme low temperatures. Freezing susceptible organisms (that die when ice forms within their body) can lower their freezing point and stabilise the supercooled state by the use of a range of cryoprotectant chemicals, including antifreezes, polyhydric alcohols and sugars, and antifreeze proteins.

Most Antarctic arthropods are freezing susceptible and have been key in the study of supercooling (Fig. 4). Freezing tolerant species of higher insect are also present, including Diptera, Coleoptera and Lepidoptera. Application of the various cold tolerance classification systems starts to break down when applied to the groups of smaller soil invertebrates that progressively assume dominance in more extreme Antarctic terrestrial habitats, including tardigrades, nematodes and rotifers. When these experience freezing immersed in water (such as would coat soil particles or vegetation) at high sub-zero temperatures, they cannot resist ice nucleation from their surroundings and can show considerable levels of freeze tolerance (Wharton 1995, Wharton and Block 1993, 1997, Convey and Worland 2000a). A similar process occurs in 'frost hardy' plants. However, when the same species are exposed to classical experimental supercooling methodologies, they can respond as freezing susceptible species, showing considerable supercooling ability but dying at the freezing point (Convey and Worland 2000b). Finally, many representatives of these groups, which typically have very little resistance to water loss, have well-developed ability to survive desiccation, showing particular development of the process of anhydrobiosis (Pickup and Rothery 1991, Wharton 1995).

The parallels between processes allowing tolerance of freezing and desiccation are well illustrated by consideration of events at the cellular level. Both involve the protection of cell membranes and organelles. The formation of ice crystals in extracellular spaces draws water out of cells by osmosis, in an analogous process to that experienced during desiccation. In both cases, cryoprotectants such as trehalose allow stabilisation of membranes and soluble proteins (Ring and Danks 1994, Block 1996). Some nematodes appear to have taken this ability one step further, and are able to survive intracellular freezing (Wharton 1995, Wharton and Ferns 1995), an ability previously thought to be restricted to certain isolated invertebrate tissues and not whole organisms. A frozen organism has no ability to actively control further water loss (Worland 1996, Worland et al. 1998). If in contact with air, water loss will inevitably continue to take place, ultimately resulting in freeze-drying. In addition to the invertebrate groups mentioned, many elements of the flora of

Antarctic ecosystems are well adapted to tolerate such desiccation processes, often repeatedly over relatively short timescales, being poikilohydrous (Green et al. 1999, Kappen and Valladares 1999).

All the ecophysiological tactics of permitting cold or desiccation tolerance referred to above involve significant physiological costs, which must reduce the finite reserves of an organism that can be made available for other activities. While there are few estimates available, it is clear that costs may be considerable (eg Green et al. 1999). Circumstantial evidence in support of a 'trade-off' between investment in stress tolerance and another ecologically important function (reproduction) is provided by a study of reproductive investment of an antarctic Oribatid mite across the natural environmental gradient between South Georgia and Marguerite Bay (c. 54 - 69°S) (Convey 1998). This found significantly greater investment in reproduction in the milder subantarctic location. Given the considerable costs involved in stress protection tactics, there is clearly likely to be a selective advantage associated with their efficient use. Thus, while variation over seasonal timescales in abilities such as cold tolerance is well known (Cannon and Block 1988, Block 1990), it is only recently that attention has been given to more rapid dynamic or responsive changes in levels of stress protection which might allow invertebrates to take maximum advantage of the opportunities provided for feeding and growth during the short austral summer season (Convey and Worland 2000b, Worland and Convey 2001, Sinclair et al. 2003c). Alternatively, in the subantarctic where thermal buffering of summer microclimate is experienced through the strong maritime influence (Convey 1996c, Danks 1999), there is reduced need to invest resources in cold survival. In the most extreme terrestrial habitats of the continental Antarctic the reverse is seen, with high levels of cold tolerance required year-round (Sømme 1986), as there is no time of the year in which freezing events are not experienced.

Antarctic terrestrial invertebrates clearly possess the features required to allow survival of the environmental stresses experienced, although few of these can be claimed to be 'true' or evolutionary adaptations and they are generally considered to be ancestral features (Norton 1994, Convey 1996b, 1997). However, it is clear that 'whole–body processes' such as growth, feeding, movement and survival, which effectively integrate the many biochemical and physiological processes occurring within the body, often have low temperature optima, which indicates the presence of adaptation even though the detail is not known (eg Marshall et al. 1995, Convey 1996b). There continues to be considerable debate over the existence or functional importance of metabolic rate elevation at low temperatures (Block 1990, Clarke 1991, 1993, Chown and Gaston 1999, Addo-Bediako et al. 2002). Although this feature has been reported in several studies of terrestrial arthropods, along with the related feature of lowered enzyme activation energies, there remain significant doubts over methodological validity or data interpretation.

Antarctic terrestrial habitats experience high natural variability in environmental stresses, with the magnitude of these variations generally far outweighing those that are linked with climate change. Therefore, as a broad generalisation, it is predicted that Antarctic terrestrial invertebrates, and their communities, are likely to benefit from current change trends (Convey 2003). They are likely to experience reduced levels of environmental stress (and hence levels of investment required to counter these stresses), greater access to resources and longer active seasons over which to develop. Indeed, it is expected that Antarctic (and Arctic) species generally will show particular sensitivity and strength of response, at least to warming: as they currently spend long periods operating near to lower thermal and other limits for activity, small increases will have a relatively much greater significance than similar increases from a higher baseline (Kennedy 1994, Strathdee et al. 1995, Freckman and Virginia 1997, Arnold and Convey 1998, Convey 2003).

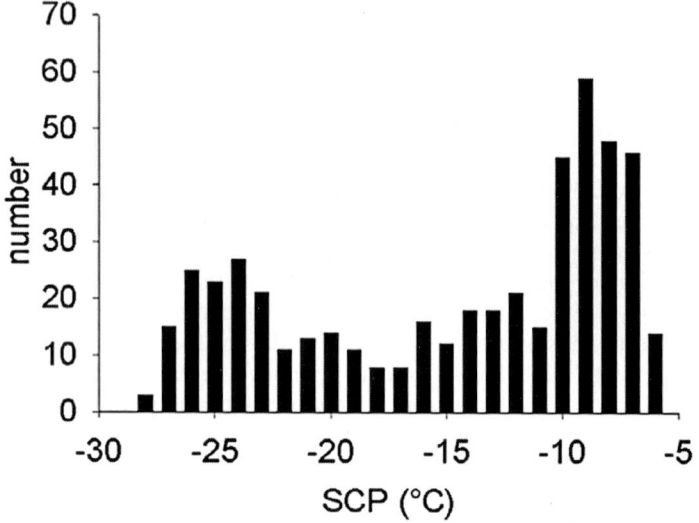

Figure 4. Typical bimodal distribution of supercooling (individual freezing) points in a natural population of the freezing-susceptible springtail, Cryptopygus antarcticus, *these two groups are formed by individuals either having relatively high sub-zero freezing points, and limited cold tolerance ability, or relatively low sub-zero freezing points and very high cold tolerance (data supplied by M.R. Worland).*

However, many interacting variables influence the biology of these invertebrates and it is clear that some circumstances can restrict the activity and distribution of some biota. Thus, the key role of water availability has been recognised (Kennedy 1993) and it is clear that some regions of the Antarctic will experience decreased precipitation and increased drought stress. Some terrestrial invertebrates, such as Collembola, while responding positively to temperature increase, are also particularly sensitive indicators of water stress and changes are likely to result in altered limits to distribution (Block and Harrisson 1995, Block and Convey 2001, Convey et al. 2003a).

In particular circumstances, it is also possible that species distributions could be limited as maximum microhabitat temperatures approach upper thermal limits. Unlike the ability shown by many invertebrates (not only polar species) to acclimate or acclimatise to low temperatures, upper thermal limits appear to show less flexibility (Addo-Bediako et al. 2000) and, thus, any increase in time spent near these temperatures may be costly. The potential importance of upper thermal limits in biological responses to climate change has been raised in a study of subantarctic weevils (Coleoptera) (van der Merwe et al. 1997).

Conclusions

Both observation and knowledge of the physiological characteristics of some antarctic organisms suggest that even the slight increase in temperature currently recorded in this region can significantly affect the organisms living there. This is also true for the expected or already observed changes in water availability. Physiological flexibility is a key factor in the ability of biota to cope with environmental change, whether caused through natural variability or anthropogenic change processes. Apparently small changes in processes (eg the limited photosynthetic response to a slight increase in mean temperature) may assume much greater significance in the context of pre-existing limitations on physiological activity in Antarctic habitats. Other metabolic processes having impacts on development, growth or reproduction may be much more sensitive to changes in environmental parameters, with significant consequences on survival, establishment and colonisation of both indigenous and introduced species. Few data are available in these areas and there is an urgent need to integrate specific studies of physiological processes at the whole organism and food web levels. Studies of physiological plasticity, linked with estimates of available genetic variability are also urgently required to assess accurately the impacts of climate change on the sensitive Antarctic terrestrial ecosystem.

As a broad generalization, maritime Antarctic biota are likely to be able to take advantage of predicted levels of environmental change, although the alternative possibility of increased limitation in specific circumstances must be recognised too. In contrast, current evidence suggests that subantarctic biota may be considerably more vulnerable. The relative importance of changes in different environmental variables (particularly thermal vs. precipitation) still needs to be assessed. At present, the great isolation of Antarctica from sources of colonizing biota acts to protect most indigenous species from the confounding effects of competition. However, the dangers of invasive species are already amply illustrated on some of the subantarctic islands, where human activity has led to the establishment of a range of vertebrates, invertebrates and plants (Frenot et al. 2005). Indigenous species are generally poor competitors (Convey 1996b) and it is far from clear how they will respond to the twin challenges presented by both climate change and alien colonization.

Finally, it is important to emphasise that studies of climate change responses need to incorporate not only individual target species, but also the possibility of interaction or synergistic links among environmental variables, and the cumulative consequences, when integrated through the food web, of apparently insignificant changes (Day 2001, Searles et al. 2001, Convey et al. 2002, Convey 2003).

References

Adamson, H. and Adamson, E. (1992) Possible effects of climate change on Antarctic terrestrial vegetation, in P. Quilty (ed.), *Impact of Climate Change on Antarctica-Australia*, Australian Government Publishing Service, Canberra, Australia, pp. 52-61.

Addo-Bediako, A., Chown, S.L. and Gaston, K.J. (2000) Thermal tolerance, climatic variability and latitude, *Proceedings of the Royal Society of London, series B* **267**, 739-745.

Addo-Bediako, A., Chown, S.L. and Gaston, K.J. (2002) Metabolic cold adaptation in insects: a large scale perspective, *Functional Ecology* **16**, 332-338.

Alberdi, M., Bravo, L.A., Gutiérrez, A., Gidekel, M. and Corcuera, L.J. (2002) Ecophysiology of Antarctic vascular plants, *Physiologia Plantarum* **115**, 479-486.

Allen S.E., Grimshaw H.M. and Holdgate M.W. (1967) Factors affecting the availability of plant nutrients on an antarctic island, *Journal of Ecology* **55**, 381-396.

Arnold, R.J. and Convey, P. (1998) The life history of the diving beetle, *Lancetes angusticollis* (Curtis) (Coleoptera: Dytiscidae), on subantarctic South Georgia, *Polar Biology* **20**, 153-160.

Aubert, S. Assard, N., Boutin, J.P., Frenot, Y. and Dorne, A.J. (1999a) Carbon metabolism in the subantarctic Kerguelen cabbage *Pringlea antiscorbutica* R. Br.: environmental controls over carbohydrates contents and relation to phenology, *Plant, Cell and Environment* **22**, 243-254.

Aubert, S., Hennion, F., Bouchereau, A., Gout, E., Bligny, R. and Dorne, A.J. (1999b) Subcellular compartmentation of proline in the leaves of the subantarctic Kerguelen cabbage *Pringlea antiscorbutica* R. Br. *In vivo* 13C-NMR study, *Plant, Cell and Environment* **22**, 255-260.

Bale, J.S. (1993) Classes of insect cold hardiness, *Functional Ecology* **7**, 751-753.

Balo, K. (1967) Some photosynthesis-ecological characteristics of forest bryophytes, *Symposia Biologica Hungarica* **35**, 125-135.

Barsig, M., Schneider, K. and Gehrke C. (1998) Effects of UV-B radiation on fine-structure, carbohydrates, and pigments in *Polytrichum commune*, *Bryologist* **101**, 357-365.

Bate, G.C. and Smith, V.R. (1983) Photosynthesis and respiration in the subantarctic tussock grass *Poa cookii*, *New Phytologist* **95**, 533-543.

Beggs, C.J. and Wellmann, E. (1994) Photocontrol of flavonoid biosynthesis, in R.E. Kendrick and G.H.M. Kronenberg (eds.), *Photomorphogenesis in Plants*, Kluwer Academic Publishers, Dordrecht, The Netherlands, pp 733-751.

Bergstrom, D.M., Selkirk, P.M., Keenan, H.M. and Wilson, M.E. (1997) Reproductive behaviour of ten flowering plant species on subantarctic Macquarie Island, *Opera Botanica* **132**: 109-132.

Bergstrom, D.M. and Chown, S.L. (1999) Life at the front: history, ecology and change on southern ocean islands, *Trends in Ecology and Evolution* **14**, 472-477.

Beyer, L., Bolter, M. and Seppelt, R.D. (2000) Nutrient and thermal regime, microbial biomass, and vegetation of Antarctic soils in the Windmill Islands region of east Antarctica (Wilkes Land), *Arctic, Antarctic and Alpine Research* **32**, 30-39.

Block, W. (1990) Cold tolerance of insects and other arthropods, *Philosophical Transactions of the Royal Society of London Series B* **326**, 613-633.

Block, W. (1996) Cold or drought - the lesser of two evils for terrestrial arthropods? *European Journal of Entomology* **93**, 325-339.

Block, W. and Convey, P. (2001) Seasonal and long-term variation in body water content of an Antarctic springtail - a response to climate change? *Polar Biology* **24**, 764-770.

Block, W. and Harrisson, P.M. (1995) Collembolan water relations and environmental change in the maritime Antarctic, *Global Change Biology* **1**, 347-359.

Bornman, J.F. and Sundby-Emanuelsson, C. (1995) Response of plants to UV-B radiation: some biochemical and physiological effects, in N. Smirnoff (ed.), *Environment and plant metabolism*, Bios Scientific Publishers, Oxford, U.K., pp. 245-262.

Buma, A.J.G., van Hannen, E.J., Roza, L., Veldhuis, M.J.W. and Gieskes, W.W.C. (1995) Monitoring Ultraviolet-B induced damage in individual diatom cells by immunofluorescent thymine dimer detection, *Journal of Phycology* **31**, 314-321.

Buma, A.G.J., De Boer, M.K. and Boelen, P. (2001) Depth distributions of DNA damage in Antarctic marine phyto- and bacterioplankton exposed to summertime ultraviolet radiation. *Journal of Phycology* **37**, 200-208.

Callaghan, T.V. and Lewis, M.C. (1971) Adaptation in the reproductive performance of *Phleum alpinum* L. at a subantarctic station, *British Antarctic Survey Bulletin* **26**, 59-75.

Cannon, R.J.C. and Block, W. (1988) Cold tolerance of microarthropods, *Biological Reviews* **63**, 23-77.

Chapuis, J.L., Hennion, F., Le Roux, V. and Le Cuziat, J. (2000) Growth and reproduction of the endemic cruciferous species *Pringlea antiscorbutica* in Kerguelen Islands, *Polar Biology* **23**, 196-204.

Chown, S.L. and Gaston, K.J. (1999) Exploring links between physiology and ecology at macro-scales: the role of respiratory metabolism in insects, *Biological Reviews* **74**, 87-120.

Clarke, A. (1991) What is cold adaptation and how should we measure it? *American Zoologist* **31**, 81-92.

Clarke, A. (1993) Seasonal acclimatisation and latitudinal compensation in metabolism: do they exist? *Functional Ecology* **7**, 139-149.

Cockell, C.S., Rettberg, P., Horneck, G., Wynn-Williams, D.D., Scherer, K. and Gugg-Helminger, A. (2002) Influence of ice and snow covers on the UV exposure of terrestrial microbial communities: dosimetric studies, *Journal of Photochemistry and Photobiology, B-Biology* **68**, 23-32.

Collins, N.J. (1977) The growth of mosses in two contrasting communities in the maritime Antarctic: measurement and prediction of net annual production, in G.A. Llano (ed.), *Adaptations within Antarctic ecosystems*, Gulf Publishing, Houston, Texas, pp. 921-933.

Convey, P. (1994) Photosynthesis and dark respiration in Antarctic mosses - an initial comparative study, *Polar Biology* **14**, 65-69.

Convey, P. (1996a) Reproduction of Antarctic flowering plants, *Antarctic Science* **8**, 127-134.

Convey, P. (1996b) The influence of environmental characteristics on life history attributes of Antarctic terrestrial biota, *Biological Reviews* **71**, 191-225.

Convey, P. (1996c) Overwintering strategies of terrestrial invertebrates from Antarctica - the significance of flexibility in extremely seasonal environments, *European Journal of Entomology* **93**, 489-505.

Convey, P. (1997) How are the life history strategies of Antarctic terrestrial invertebrates influenced by extreme environmental conditions? *Journal of Thermal Biology* **22**, 429-440.

Convey, P. (1998) Latitudinal variation in allocation to reproduction by the Antarctic oribatid mite, *Alaskozetes antarcticus*, *Applied Soil Ecology* **9**, 93-99.

Convey, P. (2001) Terrestrial ecosystem response to climate changes in the Antarctic, in G.-R. Walther, C.A. Burga and P.J. Edwards (eds.), *"Fingerprints" of climate change - adapted behaviour and shifting species ranges*, Kluwer, New York, pp 17-42.

Convey, P. (2003) Maritime Antarctic climate change: signals from terrestrial biology, in E. Domack, A. Burnett, A. Leventer, P. Convey, M. Kirby and R. Bindschadler (eds.), *Antarctic Peninsula Climate Variability: Historical and Palaeoenvironmental Perspectives*, Antarctic Research Series, Vol. 79, American Geophysical Union, Washington, D.C., pp. 145-158.

Convey, P. (2006) Antarctic climate change and its influences on terrestrial ecosystems, in D.M. Bergstrom, P. Convey, and A.H.L. Huiskes (eds.), *Trends in Antarctic Terrestrial and Limnetic Ecosystems: Antarctica as a Global Indicator*. Springer, Dordrecht (this volume).

Convey, P. and Worland, M.R. (2000a) Survival of freezing by free-living Antarctic soil nematodes, *Cryo-Letters* **21**, 327-332.

Convey, P. and Worland, M.R. (2000b) Refining the risk of freezing mortality for Antarctic terrestrial microarthropods, *Cryo-Letters* **21**, 333-338.

Convey, P., Pugh, P. J. A., Jackson, C., Murray, A. W., Ruhland, C. T., Xiong, F. S. and Day, T. A. (2002) Response of Antarctic terrestrial arthropods to multifactorial climate manipulation over a four year period. *Ecology* **83**, 3130-3140.

Convey, P., Block, W. and Peat, H.J. (2003a) Soil arthropods as indicators of water stress in Antarctic terrestrial habitats? *Global Change Biology* **9**, 1718-1730.

Convey, P., Scott, D. and Fraser, W.R. (2003b) Biophysical and habitat changes in response to climate alteration in the Arctic and Antarctic, in T.E. Lovejoy and L. Hannah (eds.), *Climate Change and Biodiversity: Synergistic Impacts, International Conservation Center for Applied Biodiversity Science, Advances in Applied Biodioversity Science* **4**, pp. 79-84.

Crittenden, P.D. (1998) Nutrient exchange in an Antarctic macrolichen during summer snowfall snow melt events, *New Phytologist* **139**, 697-707.

Danks, H.V. (1996) The wider integration of studies on insect cold-hardiness, *European Journal of Entomology* **93**, 383-403.

Danks, H.V. (1999) Life cycles in polar arthropods - flexible or programmed? *European Journal of Entomology* **96**, 83-102.

Danks, H.V., Kukal, O. and Ring, R.A. (1994) Insect cold-hardiness: insights from the Arctic, *Arctic* **47**, 391-404.

Davey, A. (1982) In situ determination of nitrogen fixation in Antarctica using a high sensitivity portable gas chromatograph, *Australian Journal of Ecology* **7**, 395-402.

Davey, M.C. (1997a) Effects of continuous and repeated dehydration on carbon fixation by bryophytes from the maritime Antarctic, *Oecologia* **110**, 25-31.

Davey, M.C. (1997b) Effects of physical factors on photosynthesis by the Antarctic liverwort *Marchantia berteroana*, *Polar Biology* **17**, 219-227.

Davey, M.C. (1997c) Effects of short-term dehydration and rehydration on photosynthesis and respiration by Antarctic bryophytes, *Environmental and Experimental Botany* **37**, 187-198.

Davey, A. and Marchant, H. (1983) Seasonal variation in nitrogen fixation in *Nostoc commune* Vaucher at the Vestfold Hills, Antarctica, *Phycologia* **22**, 377-385.

Davey, M.C. and Rothery, P. (1996) Seasonal variation in respiratory and photosynthetic parameters in three mosses from the maritime Antarctic, *Annals of Botany* **78**, 719-728.

Day, T.A. (2001) Multiple trophic levels in UV-B assessments - completing the ecosystem, *New Phytologist* **152**, 183-186.

Day, T.A., Ruhland, C.T. and Xiong, F. (2001) Influence of solar UV-B radiation on Antarctic terrestrial plants: results from a 4-year field study, *Journal of Photochemistry and Photobiology B: Biology* **62**, 78-87.

Day, T.A., Ruhland, C.T., Grobe, C.W. and Xiong, F. (1999) Growth and reproduction of Antarctic vascular plants in response to warming and UV radiation reductions in the field, *Oecologia* **119**, 24-35.

Demmig-Adams, B. and Adams, W.W. (1992) Carotenoid composition in sun and shade leaves of plants with different life forms, *Plant, Cell and Environment* **15**, 411-419.

Dorne, A.J. (1977) Analysis of the germination under laboratory and field conditions of seeds collected in the Kerguelen Archipelago, in G.A. Llano (ed.), *Adaptations within Antarctic Ecosystems*, 3rd SCAR Biology Symposium, Smithsonian, Washington DC, U.S.A., pp 1003-1013.

Dorne, AJ (1981) Etude du cycle végétatif et des premiers stades de croissance du Chou de Kerguelen, *Pringlea antiscorbutica* R.Br., *Comité National Français pour les Recherches Antarctiques* **48**, 9-22.

Dorne, A.J. and Bligny, R. (1993) Physiological adaptation to subantarctic climate by the Kerguelen cabbage, *Pringlea antiscorbutica* R. Br., *Polar Biology* **13**, 55-60.

Dorne, A.J. Joyard, J. and Douce, R. (1987) Lipid metabolism of a tertiary relic species, Kerguelen cabbage (*Pringlea antiscorbutica*), *Canadian Journal of Botany* **65**, 2368-2372.

Dufeu, M., Martin-Tanguy, J. and Hennion, F. (2003) Temperature-dependent changes of amine levels during early seedling development of the cold-adapted subantarctic crucifer *Pringlea antiscorbutica*, *Physiologia Plantarum* **118**, 164-172.

Duman, J.G., Wu, D.W., Xu, L., Tursman, D. and Olsen, M. (1991) Adaptations of insects to subzero temperatures, *Quarterly Review of Biology* **66**, 387-410.

Dunn, J. (2000) *Seasonal variation in the pigment content of three species of Antarctic bryophytes*, B.Sc.Adv.Honours Thesis, University of Wollongong, Australia.

Edwards, J.A. (1974) Studies in *Colobanthus quitensis* (Kunth) Bartl. and *Deschampsia antarctica* Desv.: VI. Reproductive performance on Signy Island, *British Antarctic Survey Bulletin* **39**, 67-86.

Edwards, J.A. and Smith, R.I.L. (1988) Photosynthesis and respiration of *Colobanthus quitensis* and *Deschampsia antarctica* from the maritime Antarctic, *British Antarctic Survey Bulletin* **81**, 43-63.

Fabiszewski, J. and Wojtun, B. (2000) Chemical composition of some dominating plants in the maritime antarctic tundra (King George Island), *Bibliotheca Lichenologica* **75**, 79-91.

Fowbert, J.A. (1996) An experimental study of growth in relation to morphology and shoot water content in maritime Antarctic mosses, *New Phytologist* **133**, 363-373.

Freckman, D.W. and Virginia, R.A. (1997) Low-diversity Antarctic soil nematode communities: distribution and response to disturbance, *Ecology* **78**, 363-369.

Frenot, Y. and Gloaguen, J.-C. (1994) Reproductive performance of native and alien colonising phanerogams on a glacier foreland, Iles Kerguelen, *Polar Biology* **14**, 473-481.

Frenot, Y., Chown, S.L., Whinam, J., Selkirk, P.M., Convey, P., Skotnicki, M. and Bergstrom, D.M. (2005) Biological invasions in the Antarctic: extent, impacts and implications, *Biological Reviews* **80**, 45-72.

Garcia-Pichel, F. and Castenholz, R.W. (1993) Occurrence of UV-absorbing Mycosporine-like compounds among cyanobacterial isolates and an estimate of their screening capacity, *Applied Environmental Microbiology* **59**, 163-169.

Gehrke, C. (1998) Effects of enhanced UV-B radiation on production-related properties of a *Sphagnum fuscum* dominated subarctic bog, *Functional Ecology* **12**, 940-947.

Gidekel, M., Destefano-Beltran, L., Garcia, P., Mujica, L., Leal, P., Cuba, M., Fuentes, L., Bravo, L.A., Corcuera, L.J., Alberdi, M., Concha, I. and Gutiérrez, A. (2003) Identification and characterization of three novel cold acclimation-responsive genes from the extremophile hair grass *Deschampsia antarctica* Desv., *Extremophiles* **7**, pp. 459-469.

George, A. L., Murray, A. W. and Montiel, P. O. (2001) Tolerance of Antarctic cyanobacterial mats to enhanced UV radiation, *FEMS Microbiology Ecology* **37**, 91-101.

George, A. L., Peat, H. J. and Buma, A. G. J. (2002) Evaluation of DNA dosimetry to assess ozone-mediated variability of biologically harmful ultraviolet radiation in Antarctica, *Photochemistry and Photobiology* **76**, 274-280.

Gimingham, C.H. and Smith, R.I.L. (1971) Growth forms and water relations of mosses in the maritime Antarctic, *British Antarctic Survey Bulletin* **25**, 1-21.

Green, T.A.G., Schroeter, B. and Sancho, L.G. (1999) Plant Life in Antarctica, in F.I. Pugnaire and F. Valladares (eds.), *Handbook of functional plant ecology*, Dekker, New York, U.S.A., pp. 495-543.

Green, T., Schroeter, B. and Seppelt, R. (2000) Effect of temperature, light and ambient UV on the photosynthesis of the moss *Bryum argenteum* Hedw. in continental Antarctica, in W. Davison, C. Howard-Williams and P. Broady (eds.), *Antarctic ecosystems: modes for wider ecological understanding*, New Zealand Natural Sciences, Christchurch, New Zealand, pp. 165-170.

Greenfield, L.G. (1992a) Retention of precipitation nitrogen by Antarctic mosses, lichens and fellfield soils, *Antarctic Science* **4**, 205-206.

Greenfield, L.G. (1992b) Precipitation nitrogen at maritime Signy Island and continental Cape Bird, Antarctica, *Polar Biology* **11**, 649-653.

Gremmen, N.J.M., Huiskes, A.H.L. and Francke, J.W. (1994) Epilithic macrolichen vegetation of the Argentine Islands, Antarctic Peninsula, *Antarctic Science* **6**, 463-471.

Hennion, F. and Bouchereau, A. (1998) Accumulation of organic and inorganic solutes in the subantarctic cruciferous species *Pringlea antiscorbutica* in response to saline and cold stresses, *Polar Biology* **20**, 281-291.

Hennion, F. and Couderc, H. (1992) Cytogenetical study of *Pringlea antiscorbutica* R. Br. and *Ranunculus moseleyi* Hook. f. from the Kerguelen Islands. *Antarctic Science* **4**, 57-58.

Hennion, F. and Couderc, H. (1993) Cytogenetical variability of *Ranunculus* species from Iles Kerguelen. *Antarctic Science* **5**, 37-40.

Hennion, F. and Martin-Tanguy, J. (1999) Amine distribution and content in several parts of the subantarctic endemic species *Lyallia kerguelensis* (Hectorellaceae), *Phytochemistry* **52**, 247-251.

Hennion, F. and Martin-Tanguy, J. (2000) Amines of the subantarctic crucifer *Pringlea antiscorbutica* are responsive to temperature conditions, *Physiologia Plantarum* **109**, 232-243.

Hennion, F. and Walton, D.W.H. (1997a) Seed germination of endemic species from Kerguelen phytogeographic zone, *Polar Biology* **17**, 180-187.

Hennion, F. and Walton, D.W.H. (1997b) Ecology and seed morphology of endemic species from Kerguelen Phytogeographic Zone, *Polar Biology* **18**, 229-235.

Hennion, F., Fiasson, J-L. and Gluchoff-Fiasson, K. (1994) Morphological and phytochemical relationships between *Ranunculus* species from Iles Kerguelen, *Biochemical Systematics and Ecology* **22**, pp. 533-42.

Holtom, A. and Greene, S.W. (1967) The growth and reproduction of Antarctic flowering plants, *Philosophical transactions of the Royal Society of London, series B* **252**, 323-337.

Honegger, R. (1986) Ultrastructural studies in lichens. II. Mycobiont and phycobiont cell wall surface layers adhering crystalline lichen products in four Parmeliaceae, *New Phytologist* **125**, 659-677.

Hovenden, M.J. and Seppelt, R.D. (1995) Uptake of water from the atmosphere by lichens in continental Antarctica, *Symbiosis* **18**, 111-118.

Hovenden, M.J., Jackson, A.E. and Seppelt, R.D. (1994) Field photosynthetic activity of lichens in the Windmill Islands oasis, Wilkes Land, continental Antarctica, *Physiologia Plantarum* **90**, 567-576.

Hoyer, K., Karsten, U., Sawall, T. And Wiencke, C. (2001) Photoprotective substances in Antarctic macroalgae and their variation with respect to depth distribution, different tissues and developmental stages, *Marine Ecology Progress Series* **21**, 117-129.

Huiskes, A.H.L., Gremmen, N.J.M. and Francke, J.W. (1997) Morphological effects on the water balance of Antarctic foliose and fruticose lichens, *Antarctic Science* **9**, 36-42.

Huiskes, A.H.L., Lud, D. and Moerdijk-Poortvliet, T.C.W. (2001) Field research on the effects of UV-B filters on terrestrial Antarctic vegetation, *Plant Ecology* **154**, 75-86.

Huiskes, A.H.L., Lud, D., Moerdijk-Poortvliet, T.C.W. and Rozema, J. (1999) Impact of UV-B Radiation on Antarctic Terrestrial Vegetation in J. Rozema (ed.), *UV-B and Terrestrial Ecosystems*, Backhuys Publishers, Leiden, The Netherlands, pp 313-333.

Hummel, I., Couée, I., El Amrani, A., Martin-Tanguy, J. and Hennion, F. (2002) Involvement of polyamines in root development at low temperature in the subantarctic cruciferous species *Pringlea antiscorbutica*, *Journal of Experimental Botany* **53**, 1463-1473.

Hummel, I., El Amrani, A., Gouesbet, G., Hennion, F. and Couée, I. (2004a) Involvement of polyamines in the interacting effects of low temperature and mineral supply on *Pringlea antiscorbutica* (Kerguelen cabbage) seedlings, *Journal of Experimental Botany* **55**, 1125-1134.

Hummel, I., Quemmerais, F., Gouesbet, G., El Amrani, A., Frenot, Y., Hennion, F. and Couée, I. (2004b) Characterization of environmental stress responses during early development of *Pringlea antiscorbutica* in the field at Kerguelen, *New Phytologist* **162**, 705-715.

Huneck, S., Sainsbury, M., Rickard, T.M.A. and Smith, R.I.L. (1984) Ecological and chemical investigations of lichens from South Georgia and the maritime Antarctic, *Journal of the Hattori Botanical Laboratory* **56**, 461-480.

Ishikura, M., Kato, C. and Maruyama, T. (1997) UV absorbing substances in zooxanthellate and a-zooxanthellate clams, *Marine Biology* **128**, 649-655.

Jackson, A.E. and Seppelt, R.D. (1997) Physiological adaptations to freezing and UV radiation exposure in *Prasiola crispa*, an Antarctic terrestrial alga, in B. Battaglia, J. Valencia and D.W.H. Walton (eds.), *Antarctic Communities: Species, Structure and Survival*, Cambridge University Press, Cambridge, U.K., pp. 226-233.

Kappen, L. (1993) Plant activity under snow and ice, with particular reference to lichens, *Arctic* **46**, 297-302.

Kappen, L. (2000) Some aspects of the great success of lichens in Antarctica, *Antarctic Science* **12**, 314-324.

Kappen, L. and Breuer, M. (1991) Ecological and physiological investigations in continental Antarctic cryptogams. II. Moisture relations and photosynthesis of lichens near Casey Station, Wilkes Land, *Antarctic Science* **3**, 273-278.

Kappen, L. and Valladares, F. (1999) Poikilohydrous autotrophs, in F.I. Pugnaire and F. Valladares (eds.), *Handbook of functional plant ecology*, Dekker, New York, U.S.A., pp. 9-80.

Kappen, L., Smith, R.I.L. and Meyer, M. (1989) Carbon dioxide exchange of two ecodemes of *Schistidium antarctici* in continental Antarctica, *Polar Biology* **9**, 415-422.

Karentz, D., Cleaver, J.E. and Michel, D.L. (1991a) Cell survival characteristics and molecular responses of Antarctic phytoplankton to UV-B radiation, *Journal of Phycology* **27**, 326-341.

Karentz, D., McEuen, F.S., Land, M.C. and Dunlap, W.C. (1991b) Survey of mycosporine-like amino acid compounds in Antarctic marine organisms: potential protection from ultraviolet exposure, *Marine Biology* **108**, 157-166.

Kennedy, A.D. (1993) Water as a limiting factor in the Antarctic Terrestrial Environment, *Arctic and Alpine Research* **25**, 308-315.

Kennedy, A.D. (1994) Simulated climate change: a field manipulation study of polar microarthropod community response to global warming, *Ecography* **17**, 131-140.

Kershaw, K.A. and Webber, M.R. (1986) Seasonal changes in the chlorophyll content and quantum efficiency of the moss *Brachythecium rutabulum*, *Journal of Bryology* **14**, 151-158.

King, J.C. and Harangozo, S.A. (1998) Climate change in the western Antarctic Peninsula since 1945: observations and possible causes, *Annals of Glaciology* **27**, 571-575.

Lange, O.L. and Kappen, L. (1972) Photosynthesis of lichens from Antarctica, in G.A. Llano (ed.) *Antarctic Terrestrial Biology*, American Geophysical Union, Washington D.C., U.S.A., pp. 83-95.

Lee, R.E. Jr. and Denlinger, D.L. (1991) (eds.) *Insects at Low Temperature*, Chapman and Hall, New York, U.S.A., 513 pp.

Leishman, M.R. and Wild, C. (2001) Vegetation abundance and diversity in relation to soil nutrients and soil water content in Vestfold Hills, East Antarctica, *Antarctic Science* **13**, 126-134.

Longton, R.E. (1974) Microclimate and biomass in communities of the *Bryum* association on Ross Island, continental Antarctica, *The Bryologist* **77**, 109-127.

Lovelock, C.E., and Robinson, S.A. (2002) Surface reflectance properties of Antarctic moss and their relationship to plant species, pigment composition and photosynthetic function, *Plant Cell and Environment* **25**, 1239-1250.

Lovelock, C.E., Osmond, C.B. and Seppelt, R.D. (1995b) Photoinhibition in the Antarctic moss *Grimmia antarctici* Card when exposed to cycles of freezing and thawing, *Plant Cell and Environment* **18**, 1395-1402.

Lovelock, C.E., Jackson, A.E., Melick, D.R. and Seppelt, R.D. (1995a) Reversible photoinhibition in Antarctic moss during freezing and thawing, *Plant Physiology* **109**, 955-961.

Lud, D., Huiskes, A.H.L., Moerdijk, T.C.W. and Rozema, J. (2001b) The effects of altered levels of UV-B radiation on an Antarctic grass and lichen, *Plant Ecology* **154**, 87-99.

Lud, D., Buma, A.G.J., Van de Poll, W., Moerdijk, T.C.W. and Huiskes, A.H.L. (2001a) DNA damage and photosynthetic performance in the Antarctic terrestrial alga *Prasiola crispa ssp antarctica* (*Chlorophyta*) under manipulated UV-B radiation, *Journal of Phycology* **37**, 459-467.

Margulis, L., Walker, J.C.G. and Rambler, M. (1976) Reassessment of roles of oxygen and ultraviolet light in precambrium evolution, *Nature* **383**, 680.

Marshall, D.J., Newton, I.P. and Crafford, J.E. (1995) Habitat temperature and potential locomotor activity of the continental Antarctic mite, *Maudheimia petronia* (Acari: Oribatei), *Polar Biology* **15**, 41-46.

Martin, C.E. (1980) Chlorophyll a/b ratios of eleven North Carolina mosses, *Bryologist* **83**, 84-87.

Martin, C.E. and Churchill, S.P. (1982) Chlorophyll concentration and a/b ratios in mosses collected from exposed and shaded habitats in Kansas, *Journal of Bryology* **12**, 297-304.

McCall, K.K. and Martin, C.E. (1991) Chlorophyll concentrations and photosynthesis in three forest understory mosses in northeastern Kansas, *Bryologist* **94**, 25-29.

Melick, D.R. and Seppelt, R.D. (1992) Loss of soluble carbohydrates and changes in freezing point of Antarctic bryophytes after leaching and repeated freeze-thaw cycles, *Antarctic Science* **4**, 399-404.

Melick, D.R. and Seppelt, R.D. (1994a) The effect of hydration on carbohydrate levels, pigment content and freezing point of *Umbilicaria decussata* at a continental Antarctic locality, *Cryptogamic Botany* **4**, 212-217.

Melick, D.R. and Seppelt, R.D. (1994b) Seasonal investigations of soluble carbohydrates and pigment levels in Antarctic bryophyte and lichens, *The Bryologist* **97**, 13-19.

van der Merwe, M., Chown, S.L. and Smith, V.R. (1997) Thermal tolerance limits in six weevil species (Coleoptera, Curculionidae) from subantarctic Marion Island, *Polar Biology* **18**, 331-336.

Meykamp, B.B., Doodeman, G. and Rozema, J. (2001) The response of *Vicia faba* to enhanced UV-B radiation under low and near ambient PAR levels, *Plant Ecology* **154**, 135-146.

Montiel, P., Smith, A. and Keiller, D. (1999) Photosynthetic responses of selected Antarctic plants to solar radiation in the southern maritime Antarctic, *Polar Research* **18**, 229-235.

Nakanishi, S. (1979) On the variation of leaf characters of an Antarctic moss, *Bryum inconnexum*. *Memoirs of the National Institute of Polar Research*, 47-57.

Nakatsubo, T. (2002) Predicting the impact of climatic warming on the carbon balance of the moss *Sanionia uncinata* on a maritime Antarctic island, *Journal of Plant Research* **115**, 99-106.

Newsham, K.K., Hodgson, D.A., Murray, A.W.A., Peat, H.J. and Smith, R.I.L. (2002) Response of two Antarctic bryophytes to stratospheric ozone depletion, *Global Change Biology* **8**, 972-983.

Newsham, K.K. (2003) UV-B radiation arising from stratospheric ozone depletion influences the pigmentation of the Antarctic moss *Andreaea regularis*, *Oecologia* **135**, 327-331.

Norton, R.A. (1994) Evolutionary aspects of oribatid mite life histories and consequences for the origin of the Astigmata, in M. Houck (ed.), *Ecological and evolutionary analyses of life-history patterns* Chapman and Hall, New York, U.S.A., pp. 99-135.

Nybakken, L., Aubert, S. and Bilger, W. (2004b) Epidermal UV-screening of arctic and alpine plants along a latitudinal gradient in Europe, *Polar Biology* **27**, 391-398.

Nybakken, L., Bilger, W., Johanson, U., Björn, L.O., Zielke, M. and Solheim, B. (2004a) Epidermal UV-screening in vascular plants from Svalbard (Norwegian Arctic), *Polar Biology* **27**, 383-390.

Pakker, H., Martins, R.S.T., Boelen, P., Buma, A.G.J., Nikaido, O. and Breeman, A.M. (2000) Effects of temperature on the photoreactivation of ultraviolet-B induced DNA damage in *Palmaria palmata* (Rhodophyta), *Journal of Phycology* **36**, 334-341.

Pannewitz, S., Schlensog, M., Green, T.G.A., Sancho, L. and Schroeter, B. (2003) Are lichens active under snow in continental Antarctica? *Oecologia* **135**, 30-38.

Pickup, J. and Rothery, P. (1991) Water-loss and anhydrobiotic survival in nematodes of Antarctic fellfields, *Oikos* **61**, 379-388.

Post, A. and Larkum, A.W.D. (1993) UV absorbing pigments, photosynthesis and UV exposure in Antarctica: Comparison of terrestrial and marine algae, *Aquatic Botany* **45**, 231-243.

Quayle, W.C., Peck, L.S., Peat, H., Ellis-Evans, J.C. and Harrigan, P.R. (2002) Extreme responses to climate change in lakes, *Science* **295**, 645.

Quayle, W.C., Convey, P., Peck, L.S., Ellis-Evans, C.J., Butler, H.G. and Peat, H.J. (2003) Ecological responses of maritime Antarctic lakes to regional climate change, in E. Domack, A. Burnett, A. Leventer, P. Convey, M. Kirby and R. Bindschadler (eds.), *Ecological responses of maritime Antarctic lakes to regional climate change. Antarctic Research Series* vol. 79, American Geophysical Union, U.S.A., pp. 159-170.

Rastorfer, J.R. (1972) Comparative physiology of four west Antarctic mosses, *Antarctic Research Series* **20**, 143-161.

Reyes, M.A., Corcuera, L.J. and Cardemil, L. (2003) Accumulation of HSP70 in *Deschampsia antarctica* Desv. leaves under thermal stress, *Antarctic Science* **15**, 345-352.

Ring, R.A. and Danks, H.V. (1994) Desiccation and cryoprotection: overlapping adaptations, *Cryo-Letters* **15**, 181-190.

Robberecht, R. and Caldwell, M.M. (1978) Leaf epidermal transmittance of ultraviolet radiation and its implications for plant sensitivity to ultraviolet radiation induced injury, *Oecologia* **32**, 277-287.

Robinson, S.A., Turnbull, J.A., and Lovelock, C.E (2005) Impact of changes in natural ultraviolet radiation on pigment composition, physiological and morphological characteristics of the Antarctic moss, *Grimmia antarctici,* Global Change Biology **11**, 476–489

Robinson, S.A., Wasley, J. and Tobin, A.K. (2003) Living on the edge - plants and global change in continental and maritime Antarctica, *Global Change Biology* **9**, 1681-1717.

Robinson, S.A., Wasley, J., Popp, M. and Lovelock, C.E. (2000) Desiccation tolerance of three moss species from continental Antarctica, *Australian Journal of Plant Physiology* **27**, 379-388.

Robinson, S.A., Wasley, J., Turnbull, J., Lovelock, C.E. (2001) Antarctic moss coping with the ozone hole, *Proceedings of the 12th International Congress on Photosynthesis*, CSIRO publishing, Canberra, Australia.

Rozema, J., R. Broekman, D. Lud, A.H.L. Huiskes, T.C.W. Moerdijk-Poortvliet, N. de Bakker, B. Meijkamp and A. van Beem (2001) Consequences of depletion of stratospheric ozone for terrestrial Antarctic ecosystems: the response of *Deschampsia antarctica* to enhanced UV-B radiation in a controlled environment, *Plant Ecology* **154**, 101-115.

Ruhland, C.T. and Day, T.A. (2000) Effects of ultraviolet-B radiation on leaf elongation, production and phenylpropanoid concentrations of *Deschampsia antarctica* and *Colobanthus quitensis* in Antarctica, *Physiologia Plantarum* **109**, 244-251.

Ruhland, C.T. and Day, T.A. (2001) Size and longevity of seed banks in Antarctica and the influence of ultraviolet-B radiation on survivorship, growth and pigment concentrations of *Colobanthus quitensis* seedlings, *Environmental and Experimental Botany* **45**, 143-154.

Schlensog, M. and Schroeter, B. (2000) Poikilohydry in Antarctic cryptogams and its influence on photosythetic performance in mesic and xeric habitats, in P. Broady (ed.), *Antarctic Ecosystems: models for wider ecological understanding*, New Zealand Natural Sciences, New Zealand, pp. 175-182.

Schlensog, M., Schroeter, B., Sancho, L.G., Pintado, A. and Kappen, L. (1997) Effect of strong irradiance on photosynthetic performance of the melt-water dependent cyanobacterial lichen *Leptogium puberulum* (Collemataceae) Hue from the maritime Antarctic, *Bibliotheca Lichenologia* **67**, 235-246.

Schroeter, B. and Scheidegger, C. (1995) Water relations in lichens at subzero temperatures: structural changes and carbon dioxide exchange in the lichen *Umbilicaria aprina* from continental Antarctica, *New Phytologist* **131**, 273-285.

Schroeter, B., Green, T.G.A., Kappen, L. and Seppelt, R.D. (1994) Carbon dioxide exchange at subzero temperatures. Field measurements on *Umbilicaria aprina* in Antarctica, *Cryptogamic Botany* **4**, 233-241.

Schroeter, B., Kappen, L., Green, T.G.A. and Seppelt, R.D. (1997) Lichens and the Antarctic environment: Effects of temperature and water availability on photosynthesis, in W.B. Lyons, C. Howard-Williams and I. Hawes (eds.), *Ecosystem processes in Antarctic ice-free landscapes*, Balkema, Rotterdam, The Netherlands, pp. 103-118.

Searles, P.S., Kropp, B.R., Flint, S.D. and Caldwell, M.M. (2001) Influence of solar UV-B radiation on peatland microbial communities of southern Argentina, *New Phytologist* **152**, 213-221.

Sinclair, B.J. (1999) Insect cold tolerance: how many kinds of frozen? *European Journal of Entomology* **96**, 157-164.

Sinclair, B.J., Addo-Bediako, A. and Chown, S.L. (2003a) Climatic variability and the evolution of insect freeze tolerance, *Biological Reviews* **78**, 181-195.

Sinclair, B.J., Vernon, P., Klok, C.J. and Chown, S.L. (2003b) Insects at low temperatures: an ecological perspective, *Trends in Ecology and Evolution* **18**, 257-262.

Sinclair, B.J., Klok, C.J., Scott, M.B., Terblanche, J.S. and Chown, S.L. (2003c) Diurnal variation in supercooling points of three species of Collembola from Cape Hallett, Antarctica, *Journal of Insect Physiology* **49**, 1049-1061.

Skvarca, P., Rack, W., Rott, H. and Ibarzábal y Donángelo, T. (1998) Evidence of recent climatic warming on the eastern Antarctic Peninsula, *Annals of Glaciology* **27**, 628-632.

Smith, R.I.L. (1984) Terrestrial Plant Biology of the Subantarctic and the Antarctic, in R.M. Laws (ed.), *Antarctic Ecology* Volume 1, Academic Press, London, pp. 61-162.

Smith, R.I.L. (1988) Recording bryophyte microclimate in remote and severe environments, In J.M. Glime (ed.), *Methods in Bryology*, Hattori Botanical Laboratory, Nichinan, Japan, pp. 275-284.

Smith, R.I.L. (1990) Signy Island as a paradigm of biological and environmental change in Antarctic terrestrial ecosystems, in K.R. Kerry and G. Hempel (eds.), *Antarctic Ecosystems, Ecological Change and Conservation*, Springer-Verlag, Berlin, Germany.

Smith, R.I.L. (1999) Biological and environmental characteristics of three cosmopolitan mosses dominant in continental Antarctica, *Journal of Vegetation Science* **10**, 231-242.

Smith, R.I.L. (2001) Plant colonization response to climate change in the Antarctic, *Folia Facultatis Scientiarum Naturalium Universitatis Masarykiana Brunensis, Geographia*, **25**, 19-33.

Smith, V.R. (2002) Climate change in the subantarctic: an illustration from Marion Island, *Climatic Change* **52**, 345-357.

Smith, V.R. (2005) Moisture, carbon and inorganic nutrient controls of soil respiration at a subantarctic island, *Soil Biology and Biochemistry* **37**, 81-91.

Smith, V.R. and Steenkamp, M. (1992) Macroinvertebrates and litter nutrient release on a sub-Antarctic island, *South African Journal of Botany* **58**, 105-116.

Solhaug, A.A. and Gauslaa, Y. (1996) Parietin, a photoprotective secondary product of the lichen *Xanthoria parietina*, *Oecologia* **108**, 412-418.

Sømme, L. (1986) Ecology of *Cryptopygus sverdrupi* (Insecta: Collembola) from Dronning Maud Land, Antarctica, *Polar Biology* **6**, 179-184.

Strathdee, A.T., Bale, J.S., Strathdee, F.C., Block, W.C., Coulson, S.J., Webb, N.R. and Hodkinson, I.D. (1995) Climatic severity and the response to temperature elevation of Arctic aphids, *Global Change Biology* **1**, 23-28.

Strullu, D.G., Frenot, Y., Maurice, D., Gloaguen, J.C. and Plenchette, C. (1999) Première contribution à l'étude des mycorhizes des îles Kerguelen, *Comptes Rendus de l'Académie des Sciences de Paris, Sciences de la Vie* **322**, 771-777.

Sullivan, J. and Rozema, J. (1999) UV-B effects on terrestrial plant growth and photosynthesis, in J. Rozema (ed.), *Stratospheric ozone depletion, the effects of enhanced UV-B radiation on terrestrial ecosystems*, Backhuys, Leiden, The Netherlands, pp. 39-57.

Swanson, A. and Fahselt, D. (1997) Effects of ultraviolet on polyphenolics of *Umbilicaria americana*, *Canadian Journal of Botany* **75**, 284-289.

Swanson, A., Fahselt, D. and Smith, D. (1996) Phenolic levels in *Umbilicaria americana* in relation to enzyme polymorphism, altitude and sampling date, *Lichenologist* **28**, 331-339.

Tarnawski, M., Melick, D., Roser, D., Adamson, E., Adamson, H. and Seppelt, R. (1992) In situ carbon dioxide levels in cushion and turf forms of *Grimmia antarctici* at Casey Station, East Antarctica, *Journal of Bryology* **17**, 241-249.

Teramura, A.H. and Ziska, L.H. (1996) Ultraviolet-B radiation and photosynthesis, in N.R. Baker (ed.), *Photosynthesis and the environment*, Kluwer Academic Publishers, Dordrecht, The Netherlands, pp. 435-450.

Verhoeven, A.S., Adams, W.W. and Demmig-Adams, B. (1998) 2 forms of sustained xanthophyll cycle-dependent energy-dissipation in overwintering *Euonymus-Kiautschovicus*, *Plant, Cell and Environment* **21**, 893-903.

Vining, E.C., Crafts-Brandner, S.J. and Day, T.A. (1997) Photosynthetic acclimation of antarctic hair grass (*Deschampsia antarctica*) to contrasting temperature regimes, *Bulletin of the Ecological Society of America* **78**, 327.

Walton, D.W.H. (1976) Dry matter production in *Acaena* (Rosaceae) on a subantarctic island, *Journal of Ecology* **64**, 399-415.

Walton, D.W.H. (1977) Studies on *Acaena* (Rosaceae) I. Seed germination, growth and establishment in *A. magellanica* (Lam.) Vahl and *A. tenera* Alboff, *British Antarctic Survey Bulletin* **45**, 29-40.

Walton, D.W.H. (1982) The Signy Island terrestrial reference sites. XV. Microclimate monitoring, 1972-4, *British Antarctic Survey Bulletin* **55**, 111-126.

Wasley, J. (2004) *The effect of Climate Change on Antarctic Terrestrial Flora.* PhD Thesis, University of Wollongong, Australia.

Wasley, J., Robinson, S.A., Popp, M., Lovelock, C.E. (2006a) Climate change manipulations show Antarctic flora is more strongly affected by elevated nutrients than water, *Global Change Biology* (in press).

Wasley, J., Robinson, S.A., Lovelock, C.E. and Popp, M. (2006b) Some like it wet – biological characteristics underpinning tolerance of extreme water events in Antarctic bryophytes, *Functional Plant Biology*, **33**, 443-455.

Webby, R.F. and Markham K.R. (1994) Isoswertiajaponin2"-O-beta-arabinopyranoside and other flavone-C-glycosides from the Antarctic grass *Deschampsia Antarctica*, *Phytochemistry* **36**, 1323-1326.

Webby, R.F., Markham, K.R. and Smith, R.I.L. (1996) Chemotypes of the Antarctic moss *Bryum algens* delineated by their flavonoid constitutents, *Biochemical Systematics and Ecology* **24**, 469-475.

Wharton, D.A. (1995) Cold tolerance strategies in nematodes, *Biological Reviews* **70**, 161-185.

Wharton, D.A. and Block, W. (1993) Freezing tolerance in some Antarctic nematodes, *Functional Ecology* **7**, 578-584.

Wharton, D.A. and Block, W. (1997) Differential scanning calorimetry studies on an Antarctic nematode (*Panagrolaimus davidi*) which survives intracellular freezing, *Cryobiology* **34**, 114-121.

Wharton, D.A. and Ferns, D.J. (1995) Survival of intracellular freezing by the Antarctic nematode *Panagrolaimus davidi*, *Journal of Experimental Biology* **198**, 1381-1387.

Wilson, M.E. (1990) Morphology and photosynthetic physiology of *Grimmia antarctici* from wet and dry habitats, *Polar Biology* **10**, 337-341.

Worland, M.R. (1996) The relationship between body water content and cold tolerance in the Arctic collembolan *Onychiurus arcticus* (Collembola: Onychiuridae), *European Journal of Entomology* **93**, 341-348.

Worland, M.R. and Convey, P. (2001) Rapid cold hardening in Antarctic microarthropods, *Functional Ecology* **15**, 515-525.

Worland, M.R., Block, W. and Oldale, H. (1996). Ice nucleation activity in biological materials with examples from Antarctic plants. *CryoLetters* **17**, 31-38.

Worland, M.R., Grubor-Lajsic, G. and Montiel, P.O. (1998) Partial desiccation induced by sub-zero temperatures as a component of the survival strategy of the Arctic collembolan *Onychiurus arcticus* (Tullberg), *Journal of Insect Physiology* **44**, 211-219.

Wynn-Williams, D.D. (1994) Potential effects of Ultraviolet radiation on Antarctic primary terrestrial colonizers: cyanobacteria, algae and cryptogams, *Antarctic Research Series* **62**, 243-257.

Xiong, F. and Day, T. (2001) Effect of solar ultraviolet-B radiation during springtime ozone depletion on photosynthesis and biomass production of Antarctic vascular plants, *Plant Physiology* **125**, 738-751.

Xiong, F.S., Mueller, E.C. and Day, T.A. (2000) Photosynthetic and respiratory acclimation and growth response of Antarctic vascular plants to contrasting temperature regimes, *American Journal of Botany* **87**, 700-710.

Xiong, F., Ruhland, C. and Day, T. (1999) Photosynthetic temperature response of the Antarctic vascular plants *Colobanthus quitensis* and *Deschampsia antarctica*, *Physiologia Plantarum* **106**, 276-286.

Zachariassen, K.E. (1985) Physiology of cold tolerance in insects, *Physiological Reviews* **65**, 799-832.

Zuñiga, G.E., Alberdi, M. and Corcuera, L.J. (1996) Non-structural carbohydrates in *Deschampsia antarctica* Desv. from South Shetland Islands, maritime Antarctic, *Environmental and Experimental Botany* **36**, 393-398.

Zuñiga, G.E., Alberdi, M., Fernandez J., Montiel P. and Corcuera, L.J. (1994) Lipid content in leaves of *Deschampsia antarctica* from the maritime Antarctic, *Phytochemistry* **37**, 669-672.

8. PLANT BIODIVERSITY IN AN EXTREME ENVIRONMENT

Genetic studies of origins, diversity and evolution in the Antarctic

M. L. SKOTNICKI

Genomic Interactions Group, Research School of Biological Sciences, Institute of Advanced Studies, Australian National University, Canberra, ACT 2601, Australia
skotnicki@rsbs.anu.edu.au

P. M. SELKIRK

Department of Biological Sciences, Macquarie University, Sydney, NSW 2109, Australia
pselkirk@rna.bio.mq.edu.au

Introduction: Plant biodiversity in the Antarctic

Plants in Antarctica survive in one of the harshest environments on Earth. Less than 2% of the 14 million km^2 that make up continental Antarctica is free of permanent ice and snow and therefore available for colonisation by plants. Vegetation is sparse and low-growing, and is dominated by mosses and lichens. Two species of flowering plants occur on the Antarctic Peninsula (Edwards and Lewis Smith 1988, Lewis Smith 2003), but none in continental Antarctica. The flora of continental Antarctica comprises 15 species of mosses (Lewis Smith 1984), one species of liverwort (Bednarek-Ochyra et al. 2000) and at least 88 taxa of lichens (Øvstedal and Lewis Smith 2001). Eight species of moss have been recorded from southern Victoria Land and Ross Island (Seppelt and Green 1998).

Continental Antarctic moss species usually grow as small colonies, but some coalesce to form turfs up to several square metres in extent and with up to 70% ground cover (Lewis Smith 2003, 2005). In a few locations, turfs can cover almost

D.M. Bergstrom et al. (eds.), Trends in Antarctic Terrestrial and Limnetic Ecosystems, 161–175.

100% of the ground over 25m^2 or more. Clumps and turfs of moss are found in areas sheltered from the strong winds common in this region: in depressions and cracks in the ground surface, drainage lines and near rocks. Mosses only grow in niches where some moisture is available in summer, such as melt water from glaciers and persistent snow banks, and melting snow accumulated amongst rocks, cracks and depressions in the ground surface. These mosses are subjected to extremes of cold, drought, wind and light, with plants south of 67°S existing for weeks or months each year in a freeze-dried state in complete darkness.

In the harsh continental Antarctic environment with its short summer growing period, mosses do not reproduce sexually. Of the moss species recorded from southern Victoria Land and Ross Island, only *Hennediella heimii* is known to produce sporophytes (Seppelt et al. 1992), although mature sporophytes and shedding spores have not been recorded. Colonies originate either from immigrant propagules from other lands (Marshall 1996, Marshall and Convey 1997) or from vegetative propagules dispersed locally. Colonisation of new locations seems to be difficult and immigration from other land masses appears infrequent.

With the annual expansion of the 'ozone hole' (Farman et al. 1985, Kennedy 1995) continental Antarctic mosses are subjected to increasing exposure to UV-B irradiation. With their haploid genomes, lack of sexual reproduction, perennial growth, and extreme isolation from colonies elsewhere around the world, these mosses appear to provide an ideal model system with the potential to reveal significant insights into colonisation, mutation and speciation. Such a combination of characteristics is not available in other plants, or even in mosses in most other parts of the world (Wyatt 1994).

On the Antarctic Peninsula, there is a wider diversity of plants, with more moss, liverwort and lichen species, and two vascular plant species, *Deschampsia antarctica* and *Colobanthus quitensis* (Lewis Smith 1984).

On the subantarctic islands, vegetation is yet more diverse, with numerous vascular plants in addition to a higher number of bryophyte and lichen species. For example, from Heard Island 12 species of vascular plants (George 1993, Turner et al. 2006) 37 species of mosses (Bergstrom and Selkirk 1997), 19 species of liverworts (Vána and Gremmen 2005) and 71 species of lichens (D. Øvestdal and N. Gremmen, unpubl. data) have been found to date. On Macquarie Island, there are 44 native species of vascular plants (George 1993), 88 species of mosses (Seppelt 2004), 51 species of liverworts and many algae and lichens (Selkirk et al. 1990). Other subantarctic islands have similarly diverse floras, especially those islands that have been subjected to human occupation and introductions of alien species (Frenot et al. 2005).

Under predicted scenarios of global climate change, the climatic constraints of the Antarctic environment are likely to be reduced and the level of biodiversity is likely to increase (Kennedy 1995). With climate change already well underway on subantarctic islands such as Macquarie and Marion Islands, Iles Kerguelen and Heard Island, on-going research can analyse the spread of species and monitor colonisation of newly deglaciated ground. Two examples of biodiversity increasing are the recent arrivals of *Poa annua* (Scott 1989) and *Leptinella plumosa* on Heard Island (Turner et al. 2006).

Uniquely in continental Antarctica, many of the short-term consequences of climate change will affect bryophyte ecosystems and colonisation; these communities may be particularly vulnerable to global change. Research into the genetic diversity and susceptibility to increased UV radiation of endemic Antarctic species may make prediction of the consequences of ozone depletion more feasible (Adamson and Adamson 1992, Kennedy 1995, Robinson et al. 2003, 2005).

An interesting question is whether the mosses found today have been present in Antarctica for a very long time, surviving periods of extensive ice cover in refugia, or whether moss populations on the Antarctic continent face regular extinction and become re-established by colonisation from outside Antarctica when conditions are favourable (Walton 1990, Marshall 1996). A promising indirect approach is to make inferences on colonisation history and processes from the current spatial (geographic and microgeographic) distribution of natural variation in particular moss genes. In addition, as bryophytes constitute one of very few successful groups with a functional haploid phase, the evolutionary processes operating in these plants are of great general interest and fundamental scientific importance (Longton 1988, 1994).

We have used techniques of molecular genetics to investigate the genetic diversity of these plants, their origins and dispersal mechanisms and their potential to respond genetically to climate change (Skotnicki et al. 2000, 2004). We have also used molecular genetics to resolve some taxonomic uncertainties (Skotnicki et al. 2001, 2002), since the extreme environment can lead to phenotypic plasticity where morphological characters can vary in response to different environmental conditions rather than being due to genetic changes (Seppelt and Selkirk 1984, Lewis Smith 1999). In some Antarctic mosses, such as *Ceratodon purpureus*, it has been suggested that the taxon is so variable that extreme phenotypes could be distinct species (Ochyra 1998). When morphological identification has been difficult or impossible, molecular techniques have proved valuable (Skotnicki et al. 1997, 2001, Bargagli et al. 2004). Molecular genetics can also assist in the taxonomic identification of potential hybrid plant species found on subantarctic islands.

Techniques of molecular genetics have the potential to reveal a wide variety of characteristics in Antarctic plants that cannot be determined by traditional morphological microscopic examination. For example, by comparing highly conserved gene sequences, it is possible to analyse the colonisation patterns of mosses in Antarctica, or dispersal of vascular plants among subantarctic islands.

Genetic diversity within and among Antarctic moss populations

The Random Amplified Polymorphic DNA (RAPD) technique has proven very useful for initial analysis of the extent of genetic variation and dispersal in several

Antarctic moss species (Selkirk et al. 1997, 1998, Skotnicki et al. 1997, 1998 a, b, c), and for similar analyses of genetic variability in temperate mosses, liverworts and ferns (Boisselier-Dubayle and Bischler 1994, Boisselier-Dubayle et al. 1995, Schneller et al. 1998). Some of the reasons for using this technique are that (a) moss growth in Antarctica is extremely slow (of the order of 1 mm per year), (b) colonies are few and far between, limiting the size of samples which can be collected consistent with minimising damage, (c) only limited genetic information is available for mosses, especially those in Antarctica (d) the gametophyte stage of mosses is haploid and so potential problems arising from amplification of codominant genes are avoided, (e) within-clump variation has already been detected, necessitating the use of single shoots, usually 3 mm or less in length, and (f) many samples can be compared relatively quickly and inexpensively.

Using the RAPD technique, our studies have clearly shown that genetic variation does occur, at relatively high levels, in Antarctic moss populations. In *Bryum argenteum, B. pseudotriquetrum, Campylopus pyriformis, Ceratodon purpureus, Hennediella heimii* and *Sarconeurum glaciale,* variation can be detected among populations, within isolated populations and even within individual colonies (Selkirk et al. 1997, 1998, Dale et al. 1999, Skotnicki et al. 1998b, c, 2004, Seppelt et al. 1999).

DNA sequencing of highly conserved genes such as the nuclear ribosomal DNA internal transcribed spacer region (ITS1–5.8S–ITS2) has also become a widely-used technique for phylogenetic analyses in a range of plant species (Baldwin 1992). DNA sequencing of this region has recently been applied to phylogenetic investigations of temperate bryophytes (Bopp and Capesius 1996, Patterson et al. 1998, Chiang and Schaal 1999, Longton and Hedderson 2000, Shaw 2000a, Vanderpoorten et al. 2001, Boisselier-Dubayle et al. 2002). In a study of a very isolated population of *Pohlia nutans* on heated ground on Mt Rittmann in Northern Victoria Land, the RAPD and DNA sequencing techniques proved to be complementary, with DNA sequencing confirming and extending the results of RAPD analyses: the population has little genetic variability and appears to be derived from a single immigration event. Although some colonies did exhibit some RAPD band differences, these were small changes in comparison with specimens of the same species from islands near the Antarctic Peninsula (Skotnicki et al. 2002).

Overall, DNA sequencing experiments have completely complemented and extended genetic diversity results previously obtained using the RAPD technique in Antarctic mosses (Skotnicki et al. 2004). While RAPDs can facilitate a quick, simple and inexpensive overall analysis of a plant population, this technique can have problems of reproducibility in some laboratories (Jones et al. 1997). DNA sequencing of the ITS region reveals similar levels of genetic variability but at a much finer and more detailed scale.

Recently, several other genes including the three rps4, rbcL and trnF-trnL plastid gene regions have been analysed in mosses and liverworts (Lewis et al. 1997, Buck et al. 2000, Cox et al. 2000, De Luna et al. 2000, Newton et al. 2000, Shaw 2000b,

Cox and Hedderson 2003, Magombo 2003, Pederson et al. 2003, Virtanen 2003), as they are more highly conserved than the nuclear ITS region and more useful for intergeneric comparisons. For some of the Antarctic mosses, we have also sequenced these three plastid regions in addition to the nuclear ITS region, to compare the Antarctic genera with those from elsewhere. The plastid genes sequenced in Antarctic mosses have so far simply confirmed the ITS results obtained.

Dispersal and colonisation by Antarctic mosses

Molecular genetics techniques have been useful to document dispersal of Antarctic mosses over distances of a few metres to hundreds of kilometres. The RAPD technique was used to analyse genetic variation in populations of mosses in the vicinity of and along drainage channels in the Garwood Valley and at Cape Chocolate Southern Victoria Land. The results from the Garwood Valley populations were consistent with dispersal of *Bryum argenteum* propagules principally by overland flow of water along meltwater drainage channels some 50m long, with some wind dispersal that would account for the occasional appearance in one channel of a genotype with close affinities to plants in another channel. The uniform spread of variation in *Hennediella heimii* at the same locality was consistent with dispersal of propagules by wind (Selkirk et al. 1998). In the Cape Chocolate populations, the results were consistent with initial colonisation of a site towards the top of the drainage channel by a single wind borne propagule and subsequent downslope dispersal within the channel via water. Wind borne propagules that may have landed outside the protection of the channel sides in this exposed area presumably were not able to survive and establish moss colonies (Skotnicki et al. 1998c). Local dispersal by overland flow of water has been inferred also for *Bryum pseudotriquetrum* near Casey Station in Wilkes Land, Antarctica (Selkirk and Seppelt 1987).

Near the summit of Mt Melbourne in Victoria Land, small patches of *Campylopus pyriformis* occur on several areas of geothermally heated ground. RAPD studies of single shoots from each of 26 clumps showed three identical samples and very little genetic variation amongst the others. This extremely isolated population of *C. pyriformis* appears to have been derived from a single intercontinental immigration event, followed by local wind dispersal (over a few to tens of metres) and some mutational change. Further dispersal from Mt Melbourne to heated ground on Mt Erebus, 300km to the south, is indicated, since there is a close genetic relationship between the protonemal *C. pyriformis* on Mt Erebus and the mature gametophyte plants on Mt Melbourne (Skotnicki et al. 2001).

Results of RAPD genetic analyses with *Sarconeurum glaciale* at Arrival Heights on Ross Island showed that colonies within drainage and frost heave channels are more closely related to each other than to other colonies from nearby and distant sites. Combined with the observation of snow drifts collecting and melting in these channels, it was clear that water plays a role in local dispersal of this species over distances of between 1 and 30m (Skotnicki et al. 1999).

Investigations using the RAPD technique to examine colonisation patterns on recently exposed ground near the front of retreating glaciers has provided evidence for probable long-distance dispersal. A single, very isolated colony of *Bryum pseudotriquetrum* sampled near the northern face of Crescent Glacier on the southern slope of Taylor Valley appeared very different from the other samples collected near the southern shore of Lake Fryxell and the southern face of Canada Glacier, 4km away on Taylor Valley floor. The Crescent Glacier colony however gave an identical RAPD banding pattern to a sample from an extensive turf from Cape Chocolate 45km to the south (Skotnicki et al. 1998c) with this turf being the potential source population.

Recolonisation of anthropogenically disturbed sites is an important issue in Antarctica where the conservation of the small areas of natural vegetation is crucial. Adherence to strategies that ensure minimal long-term environmental disturbance is an important component of any human activity. The origins of recolonising propagules from moss colonies growing near or on a bulldozed track near Scott Base were compared with samples from nearby undisturbed colonies of the same species. Nearby colonies in the general area were shown to be the most likely source of propagules for recolonisation, rather than those from further away, despite the prevailing strong winds (Skotnicki et al. 1999). It seems that fragments of whole colonies, up to several centimetres in diameter, in addition to smaller propagules (perhaps even leaf tip gemmae) can initiate growth of new colonies. However, recolonisation does not occur rapidly or easily and it is clear that promoting recolonisation of disturbed ground in Antarctica will need considerable attention, time and care.

Mutation in Antarctic moss populations

RAPD analysis of *Ceratodon purpureus* specimens from a range of Antarctic, subantarctic and temperate populations showed that significant levels of genetic variation occurred within populations (Skotnicki et al. 1998b). In addition, using RAPDs we have demonstrated that mutation in *C. purpureus* occurs within single colonies, with the number of mutations being directly correlated with the distance apart shoots are sampled within a colony: closer shoots are less likely to have genetic differences, whereas those separated by 5mm or more (presumably equivalent to many years' growth) are much more likely to exhibit genetic changes in their RAPD profile (Skotnicki et al. 2004).

This has now been confirmed and extended with sequencing of the 18S – 26S nuclear ribosomal Internal Transcribed Spacer (ITS) region indicate that Antarctic and subantarctic isolates form one clade and Australian isolates another. The differences detected may be the beginning of cryptic speciation (Longton and Hedderson 2000, Shaw 2000a, 2001, Shaw et al. 2002) although the changes are not yet sufficient to warrant their separation into a separate species as has been proposed by Burley and Pritchard (1990).

Some individual colonies of *Ceratodon purpureus* at Granite Harbour in southern Victoria Land have very long shoots (up to 10 cm in length) and are likely to be around 100 years old. Preliminary experiments on single long shoots cut into 50 or more segments, have very occasionally detected mutation along the length of such shoots using RAPDs (Skotnicki et al. 2004). This fascinating and important finding, that mutation appears to be responsible for the high levels of genetic diversity observed in Antarctic moss populations, can now be investigated further and additional experiments are planned on these mosses. Cryptic speciation in Antarctic mosses, as in temperate mosses (McDaniel and Shaw 2003), is probably increasingly prevalent as exposure to mutagenic UVB radiation becomes more prolonged (Skotnicki et al. 2004).

Robinson et al. (2005) reported detrimental physiological effects, morphological irregularities and potential DNA damage in plants of the Antarctic endemic moss *Schistidium antarctici* growing in ambient UV–B conditions. Plants growing under screens that reduced the levels of exposure to UV–B did not express similar effects. Using a combination of RAPDs, DNA sequencing of individual genes and biochemical assays for rates of DNA repair, it is now possible to make detailed investigations of mutation in Antarctic populations compared with temperate populations where colonies are not yet exposed to such high levels of mutagenic UV–B irradiation.

Antarctic mosses provide excellent model systems for measuring mutation (and cryptic speciation) within living plants, with the possibility of assessing the effect of increased UV-B exposure on mutation rates over time. In addition, the finding that useful DNA can satisfactorily be extracted and analysed from very old dry herbarium specimens is valuable in this context.

Origins of Antarctic moss populations

For *Bryum argenteum* populations in Southern Victoria Land, our study of the geographic distribution of isozyme and RAPD markers showed one common genotype occurring at high frequency in each of the five locations studied and other less frequent genotypes with more restricted distributions. These results are consistent with the present populations of *B. argenteum* in this area being remnants of genetically variable indigenous populations, or being the result of immigration by a range of genetically diverse propagules, or a combination of these (Adam et al. 1997). Recently, it has been suggested that *Bryum subrotundifolium* may also occur in some of these populations and further genetic analysis should help to determine whether two distinct species do exist there.

Lewis Smith (1993) has experimentally demonstrated the presence of a moss-propagule bank in Antarctic soils, capable of developing into plants when conditions become appropriate and the capacity for oases (large and small) to act as sources of moss propagules for local dispersal (Lewis Smith 1997). Using the RAPD technique we have provided evidence for dispersal of propagules over distances of a few metres to tens of kilometres for *Bryum argenteum, B. pseudotriquetrum, Sarconeurum glaciale* and *Hennediella heimii* (Selkirk et al. 1998, Skotnicki et al. 1998b, 1999). Using the RAPD techniques and DNA sequencing we infer dispersal over hundreds to thousands of kilometres for *Campylopus pyriformis* and *Pohlia nutans* (Skotnicki et al. 2001, 2002, Bargagli et al. 2004). Muñoz et al. (2004) have shown stronger correlation of floristic similarities among locations with wind-route connections than between geographically close locations, supporting the idea that wind is an intercontinental dispersal agent for many organisms in the Southern Hemisphere.

It seems probable that the genetically variable populations of mosses now in continental Antarctica have multiple origins:

(1) some species have long been established there, with genetic diversity within their populations increasing over time by a combination of mutation, immigration of propagules of different genetic makeup and local dispersal,

(2) some species are relatively recently established in habitats appropriate for colonisation by the species,

(3) cryptic species, developed by mutation from their ancestral species, that may in future become morphologically or physiologically distinct species, meanwhile contributing to the genetic variation within populations of Antarctic mosses.

Antarctic populations of *B. argenteum* appear to be long-established (Adam et al. 1997), but whether this moss has colonised Antarctica rarely in the past and spread to other parts of the continent since then, or whether many colonisation events have occurred, is not known. Similarly, it is not known whether this species has adapted genetically to survive and spread in the harsh Antarctic climate; transplantation experiments with plants from different origins provided little evidence for specialised adaptations by Antarctic plants to their severe environment (Longton and MacIver 1977). *Campylopus pyriformis* and *Pohlia nutans* are cosmopolitan moss species that occur in Victoria Land only on geothermally heated ground and are not known from elsewhere in continental Antarctica (Skotnicki et al. 2001, 2002). It is likely that their propagules have fallen elsewhere in Victoria Land and Antarctica, but have not become established in ambient conditions that were unsuitable. There is the potential, with global climate change and other environmental change, including anthropogenic modifications, for conditions in the future to allow establishment of additional temperate plant species whose propagules may well have reached the Antarctic but not become established in the past (Lewis Smith 1993).

Plant pathogens in the Antarctic

Where there are plants, there are likely to be plant pathogens and the Antarctic is proving to be no exception. In continental Antarctica pathogenic fungus causing rings of dead shoots on colonies of the moss *Bryum argenteum* have been observed and the fungus has been identified as a new *Embellisia* species (Bradner et al. 2000). In the absence of fruiting material on these Antarctic fungi due to the climate, it appears that molecular genetics may be the quickest and simplest method of initial identification.

Fungal rings are relatively common on several other moss species in the maritime Antarctic and also on *Ceratodon*, *Syntrichia principes* and *Schistidium antarctici* in Victoria land (R. Lewis Smith, pers. comm.). Similarly, we are characterising a fungus that causes disease and death of the moss *Dicranoweisia brevipes* on Heard Island, with rings of dead shoots in the centre of an infected patch. Symptoms were observed on large colonies of this moss at several locations around Heard Island. The results so far indicate this will also prove to be a new species of fungus.

Molecular genetic techniques, combined with electron microscopy, have clearly demonstrated the presence of a new plant badnavirus in the vascular species *Stilbocarpa polaris* on Macquarie Island, which we have named *Stilbocarpa* mosaic bacilliform virus (SMBV, Skotnicki et al. 2003). Although algal viruses have been detected in various Antarctic aquatic ecosystems (Kepner et al. 1998, Pearce and Wilson 2003) this virus in a terrestrial plant native to Macquarie Island is the first report of a virus in a vascular plant anywhere in the southern polar regions. The presence of yellow mosaic symptoms, virus particles in low concentrations in the plant sap and badnavirus DNA in diseased plants are consistent with infection of *Stilbocarpa* by a badnavirus. Although SMBV has only recently been reported from Macquarie Island, it is not known how long it has been on the island, but recent results indicate it has most probably been present for a very long time. Although *Stilbocarpa polaris* also occurs on other Southern Ocean islands Campbell Island, Auckland Islands and the Snares, no virus infection has been reported from them.

The mechanism of transmission of SMBV in *Stilbocarpa* on Macquarie Island is not yet known, but it appears that the extensive rabbit grazing of *Stilbocarpa* on this island is not sufficient to transmit badnavirus infection. The restriction of disease symptoms to small isolated foci of infection, with up to 20-30% of plants having mosaic symptoms at some sites, also indicates that transmission among these perennial plants occurs infrequently.

Changing patterns of atmospheric circulation may bring propagules of new species to the island: other known badnaviruses occur in warmer climates. Potential links between documented climate change on Macquarie Island and the introduction and spread of SMBV will be investigated. This may give an indication of how plant viruses will respond to climate change elsewhere, with the potential to correlate the spread of the virus along the island with climate warming. Further research on this new plant virus from a very isolated location, including the sequencing of the complete viral genome (allowing analysis of its relationship to other badnaviruses) will yield information of practical importance to management and conservation of the World Heritage Macquarie Island ecosystem, in addition to fascinating information about evolution and dispersal of plant viruses.

Snow algae and endolithic algae

ITS gene sequences have been determined for endolithic algae from southern Victoria Land, Antarctica (Friedmann et al. 1988, de la Torre et al. 2003, M. Skotnicki et al. unpubl. data). These sequences should assist in identification of novel isolates of these algae. Similarly, we have sequenced the ITS region of several species of snow algae mainly from the Windmill Islands. The sequences obtained were species-specific (M. Skotnicki et al. unpubl. data), and the availability of known species together with gene sequences should aid in the identification of new isolates, even in the absence of morphological and life history characteristics previously necessary for taxonomic identification (Ling 2001).

Maritime Antarctic and subantarctic vascular plants

Populations of the only two flowering plants on the Antarctic Peninsula have recently been increasing in size and number, and this has been related to climate warming (Lewis Smith 1994, Convey this volume). In a study using Amplified Fragment Length Polymorphisms (AFLPs), investigating 10 populations of *Deschampsia antarctica* spanning a distance of 1350km, little genetic diversity was observed (Holdregger et al. 2003). However, populations from the northern and southern extents of the study were deemed to be genetically distinct from each other and it appeared that there were low levels of gene flow between them. It was concluded that populations of *D. antarctica* in the maritime Antarctic were founded by one or few individuals and that the species mainly reproduced by selfing or vegetative propagation. A database of gene sequences of subantarctic *Deschampsia* and *Colobanthus* species will facilitate comparison with specimens from the Antarctic Peninsula and allow analysis of their origins and dispersal patterns (van de Wouw 2003).

Molecular genetics techniques are also being used to investigate the diversity and dispersal of vascular plants on subantarctic islands. For the following dicotyledon species, sequences have been obtained for the ITS region, enabling comparison of populations within islands and among islands where appropriate: from Macquarie Island *Stilbocarpa polaris, Pleurophyllum hookeri, Crassula moschata, Callitriche antarctica, Cardamine corymbosa,* from Heard Island and Iles Kerguelen *Crassula moschata* and *Callitriche antarctica.* This study will extend to other subantarctic islands using species common to all or some of these islands. Sequences of the ITS region from a wide range of the grasses found on Heard Island, Macquarie Island and Iles Kerguelen and nearby temperate localities, including the rapidly spreading non-indigenous species *Poa annua* (Scott 1989), is enabling comparison and inferences about their dispersal patterns (M. Skotnicki et al. unpubl. data). Genetic studies have the potential to enable identification of dispersal routes in a way not possible by any other techniques.

Overall utility of genetics

The results presented here have shown that levels of ITS sequence variation are high in some Antarctic moss species. As the mosses appear to reproduce solely by vegetative means (Seppelt et al. 1992), such variation appears to have arisen either by infrequent multiple colonisation events, or by *de novo* mutation within the haploid plants.

We are also currently sequencing several more genes from these Antarctic mosses, including the *trn*L-F, *rps*4 and *rbc*L chloroplast genes (De Luna et al. 2000, Pedersen et al. 2003, Virtanen 2003), to assist with determining the extent of genetic variability, origins and evolution of moss species in this isolated and extreme location. However, the ITS region appears to be excellent for analysis of Antarctic mosses, enabling identification and analysis of some genetic variability among populations.

Although the extent of genetic diversity is much lower in vascular plants than in mosses, the results obtained so far indicate that the ITS region will also be useful in determining the origins and dispersal routes of plants on subantarctic islands, including the grasses and other flowering plants. Molecular genetics has a major and valuable role to play in understanding how plants colonise and evolve to survive in the extremely isolated and harsh environments of Antarctica and the subantarctic.

References

Adam K.D., Selkirk, P.M., Walsh, S.M. and Connett, M.B. (1997) Genetic variation and patterns of colonisation in the moss *Bryum argenteum* in East Antarctica, in B. Battaglia, J. Valencia and D.W. H. Walton, (eds.), *Antarctic Communities, Species, Structure and Survival,* Cambridge University Press pp 33-38.

Adamson, H. and Adamson, E. (1992) Possible effects of global climate change on Antarctic terrestrial vegetation, in P.G. Quilty (ed.), *Impact of Climate Change Australia Antarctica.* Australian Government Publishing Service, Canberra, pp 52-62.

Baldwin, B.G. (1992) Phylogenetic utility of the internal transcribed spacers of nuclear ribosomal DNA in plants: an example from the Compositae, *Molecular Phylogenetics and Evolution* **1**, 3-16.

Bargagli, R., Skotnicki, M.L., Marri, L., Pepi, M., Mackenzie, A. and Agnorelli, C. (2004) New record of moss and thermophilic bacteria species and physico-chemical properties of geothermal soils on the northwest slope of Mt. Melbourne (Antarctica), *Polar Biology,* **27**, 423-431.

Bednarek-Ochyra, H., Vána, J., Ochyra, R. and Lewis Smith, R.I. (2000) *The Liverwort Flora of Antarctica*, Polish Academy of Sciences, Institute of Botany, Cracow. 238 pp.

Bergstrom, D.M. and Selkirk, P.M. (1997) Distribution of bryophytes on subantarctic Heard Island, *The Bryologist* **100**, 349-355.

Boisselier-Dubayle, M. C. and Bischler, H. (1994) A combination of molecular and morphological characters for delimitation of taxa in European *Porella, Journal of Bryology* **18**, 1-11.

Boisselier-Dubayle, M.C., Lambourdière, J. and Bischler, H. (2002) Molecular phylogenies support multiple morphological reductions in the liverwort subclass Marchantiidae (Bryophyta), *Molecular Phylogenetics and Evolution* **24**, 66-77.

Boisselier-Dubayle, M.C., Jubier, M.F., Lejeune, B. and Bischler, H. (1995) Genetic variability in the three subspecies of *Marchantia polymorpha* (Hepaticae): isozymes, RFLP and RAPD markers, *Taxon* **44**, 363-376.

Bopp, M. and Capesius, I. (1996) New aspects of bryophyte taxonomy provided by a molecular approach, *Botanica Acta* **109**, 368-372.

Bradner, J.R., Sidhu, R.K., Yee, B., Skotnicki, M.L., Selkirk, P.M. and Nevalainen, K.M.H. (2000) A new microfungal isolate, *Embellisia* sp., associated with the Antarctic moss *Bryum argenteum, Polar Biology* **23**, 730-732.

Buck, W. R., Goffinet, B. and Shaw, A.J. (2000) Testing morphological concepts of orders of pleurocarpous mosses (Bryophyta) using phylogenetic reconstructions based on *trn*L-*trn*F and *rps*4 Sequences, *Molecular Phylogenetics and Evolution* **16**, 180-198.

Burley, I.S. and Pritchard, N.M. (1990) Revision of the genus *Ceratodon* (Bryophyta), *Harvard Papers in Botany* **2**, 368-372.

Chiang, T.Y. and Schaal, B.A. (1999) Phylogeography of North American populations of the moss species *Hylocomium splendens* based on the nucleotide sequence of internal transcribed spacer 2 of nuclear ribosomal DNA, *Molecular Ecology*, **8**, 1037-1042.

Convey, P. (2006) Antarctic climate change and its influences on terrestrial ecosystems, in D.M. Bergstrom, P. Convey, and A.H.L. Huiskes (eds.), *Trends in Antarctic Terrestrial and Limnetic Ecosystems: Antarctica as a Global Indicator*, Springer, Dordrecht (this volume).

Cox, C.J. and Hedderson, T.A.J. (2003) Phylogenetic relationships within the moss family Bryaceae based on chloroplast DNA evidence, *Journal of Bryology* **25**, 31-40.

Cox, C.J., Goffinet, B., Newton, A.E., Shaw, A.J. and Hedderson, T.A.J. (2000) Phylogenetic relationships among the diplolepideous-alternate mosses (Bryidae) inferred from nuclear and chloroplast DNA sequences, *The Bryologist* **103**, 224-241.

Dale, T.M., Skotnicki, M.L., Adam, K.D. and Selkirk, P.M. (1999) Genetic diversity in the moss *Hennediella heimii* in Miers Valley, southern Victoria Land, Antarctica, *Polar Biology* **21**, 228-233.

de la Torre, J.R., Goebel, B.M., Friedmann, I. and Pace, N.R. (2003) Microbial diversity of cryptoendolithic communities from the McMurdo Dry Valleys, Antarctica, *Applied and Environmental Microbiology* **69**, 3858-3867.

De Luna, E., Buck, W.R., Akiyama, H., Arikawa, T., Tsubota, H., González, D., Newton, A.E. and Shaw, A.J. (2000) Ordinal phylogeny within the hypnobryalean pleurocarpous mosses inferred from cladistic analyses of three chloroplast DNA sequence data sets: *trn*L-F, *rps*4, and *rbc*L, *The Bryologist* **103**, 242-256.

Edwards, J.A. and Smith, R.I.L. (1988) Photosynthesis and respiration of *Colobanthus quitensis* and *Deschampsia antarctica* from the maritime Antarctic, *British Antarctic Survey Bulletin* **81**, 43-63.

Farman, J.C., Gardiner, B.G. and Shanklin, J.D. (1985) Large losses of total ozone in Antarctica reveal seasonal ClO$_x$/No$_x$ interaction, *Nature* **315**, 207-210.

Frenot, Y., Chown, S., Whinam, J., Selkirk, P.M., Convey, P., Skotnicki, M., and Bergstrom D.M. (2005) Biological invasions in the Antarctic: extent, impacts and implications. *Biological Reviews* **80**, 45-72.

Friedmann, E.I., Hua, M. and Ocampo-Friedmann, R. (1988) Cryptoendolithic lichen and cyanobacterial communities of the Ross Desert, Antarctica, *Polarforschung* **58**, 251-259.

George, A.S. (ed.) (1993) *Flora of Australia Volume 50, Oceanic islands 2,* Australian Government Publishing Service, Canberra.

Holderegger, R., Stehlik, I.S., Smith, R.I.L. and Abbott, R.J. (2003) Populations of Antarctic hairgrass (*Deschampsia antarctica*) show low genetic diversity, *Arctic, Antarctic and Alpine Research* **35**, 214-217.

Jones, C.J., Edwards, K.J., Castaglione, S., Winfield, M.O., Sala, F., van de Wiel, C., Bredemeijer, G., Vosman, B., Matthes, M., Daly, A., Brettschneider, R., Bettini, P., Buiatti, M., Maestri, E., Malcevschi, A., Marmiroli, N., Aert, R., Volckaert, G., Rueda, J. and Linacer, R. (1997) Reproducibility testing of RAPD, AFLP and SSR markers in plants by a network of European laboratories. *Molecular Breeding* **3**, 381-390.

Kennedy, A.D. (1995) Antarctic terrestrial ecosystem response to global environmental change, *Annual Review of Ecology and Systematics* **26**, 683-704.

Kepner, R.L., Wharton, R.A. and Suttle, C.A. (1998) Viruses in Antarctic lakes. *Limnology and Oceanography* **43**, 1754-1761.

Lewis, L.A., Mishler, B.D. and Vilgalys, R. (1997) Phylogenetic relationships of the Liverworts (Hepaticae), a basal Embryophyte lineage, inferred from nucleotide sequence data of the chloroplast gene rbcL, *Molecular Phylogenetics and Evolution,* **7**, 377-393.

Lewis Smith, R.I. (1984) Terrestrial plant biology of the sub-Antarctic and Antarctic, in R.M. Laws (ed.), *Antarctic Ecology,* Academic Press, London, pp 61-162.

Lewis Smith, R.I. (1993) The role of bryophyte propagule banks in primary succession: case-study of an Antarctic fellfield soil, in J. Miles and D.W.H. Walton (eds.), *Primary Succession on Land,* Special Publication Number 12 of the British Ecology Society, Blackwell, Oxford, pp 55-77.

Lewis Smith, R.I. (1994) Vascular plants as bioindicators of regional warming in Antarctica, *Oecologia,* **99**, 322-328.

Lewis Smith, R.I. (1997) Oases as centres of high plant diversity and dispersal in Antarctica, in W.B. Lyons, C. Howard-Williams and I. Hawes (eds.), *Ecosystem Processes in Antarctic Ice-free Landscapes.* Balkema, Rotterdam, pp 119-128.

Lewis Smith, R.I. (1999) Biological and environmental characteristics of three cosmopolitan mosses dominant in continental Antarctica, *Journal of Vegetation Science* **10**, 231-242.

Lewis Smith, R.I. (2003) The enigma of *Colobanthus quitensis* and *Deschampsia antarctica* in Antarctica, in A.H.L. Huiskes, W.W.C. Gieskes, J. Rozema, R.M.L. Schorno, S.M. van der Vies, W.J. Wolf, (eds.) *Antarctic Biology in a Global Context.* Leiden, Backhuys, pp 234-239.

Lewis Smith, R.I. (2005) Bryophyte diversity and ecology of two geologically contrasting Antarctic islands, *Journal of Bryology* **27**, 195-206.

Ling, H.U. (2001) Snow algae of the Windmill Islands, continental Antarctica, *Desmotetra aureospora* sp. nov. and *D. antarctica,* comb. nov., Chlorophyta, *Journal of Phycology* **37**, 160-174.

Longton, R.E. (1988) *The Biology of Polar Bryophytes and Lichens,* Cambridge University Press, Cambridge, 391 pp.

Longton, R.E. (1994) Reproductive biology in bryophytes–the challenge and the opportunities, *Journal of the Hattori Botanical Laboratory* **76**, 159-172.

Longton, R.E. and Hedderson, T.A. (2000) What are rare species and why conserve them? *Lindbergia,* **25**, 53-61.

Longton, R.E. and MacIver, M.A. (1977) Climate relationships in Antarctic and Northern Hemisphere populations of a cosmopolitan moss, *Bryum argenteum* Hedw., in G.A. Llano (ed.) *Adaptations within Antarctic Ecosystems.* Smithsonian Institution, Washington. pp. 899-919.

Magombo, Z.L.K. (2003) The phylogeny of basal peristomate mosses: evidence from cpDNA, and implications for peristome evolution, *Systematic Botany* **28**, 24-38.

Marshall, W.A. (1996) Biological particles over Antarctica, *Nature,* **383**, 680.

Marshall, W.A. and Convey, P. (1997) Dispersal of moss propagules on Signy Island, maritime Antarctic, *Polar Biology* **18**, 376-383.

McDaniel, S.F. and Shaw, A. J. (2003) Phylogeographic structure and cryptic speciation in the trans-Antarctic moss *Pyrrhobryum mniodes, Evolution* **57**, 205-215.

Muñoz, J., Felicísimo, Á.M., Cabezas, F., Burgaz, A.R. and Martínez, I. (2004) Wind as a long-distance dispersal vehicle in the Southern Hemisphere, *Science,* **304**, 1144-1147.

Newton, A.E., Cox, C.J., Duckett, J.G., Wheeler, J.A., Goffinet, B., Hedderson, T.A.J. and Mishler, B.D. (2000) Evolution of the major moss lineages: phylogenetic analyses based on multiple gene sequences and morphology, *The Bryologist* **103**, 187-211.

Ochyra, R. (1998) *The Moss Flora of King George Island Antarctica*, Polish Academy of Sciences, Cracow. 278pp.

Øvstedal, D.O. and Lewis Smith, R.I. (2001) *Lichens of Antarctica and South Georgia A Guide to their Identification and Ecology*, Cambridge University Press, Cambridge, 411 pp.

Patterson, E., Boles, S.B. and Shaw, A.J. (1998) Nuclear ribosomal DNA variation in *Leucobryum glaucum* and *L. albidum* (Leucobryaceae): a preliminary investigation, *The Bryologist* **101**, 272-277.

Pearce, D.A. and Wilson, W.H. (2003) Viruses in Antarctic ecosystems, *Antarctic Science* **15**, 319-331.

Pedersen, N., Cox, C.J. and Hedenäs, L. (2003) Phylogeny of the moss family Bryaceae inferred from chloroplast DNA sequences and morphology, *Systematic Botany* **28**, 471-482.

Robinson, S.A, Turnbull, J.A, and Lovelock, C.E (2005). Impact of changes in natural ultraviolet radiation on pigment composition, physiological and morphological characteristics of the Antarctic moss, *Grimmia antarctici*, *Global Change Biology* **11**, 476-489.

Robinson, S.A., Wasley, J. and Tobin, A.K. (2003) Living on the edge — plants and global change in continental and maritime Antarctica, *Global Change Biology* **9**, 1681-1717.

Schneller, J., Holderegger, R., Gugerli, F., Eichenberger, K. and Lutz, E. (1998) Patterns of genetic variation detected by RAPDs suggest a single origin with subsequent mutations and long distance dispersal in the apomictic fern *Dryopteris remota* (Dryopteridaceae), *American Journal of Botany*, **85**, 1038-1042.

Scott, J.J. (1989) New records of vascular plants from Heard Island, *Polar Record*, **25**, 37-42.

Selkirk, P.M., and Seppelt, R.D. (1987) Species distribution within a mossbed in Greater Antarctica, *Symposia Biologica Hungarica* **35**, 279-284.

Selkirk, P.M., Seppelt, R.D. and Selkirk, D.R. (1990) *Subantarctic Macquarie Island. Environment and Biology*. Studies in Polar Research Cambridge University Press, Cambridge, 285 pp.

Selkirk, P M., Skotnicki, M.L., Adam, K.D., Connett, M.B., Dale, T., Joe, T.W. and Armstrong, J. (1997) Genetic variation in Antarctic populations of the moss *Sarconeurum glaciale*, *Polar Biology* **18**, 344-350.

Selkirk, P. M., Skotnicki, M.L., Ninham, J.A., Connett, M.B. and Armstrong, J. (1998) Genetic variation and dispersal of *Bryum argenteum* and *Henediella heimii* populations in the Garwood Valley, Southern Victoria Land, Antarctica, *Antarctic Science* **10**, 423-430.

Seppelt, R.D. (2004) *The Moss Flora of Macquarie Island*, Australian Antarctic Division, Kingston, Tasmania, 328 pp.

Seppelt, R.D. and Green, T.G.A. (1998) A bryophyte flora for southern Victoria Land, Antarctica, *New Zealand Journal of Botany* **36**, 617-635.

Seppelt, R.D. and Selkirk, P.M. (1984) Effects of submersion on morphology and implications of induced environmental modification on the taxonomic interpretation of selected Antarctic moss species, *Journal of the Hattori Botanical Laboratory* **55**, 273-279.

Seppelt, R.D., Green, T.G.A. and Skotnicki, M.L. (1999) Notes on the flora, vertebrate fauna and biological significance of Beaufort Island, Ross Sea, Antarctica, *Polarforschung* **66**, 53-59.

Seppelt, R.D., Green, T.G.A., Schwartz, A.J. and Frost, A. (1992) Extreme Southern locations for moss sporophytes in Antarctica, *Antarctic Science* **4**, 37-39.

Shaw, A.J. (2000a) Molecular phylogeography and cryptic speciation in the mosses, *Mielichhoferia elongata* and *M. mielichhoferiana* (Bryaceae), *Molecular Ecology* **9**, 595-608.

Shaw, A.J. (2000b) Phylogeny of the Sphagnopsida based on chloroplast and nuclear DNA sequences, *Bryologist* **103**, 277-306.

Shaw, A.J. (2001) Biogeographic patterns and cryptic speciation in bryophytes, *Journal of Biogeography* **28**, 253-261.

Shaw, A.J., McDaniel, S.F., Werner, O. and Ros, R.M. (2002) New frontiers in bryology and lichenology: phylogeography and phylodemography, *The Bryologist* **105**, 373-383.

Skotnicki, M.L., Bargagli, R. and Ninham, J.A. (2002) Genetic diversity in the moss *Pohlia nutans* on geothermal ground of Mount Rittmann, Victoria Land, Antarctica, *Polar Biology* **25**, 771-777.

Skotnicki, M.L., Mackenzie, A.M. and Selkirk, P.M. (2004) Mosses surviving on the edge: origins, genetic diversity and mutation in Antarctica, in Goffinet, B. and Magill, R. (eds.) *Molecular Systematics of Bryophytes: Progress, problems and perspectives,* Monographic Series in Botany, Missouri Botanical Gardens Press, St. Louis, 98: 388-403.

Skotnicki, M L., Ninham, J.A. and Selkirk, P.M. (1998b) Genetic diversity in the moss *Bryum argenteum* in Australia, New Zealand and Antarctica, *The Bryologist* **101**, 412-421.

Skotnicki, M.L., Ninham, J.A. and Selkirk, P.M. (1999) Genetic diversity and dispersal of the moss *Sarconeurum glaciale* on Ross Island, East Antarctica, *Molecular Ecology* **8**, 753-762.

Skotnicki, M. L., Ninham, J.A. and Selkirk, P.M. (2000) Genetic diversity, mutagenesis and dispersal of Antarctic mosses - a review of progress with molecular studies. *Antarctic Science* **12**, 363-373.

Skotnicki M.L., Selkirk, P.M. and Beard, C. (1998a) RAPD profiling of genetic diversity in Antarctic populations of the moss *Ceratodon purpureus, Polar Biology* **19**, 172-176.

Skotnicki, M.L., Selkirk, P.M. and Dale, T.M. (1997) RAPD profiling of Antarctic mosses, in W.B. Lyons, C. Howard-Williams and I. Hawes (eds.), *Ecosystem Processes in Antarctic Ice-Free Landscapes,* Balkema, Rotterdam, pp129-136.

Skotnicki, M.L., Selkirk, P.M. and Ninham, J.A. (1998c) RAPD analysis of genetic variation and dispersal of the moss *Bryum pseudotriquetrum* from Southern Victoria Land, Antarctica, *Polar Biology* **20**, 121-126.

Skotnicki, M.L., Selkirk, P.M., Broady, P., Adam K.D. and Ninham, J.A. (2001) Dispersal of the moss *Campylopus pyriformis* on geothermal ground near the summits of Mount Erebus and Mount Melbourne, Victoria Land, Antarctica, *Antarctic Science* **13**, 280-285.

Skotnicki, M.L., Selkirk, P.M., Kitajima, E., McBride, T.P., Shaw, J. and Mackenzie, A. (2003) The first subantarctic plant virus report: Stilbocarpa mosaic bacilliform badnavirus (SMBV) from Macquarie Island, *Polar Biology* **26**, 1-7.

Turner P.A.M., Scott J.J. and Rozefelds A. (2006) Probable long distance dispersal of *Leptinella plumosa* Hook. f. to Heard Island: habitat, status and discussion of its arrival, *Polar Biology,* **29**, 160-168.

Vána, J. and Gremmen, N.J.M (2005) Hepatics of Heard Island. *Cryptogamie Bryologie Lichenologie*

Vanderpoorten, A., Shaw, A.J. and Goffinet, B. (2001) Testing controversial alignments in *Amblystegium* and related genera (Amblystegiaceae: Bryopsida). Evidence from rDNA ITS sequences, *Systematic Botany* **26** 470-479.

Van de Wouw, M. (2003) Population increase in *Deschampsia antarctica*: genotypic variation or phenotypic plasticity? in: *Regional Sensitivity to Climate Change in Antarctic Terrestrial and Limnetic Ecosystems (RiSCC),* Report on the Fifth Workshop, Varese, Italy, 2-8 July 2003.

Virtanen, V. (2003) Phylogeny of the Bartramiaceae (Bryopsida) based on morphology and on *rbc*L, *rps*4, and *trn*L-*trn*F sequence data, *The Bryologist* **106**, 280-296.

Walton, D.W.H. (1990) Colonization of terrestrial habitats—organisms, opportunities and occurrence, in K.R. Kerry and G. Hempel (eds.), *Antarctic Ecosystems, Ecological Change and Conservation,* Springer-Verlag, Berlin, pp 51-60.

Wyatt, R. (1994) Population genetics of bryophytes in relation to their reproductive biology, *Journal of the Hattori Botanical Laboratory* **76** 147-157.

9. THE MOLECULAR ECOLOGY OF ANTARCTIC TERRESTRIAL AND LIMNETIC INVERTEBRATES AND MICROBES

M. I. STEVENS
Allan Wilson Centre for Molecular Ecology and Evolution,
Massey University
Private Bag 11-222, Palmerston North, New Zealand
&
Department of Genetics, La Trobe University
Bundoora, 3083 Victoria, Australia
M.I.Stevens@massey.ac.nz

I. D. HOGG
Centre for Biodiversity and Ecology Research, University of Waikato
Private Bag 3105, Hamilton, New Zealand
hogg@waikato.ac.nz

Introduction

The Antarctic landscape has been dominated by long-term habitat fragmentation, with less than 1% of the 14 million km^2 of the continent ice-free today, and more than 10 major glacial cycles over the last one million years (Hays et al. 1976, Lawver and Gahagan 2003, Roberts et al. 2003). The continental landscape is well known as an extreme environment, but there remains debate about 'when' these conditions became extreme (Miller and Mabin 1998, Roberts et al. 2003). Climate cooling and subsequent glaciation of Antarctica was not possible until both the South Tasman Rise cleared the Oates Land coast of East Antarctica (~32 MYA) and the opening of the Drake Passage to deep water circulation (~28 MYA) (Lawver and Gahagan 2003). However, while the circum-Antarctic currents were sufficient to isolate Antarctica from other continental landmasses, glaciation of Antarctica was not immediate (see also Bergstrom et al. this volume). It is clear from palaeological

177

D.M. Bergstrom et al. (eds.), Trends in Antarctic Terrestrial and Limnetic Ecosystems, 177–192.
© *2006 Springer.*

evidence that a *Nothofagus*-herb-moss tundra vegetation replaced the Gondwanan forest and persisted in the region until possibly the Pliocene (~2-5 MYA) or late Miocene (~5-12 MYA) (Ashworth and Preece 2003). Furthermore, fossil records suggest that some elements of the terrestrial (eg weevils), and limnetic faunas (eg lymnaeid gastropods, the bivalve *Pisidium* and at least one species of fish) persisted throughout the Trans-Antarctic Mountains until their extinction in the Pliocene (Ashworth and Kuschel 2003, Ashworth and Preece 2003). This extinct continental fauna evolved during or before the Jurassic, with most achieving global distributions (including Antarctica) by the Cretaceous (Ashworth and Preece 2003). Accordingly, the present-day invertebrate taxa may be relics of this once more abundant and widespread fauna present on the Gondwanan supercontinent.

On the Antarctic continent the terrestrial invertebrate fauna is taxonomically limited and consists largely of arthropods, particularly the springtails (Collembola) and mites (Acari) (Hogg and Stevens 2002). However, various flies (Diptera), beetles (Coleoptera), aphids (Homoptera), copepods, isopods, amphipods, annelids and planarians have also been recorded from the maritime regions of the Antarctic Peninsula and subantarctic islands (Gressitt 1964, Balfour-Browne and Tilbrook 1966, Wirth and Gressitt 1967, Gressitt 1970, Richardson and Jackson 1995, Convey and Block 1996, Vernon et al. 1997, Hullé et al. 2003, Winsor and Stevens 2005). Exceptions to this generalisation particularly in limno-terrestrial (*sensu* McInnes and Pugh 1998) and limnetic systems are the smaller protozoans, tardigrades, rotifers and nematodes (Gressitt 1965, Utsugi and Ohyama 1989, Bullini et al. 1994 Vincent 2000, Treonis et al. 2002, de la Torre et al. 2003, Moorhead et al. 2003, Lawley et al. 2004). Indeed, Wharton (2003) has suggested that nematodes may be the most diverse and abundant invertebrates in both the maritime and continental Antarctic regions. Much of the available information is based on studies documenting and/or describing new species (eg Womersley and Strandtmann 1963, Wise 1967, 1971, Dastych 1984, Greenslade and Wise 1984, 1986, Miller et al. 1988, Greenslade 1995). From the earliest Antarctic expeditions, several reports have described, added to, and revised the inventories of terrestrial, limno-terrestrial and limnetic invertebrates (eg Carpenter 1902, Hunter 1967, Strandtmann 1967, Wise 1967, 1971, Jennings 1976, Miller et al. 1988, Shishida and Ohyama 1989, Utsugi and Ohyama 1989, Potapov 1991). Such inventories have relied on morphology-based taxonomy, which in many cases may not adequately reflect true levels of diversity, or fully scrutinize levels of endemicity, particularly since morphological conservatism is a prevalent feature among many invertebrate taxa (eg Witt et al. 2003). However, until recently these inventories remained the only approach to studying the diversity and taxonomic affinities of Antarctic invertebrates. Fortunately, with increased access to molecular techniques (Gaffney 2000, Sunnucks 2000), the diversity of Antarctic invertebrates can now be assessed at levels previously unattainable. Furthermore, these same techniques can be used to test hypotheses related to connectivity (ie gene flow), and can reveal processes and historical events that shaped the pattern of genetic diversity among populations (phylogeography), in addition to their evolutionary history and relationships to other taxa (phylogeny).

This chapter discusses the application of molecular techniques for species identifications and the role of population genetic studies in assessing levels of gene flow/dispersal among habitats. We then continue with a discussion of the phylogenetic relationships of some of the Antarctic fauna and conclude with a summary of future research needs. In keeping with the theme of this volume, we include the terrestrial and limnetic environments. We consider the 'Antarctic' to include the main continental landmass, the Antarctic Peninsula and associated islands and archipelagos (South Shetland, South Orkney, South Sandwich Islands, Bouvetøya), and the ring of 'subantarctic' oceanic islands surrounding the continent at relatively high latitude in the Southern Ocean (Marion Island, Iles Crozet, Iles Kerguelen, Heard Island, Macquarie Island). Accordingly, three main biogeographical zones are recognized - continental, maritime and subantarctic (Huiskes et al. this volume).

Molecular Techniques as an Aid to Taxonomic Identification

The ability to accurately identify taxa has long been an arduous task, even on the Antarctic continent, which lacks the species richness found in other regions (Chown and Convey this volume, Gibson et al. this volume). However, this task may be revolutionised with the application of molecular approaches to taxonomy. Such approaches typically utilize short (400-700 base pair) sequences of mitochondrial or ribosomal DNA (mtDNA and rDNA, respectively), to discriminate among species. More recently, Hebert et al. (2003b) have suggested that DNA "barcoding" using the mitochondrial cytochrome *c* oxidase subunit I (COI) gene, could be used to provide accurate identifications for all animal life. This system would operate with DNA sequences from reference specimens (morphologically confirmed by an appropriate expert) being added to a global database. Any future specimens could then be 'matched' to confirm their taxonomic identities (eg Hebert et al. 2003a, Hogg and Hebert 2004). This system could provide an immediate and long-term solution to tackling the obvious logistical problems associated with species identifications (Hebert et al. 2003b). Although only limited COI data exist for Antarctic invertebrates (Stevens and Hogg 2003, in press, Stevens et al. 2006), additional analyses are ongoing (I.D. Hogg, M.I. Stevens unpubl. data). DNA barcoding may also assist in discriminating morphologically similar (cryptic) species and in determining if any morphological variation is the result of phenotypic plasticity (within-species variation) or genuine species-level differences.

Other molecular approaches have used 16S rDNA to examine microbial diversity. For example, soil samples collected on Mt. Melbourne, Victoria Land revealed one new thermophilic species, in addition to a number of others with specific identities ranging up to 98% in comparison to other known species using BLAST sequence similarity searches in the GenBank database (Bargagli et al. 2004). Similarly, Lawley et al. (2004) examined soil sites along a latitudinal gradient from continental Antarctica (La Gorce Mts), through the Antarctic Peninsula, to Signy Island. They found a very limited overlap using the SSU rRNA gene between the eukaryotic biota among six sites, with generally low relatedness to existing sequence databases. Lawley et al. (2004) found that most sequences were specific to a particular site, and that most of the sequences could be identified to the species or genus levels, while others could only be loosely identified at a family or order level. This comparison among sites suggests high levels of isolation and possibly endemicity. However, even bipolar gene flow has been inferred for microbial taxa (Darling et al. 2000), suggesting that unidentified DNA sequences may simply indicate a lack of known species sequences in such databases.

Whilst Lawley et al. (2004) may have detected the presence of known Antarctic species it was not possible to match sequences to known morphologically characterised taxa. Vincent (2000) suggests that if microbial endemism is possible, Antarctica should be amongst the most likely places in which such organisms may be found. However, contrary to this notion, there is currently limited evidence for microbial endemism in Antarctic terrestrial, or indeed any, polar environment (Priscu et al. 1999, Vincent 2000, Finlay 2002, Lawley et al. 2004). Most surprising is that molecular (16S rDNA) profiling of microbes from accreted ice from cores taken above Lake Vostok (to a depth of 3603 m) show a close agreement with present-day surface microbiota (Priscu et al. 1999, Siegert et al. 2003), although their isolation (approx. 14 million years) is perhaps too recent to expect an evolutionarily distinct biota considering that species-level divergence in prokaryotes may require up to 100 million years (Lawrence and Ochman 1998). Regardless of whether sequence uniqueness is indicative of endemism, or simply a lack of molecular data in other environments, Lawley et al. (2004) have demonstrated that there is very little effective transfer of biota among the study sites they examined.

Gordon et al. (2000) examined 16S rDNA from terrestrial cyanobacterial mats, and from lake ice communities in Taylor Valley, southern Victoria Land. They demonstrated the presence of a diverse microbial community dominated by cyanobacteria in both the terrestrial and lake samples. The lake ice microbial community appeared to be dominated by organisms that are not uniquely adapted to the lake ice ecosystem, but instead are species that originate elsewhere in the surrounding region and opportunistically colonise the unusual habitat provided in the lake (Gordon et al. 2000). Numerous studies (eg Taton et al. 2003, Hirsch et al. 2004, Van Trappen et al. 2004) from a diverse range of environments (ie microbial mats in Lake Fryxell, Taylor Valley, maritime Antarctic lakes and Antarctic sandstone) have revealed several new genera and species. Furthermore, microbiological analyses of deep glacial ice cores above Lake Vostok have revealed the presence of viable species including an actinomycete (*Nocardiopsis antarcticus*) not found elsewhere (Abyzov 1993). Even the lake ice contains viable bacteria (Karl et al. 1999), and 16S rDNA analysis has shown that this bottom assemblage is dominated by the genera *Actinomyces*, *Acidovorax*, *Comomonas* and *Afipia* (Priscu et al. 1999). Affinities to temperate organisms suggest that Antarctica has been open to colonisation by long range transport for a considerable period. However, problems associated with accurate identification have been demonstrated by Pearce et al. (2003) using the bacterioplankton community of a maritime Antarctic freshwater lake on Signy Island. These authors obtained significantly different results using both cultivation-dependent versus cultivation-independent techniques. In particular, the phylogenetic diversity represented by the cultured bacteria differed from that of the 16S rDNA clone library. Nevertheless, they found no evidence of prokaryotes endemic to Antarctica.

With the increasing use of molecular approaches for taxonomic identification of polar invertebrates, (eg Valbonesi et al. 1994, Hogg and Hebert 2004), faunal lists will undoubtedly prove incomplete and will continue to be modified. DNA-based identification systems will also allow for independent molecular verification of 'new' species and/or records that initially appear biogeographically unusual. For example, three species of springtails were recorded from King George Island in the South Shetland Islands (Yue and Tamura 2001). One of the species recorded was *Tullbergia mediantarctica*, previously recorded only from Shackleton Glacier, Queen Maud Mountains. However, the closely related congeneric *Tullbergia mixta*, which has been previously recorded throughout the western Antarctic region, including King George Island (Wise 1967, Convey et al. 1996) was not mentioned. One new species was described, *Cryptopygus nanjiensis*, and again no record of the congeneric *C. antarcticus*, with previous records on King George Island (eg Wise 1967, 1971, Weiner and Hajt 1994, Greenslade 1995, Convey et al. 1996). It has since been suggested that these newer records were erroneous and the record of *Tullbergia mediantarctica* from King George Island was *T. mixta* and *Cryptopygus nanjiensis* was in fact *C. antarcticus* (P. Greenslade unpubl. data). In addition, a study of springtail species distribution and abundance in northern Victoria Land details three previously known species, but also provides details of a new species, *Folsomia antarctica* (Frati et al. 1997). However, in later publications (eg Frati et al. 2000, Frati et al. 2001, Fanciulli et al. 2001) this new species is not included and it would appear that *Folsomia antarctica* may be synonymous with *Cryptopygus* (P. Greenslade unpubl. data).

Interhabitat Dispersal and Population Genetic Structure

Much of what is known of potential dispersal of Antarctic invertebrates is based on anecdotal evidence and/or casual observation. For example, air currents are thought to be one agent of passive dispersal (Pryor 1962, Strong 1967, Greenslade et al. 1999, Muñoz et al. 2004). However, on the continent this mode of transport may not be particularly effective for larger invertebrates (eg springtails, mites, spiders, dipterans) due to a high risk of desiccation and lack of an anhydrobiotic dispersal stage (Marshall and Pugh 1996). By contrast, such dispersal may be more likely in maritime or subantarctic regions due to the relatively more humid environment (Pugh 2003, 2004). Alternative methods include "rafting" along melt-water streams, and subsequently on seawater (Gressitt 1967, Moore 2002, and see Nolan et al. in press for a molecular perspective), or 'hitchhiking' on other animals (eg birds). For taxa that have a specialised dispersal life-stage, such as the nematodes, tardigrades and rotifers, they may possess a much greater potential for dispersal via wind and water that may act as a homogenising force over large distances. These taxa have been recorded widely on the Antarctic continent and in the subantarctic and maritime regions, even though they appear to be restricted primarily to lakes, streams and areas of high moisture content (Pryor 1962, Tilbrook 1967, McInnes and Pugh 1998, Moorehead et al. 2003). For these organisms, geographic barriers may be of limited importance. Although these observations allow speculation on potential dispersal methods, they rarely provide information on the actual frequency or success of such events (Bilton et al. 2001). One method to address this issue is to evaluate population genetic structures to quantify rates of dispersal among habitats.

Despite the usefulness of molecular studies in an Antarctic context (Gaffney 2000), limited data exist for the Antarctic invertebrate fauna. Studies to date include work on springtails, *Desoria klovstadi* and *Gressittacantha terranova* (Collembola, Isotomidae) in northern Victoria Land (eg Frati et al. 1997, 2000, 2001, Fanciulli et al. 2001, Stevens et al. in press) and *Gomphiocephalus hodgsoni* (Collembola, Hypogastruridae) in southern Victoria Land (Stevens and Hogg 2003, Nolan et al. in press). Work has also been undertaken on nematodes (Courtright et al. 2000), dipterans (Vernon et al. 1997), mites (Hayward 2002, Stevens and Hogg in press) and the microbiota (eg Vincent et al. 2000, Lawley et al. 2004).

Courtright et al. (2000) examined the phylogeography of the nematode *Scottnema lindsayae* across southern Victoria Land (including Terra Nova Bay). They found 12 mitochondrial haplotypes from 188 nematodes, with only 11 nucleotide substitutions identified over 10 positions. In addition, *S. lindsayae* exhibited no phylogeographic pattern of mitochondrial or ribosomal haplotypes which were genetically homogeneous among six locations examined — their data did not support the existence of long-term barriers among populations (Courtright et al. 2000). Some geographic patterns were revealed when analyses focussed only on Taylor and Wright Valleys and may suggest local re-colonisation events. Similar patterns were identified in Taylor Valley for the springtail *Gomphiocepahlus hodgsoni* (Stevens and Hogg 2003, Nolan et al. in press), which may reflect the establishment of local populations by a low number of founding individuals from few glacial refuges since the last glacial maximum (<17 000 years). Similar levels of divergence were found throughout the same geographic region for *G. hodgsoni*. However, unlike *S.lindsayae*, phylogeographic patterns across southern Victoria Land were identified for *G. hodgsoni* (Stevens and Hogg 2003). Across 45 *G. hodgsoni* sequences, there were 14 mtDNA haplotypes, with 22 variable nucleotide substitutions. The number of nucleotide substitutions between each *G. hodgsoni* haplotype ranged from 1 to 12. Stevens and Hogg (2003) utilised both allozymes and mitochondrial data to reveal likely refuges and local colonisation events.

The patterns found for *G. hodgsoni* are supported by a comparative phylogeographic study using the mtDNA (COI) gene for the prostigmatic mite *Stereotydeus mollis* (Stevens and Hogg in press). This work has revealed concordant phylogeographic patterning, and suggests that *G. hodgsoni*, *S. mollis*, and *S. lindsayae* have shared a common, local geological history, although long distance dispersal has been far more effective in the latter species. In particular, all three studies in southern Victoria Land indicate that Taylor Valley not only harbours unique genetic haplotypes, but also accounts for much of the haplotype diversity for each species found elsewhere in southern Victoria Land. Interestingly, the most divergent mitochondrial haplotypes identified for *G. hodgsoni* (Taylor Valley and Beaufort Island, see Stevens and Hogg 2003) were also identified as the most divergent for *S. mollis* (Stevens and Hogg in press). Such phylogeographic concordance suggests that these terrestrial arthropods have been exposed to common geological and glacial effects that have dominated the Victoria Land terrestrial landscape since the Pliocene.

Northern Victoria Land has also been the site of two recent population genetic studies. The northern Victoria Land springtail *Desoria klovstadi* revealed 18 different haplotypes from 40 individuals, ranging from three to seven haplotypes among each of four continental populations surveyed (Frati et al. 2001). Although Frati et al. (2001) found only a single mtDNA (COII) haplotype in more than one population, other similarities to *G. hodgsoni* and *S. lindsayae* are notable. These species are geographically separated along the Trans-Antarctic Mountains, yet *D. klovstadi* was found to have a similar number of variable nucleotide substitutions (26), with 1 to 13 nucleotide substitutions between each haplotype (Frati et al. 2001). Fanciulli et al. (2001) found a similar pattern with high levels of differentiation and heterozygosity among 22 populations for the northern Victoria Land springtail *Gressittacantha terranova*. These populations appeared to be reproductively isolated from one another, as indicated by a high level of genetic differentiation ($F_{ST} = 0.31$). The populations grouped into three main geographical regions divided by the Aviator and Campbell Glaciers, which appeared the most likely primary physical barriers to dispersal, with the exception of one population which has probably been influenced by immigrants from others. Fanciulli et al. (2001) showed fixed allelic differences (non-shared alleles) at sites within 100km suggesting that no current gene flow was occurring among these locations.

Our results too, have suggested limited gene flow among sites on Ross Island and those of southern Victoria Land on the Antarctic continent where we found fixed allelic differences at two of the 10 allozyme loci examined (Stevens and Hogg 2003). Furthermore, *Gomphiocephalus hodgsoni* populations were characterised by high levels of allelic diversity throughout the continental sites but lower levels for populations on Ross Island (Stevens and Hogg 2003). Limited gene flow ($F_{ST} = 0.55$) was found across all populations for *G. hodgsoni*. However, a moderate level of differentiation for the continental populations ($F_{ST} = 0.27$) contrasted with the low level found across the island populations ($F_{ST} = 0.05$). Lower allelic variability, and high genetic similarity for the island populations (in contrast to continental populations), may indicate that the consequences of bottleneck and/or founder effects (recent or historic) have been more pronounced for the island populations. Accordingly, these data suggest that gene flow and hence dispersal for Antarctic taxa may be limited to local events and that species distributions were restricted to very few refugia during the last glacial maximum. Such patterns of isolation among fragmented habitats and subsequent re-colonisation following availability of habitat may be a common feature for the Antarctic continental terrestrial and limno-terrestrial faunas.

In the maritime Antarctic, Hayward (2002) examined population genetic structure using allozymes for seven populations of the terrestrial mite *Alaskozetes antarcticus*. Levels of polymorphic loci (across eight loci), heterozygosity, and F_{ST} (0.105), from populations ranging in geographic distance from 1.8km (Signy Island) to 1300km (Ryder Bay, Antarctic Peninsula), suggest limited genetic differentiation likely resulting from dispersal and/or recent colonisation (<6000yr bp). Comparisons to allozyme studies on other terrestrial invertebrates in Antarctica (*Gomphiocephalus hodgsoni* and *Gressittacantha terranova*) indicate that within and among population variability was lower in *A. antarcticus* populations.

Work on the subantarctic islands is limited to two studies, one on Diptera from Iles Crozet (Vernon et al. 1997), and one on Coleoptera, introduced to South Georgia from the Falkland Islands (Ernsting et al. 1995). Ernsting et al. (1995) analysed eight allozyme loci for *Trechisibus antarcticus* (Coleoptera: Carabidae), collected on South Georgia and the Falkland Islands. These data showed an absence of rare alleles in the South Georgia population, with a high similarity compared to the Falkland Islands population, which supports the recent arrival of this species on South Georgia. Vernon et al. (1997) found that the Ile aux Pingouins population was clearly separated from all Ile de la Possession populations. Heterozygosity levels were also found to be slightly lower among the subantarctic populations, compared to those reported for Diptera elsewhere. However, Vernon et al. (1997) have questioned whether this may be confounded by the taxonomic status of the two morphologically recognised species. Specifically, morphologically distinct individuals were found to be genetically similar (based on an allozyme analysis), whereas morphologically similar individuals were found to be genetically distinct. This study highlights the discrepancies which can exist between biochemical and morphological features. Collectively, studies on the endemic fauna suggest that the extensive climate change during the Pleistocene, in conjunction with limited dispersal opportunities, appear to have promoted isolation and divergence among Antarctica's fragmented habitats.

Phylogenetic Relationships of Southern Hemisphere Fauna

To date, only a single study has the Gondwanan relationships of terrestrial invertebrates in Antarctica (Stevens et al. 2006). This is unfortunate as they are the only extant animals to have survived the glaciation of the Antarctic continent (see Lawver and Gahagan 2003 for a review of the glacial evolution of Antarctica). The recolonisation of Antarctica is possible for species that can disperse over vast oceanic distances. However, springtails (Wise 1971, Hogg and Stevens 2002, Stevens and Hogg 2003, Stevens et al. 2006), free-living mites (Pugh 2003), dipterans (Vernon et al. 1997) and spiders (Pugh 2004) in Antarctica do not have these capabilities and therefore their origins are likely to be pre-Pliocene. Accordingly, testing phylogenetic hypotheses on the continental Antarctic springtails and mites should provide an unparalleled contribution to our understanding of the evolutionary basis of species' distributions and genetic diversity.

From the Devonian period, these taxa have followed the geological evolution of continents, with great adaptive capacity to different climates/environments and without undergoing any great morphological changes. The proportion of endemic genera and species is high in Antarctica, although the majority occur in one region, the Trans-Antarctic Mountains. This region including Victoria Land and the more southern Queen Maud Mountains contains a unique springtail assemblage where six of 10 genera are endemic, and nine of 10 species are endemic (Wise 1967, Greenslade 1995). Some, like *Biscoia sudpolaris* (Hypogastruridae), found only in some of the most southerly isolated soils of the Trans-Antarctic Mountains (Wise 1967), may be relics of once widespread taxa on the supercontinent Gondwana (Rapoport 1971). Other endemic springtails (eg *Antarctophorus subpolaris, Antarcticinella monoculata, Neocryptopygus nivicolus, Gressittacantha terranova,* and *Isotoma klovstadi*) and mites (eg *Stereotydeus* spp.) are now forming the basis of Southern Hemisphere phylogenetic comparisons to examine the relationships and evolutionary history of the Antarctic terrestrial invertebrates (eg Stevens et al. 2006).

Frati and Carapelli (1999) assessed the usefulness of nuclear (large ribosomal RNA subunit (D3), and Elongation Factor-1$\propto$) and mitochondrial (COII) genes in resolving phylogenetic relationships among springtails. They examined the phylogenetic position of *Desoria klovstadi* from northern Victoria Land, using three genera of Isotomidae and three Entomobryidae in their analyses. They found that mtDNA (COII) and nuclear EF-1$\propto$ genes were the most informative, but their preliminary study lacked the relevant taxa that would have allowed an assessment of the evolutionary history of the Antarctic species. Frati et al. (2000) further examined the phylogenetic position of *D. klovstadi*, in comparison with *Isotomurus maculatus* and *Tetracanthella* sp. and again found that mtDNA (COII) was the most informative for phylogenetic reconstructions.

Using the COII gene Frati et al. (2000) also examined the morphological hypothesis that the subfamily Pseudachorutinae (represented by *Anurida maritima*) is the next sister-group to the Neanurinae (12 taxa) and that Frieseinae (using *Friesea grisea* from northern Victoria Land) is their closest sister taxon. They found that a phylogenetic reconstruction contradicted morphological evidence among the Neanuridae, by suggesting that the subfamilies Frieseinae and Pseudachorutinae were sister taxa, with the third subfamily Neanurinae being their sister-group (Frati et al. 2000). D'Haese (2002) has perhaps provided the most extensive phylogenetic study that includes Antarctic species using the D1 and D2 regions of the 28S rDNA. Indeed, springtails are often one of the key taxa used to assess arthropod relationships, particularly among the hexapods (see also Delsuc et al. 2003). D'Haese (2002) included in his analyses three Antarctic springtail species from King George Island, South Shetland Islands (*Cryptopygus antarcticus, Friesea grisea* and *Tillieria penai*) and inferred relationships among the orders Symphypleona, Entomobryomorpha, and Poduromorpha and more specifically among the subfamilies within the Poduromorpha. Furthermore, D'Haese (2002) challenges morphological interpretations, in particular, that *Podura aquatica* has been regarded as 'primitive', and thus representing ancestral characteristics.

D'Haese (2002) clearly shows that *P. aquatica* is not basal, or 'primitive', but well nested in the Poduromorpha.

Comparatively more work has been undertaken on the microbiota of Antarctica. However, interpreting these results is confounded by several issues. These issues appear to be centred on the assumption that classic taxonomic criteria do not provide an accurate guide to the extent of genetic divergence (Vincent 2000). Franzmann (1996) undertook the first detailed molecular analysis of Antarctic microbiota to address the question of prokaryotic divergence in the southern polar region. Ten Antarctic bacteria (mostly from saline lakes in the Vestfold Hills) were compared using 16S rDNA sequences relative to the most closely related bacteria available in culture from temperate latitudes. The Antarctic species sequence dissimilarity relative to the temperate strains was 4.5%. Assuming an evolutionary rate of 1% divergence in 16S rDNA in 25 million years (Franzmann 1996), this would equate to phylogenetic divergence of the Antarctic taxa from those in temperate latitudes beginning > 100 million years ago, well before the isolation and cooling of Antarctica (Vincent 2000). These results imply that Antarctica contains an unusual subset of prokaryotes that are poorly represented at lower latitudes. However, Franzmann et al. (1997) note the incompleteness of the database and that the branching patterns may change substantially as more microbiota are discovered and sequenced within and outside Antarctica. For example, the 4.5% dissimilarity value might reflect the possibility that more closely related strains from temperate regions have yet to be sequenced, a view shared by others (eg Lawley et al. 2004). An analysis of high latitude picocyanobacteria showed that there were considerable differences between Arctic and Antarctic strains (Vincent et al. 2000). Three *Synechococcus* isolates from saline lakes in the Vestfold Hills, East Antarctica, were mostly related (96% similar) to *Prochlorococcus marinus* and formed a distinct cluster relative to all other known picocyanobacteria (Vincent et al. 2000). Sequences of isolates from a much broader range of habitats are required to assess the evolutionary divergence of Antarctic cyanobacteria and this approach would be useful to examine other groups of microphototrophs such as diatoms and phytoflagellates.

Many new species of bacteria and protozoa have been described from Antarctica (detailed in Vincent 2000), suggesting that the level of endemism could be considerable. Similarly, the application of 16s rDNA analysis to Antarctic bacteria implies not only the presence of unusual species, but also novel taxa at the genus, family and even higher levels (Vincent 2000). However, most critical to the interpretation of these results is whether the molecular marker is suitable for the time-scale being examined. The 16s rDNA sequences may be too highly conserved (Lawrence and Ochman 1998) to allow the identification of taxa that are endemic to Antarctica. Studies employing a "polyphasic" approach (eg Pearce et al. 2003), which combine several analytical techniques (eg morphological analysis, pigment analysis, rDNA gene sequencing), are likely to provide additional insights into the genetic diversity and evolutionary relationships of the Antarctic fauna.

Further Research

Continental and maritime Antarctica, and the subantarctic islands, and their relationships to other Southern Hemisphere land-masses, provide a wealth of research opportunities including population genetics, phylogeography and phylogenetics. Here, we outline three research areas that would benefit from immediate attention and would greatly enhance our understanding:

1) Several invertebrate groups have revealed large-scale biogeographic distributions across Antarctica and the subantarctic islands (including the lower latitude islands, for example, Campbell Island and Auckland Islands). This is true for the springtails (eg *Friesea grisea, Cryptopygus antarcticus, Tullbergia bisetosa*) and mites (eg *Alaskozetes antarcticus, Halozetes crozetensis, Halozetes belgicae, Nanorchestes antarcticus, Globoppia intermedia longiseta, Austroppia crozetensis and Maudheimia* spp.), which we assume have very little dispersal capability. Hence, an examination on a wider scale across Antarctica, including the subantarctic islands, would test assumptions of dispersal and widespread endemicity.

2) Further allozyme work, in conjunction with other genetic markers (eg mtDNA, ncDNA), for dipteran species and populations from the Iles Crozet, Iles Kerguelen and Heard Island, would further our understanding of the historical processes by which these islands were colonised. Such studies would provide information on the long-term survival of terrestrial invertebrates on these subantarctic islands. For example, it seems extremely unlikely that terrestrial life would have continued to survive on Heard Island throughout the last glaciation in the absence of adequate terrestrial habitat. In fact, Gressitt and Temple (1970) have suggested that much of the Heard Island terrestrial fauna were colonisers from Iles Kerguelen since the glacial maximum. Furthermore, Gressitt (1970) has suggested air dispersal across the southern seas providing colonisation of many of the subantarctic islands. Molecular markers have only recently been applied to address such relationships on vertebrates and plants (eg Muñoz et al. 2004, Ritchie et al. 2004, Sanmartin and Ronquist 2004).

3) Studies testing the utility of DNA barcoding methods (*sensu* Hebert et al. 2003a), for the Antarctic fauna would allow accurate assessment of true levels of (reproductively isolated) species diversity within and among habitats. These data in conjunction with further allozyme (or other nuclear markers) will also provide estimates of gene flow among populations, thus allowing assessment of interhabitat dispersal, in addition to providing data on levels of genetic variability within and among populations. This in turn will be important for targeting management decisions towards the conservation of Antarctica's unique fauna. Multidisciplinary studies in continental, maritime and subantarctic regions would be useful for assessing local dispersal dynamics and evolutionary persistence of Antarctic taxa. No region is more suited for such studies than the Trans-Antarctic Mountains with high species richness and generic endemicity for springtails and mites. Studies so far are revealing much higher levels of endemism than previously thought and have conservation implications in addition to better understanding of the extent and persistence of glacial refugia.

In this chapter, we have presented an overview of molecular studies on the terrestrial and limno-terrestrial invertebrates from the maritime and continental Antarctic regions. It is clear that knowledge of this important component of the Antarctic fauna is incomplete and ongoing research will no doubt enhance our understanding considerably.

Acknowledgements

We thank Trish McLenachan, David Penny, Penny Greenslade and two anonymous reviewers for their thorough and helpful comments on the manuscript. We are grateful to Antarctica New Zealand for their logistical support to IDH, the Australian Antarctic Division for financial and logistic support to MIS through ASAC grants 2355 and 2397 to Paul Sunnucks and to financial support from David Penny.

References

Abyzov, S.S. (1993) Microorganisms in the Antarctic ice, in E.I. Friedmann (ed.), *Antarctic microbiology*, Wiley-Liss, New York, pp. 265-295.

Ashworth, A.C. and Kuschel, G. (2003) Fossil weevils (Coleoptera: Curculionidae) from latitude 85°S Antarctica. *Palaeogeography, Palaeoclimatology, Palaeoecology* **191**, 191-202.

Ashworth, A.C. and Preece, R.C. (2003) The first freshwater molluscs from Antarctica. *Journal of Molluscan Studies* **69**, 97-100.

Balfour-Browne, J. and Tilbrook, P.J. (1966) Coleoptera collected in the South Orkney and South Shetland Islands, *British Antarctic Survey Bulletin* **9**, 41-43.

Bargagli, R., Skotnicki, M.L., Marri, L., Pepi, M., Mackenzie, A. and Agnorelli, C. (2004) New record of moss and thermophilic bacteria species and physico-chemical properties of geothermal soils on the northwest slope of Mt. Melbourne (Antarctica), *Polar Biology* **27**, 423-431.

Bergstrom, D.M., Hodgson, D.A. and Convey, P. (2006) The physical setting of the Antarctic, in D.M. Bergstrom, P. Convey, and A.H.L. Huiskes (eds.), *Trends in Antarctic Terrestrial and Limnetic Ecosystems: Antarctica as a Global Indicator*, Springer, Dordrecht (this volume).

Bilton, D.T., Freeland, J.R. and Okamura, B. (2001) Dispersal in freshwater invertebrates. *Annual Review of Ecology and Systematics* **32**, 159-181.

Bullini, L., Arduino, P., Cianchi, R., Nascetti, G., D'Amelio, S., Mattiucci, S., Paggi, L. and Orecchia, P. (1994) Genetic and ecological studies on nematode endoparasites of the genera *Contracaecum* and *Pseudoterranova* in the antarctic and arctic-boreal regions, in B. Battaglia, P.M. Bisol, and V. Varotto (eds.), Proceedings of the 2nd meeting on Antarctic Biology, Edizioni Universitarie Patavine, Padova, Italy, pp. 131-146.

Carpenter, G. (1902) Aptera: Collembola. Chapter 9 Insecta, in *The Report on the Collections of Natural History made in the Antarctic Regions during the voyage of the Southern Cross*, British Museum (Natural History), London, pp. 221-223.

Chown, S.L. and Convey, P. (2006) Biogeography, in D.M. Bergstrom, P. Convey, and A.H.L. Huiskes (eds.), *Trends in Antarctic Terrestrial and Limnetic Ecosystems: Antarctica as a Global Indicator*, Springer, Dordrecht (this volume).

Convey, P. and Block, W. (1996) Antarctic Diptera: Ecology, physiology and distribution, *European Journal of Entomology* **93**, 1-13.

Convey, P., Greenslade, P., Richard, K.J. and Block, W. (1996) The terrestrial arthropod fauna of the Byers Peninsula, Livingston Island, South Shetland Islands – Collembola, *Polar Biology* **16**, 257-259.

Courtright, E.M., Wall, D.H., Virginia, R.A., Frisse, L.M., Vida, J.T. and Thomas, W.K. (2000) Nuclear and mitochondrial DNA sequence diversity in the Antarctic nematode *Scottnema lindsayae*, *Journal of Nematology* **32**, 143-153.

Darling, K.F., Wade, C.M., Stewart, I.A., Kroon, D., Dingle, R. and Brown, A.J.L. (2000) Molecular evidence for genetic mixing of Arctic and Antarctic subpolar populations of planktonic foraminifers, *Nature* **405**, 43-47.

Dastych, H. (1984) The Tardigrada from the Antarctic with descriptions of several new species, *Acta Zoologica Cracoviensia* **27**, 377-436.

Delsuc, F., Phillips, M.J. and Penny, D. (2003) Comment on 'Hexapod origins: monophyletic or paraphyletic?' *Science*, **301**, 1482d.

D'Haese, C.A. (2002) Were the first springtails semi-aquatic? A phylogenetic approach by means of 28S rDNA and optimization alignment, *Proceedings of the Royal Society of London B* **269**, 1143–1151.

Ernsting, G., van Ginkel, W. and Menken, S.B.J. (1995) Genetical population structure of *Trechisibus antarcticus* (Coleoptera, Carabidae) on South Georgia and on the Falkland Islands, *Polar Biology* **15**, 523-539.

Fanciulli, P.P., Summa, D., Dallai, R. and Frati, F. (2001) High levels of genetic variability and population differentiation in *Gressittacantha terranova* (Collembola, Hexapoda) from Victoria Land, Antarctica, *Antarctic Science* **13**, 246-254.

Finlay, B.J. (2002) Global dispersal of free-living microbial eukaryote species. *Science* **296**, 1061-1063.

Franzmann, P.D. (1996) Examination of Antarctic prokaryotic diversity through molecular comparisons, *Biodiversity and Conservation* **5**, 1295-1305.

Franzmann, P.D., Dobson, S.J., Nichols, P.D. and McMeekin, T.A. (1997) Prokaryotic microbial diversity, in B. Battaglia, J. Valencia, and D.W.H. Walton (eds.), *Antarctic communities: species, structure and survival*, Cambridge University Press, Cambridge, pp. 51-56.

Frati, F. and Carapelli, A. (1999) An assessment of the value of nuclear and mitochondrial genes in elucidating the origin and evolution of *Isotoma klovstadi* Carpenter (Insecta, Collembola), *Antarctic Science* **11**, 160-174.

Frati, F., Spinsanti, G. and Dallai, R. (2001) Genetic variation of mtCOII gene sequences in the collembolan *Isotoma klovstadi* from Victoria Land, Antarctica: evidence for population differentiation, *Polar Biology* **24**, 934-940.

Frati, F., Fanciulli, P.P., Carapelli, A., De Carlo, L. and Dallai, R. (1997) Collembola of northern Victoria Land: distribution, population structure and preliminary molecular data to study origin and evolution of Antarctic Collembola, in Proceedings of the 3rd meeting on Antarctic Biology, Santa Margherita Ligure, Italy, 1996, G. di Prisco, S. Focardi, and P. Luporini (eds.), Camerino University Press, pp. 321-330.

Frati, F., Fanciulli, P.P., Carapelli, A., Dell'ampio, E., Nardi, F., Spinsanti, G. and Dallai, R. (2000) DNA sequence analysis to study the evolution of Antarctic Collembola, *Italian Journal of Zoology, Supplement* **1**, 133-139.

Gaffney, P.M. (2000) Molecular tools for understanding population structure in Antarctic species, *Antarctic Science* **12**, 288-296.

Gibson, J.A.E., Wilmotte, A., Taton, A., Van De Vijver, B., Beyens, L. and Dartnall, H.J.G. (2006) Biogeographic trends in Antarctic lake communities, in D.M. Bergstrom, P. Convey, and A.H.L. Huiskes (eds.), *Trends in Antarctic Terrestrial and Limnetic Ecosystems: Antarctica as a Global Indicator*, Springer, Dordrecht (this volume).

Gordon, D.A., Priscu, J. and Giovannoni, S. (2000) Origin and phylogeny of microbes living in permanent Antarctic lake ice. *Microbial Ecology* **39**, 197-202.

Greenslade, P. (1995) Collembola from the Scotia Arc and Antarctic Peninsula including descriptions of two new species and notes on biogeography, *Polskie Pismo Entomologiczne* **64**, 305-319.

Greenslade, P. and Wise, K.A.J. (1984) Additions to the collembolan fauna of the Antarctic, *Transactions of the Royal Society of South Australia* **108**, 203-205.

Greenslade, P. and Wise, K.A.J. (1986) Collembola of Macquarie Island. *Records of the Auckland Institute Museum* **23**, 67-97.

Greenslade, P., Farrow, R.A. and Smith J. M.B. (1999) Long distance migration of insects to a subantarctic island, *Journal of Biogeography* **26**, 1161-1167.

Gressitt, J.L. (1964) Ecology and biogeography of land arthropods in Antarctica, in R. Carrick, M.W. Holdgate, and J. Prévost (eds.), *Biologie Antarctique*, Hermann, Paris pp. 211-222.

Gressitt, J.L. (1965) Terrestrial Animals, in T. Hatherton (ed.), *Antarctica*, Methuen, London pp. 351-371.

Gressitt, J.L. (1967) Notes on arthropod populations in the Antarctic Peninsula – South Shetland Islands – South Orkney Islands area, *Antarctic Research Series* **10**, 373-391.

Gressitt, J.L. (1970) Subantarctic entomology and biogeography, *Pacific Insects Monograph* **23**, 295-374.

Gressitt, J.L. and Temple, P. (1970) Introduction to Heard Island, *Pacific Insects Monograph* **23**, 17-30.

Hays, J.D., Imbrie, J. and Shackleton, N.J. (1976) Variations in the Earth's orbit: pacemaker of the ice ages. *Science* **194**, 1121-1132.

Hayward, S.A.L. (2002) The Functional Ecology of Polar Terrestrial Invertebrates. Unpublished Ph.D. Thesis, Birmingham University, UK.

Hebert, P.D.N., Cywinska, A., Ball, S.L. and deWaard, J.R. (2003a) Biological identifications through DNA barcodes, *Proceedings of the Royal Society of London B* **270**, 313-322.

Hebert, P.D.N., Ratnasingham, S. and de Waard, J.R. (2003b) Barcoding animal life: cytochrome *c* oxidase subunit 1 divergences among closely related species, *Proceedings of the Royal Society of London B* **270** (Suppl.), S96-99.

Hirsch, P., Mevs, U., Kroppenstedt, R.M., Schumann, P. and Stackebrandt, E. (2004) Cryptoendolithic Actinomycetes from Antarctic Sandstone Rock Samples: *Micromonospora endolithica sp. nov.* and two Isolates Related to *Micromonospora coerulea* Jensen 1932. *Systematic and Applied Microbiology* **27**, 166-174.

Hogg, I.D. and Hebert, P.D.N. (2004) Biological identification of springtails (Collembola: Hexapoda) from the Canadian Arctic, using mitochondrial DNA barcodes, *Canadian Journal of Zoology* **82**, 82, 749-754.

Hogg, I.D. and Stevens, M.I. (2002) Soil Fauna of Antarctic Coastal Landscapes, in L. Beyer and M. Bölter (eds.), *Geoecology of Antarctic Ice-Free Coastal Landscapes, Ecological Studies Analysis and Synthesis*, Springer-Verlag, Berlin, Volume 154, pp. 265-278.

Huiskes, A.H.L., Convey, P. and Bergstrom, D.M. (2006) Trends in Antarctic terrestrial and limnetic ecosystems: Antarctica as a global indicator in D.M. Bergstrom, P. Convey, and A.H.L. Huiskes (eds.), *Trends in Antarctic Terrestrial and Limnetic Ecosystems: Antarctica as a Global Indicator*, Springer, Dordrecht (this volume).

Hullé, M., Pannetier, D., Simon, J.-C., Vernon, P. and Frenot, Y. (2003) Aphids of sub-Antarctic Îles Crozet and Kerguelen: species diversity, host range and spatial distribution, *Antarctic Science* **15**, 203-209.

Hunter, P.E. (1967) Mesostigmata: Rhodacaridae, Laelapidae (Mesostigmatic mites), *Antarctic Research Series* **10**, 35-39.

Jennings, P.G. (1976) The Tardigrada of Signy Island, South Orkney Islands, with a note on the Rotifera, *British Antarctic Survey Bulletin* **44**, 1-25.

Karl, D.M., Bird, D.F., Björkman, K., Houlihan, T., Shackelford, R. and Tupas, L. (1999) Microorganisms in the accreted ice of Lake Vostok, Antarctica. *Science* **286**, 2144-2147.

Lawley, B., Ripley, S., Bridge, P. and Convey, P. (2004) Molecular analysis of geographic patterns of eukaryotic diversity in Antarctic soils, *Applied and Environmental Microbiology* 70, 5963-5972.

Lawrence, J.G. and Ochman, H. (1998) Molecular archaeology of the *Escherichia coli* genome, *Proceedings of the National Academy of Science* **95**, 9413-9417.

Lawver, L.A. and Gahagan, L.M. (2003) Evolution of Cenozoic seaways in the circum-Antarctic region, *Palaeogeography, Palaeoclimatology, Palaeoecology* **198**, 11-37.

Marshall, D.J. and Pugh, P.J.A. (1996) Origin of the inland Acari of continental Antarctica with particular reference to Dronning Maud Land, *Zoological Journal of the Linnean Society* **118**, 101-118.

McInnes, S.J. and Pugh, P.J.A. (1998) Biogeography of limno-terrestrial Tardigrada, with particular reference to the Antarctic fauna. *Journal of Biogeography* **25**, 31-36.

Miller, J.D., Horne, P., Heatwole, H., Miller, W.R. and Bridges, L. (1988) A survey of the terrestrial Tardigrada of the Vestfold Hills, Antarctica, *Hydrobiologia* **165**, 197-208.

Miller, M.F. and Mabin, M.C.G. (1998) Antarctic neogene landscapes – in the refrigerator or in the deep freeze? *GSA Today* **8**, 1-3.

Moore, P.D. (2002) Springboards for springtails, *Nature* **418**, 381.

Moorhead, D.L., Barrett, J.E., Virginia, R.A., Wall, D.H. and Porazinska, D. (2003) Organic matter and soil biota of upland wetlands in Taylor Valley, Antarctica, *Polar Biology* **26**, 567-576.

Muñoz, J., Felicisima, Á.M., Cabezas, F., Burgaz, A.R. and Martinez, I. (2004) Wind as a long-distance vehicle in the Southern Hemisphere, *Science* **304**, 1144-1147.

Nolan, L., Hogg, I.D., Stevens, M.I. and Haase, M. (in press) Molecular support for a secondary contact zone among late Pleistocene glacial refugia for *Gomphiocephalus hodgsoni* (Collembola: Hypogastruridae) in Taylor Valley, continental Antarctica, *Polar Biology*.

Pearce, D.A., van der Gast, C.J., Lawley, B. and Ellis-Evans, J.C. (2003) Bacterioplankton community diversity in a maritime Antarctic lake, determined by culture-dependent and culture-independent techniques. *FEMS Microbiology Ecology*, **45**, 59-70.

Potapov, M. (1991) *Antarctophorus* – a new genus of Isotomidae (Collembola) from Antarctica, *Revue d'Ecologie et de Biologie du Sol* **28**, 491-495.

Priscu, J.C., Adams, E.E., Lyons, W.B., Voytek, M.A., Mogk, D.W., Brown, R.L., McKay, C.P., Takacs, C.D., Welch, K.A., Wolf, C.F., Kirshtein, J.D. and Avci, R. (1999) Geomicrobiology of subglacial ice above Lake Vostok, Antarctica. *Science* **286**, 2141-2144.

Pryor, M.E. (1962) Some environmental features of Hallett Station, Antarctica, with special reference to soil arthropods, *Pacific Insects* **4**, 681-728.

Pugh, P.J.A. (2003) Have mites (Acarina: Arachnida) colonised Antarctica and the islands of the Southern Ocean via air currents? *Polar Record* **39**, 239-244.

Pugh, P.J.A. (2004) Biogeography of spiders (Araneae: Arachnida) on the islands of the Southern Ocean, *Journal of Natural History* **38**, 1461-1488.

Rapoport, E.H. (1971) The geographical distribution of Neotropical and Antarctic Collembola, *Pacific Insects Monograph* **25**, 99-118.

Richardson, A.M.M. and Jackson, J.E. (1995) The first record of a terrestrial landhopper (Crustacea: Amphipoda: Talitridae) from Macquarie Island, *Polar Biology*, **15**, 419-422.

Ritchie, P.A., Millar, C.D., Gibb, G.C., Baroni, C. and Lambert, D.M. (2004) Ancient DNA enables timing of the Pleistocene origin and Holocene expansion of two Adélie penguin lineages in Antarctica, *Molecular Biology and Evolution* **21**, 240–248.

Roberts, A.P., Wilson, G.S., Harwood, D.M. and Verosub, K.L. (2003) Glaciation across the Oligocene-Miocene boundary in southern McMurdo Sound, Antarctica: new chronology from the CIROS-1 drill hole, *Palaeogeography, Palaeoclimatology, Palaeoecology* **198**, 113-130.

Sanmartin, I. and Ronquist, F. (2004) Southern Hemisphere biogeography inferred by event-based models: plant versus animal patterns, *Systematic Biology* **53**, 216-243.

Siegert, M.J., Tranter, M., Ellis-Evans, C.J., Priscu, J.C. and Lyons, W.B. (2003) The hydrochemistry of Lake Vostok and the potential for life in Antarctic subglacial lakes. *Hydrological Processes* **17**, 795-814.

Shishida, Y. and Ohyama, Y. (1989) A note on the terrestrial nematodes around Palmer Station, Antarctica (extended abstract). *Proceedings NIPR Symp Polar Biology* **2**, 223-224.

Stevens, M.I. and Hogg, I.D. (2003) Long-term isolation and recent range expansion from glacial refugia revealed for the endemic springtail *Gomphiocephalus hodgsoni* from Victoria Land, Antarctica. *Molecular Ecology* **12**, 2357-2369.

Stevens, M.I. and Hogg, I.D. (in press) Contrasting levels of mitochondrial DNA variability between mites (Penthalodidae) and springtails (Hypogastruridae) from the Trans-Antarctic Mountains suggest long-term effects of glaciation and life history on substitution rates, and speciation processes, *Soil Biology and Biochemistry*.

Stevens, M.I., Greenslade, P., Hogg, I.D. and Sunnucks, P. (2006) Examining Southern Hemisphere springtails: could any have survived glaciation of Antarctica? *Molecular Biology and Evolution* **23**, 874-882.

Stevens, M.I., Fjellberg, A., Greenslade, P., Hogg, I.D. and Sunnucks, P. (in press) Redescription of the Antarctic springtail *Desoria klovstadi* using morphological and molecular evidence. *Polar Biology*.

Strandtmann, R.W. (1967) Terrestrial Prostigmata (Trombidiform mites), *Antarctic Research Series* **10**, 51-95.

Strong, J. (1967) Ecology of terrestrial arthropods at Palmer station, Antarctic Peninsula, *Antarctic Research Series* **10**, 357-371.

Sunnucks, P. (2000) Efficient genetic markers for population biology, *Trends in Ecology and Evolution* **15**, 199-203.

Taton, A., Grubisic, S., Brambilla, E., De Wit, R. and Wilmotte, A. (2003) Cyanobacterial diversity in natural and artificial microbial mats of Lake Fryxell (McMurdo Dry Valleys, Antarctica): a morphological and molecular approach. *Applied and Environmental Microbiology* **69**, 5157-5169.

Tilbrook, P.J. (1967) Arthropod ecology in the maritime Antarctic, *Antarctic Research Series* **10**, 331-356.

de la Torre, J.R., Goebel, B.M., Friedmann, E.I. and Pace, N.R. (2003) Microbial diversity of cryptoendolithic communities from the McMurdo Dry Valleys, Antarctica, *Applied and Environmental Microbiology* **69**, 3858-3867.

Treonis, A.M., Wall, D.H. and Virginia, R.A. (2002) Field and microcosm studies of decomposition and soil biota in a cold desert soil, *Ecosystems* **5**, 159-170.

Utsugi, K. and Ohyama, Y. (1989) Antarctic Tardigrada, *Proceedings NIPR Symp. Polar Biology* **2**, 190-197.

Valbonesi, A., Ballarini, P., Di Giuseppe, G., Miceli, C., Felici, A., Ortenzi, C. and Luporini, P. (1994) Speciation and adaptive biology of antarctic ciliated protozoa, in B. Battaglia, P.M. Bisol, and V. Varotto (eds.), Proceedings of the 2nd meeting on Antarctic Biology, Edizioni Universitarie Patavine, Padova, Italy, pp. 111-120.

Van Trappen, S., Vandecandelaere, I., Mergaert, J. and Swings, J. (2004) *Gillisia limnaea* gen. nov., sp. nov., a new member of the family Flavobacteriaceae isolated from a microbial mat in Lake Fryxell, Antarctica, *International Journal of Systematic and Evolutionary Microbiology* **54**, 445-448.

Vernon, P., Cariou, M.L. and Deunff, J. (1997) Genetic variability in the wingless subantarctic genus *Anatalanta* (Diptera, Sphaeroceridae): a preliminary approach, *Polar Biology* **18**, 384-390.

Vincent, W.F. (2000) Evolutionary origins of Antarctic microbiota: invasion, selection and endemism, *Antarctic Science* **12**, 374-385.

Vincent, W.F., Bowman, J., Powell, L. and McMeekin, T. (2000) Phylogenetic diversity of picocyanobacteria in Arctic and Antarctic ecosystems, in M. Brylinsky, C. Bell, and P. Johnson-Green (eds.), Microbial Biosystems: New Frontiers, Proceedings of the 8th International Symposium on Microbial Ecology, Atlantic Canada Society for Microbial Ecology, pp 317-322.

Weiner, W.M. and Najt, J. (1994) Une nouvelle espèce de Tillieria (Collembola, Tullbergiinae) de l'île King George. *Polskie pismo entomologiczne* **63**, 17-21.

Wharton, D.A. (2003) The environmental physiology of Antarctic terrestrial nematodes: a review. *Journal of Comparative Physiology B* **173**, 621-628.

Winsor, L. and Stevens, M. (2005) Terrestrial flatworms (Platyhelminthes: Tricladida: Terricola) from subantarctic Macquarie Island. *Kanunnah*, **1**, 17-32.

Wirth, W.W. and Gressitt, J.L. (1967) Diptera: Chironomidae (midges), *Antarctic Research Series* **10**, 197-203.

Wise, K.A.J. (1967) Collembola (Springtails), *Antarctic Research Series* **10**, 123-148.

Wise, K.A.J. (1971) The Collembola of Antarctica, *Pacific Insects Monograph* **25**, 57-74.

Witt, J.D.S., Blinn, D.W., Hebert, P.D.N. (2003) The recent evolutionary origin of the phenotypically novel amphipod *Hyalella montezuma* offers an ecological explanation for morphological stasis in a closely allied species complex. *Molecular Ecology* **12**, 405-413.

Womersley, H. and Strandtmann, R.W. (1963) On some free living prostigmatic mites of Antarctica, *Pacific Insects* **5**, 451-472.

Yue, Q.Y. and Tamura, H. (2001) Three species of Collembola from Antarctic, *Entomologia Sinica* **8**, 1-7.

10. BIOLOGICAL INVASIONS

P. CONVEY
British Antarctic Survey, Natural Environment Research Council
High Cross, Madingley Road
Cambridge CB3 0ET, United Kingdom
p.convey@bas.ac.uk

Y. FRENOT
UMR 6553 CNRS-Université de Rennes
&
French Polar Institute (IPEV)
Station Biologique F-35380 Paimpont France
yves.frenot@univ-rennes1.fr

N. GREMMEN
Data-Analyse Ecologie
Hesselsstraat 11
7981 CD Diever
The Netherlands
gremmen@wxs.nl

D. M. BERGSTROM
Department of Environment and Heritage
Australian Government Antarctic Division
203 Channel Highway
Kingston, Tasmania 7050, Australia
dana.bergstrom@agad.gov.au

Introduction

At first sight, and certainly in comparison with most other land areas worldwide, Antarctica appears exceptionally well protected against the dangers of invasion by

D.M. Bergstrom et al. (eds.), Trends in Antarctic Terrestrial and Limnetic Ecosystems, 193–220.
© *2006 Springer.*

non-indigenous (alien) species. It is geographically isolated from other Southern Hemisphere continents and smaller landmasses, historically lacks indigenous human populations or contact, and presents extreme environmental challenges that must be survived both during any transfer process and after establishment at an Antarctic location. Despite this, it is clear that biological invasions have taken place, and have led to serious impacts on indigenous biota, ecosystems, and ecosystem functions, posing a serious risk to the Antarctic region (Dingwall 1995, Smith 1996, Chown et al. 2001, Greenslade 2002, Frenot et al. 2005).

In common with elsewhere, parts of the Antarctic have been experiencing a period of very rapid environmental change, relating to a number of significant variables, over the last 50 or more years (Huiskes et al. this volume, Convey this volume, Lyons et al. this volume). In addition to confirming the significant influence of some existing biological invasions, Frenot et al. (2005) have identified that rapid climate change, in combination with increased human activity, is likely to increase the frequency and significance of future invasions, and increase the impacts of alien biota that are already established.

Worldwide, biological invasions are one of the most important threats to biodiversity (McKinney and Lockwood 1999, Sala et al. 2000, Courchamp et al. 2003) and ecosystem processes (Heywood 1989, d'Antonio and Dudley 1995, Mack et al. 2000). In an Antarctic context, these threats are serious. The subantarctic islands, continental margin, packice and surrounding seas are home to spectacular concentrations of marine megafauna, including a large proportion of the world's seabird species and marine mammals. Life on land, while species poor and less visually spectacular (Gressitt 1970, Chown et al. 1998, Vernon et al. 1998, Convey 2001) is no less significant, and terrestrial biotas often include a particularly high proportion of endemic taxa (as illustrated by lichens, liverworts, flowering plants, arthropods and nematodes).

Antarctic terrestrial habitats are typified by low species richness and the absence of many functional groups that are present elsewhere. This itself may be sufficient to render the sub- and maritime Antarctic islands, and the ice-free islands of exposed land on the continent, susceptible to alien invasion (Bergstrom and Chown 1999, Chown et al. 2000). Furthermore, island biotas may be more susceptible to invasion as indigenous species are less able to cope with the associated change (d'Antonio and Dudley 1995, Vermeij 1996, Williamson 1996, Bowen and van Vuren 1997).

Although human contact with the Antarctic has occurred only over the last two centuries, our influence has increased rapidly. Initially, effort was almost exclusively focused on economic activity. On land, this related to the support requirements that were necessary to allow the excessive commercial exploitation of marine resources (whales, seals, penguins) from the Southern Ocean. In parallel, some farming, social and recreational development also occurred, resulting in many of the introductions of grazing and predatory vertebrates that remain on most of the subantarctic islands today. Throughout this period, concerns over human impact on indigenous Antarctic biota received scant attention, even in the context of the virtual extermination of successive target industrial species.

The second phase of human impact in the Antarctic became apparent through the importance attached to scientific research as being integral to the expeditions of the 'heroic age' of exploration of the early 20th Century. The initially piecemeal and competitive development of scientific activities eventually led to a large and coordinated Antarctic contribution to the International Geophysical Year (1958) and, soon after, to the development of the Antarctic Treaty System (Hull and Bergstrom 2006). Subsequently, research stations have been established by over 30 nations across the Antarctic and approximately 45 are now signatories to the Treaty. In recognition of the need to protect the Antarctic environment in a comprehensive and legally binding form, in 1991 the Madrid Protocol was established. Within the protocol, in the context of this paper, are included the prohibition of introduction of fauna or flora and the establishment of protected areas in Antarctica.

A third phase of human impact on the continent – tourism – developed during the latter decades of the 20th Century, with tourists (mostly arriving on specially designed cruise ships) now numerically outnumbering the scientific and associated logistical operations of national operators by a factor of four to five, and numbers continuing to increase rapidly (Frenot et al. 2005). In the austral summer season of 2004/05 just under 50 000 people (tourists and crew) visited the region with 27 950 tourists on expeditions that included a landing component (http://www.IAATO.org).

Current significance of invasive species in the Antarctic

Chown et al. (1998) studied the correlates of successful invasions on Southern Ocean islands by investigating the relationships among several abiotic and biotic variables and the richness of alien vascular plants, insects, birds and mammals. Their data were subsequently re-analysed by Selmi and Boulinier (2001) to take into account spatial auto-correlations, resulting in similar conclusions. Island size was a significant contributing factor, with larger islands having more alien vascular plants because of both greater habitat heterogeneity and human populations. Temperature was also important, with cold islands being less susceptible than warm ones. Similar relationships were found for insects, with the additional contributing factor of indigenous plant species richness. Human occupancy and temperature were the main correlates for alien mammals. Studies on Marion Island (Chown et al. 1998, Gabriel et al. 2001) suggest that the interactions of alien invertebrates with indigenous biota are less important to the success of the invasion process than the direct impacts of local (micro) climate. These findings support the proposal that climate matching rather than biotic resistance (competition) (Lee 2002) is a major determinant of invasion success, although this remains a subject of considerable debate.

Frenot et al. (2005) provide an up-to-date and comprehensive review and literature resource of the current status of invasive species across the Antarctic continent and subantarctic islands. Here, we do not seek simply to repeat the detail of this review, rather drawing upon it to provide a concise overview of the current and likely future significance of biological invasions in this region (Table 1).

Table 1. Total number of alien species currently established on the main subantarctic islands. 'nd' indicates no data available

Island	Plants	Invertebrates	Vertebrates:		
			mammals	fishes	birds
Iles Crozet:					
Cochons 46.10°S 50.23°E	nd	nd	3	0	0
Possession 46.42°S 51.50°E	59	14	1	2	0
Est 46.43°S 52.20°E	nd	nd	1	0	0
Pingouins 46.50°S 50.40°E	0	0	0	0	0
Prince Edward I 46.63°S 37.95°E	3	1	0	0	0
Marion I 46.90°S 36.75°E	12	18	1	0	0
Iles Kerguelen 49.37°S 69.50°E	69	30	7	5	0
McDonald I 53.03°S 72.60°E	0	0	0	0	0
Heard I 53.10°S 73.50°E	1	3	0	0	0
South Georgia 54.25°S 37.00°W	33	12	3	0	0
Macquarie I 54.62°S 158.90°E	3	28	3	0	3

PLANTS

Virtually all non-indigenous plant species known in the Antarctic are found on the subantarctic islands (Table 1, Fig. 1a, b). To date, all are higher plants, with no confirmed examples of cryptogams (bryophytes, lichens), despite these being the dominant native vegetation type across most of Antarctica, including large parts of some subantarctic islands. This dichotomy is unlikely to reflect the true situation, rather being a function of the focus of past scientific research efforts - partly through poor knowledge of the worldwide distribution for many species, while Antarctic data are also patchy. Little is known about invasive bryophytes, diatoms and other lower plant groups. Thus, there is insufficient basis to assess whether an occurrence on a single island is a true disjunctive distribution, indicating the possibility of introduction, or simply the first record in a very incompletely surveyed area. Furthermore, the local (within-island) distribution patterns of most Antarctic cryptogams are also hardly known. A distribution centering on a locality of human occupation and spreading out from there can be interpreted as a strong indication of introduction, but such data are generally lacking. Lindsay (1973) considered a number of lichens on South Georgia to be probably introduced and Ochyra et al. (2003) suggest that the moss *Thuidium delicatulum* (Hedw.) Schimp. may have been introduced to Marion Island.

Figure 1. Examples of highly visible invasive plants and vertebrates on the subantarctic islands. (a) Large stand of the alien grass Agrostis stolonifera *in an area originally covered by* Acaena magellanica *dominated vegetation, Marion Island, 1998 (photo: N. Gremmen), (b) Well-drained slope on Ile Australia, Golfe du Morbihan, Iles Kerguelen, invaded by* Taraxacum spp. *and* Senecio vulgaris, *January 2004 (photo: N. Gremmen), (c) Mouflon on Ile Haute, Iles Kerguelen, during winter 1991 (photo: D. Réale), (d) Reindeer grazing on a sward of the introduced grass* Poa annua, *South Georgia (Photo: D. Bone).*

Frenot et al. (2005) list 108 species of non-indigenous vascular plant currently present in the subantarctic, providing a striking comparison with the two species known from single locations in the maritime zone (Smith 1996) and none from the continental Antarctic (although in the late 1990s, a flowering grass and a daisy were found growing and removed from the vicinity of Progress Station, in addition to seeds of many angiosperm species being found within the station, M. Riddle, pers. comm.). It is clear that a range of plants from lower southern latitudes, and from the Northern Hemisphere, can survive and in some cases reproduce under the conditions of the sub- and even maritime Antarctic and that the challenges of long-distance dispersal and establishment are greater than survival alone.

Most of the higher plants established in the Antarctic belong to common and widely distributed families that are often invasive at a global scale (Pyšek 1998). These include Poaceae (39 species), Asteraceae (20), Brassicaceae (8) and Juncaceae (7) (Frenot et al. 2005). In terms of simple species numbers, alien plants contribute a considerable proportion of the contemporary biodiversity on some subantarctic islands – approaching 50% on South Georgia and 70% on Iles Kerguelen. However, there is little commonality at the species level among the different islands, with only one species (the grass *Poa annua*) present on all major islands, and a further five with wide distributions. Indeed, most alien plant species are found only on one or at most two islands.

The contemporary impacts of most species are also small, as they can be classified as persistent rather than invasive (as defined by Frenot et al. 2005), with very restricted distributions (in some cases even limited to single plants). For instance, only 7/69 alien species on Iles Kerguelen and 7/59 on Possession are invasive and more widely distributed away from sites of human activity (Frenot et al. 2001). The impact of these low numbers of species can be severe, with native species being displaced (Fig. 1).

Established alien species are generally long-lived, with 75% of species being perennial. Possession of obligate annual or biennial life cycles may not be a viable strategy for plants with the opportunity of colonising the subantarctic, with data indicating that two-thirds of transient species recorded on Iles Kerguelen and Possession were of this type (Frenot et al. 2001). Even the normally annual *Poa annua* can adopt a perennial life cycle in certain habitats or circumstances in the subantarctic (Frenot and Gloaguen 1994, Smith and Steenkamp 2001).

VERTEBRATES

No alien vertebrates (other than the permanent human presence) have become established in the maritime or continental Antarctic. Historically, commercial concerns, research stations and exploring expeditions have imported a range of mammals and birds for logistic (dogs, ponies), food (cattle, reindeer, mouflon, pigs, rabbits, hens) or companionship or pest control (cats) purposes. Such activities are no longer permitted on the continent or Antarctic Peninsula under the terms of the Madrid Protocol. Alien mammals have received considerable study on several of the subantarctic islands and receive the highest profile in public awareness. It is important to realise that a unique feature of Antarctic terrestrial ecosystems is that they naturally lack terrestrial mammalian herbivores or carnivores. Various introductions of these groups (both deliberate and accidental) clearly have had and continue to have considerable impacts on native ecosystems (Bonner 1984, Leader-Williams 1988, Chapuis et al. 1994, Bester et al. 2002). Large and visible vertebrates that are present in spatially defined locations are, in principle at least, potential targets for effective eradication measures and some such have been attempted or completed (Micol and Jouventin 1995, Myers et al. 2000, Bester et al. 2002, Chapuis et al. 2001, Copson and Whinam 2001, Whinam et al. this volume).

Eight invasive mammals are currently established on subantarctic islands (Frenot et al. 2005) (Table 1, Fig. 1c, d). Other than rodents, the remainder were originally deliberately introduced by humans. The Iles Kerguelen hosts the highest number of alien mammals (7), and a number of alien freshwater fish. Mice and rats are the most widespread species (Cumpston 1968, Chapuis et al. 1994, Pye et al. 1999). The selective herbivorous nature of rabbits has resulted in significant ecosystem modification on all islands where they have been released (Copson and Whinam 2001). However, alien predators have had the greatest impacts on the native fauna – cats are responsible for drastic reductions in some seabird populations and local extinctions of several species (Pascal 1980, van Aarde 1980, Bonner 1984, Brothers 1984), while rats have also had major impacts on burrow-nesting bird species (Jouventin et al. 1984) and the endemic South Georgia pipit (Pye and Bonner 1980), the only passerine occurring naturally anywhere in the Antarctic. Rats and mice also have considerable impacts on endemic invertebrate (Pye and Bonner 1980, Chown and Smith 1993, Smith et al. 2002) and plant populations (Shaw et al. 2005). Cats have been the target of successful eradication programmes on Marion and Macquarie Islands (Copson and Whinam 2001, Bester et al. 2002). No concerted efforts have yet been attempted to eradicate rats from any large subantarctic island, although trials have commenced on some small coastal islets off South Georgia in addition to on two islands of Iles Kerguelen. Successful eradication of rodents has been achieved on some of the more northerly cold temperate Southern Ocean Islands including St Paul Island (Micol and Jouventin 2002), and Campbell and Enderby Islands (Torr 2002, see also http://www.doc.govt.nz/Conservation/Offshore-Islands/Campbell-Island-Rat-Eradication.asp).

Few alien birds are established on the subantarctic islands (Table 1) and little is known about their biology and impact. None are established on the continent. Those in the subantarctic are representatives of families with the highest success of introduction globally (Lockwood 1999) and include the Palaearctic mallard (*Anas platyrhynchos*), redpoll (*Carduelis flammea*) and starling (*Sturnus vulgaris*) on Macquarie Island. Species such as the starling (originally of European origin) can probably reach the subantarctic relatively frequently, and without human assistance, from southern temperate locations where they have been long established. Indeed, the subantarctic islands and more southerly locations record vagrant birds regularly (eg Burger et al. 1980, Gauthier-Clerc et al. 2002). Humans did however introduce wekas (*Gallirallus australis scotti*), flightless birds native to New Zealand, as a food source on Macquarie Island in the 1870s (Copson and Whinam 2001). Taylor (1979) credited the extinction of the endemic subspecies of the Macquarie Island rail (*Rallus phillipensis macquariensis*) and the Macquarie Island parakeet (*Cyanoramphus novaezelandiae erythrotis*) to predation from both wekas and cats. Wekas have subsequently died out on Macquarie Island, partly resulting from the efforts of an eradication program, but more importantly through increased predation by cats, a secondary impact of a rabbit eradication program (Copson and Whinam 2001).

INVERTEBRATES

Our ability to provide an overview of the current status or impacts of non-indigenous invertebrates in the Antarctic is compromised by considerable variation in the level of knowledge available, both among taxonomic groups and across locations. The highest numbers (Table 1) are known from the subantarctic Iles Kerguelen (30 species) and Macquarie Island (28 species) (Frenot et al. 2005). Some islands, notably McDonald, Pingouins and Apôtres Islands, are largely non-impacted, while Macquarie, Iles Kerguelen, Possession, Marion Islands and South Georgia are those with the highest numbers. However, even on these much larger islands, the distribution of alien species is very patchy, with sometimes considerable areas (eg between glaciers, offshore islands) currently remaining pristine. Only two species of invertebrate (a dipteran and an enchytraeid worm) are confirmed to have established and remain persistent in the maritime Antarctic, with both introductions associated with human activity on Signy Island in the South Orkney Islands (Block et al. 1984) and none are proved to be resident in the continental Antarctic. A small fly (*Lycoriella* sp.) has been known from within station buildings at the continental station Casey (Hughes et al. 2005), illustrating a frequently observed feature of alien species being able to co-exist synanthropically but (currently at least) unable to expand beyond the confines of such situations. An eradication program on this fly was conducted at Casey station in April 2005, and at the beginning of 2006 appeared to have been successful.

Most studies of alien invertebrates in the Antarctic have focused on the physically larger groups of molluscs and arthropods, with little attention yet paid to smaller microscopic groups or the soil fauna. For instance, the tardigrade, rotifer and nematode species known from subantarctic islands generally are thought to have cosmopolitan distributions, but both detailed taxonomic studies and distributional data are lacking.

Most aliens are drawn from three groups of insects - Diptera, Hemiptera and Coleoptera. It is no coincidence that the two most widely distributed species are both capable of parthenogenetic reproduction (see Crafford et al. 1986) and very commonly associated with horticultural activities and food transport, these being *Psychoda parthenogenetica* (Diptera, Psychodidae) and *Rhopalosiphum padi* (Hemiptera, Aphididae). Indeed, many of the alien invertebrates recorded both as being established on subantarctic islands (Frenot et al. 2005) and anecdotally from maritime and continental Antarctic research stations (Hughes et al. 2005) are known to have been imported amongst general and food stores. As concluded by Pugh (1994) for mites (Acari), many species introduced to the region will have been imported with live vegetation, litter or soil.

Even though well documented for some islands (eg Ile de la Possession and Iles Kerguelen, Bouché 1982, Frenot 1985), some large and visible groups such as earthworms have not been well surveyed elsewhere. Likewise, although alien species of slug are known from three subantarctic islands (Iles Kerguelen, Macquarie Island, Marion Island), little is known of their biology or impacts. Other groups, such as non-marine Crustacea, while small in representation, have the potential for considerable impacts. The single species established on Marion Island (*Porcelio scaber*) is also already an invasive species on the South Atlantic cold temperate Gough Island (Jones et al. 2003). If the same situation were to develop on Marion Island the species could substantially alter nutrient cycling by reducing the bottleneck currently imposed by native lepidopteran larvae and earthworms (Slabber and Chown 2002, see Smith and Steenkamp 1992a,b).

Alien springtails (Collembola) are known from several subantarctic islands. The only major island from which none are known is Heard Island, and the group contributes 10% of the springtail fauna on South Georgia, >15% on Macquarie, 17% on Iles Kerguelen, 21% on Iles Crozet and 38% on Marion Island (Frenot et al. 2005). The alien faunas include several widespread invasives, particularly in the genus *Hypogastrura*. A member of this genus, *Hypogastrura viatica*, has been reported from the northern maritime Antarctic South Shetland Islands (Wise 1971) and from Léonie Island in the southern maritime Antarctic (Greenslade 1995), although the current status at either location is unconfirmed. Also in the South Shetland Islands, the presence of geothermal activity and heated ground on Deception Island may provide assistance for any alien species accidentally imported, such as the records of *Folsomia candida* and *Protaphorura* sp. reported by Greenslade and Wise (1984). Extensive geothermal activity on the maritime Antarctic South Sandwich Islands is also proposed to explain the presence of a range of subantarctic bryophytes and arthropods (Convey et al. 2000a,b), although human influence at this most isolated and extreme of locations remains so minimal that no evidence of anthropogenic introductions has been suggested.

MICROBIAL GROUPS

In much of the world, and despite the recognition that soil communities are key to the overall maintenance of ecosystem processes, remarkably little attention has been paid to their understanding. This is particularly true of the Antarctic, where many soils are relatively barren of both multicellular autotrophs and metazoans (Freckman and Virginia 1997, Convey and McInnes 2005), while decomposition pathways dominate nutrient and energy flows. Little is known about levels of endemism in most of the microbial flora (Lawley et al. 2004) or, alternatively, cosmopolitanism, although the algal flora is thought to be largely cosmopolitan (Broady 1996). However, if the 'global ubiquity hypothesis' (Finlay and Clarke 1999) does apply even in part to the Antarctic microbial flora, it is clear that the constraints to microbial dispersal and biodiversity will be fundamentally different to those applying to multicellular organisms.

There are virtually no data available on the presence of alien microbial species in the Antarctic, with the exception of a number of yeasts, fungi (see Wynn-Williams 1996a, Downes 2003) and algae (Broady and Smith 1994). The lack of detailed Antarctic studies is further complicated by a lack of comparable data from elsewhere including, in contemporary molecular studies, a lack of both Antarctic and non-Antarctic sequence data. However, there is evidence for at least some of the Antarctic prokaryote and eukaryote microbial floras being distinct (Franzmann 1996, Lawley et al. 2004).

Notwithstanding the overall lack of knowledge, the risk of importation into the Antarctic has been recognised (Smith 1996, Wynn-Williams 1996b). It is already clear, as with the macroscopic biota discussed above, that human-mediated imports play a significant role. Examples include the discovery of spores of a *Penicillium* species at Mt. Howe, of human pathogens in soil close to McMurdo station (Wynn-Williams 1996b), the isolation of fungi from historic sites on Ross Island (Minasaki et al. 2001) and, in the subantarctic, the infection of stands of *Pringlea antiscorbutica* by *Botryotinia fuckeliana* on Marion Island (Kloppers and Smith 1998) and *Albugo candida* on Iles Kerguelen (Y. Frenot and F. Hennion unpubl. data, Fig. 2a), both fungus species probably transferred from fresh vegetables. Circumstantial evidence of human introduction is also provided by diversity studies that report a proportion of taxa restricted to sites of human activity, as found by Azmi and Seppelt (1998) near Casey Station in the Windmill Islands and Kerry (1990) in the Vestfold Hills and Mac.Robertson Land.

Figure 2. Consequential impacts of some invasive species. (a) Infection of Pringlea
antiscorbutica *by the fungus* Albugo candida *on Ile Australia, Iles Kerguelen, February 2000
(Photo: Y. Frenot), (b) Damage to* P. antiscorbutica *caused by rats on Ilôt Colbeck, Iles
Kerguelen (Photo: J.-L. Chapuis), (c) Cat in a king penguin colony, Iles Kerguelen (Photo:
J.-L. Chapuis), (d) The predatory carabid beetle* Trechisibus antarcticus *introduced to South
Georgia (Photo: P. Bucktrout/British Antarctic Survey).*

A potentially significant, and certainly widely publicised, risk associated with human activity in Antarctica lies in the potential introduction or activation of diseases to which regional wildlife are susceptible (Kerry et al. 1999). Although the risk is clear, such a causal link has yet to be documented, and categorical proof is confounded by the fact that many Antarctic birds and mammals interact with humans well beyond the boundaries of the region. Nevertheless, there is considerable evidence of exposure to various pathogens (see reviews by Clarke and Kerry 1993, Kerry et al. 1999, Frenot et al. 2005). Furthermore, mass mortality events, such as the deaths of hundreds of chinstrap penguins from what is believed to be avian cholera (*Pasturella multocida*) at a tourist visitation locality on South Georgia, have now been reported (S. Harvey pers. comm. http://www.sgisland.org/pages/main/news16.htm).

Some human activities carry with them a virtual certainty of some form of contamination of the Antarctic environment which, while they can be controlled by reasonable measures (eg Hughes and Blenkharn 2003), could realistically only be removed completely by ending human contact with the continent. One such is the introduction of microorganisms associated with sewage. Although, again, few detailed studies have been completed, these organisms have been located in the marine and sea-ice environments near to McMurdo Station (Ross Sea) (Smith and McFeters 1999, Edwards et al. 1998) and in nearshore and coastal locations near to Rothera Station (Adelaide Island) (Hughes 2003a,b), in the continental and maritime Antarctic, respectively. Smith and McFeters (1999) further identified a risk that indigenous microbiota may be susceptible to the transfer of harmful genetic features from pathogenic microorganisms in untreated sewage, with potential knock-on consequences for other wildlife.

FRESHWATER AND MARINE HABITATS

There have been no reports of non-indigenous species in freshwater habitats (lakes, pools, streams) of the continental and maritime Antarctic. Pugh et al. (2002) conclude that anthropogenic dispersal is a likely explanation for at least some records of non-marine (ie terrestrial and freshwater) Crustacea on subantarctic islands. However, their suggestion that the presence of the copepod *Boeckella poppei* in Beaver Lake and adjacent lakes (Amery Oasis, continental Antarctica) is linked with human transfer is now known to be incorrect, as the species is found in sediment cores from these lakes dating back at least several thousand years (see Gibson et al. this volume).

Antarctica has no native freshwater fish. Several salmonids were introduced and survive on the subantarctic Iles Kerguelen and Iles Crozet (Davaine and Beall 1997), but their current status has not been researched in detail. Brown trout introduced to Marion Island were restricted to a single river system and are now considered to be extinct (Cooper et al. 1992), as are rainbow trout introduced to some pools. Although these probably have had substantial local impact on invertebrate populations no detailed studies are available. Trout were also introduced to some lakes on South Georgia near to occupied whaling stations (Headland 1984), but have long been extinct, with no information available on any impacts on these ecosystems. The subantarctic islands also have few native freshwater birds, these being restricted to two species of duck on South Georgia and Iles Kerguelen.

Palaearctic mallard (*Anas platyrhynchos*) is a non-indigenous species of duck first recorded on Macquarie Island in 1950s. The species is to be expected to have impacted the freshwater ecosystems of this island, though no specific data appear to exist.

Although marine ecosystems do not fall within the scope of this volume, brief mention is warranted given the acknowledged significance of human activity in dispersal and alien colonisation in marine habitats worldwide. Despite this attention, there are few studies or even anecdotal observations of alien marine taxa in the Antarctic region. Two clear records have been published, both from the South Shetland Islands. Mats of a non-indigenous green alga (*Enteromorpha intestinalis*) have been found in the intertidal zone at Half Moon Island (62°37'W 59°57'S), which Clayton et al. (1997) suggested may have been introduced on the hulls of ships or yachts. Most recently, Tavares and de Melo (2004) reported a North Atlantic species of spider crab from a trawl collection off the Antarctic Peninsula, to date the only record of a non-indigenous marine species in Antarctic seas.

Human activity again presents clear opportunities for transport of alien marine species into the Antarctic. The most direct route (Lewis et al. 2003) involves the ships used in Antarctic science and tourism activities, with opportunities presented by the transport of ballast water and by hull fouling assemblages. Lewis et al. (in press) reported how a barge loaded onto a supply ship had the potential to introduce an entire temperate epibenthic community to Macquarie Island (the barge's deployment was stopped once the biosecurity hazard had been identified). A second, more indirect, route exists through the potential for transfer on anthropogenic marine debris (Barnes 2002, Barnes and Fraser 2003), a route that also may provide opportunities for transfer of terrestrial and intertidal taxa (Hughes et al. this volume).

Case studies

PLANT COMMUNITIES

Prince Edward Islands
The recorded vascular flora of Marion Island consists of 23 native, 18 alien and three species of uncertain status. Of the 18 introductions, six have died out or have been eradicated. Despite the imposition of strict quarantine measures, even in recent years new species have reached the islands, with four becoming established the 1990s (Gremmen and Smith 1999). Several of the alien species introduced to Marion Island at the time of construction of the weather station in 1948 have spread rapidly. Gremmen and Smith (1999) and Ryan et al. (2003) estimated continuous rates of spread of 200 - 600 m.year^{-1}. These rates exclude occasional 'jumps' of several kilometres by some species. Of the alien species that have become established on Marion Island, two (*Cerastium fontanum* and *Sagina procumbens*) have subsequently spread to Prince Edward Island without apparent human assistance (Bergstrom and Smith 1990, Gremmen and Smith 1999, Ryan et al. 2003).

Of the 12 alien species presently occurring on Marion Island, eight are grasses. Nine species are able to become dominant in the communities they invade, of which seven are grasses (Gremmen 1997, Gremmen and Smith 1999, Fig. 1a). One other species, *Rumex acetosella*, may locally also become dominant, but not in the sense that it crowds out many native species. Six species are presently widespread, three of which are grasses, and at least one other grass, the recently introduced *Agrostis gigantea*, is expected to become widespread if unchecked. A program to eradicate this latter species is presently underway.

The impact of *Agrostis stolonifera*, which now dominates many plant communities of slope drainage lines and riverbanks, has been studied in detail by Gremmen et al. (1998). This species has invaded 19 of 22 major plant communities on Marion Island, and in seven of these, it has become dominant, at least locally. In the drainage line communities it has invaded, vegetation structure and community species composition and species richness have been significantly affected. The vegetation has changed from a deciduous *Acaena magellanica* dwarf-shrub dominated community into a permanent dense sward of high grass. Total standing crop has not changed significantly, but the percentage of bryophyte matter in the standing crop has diminished from 15% to 1%, and the total number of native plant species (vascular plants and bryophytes) from 7.4 to 3.6 per 4m^2 sample plot. A comparison of estimated total species richness of all invaded *vs.* unaffected drainage lines led to a prediction of a reduction in total species richness in this habitat by some 20%, once *A. stolonifera* has spread throughout the island. The impacts of *A. stolonifera* also extended to soil macro- and meso-invertebrates, with significant consequences detected in terms of overall species composition and population densities of individual species. Changes of a similar magnitude are also associated with invasion by a number of other alien species, including *Poa annua*, *Poa pratensis* and *Sagina procumbens*, although rigorous studies have not been completed (Gremmen 1997). On South Georgia, the different nutritional qualities of alien (*P. annua*) over native grasses have been shown to impact the biology of endemic herbivorous/detritivorous beetles (Chown and Block 1997).

The impacts of alien plants on their host indigenous plant communities appears to be related to their ability to spread vegetatively, thus forming dense colonies, rather than simply to their ability to disperse rapidly. For instance, some grasses with a very restricted distribution on Marion Island have a massive impact on the original communities of the sites they have invaded. Examples include *Agropyron repens* (1 site) and *Festuca rubra* (2 sites), both of which have almost totally replaced the native species at the locations they have invaded. The impact of alien species can be measured in different ways. In addition, as above, to quantifying the changes introduced in individual invaded communities, an alternative or complementary approach is to consider the total invaded area. A combination of both approaches can then be used to provide a better overall assessment of impacts. Following this approach it can be seen that, although several alien plants on Marion and Prince Edward Islands have large impacts at a local scale, it is also the case that, at present, only a low percentage of the total area of the islands has been invaded. Most alien plants are patchily distributed, relating to their mode of reproduction and dispersal (often vegetative) and to the patchiness of occurrence of suitable habitats. This contrasts markedly with distribution patterns of some invasive animals (eg mice), which rapidly become much more ubiquitous.

Heard Island

On Heard Island until recently, only a single alien vascular species was known to occur, *Poa annua*, with the first report in 1987. Subsequent to this, the species' density and abundance have increased (Scott and Kirkpatrick 2005). Its mode of introduction or arrival is unknown, and its centre of distribution is not related to the main areas of human activities on the island, although the fact that it is a Northern Hemisphere species, widely introduced to most of the Southern Hemisphere, makes its alien status unequivocal. On Heard Island, *P. annua* is most common in open communities on moraine soils, where it usually achieves low cover but can reach 75% over small areas (measured in $1m^2$ sample plots). It is not clear if *P. annua* simply occupies open space in these communities, or replaces any of the native species, as the total native species richness of sample plots in invaded areas is identical to comparable, non-invaded areas (N.J.M. Gremmen unpubl. data). Quadrat survey data obtained in February 2001 indicated an average density of $18\ 908 \pm 880$ plants.m^{-2} and a mean of $10\ 502 \pm 825$ inflorescences.m^{-2} (D.M. Bergstrom unpubl. data).

Iles Kerguelen

A number of alien vascular plants are widespread on Iles Kerguelen and locally reach dominance in the vegetation. In a study of islands in the Golfe du Morbihan, aliens were found to invade mainly well-drained lowland slopes and other well-drained lowland habitats (Fig. 1b). Again, some grasses reach local dominance, spreading vegetatively and replacing most of the native plants in the areas they invade. Several non-graminoid species, however, notably *Taraxacum officinale*, *T. erythrophyllum* and *Senecio vulgaris*, also reach high cover values, assisted by the abundant production of wind-borne seeds. In general, sites in which alien species have reached high cover values contain fewer native species than comparable pristine sites (N. Gremmen unpubl. data).

One of the notable ecosystem interactions of alien plant species on islands of the Iles Kerguelen is that with native seabirds and alien rats. On Ile Australia, rats have had a negative impact on burrowing bird populations. The stature and cover of alien plant species such as *Poa pratensis*, *Vulpia bromoides* and *Taraxacum officinale* is substantially less on Ile Australia than on the rat free Ile Mayes, which thus supports substantial burrowing bird populations. It is believed that the alien plants are responding to increased levels of nutrient input from bird guano on Ile Mayes, although further research to substantiate this hypothesis is needed.

Pattern of colonisation along tracks on Iles Crozet

On Possession Island, the last six years have been marked by significant increases in the distribution of the dominant alien plant species established away from the immediate vicinity of the research station (M. Lebouvier and Y. Frenot, unpubl.

data). *Sagina procumbens*, for example, was first recorded in 1978 at Alfred Faure Station. By the mid 1990s, it was present in the vicinity of all the field huts on the island and, by 2002, its distribution had expanded considerably, not only around sites of most human activity, but also along the tracks linking these sites. Similar trends are apparent in *Taraxacum erythrospermum*, a species located only at the research station until 1996. These two examples demonstrate the obvious role of humans in the spread of alien species in the subantarctic islands and the significance of tracks used by walkers in assisting alien establishment and spread.

VERTEBRATES

The house mouse (*Mus musculus domesticus*) is the most widespread alien vertebrate species, introduced to several subantarctic islands. On Iles Kerguelen (Le Roux et al. 2002) and Marion Island (Smith et al. 2002) this opportunistic rodent includes a variety of items in its diet: earthworms, larvae of a flightless and endemic moth, weevil adults and larvae, seeds of grasses and *Acaena magellanica* and floral parts of the alien dandelion (*Taraxacum officinale*). At Iles Kerguelen, Le Roux et al. (2002) showed that mice had a marked preference for plants in the summer months (January and February), whereas invertebrates formed nearly 100% of the prey items taken in winter (July). Pye and Bonner (1980) reported analogous sequential changes in dietary composition of rats (plant material, invertebrates, vertebrates) over the seasonal cycle on South Georgia. On Marion Island mice have a significant impact on *Azorella* cushions, which they destroy by burrowing into them, often eating through the main roots. Mice also selectively collect *Uncinia compacta* seedheads. As a result, ripe *Uncinia* seeds are now rarely seen on Marion Island, in contrast with the situation on Prince Edward Island, which is mouse-free and where ripe seeds are produced in abundance (Smith and Steenkamp 1990). Rats were seen to destroy *Azorella* cushions in a similar fashion on Ile Australia and Iles Kerguelen (D.M. Bergstrom, pers. obs., Fig. 2b). It is clear that alien rodents can affect several different components of subantarctic terrestrial food webs. Furthermore, on Marion Island Huyser et al. (2000) identified mice as playing an important role in the decrease in lesser sheathbill (*Chionis minor*) populations, an indigenous bird that relies on terrestrial invertebrates for winter survival.

The cat is the only mammalian predator present on some subantarctic islands, introduced as a pet or for the control of alien rodent populations. Food sources include other alien mammals (rats, mice and rabbits) and native birds. Say et al. (2002) estimated that the population of cats at Iles Kerguelen approached 7000 individuals on the main island, while Pontier et al. (2002) showed that the diet of feral cats varied markedly among different sites. Rabbits were the most common dietary item (about 90% throughout the year) in sites remote from bird colonies, whereas near a king penguin colony (Fig. 2c) and a large black-browed albatross colony rabbits, mice and birds were similarly represented (about 30%). These results contrast strongly with those of an earlier study (Derenne 1976): in 1976, 66.3% of stomachs examined contained birds and 35.0% contained rabbits whereas in 1998/99, scats consisted of 7.3% bird remains and 84.2% rabbits. While the methodologies used differed, it is known that rabbit populations did not fluctuate significantly between the two studies and, therefore, the large difference is interpreted as strongly suggesting that bird availability is now lower than in 1976 and that cats have had a major impact on the Kerguelen avifauna over a time period of less than 50 years (Pontier et al. 2002).

The impact of rabbits has been disastrous for most of the sensitive subantarctic islands to which the species has been introduced, causing soil erosion and rapid decreases in the native vegetation. At Iles Kerguelen, some plant species normally dominant in native communities, such as the Kerguelen cabbage *Pringlea antiscorbutica* and the cushion plant *Azorella selago* have became rare, replaced by almost monospecific communities of *Acaena magellanica* (Chapuis et al. 1994). Changes in plant cover have indirect consequential impacts on invertebrate communities (Chapuis et al. 1991) and the breeding potential of some burrow-nesting birds (Weimerskirch et al. 1989). Rabbits were also the main winter food resource for cats and ensured their survival during the less favourable months (Chapuis et al. 2004). Consequently, the rabbit can now be considered as a keystone species at Iles Kerguelen.

Similar ecosystem damage and impacts have occurred on Macquarie Island, where rabbits have demonstrated selective grazing of approximately 50% of native vascular plant species as well altering habitat for burrowing birds (Copson and Whinam 1998). Modifications to the plant communities and bird habitats may also have impacted on the invertebrate fauna and caused alterations to edaphic processes (Copson and Whinam 2001). Recent analysis of native plant species with restricted populations on the island suggests that rabbits may be limiting the potential expansion of the tussock grass *Poa littorosa*, with grazing observed at all four known populations. Rabbits may therefore be affecting ecosystem evolution (Bergstrom et al. in press).

Norwegian reindeer were introduced to South Georgia on three occasions between 1911 and 1925, in the vicinity of whaling stations. Archival information is summarized by Leader-Williams (1988). There are currently two genetic stocks and the total number of reindeer is estimated to about 2 - 4000. Population densities are far greater than those typical for the source locations in Scandinavia. Ten animals from Sweden were introduced to Iles Kerguelen in 1955/56 (Lésel 1967). In 1972, Pascal (1982) estimated the population at 2000 and there is no indication of any subsequent decrease (J.-L. Chapuis, pers. comm.). Reindeer on the Iles Kerguelen are restricted to Grande Terre, with the reported presence on the pristine Ile Foch (Chapuis et al. 1994) now appearing doubtful. On South Georgia (Leader-Williams 1988), snow cover limits the choice of forage available for reindeer to tussock grass (*Paradiochloa flabellata*) for up to three months of winter and other forage species (*Deschampsia antarctica, Acaena magellanica*) remain unavailable for up to six months. This dependence upon a single grass species in winter is mainly responsible for the overgrazing of the tussock grassland and to a lesser extent, of mesic meadow, dwarf-shrub sward and dry meadow. Overgrazing has also resulted in soil erosion. In addition to altering the structure and composition of plant communities (Leader-Williams et al. 1987), a further consequence of the activities of reindeer on South

Georgia lies in their assisting the rapid spread of the alien grass *Poa annua* (Fig. 1d), whose nutritional characteristics are different to those of native grasses, leading to impacts on invertebrate communities (Vogel et al. 1984, Chown and Block 1997). No accurate study on the impact of reindeer on the vegetation of Iles Kerguelen has been carried out but some effects are obvious, for example, many cushions of *Azorella selago* are turned over (Chapuis et al. 1994) and trampling causes severe damage in wet areas (Y. Frenot, pers. obs.).

The eradication of alien vertebrates is likely to be practicable, at least for the larger mammals. At first sight, such action provides a visible and probably useful response limiting or mitigating the damage caused by certain alien species. However, management approaches must be carefully planned and executed and require a multidisciplinary approach. There is no doubt that eradication campaigns removing rabbits from Macquarie Island (Copson and Whinam 2001) or cats from Marion Island (Bloomer and Bester 1991) have been successful and have led to positive consequences for native ecosystems. However, several examples are available in the literature demonstrating that undesirable cascade or consequential effects may ensue. For instance, van Aarde et al. (1996) pointed out the significance of the regulatory role of cats on mouse populations on Marion Island. Similarly, eradication of rats on St. Paul Island led to a large increase in the abundance of mice, most probably with an increase in their impact on native invertebrate communities. The eradication of rabbits on three islands of the Iles Kerguelen also produced unexpected results. Chapuis et al. (2004) observed a consequential decrease in cover of some plant species (primarily *A. magellanica*) and an increase in the abundance of other alien species such as *Taraxacum officinale*. While this trend is most closely linked to the removal of grazing, recent changes in climatic conditions (warming and increased drought stress) are also influential in the decrease or slow recovery of native species and the success of alien species (Convey this volume). In these contexts, the decisions over whether to attempt eradication are not trivial and, as emphasized by Myers et al. (2000), alternative approaches may be preferable.

INVERTEBRATES

In comparison with vertebrates, the consequences of most invertebrate introductions have received scant attention, indeed it is likely that many more examples exist than are currently recognised, at least on the subantarctic islands. Where terrestrial vertebrates are lacking, invertebrates play a correspondingly greater role in ecosystem processes. Ecosystems of the subantarctic and more extreme Antarctic regions are also simple in terms of both biodiversity and trophic complexity, featuring reduction in the number of higher taxonomic groups and trophic levels present. In the maritime and continental Antarctic, it is doubtful whether true herbivory is currently sustainable on energetic grounds and a large majority of the invertebrates present are regarded as detrivores or microbivores. While some predatory invertebrates are present (mites, springtails, nematodes, tardigrades), their impact currently appears to be low or undetectable (Lister et al. 1988, Convey

1996a), although few studies have been attempted. The two invertebrates known to have been introduced to the maritime Antarctic by human activity (Block et al. 1984) are both detritivores. Although they have local distributions that are gradually expanding, they have not been the subjects of detailed investigation and any impacts on indigenous communities remain unknown.

Some invertebrates introduced to the subantarctic islands have the potential to markedly accelerate the rate at which ecological processes related to nutrient cycling can take place. Thus, on Marion Island, the woodlouse *Porcelio scaber* may in future remove a bottleneck in decomposition processes as discussed above (Slabber and Chown 2002). On the same island, the alien midge *Limnophyes minimus* may have comparable trophic impact as an indigenous and endemic lepidopteran (Hänel and Chown 1998). Others, such as the fly *Calliphora vicina* introduced to Iles Kerguelen, are thought likely to compete strongly with native dipterans (in this example, *Anatalanta aptera*) (Chevrier et al. 1997).

Particularly significant introductions are likely to be those that involve new feeding guilds or trophic levels. In this context, the introductions of two predatory carabid beetles (*Trechisibus antarcticus* and *Oopterus soledadinus*) to South Georgia and one (*O. soledadinus*) to Iles Kerguelen have been studied in detail (Ernsting 1993, Ernsting et al. 1995, 1999, Todd 1996, Chevrier et al. 1997, Brandjes et al. 1999) (Fig. 2d). These introductions are relatively unusual, involving Southern Hemisphere species and most likely taking place during the mid to late 20th Century (Ernsting 1993, Chevrier et al. 1997) and provide an illustration of the potentially rapid consequential changes experienced in ecosystems. They also provide a form of 'natural experiment' that would otherwise be impossible elsewhere in the world, in which fundamental ecological questions and theories can be tested relating to the consequences of introducing new trophic levels into pre-existing natural ecosystems. On South Georgia, where the most detailed studies have been completed, major consequences include considerable reductions in populations of endemic herbivorous perimylopid beetles, whose larvae form a major prey item, combined with accelerated rates of development of prey larvae permitting a less vulnerable size to be reached more rapidly. On both islands, the two carabids are also thought to be restricted (to different extents) by the low temperatures of their habitats and hence to be sensitive to any increase in availability of thermal energy brought about by climate warming.

The introduction of alien earthworms to many of the major subantarctic islands provides a further example of the potential impacts of a new functional group within an existing ecosystem. The native earthworms of the subantarctic islands belong to the family Acanthodrilisae and the genus *Microscolex*. However, earthworms of the family Lumbricidae originating from the Northern Hemisphere are now present. Their introduction is generally related to the importation of soil for use in glasshouses or to the rinsing of water barrels by sealers during the 19th Century. Until recently, all alien Lumbricidae known from the subantarctic were epigeous and humus feeders, as are the native *Microscolex* species. The most widespread alien species, *Dendrodrilus rubidus tenuis*, has similar digestive capabilities to the indigenous *M. kerguelensis* (Prat et al. 2002), characterized by a low cellulose decomposition capacity, a feature that largely reduces their role to the fragmentation of soil organic matter. In February 2004, a new lumbricid species (*Allolobophora chlorotica chlorotica*, albinic form) was found at Iles Kerguelen, in a location close to the glasshouses at Port-aux-Français (M. Lebouvier and Y. Frenot unpubl. data). This observation is significant because this species shares ecological characteristics of the anecic and endogeous earthworms (as defined by Bouché 1972), meaning that *A. chlorotica chlorotica* is not only detritivorous and a litter feeder, but also geophagous. It contributes, therefore, to the burying of organic matter in soil and its presence in a subantarctic location could drastically alter the processes of pedogenesis in this region.

Future scenarios

The term 'climate change' is often equated with 'climatic amelioration' – thereby carrying an implicit assumption that change will automatically lead to less stressful and hence 'better' conditions for biota. This is a particularly tempting assumption in the context of Antarctic biology, where change is often perceived simply as an increase in temperature, hence in availability of thermal energy and a reduction in the risks and costs of freezing stress. In many locations and circumstances, this simplistic interpretation does carry some weight, but it is also clear that patterns and consequences of climate change are far more complex and, particularly if changes in patterns of water availability are seen, can result in considerable increases in stress (Kennedy 1995, Block and Convey 2001, Convey et al. 2002, Convey this volume).

Where climate amelioration occurs, it is likely to enhance the ability of both natural long-distance colonists and human-assisted aliens to complete successfully the two key stages of long distance transport and establishment, particularly in the subantarctic (Kennedy 1995, Bergstrom and Chown 1999, Frenot et al. 2005). Whinam et al. (2005) has identified transport routes for propagules in association with national logistic programs. However, to date there are only a few documented instances of long-distance colonization of new sites within the Antarctic even by indigenous species and no new records of alien species that can be linked clearly with climate amelioration as distinct from direct human intervention. Founder populations of four flowering plant species (two species on each island) have been identified as being less that 200 years old on subantarctic Macquarie and Heard Islands (Bergstrom et al. in press) and another single plant found on Heard Island in 2004 (*Leptinella plumosa*) provides a very recent possible example (Turner et al. 2006). Local colonisation and rapid population expansion has been well documented in the two native plant species in the maritime Antarctic (Fowbert and Smith 1994, Smith 1994) and linked with regional climate warming. Similar effects might be predicted for plant species across the Antarctic, but no rigorous studies of other species over time have yet been attempted.

In the subantarctic, where many alien plant and animal species are already established, climatic amelioration is likely to have two main effects. First, species already established but only of persistent status may be able to switch to a more aggressively invasive status. Recent changes in status of some alien plants on Marion Island (Gremmen and Smith 1999) might already illustrate such a response. In this context, it has also been proposed that indigenous subantarctic invertebrates have inherently slower life cycles and less ability to respond to temperature increase than their alien competitors (Barendse and Chown 2000). However, some indigenous species may be able to respond more rapidly. This seems to be true across the region. The diving beetle *Lancetes angusticollis* is the top predator in lake ecosystems on South Georgia. It currently undergoes a typically biennial life cycle, with its development limited by the thermal energy currently available in its habitats and the possession of a temperature mediated larval diapause stage (Nicolai and Droste 1984). This species may show a very rapid response to warming, with an increase of only 1°C in lake temperature providing enough additional thermal energy to allow development to be completed on an annual rather than a biennial timescale. Such a change in population dynamics of the top predator is likely to have considerable impacts on the trophic dynamics of these lake food webs (Arnold and Convey 1998).

Changes in colonisation patterns may also be facilitated by alterations in the relative success of asexual and sexual reproduction. Studies of the two native maritime Antarctic flowering plants have indicated a greater frequency of successful seed maturation with climate warming (Convey 1996b), leading to greater opportunities for more rapid and distant dispersal. Such a process may also allow an established alien species to colonize nearby locations by natural means, as appears to have occurred at the carefully protected Prince Edward Islands (Gremmen and Smith 1999). A general consequence of increasing success in sexual reproduction will be to generate increased genetic diversity, with the attendant possibility of offspring better capable of surviving the environmental extremes of Antarctic habitats. Although untested, such an explanation may underpin the success of the invasive microlepidopteran *Plutella xylostella* on Marion Island (Chown and Avenant 1992, Chown and Language 1994), as wider literature on this otherwise cosmopolitan pest species suggests it does not possess appropriate ecophysiological abilities to survive the subantarctic climate (Convey 2005).

Summary

A wide range of non-indigenous plants, invertebrates and vertebrates occur on most of the subantarctic islands and on some much more restricted parts of the Antarctic continent. The same is likely to hold true for microbial groups, although as yet there are few explicit demonstrations or even ongoing studies. Their arrival in the Antarctic region has been closely linked with human activity, which commenced only two centuries ago and exerts ever increasing pressure today. To date, despite the Southern Hemisphere location of the Antarctic, a large majority of the alien species known are European in origin. Their impacts on the functioning of Antarctic terrestrial ecosystems are diverse, but include examples of substantial loss of local biodiversity and changes to ecosystem processes.

Human activity essentially circumvents one of the key stages required for successful colonisation of Antarctic terrestrial habitats by biota from lower latitudes – that of long distance transport of organisms or propagules from source locations and the associated requirement for survival of the stresses experienced *en route*. Given this assistance, it is already known that a range of biota possess appropriate life history and ecophysiological capabilities to allow longer term establishment in the Antarctic. On their own, the rapid climate changes that are occurring in some parts of Antarctica are likely to reduce the barriers to successful survival of long distance transport by propagules and to increase the chance of successful establishment on arrival at an Antarctic location. However, such natural colonisation events are likely to be far outweighed in frequency by instances of human-assisted transport. Thus, climate change and human assistance in tandem are expected to result in an elevated frequency of introduction and establishment events.

Consequent increases in impacts on ecosystems are hard to forecast in detail, or for specific locations. The large majority of examples existing today are drawn from the subantarctic, where most islands already host a range of non-indigenous species. The most obvious impacts of these include (i) extensive habitat modification or destruction by grazing vertebrates, (ii) local population reduction or extinction of indigenous bird species by alien predatory vertebrates, (iii) the introduction of new trophic levels (vertebrate and invertebrate predators) into invertebrate communities where they were previously not present, with consequential impacts on prey, and (iv) alterations in the levels of competition faced in plant and invertebrate communities. Of course, the true picture is far more complex, with many direct, indirect and synergistic effects implicated across different communities and ecosystems.

Acknowledgements

The authors acknowledge the research resources of the Australian Antarctic Division, the British Antarctic Survey (BIRESA Project), the French Polar Institute (Programmes 136 and 272) and the CNRS (Zone-atelier de recherches sur l'environnement antarctique et subantarctique). Fieldwork on Heard Island and Macquarie Island was supported by the Australian Antarctic Program (Project 1015), Australian Government, Department of Environment and Heritage.

References

van Aarde, R.J. (1980) The diet and feeding behaviour of feral cats *Felis catus* on Marion Island, *South African Journal of Wildlife Research* **10**, 123-128.

van Aarde, R., Ferreira, S., Wassenaar, T. and Erasmus, D.G. (1996) With the cats away the mice may play, *South African Journal of Science* **92**, 357-358.

d'Antonio, C.M. and Dudley, T.L. (1995) Biological invasions as agents of change on islands versus mainlands, in P.M. Vitousek, L.L. Loope and H. Adersen (eds.), *Islands. Biological diversity and ecosystem function*, Springer, Berlin, pp. 103-121.

Arnold, R.J. and Convey P. (1998) The life history of the diving beetle, *Lancetes angusticollis* (Curtis) (Coleoptera: Dytiscidae), on subantarctic South Georgia, *Polar Biology* **20**, 153-160.

Azmi, O.R. and Seppelt, R.D. (1998) The broad-scale distribution of microfungi in the Windmill Islands region, Antarctica, *Polar Biology* **19**, 92-100.

Barendse, J. and Chown, S.L. (2000) The biology of *Bothrometopus elongatus* (Coleoptera, Curculionidae) in a mid-altitude fellfield on sub-Antarctic Marion Island, *Polar Biology* **23**, 346-351.

Barnes, D.K.A. (2002) Invasions by marine life on plastic debris, *Nature* **416**, 808-809.

Barnes, D.K.A. and Fraser, K.P.P. (2003) Rafting by five phyla on man-made flotsam in the Southern Ocean, *Marine Ecology Progress Series* **262**, 289-291.

Bergstrom, D.M. and Chown, S.L. (1999) Life at the front: history, ecology and change on southern ocean islands, *Trends in Ecology and Evolution* **14**, 472-477.

Bergstrom, D.M. and Smith, V.R. (1990) Alien vascular flora of Marion and Prince Edward Islands: new species, present distribution and status, *Antarctic Science* **2**, 301-308.

Bergstrom, D.M., Turner, P.A.M., Scott, J.J. Copson G. and Shaw J. (in press) Restricted plant species on subantarctic Macquarie and Heard Islands, *Polar Biology*, DOI 10.1007/s00300-005-0085-2.

Bester, M.N., Bloomer, J.P., Van Aarde, R.J., Erasmus, B.H., Van Rensburg, P.J.J., Skinner, J.D., Howell, P.G. and Naude, T.W. (2002) A review of the successful eradication of feral cats from sub-Antarctic Marion Island, Southern Indian Ocean, *South African Journal of Wildlife Research* **32**, 65-73.

Block, W. and Convey, P. (2001) Seasonal and long-term variation in body water content of an Antarctic springtail - a response to climate change? *Polar Biology* **24**, 764-770.

Block, W., Burn, A.J. and Richard, K.J. (1984) An insect introduction to the maritime Antarctic, *Biological Journal of the Linnean Society* **23**, 33-39.

Bloomer, J.P. and Bester, M.N. (1991) Effects of hunting on population characteristics of feral cats on Marion Island, *South African Journal of Wildlife Research* **21**, 97-102.

Bonner, W.N. (1984) Introduced mammals, in R.M. Laws (ed.), *Antarctic Ecology* Vol. 1., Academic Press, London, pp. 237-278.

Bouché, M.B. (1972) *Lombriciens de France. Ecologie et Systématique*, I.N.R.A. (Ann. Zool.- écol. anim. numéro hors série 72/2), Paris, 671pp.

Bouché, M.B. (1982) Les Lombriciens (Oligochaeta) des Terres Australes Françaises, *Comité National Français des Recherches Antarctiques* **51**, 175-180.

Bowen, L. and van Vuren, D. (1997) Insular endemic plants lack defenses against herbivores, *Conservation Biology* **11**, 1249-1254.

Brandjes, G.J. Block, W. and Ernsting, G. (1999) Spatial dynamics of two introduced species of carabid beetles on the sub-Antarctic island of South Georgia, *Polar Biology* **21**, 326-334.

Broady, P.A. (1996) Diversity, distribution and dispersal of Antarctic terrestrial algae, *Biodiversity and Conservation* **5**, 307-1335.

Broady, P.A. and Smith, R.A. (1994) A preliminary investigation of the diversity, survivability and dispersal of algae introduced into Antarctica by human activity, *Proceedings of the NIPR Symposium on Polar Biology* **7**, 185-197.

Brothers, N.P. (1984) Breeding distribution and status of burrow-nesting petrels on Macquarie Island, *Australian Wildlife Research* **11**, 113-131.

Burger, A.E., Williams, A.J. and Sinclair, J.C. (1980) Vagrants and the paucity of land bird species at the Prince Edward Islands, *Journal of Biogeography* **7**, 305-310.

Chapuis, J.L., Boussès, P. and Barnaud, G. (1994) Alien mammals, impact and management in the French subantarctic islands, *Biological Conservation* **67**, 97-104.

Chapuis J.-L., Frenot Y. and Lebouvier M. (2004) Recovery of native plant communities after eradication of rabbits from the subantarctic Kerguelen Islands, and influence of climate change, *Biological Conservation* **117**, 167–179.

Chapuis, J.-L., Vernon, P. and Frenot, Y. (1991) Fragilité des peuplements insulaires: exemple des îles Kerguelen, archipel subantarctique, in Z. Massoud (ed.), *Réactions des êtres vivants aux changements de l'environnement*, Actes des Journées de l'Environnement du Centre National de la Recherche Scientifique, 30 nov.-1er déc. 1989, Paris, pp. 235-248.

Chapuis, J.-L., Le Roux, V., Asseline, J., Lefèvre, L. and Kerleau, F. (2001) Eradication of rabbits (*Oryctolagus cuniculus*) by poisoning on three islands of the subantarctic Kerguelen Archipelago, *Wildlife Research* **28**:323-331.

Chevrier, M., Vernon, P. and Frenot, Y. (1997) Potential effects of two alien insects on a subantarctic wingless fly in the Kerguelen islands, in B. Battaglia, J. Valencia and D.W.H. Walton (eds.), *Antarctic Communities: Species, Structure and Survival*, Cambridge University Press, Cambridge, UK, pp. 424-431.

Chown, S.L. and Avenant, N. (1992) Status of *Plutella xylostella* at Marion Island six years after its colonisation, *South African Journal of Antarctic Research* **22**, 37-40.

Chown, S.L. and Block, W. (1997) Comparative nutritional ecology of grass-feeding in a sub-Antarctic beetle: the impact of introduced species on *Hydromedion sparsutum* from South Georgia, *Oecologia* **111**, 216-224.

Chown, S.L. and Language, K. (1994) Recently established Diptera and Lepidoptera on sub-Antarctic Marion Island, *African Entomology* **2**, 57-60.

Chown S.L. and Smith, V.R. (1993) Climate change and the short-term impact of feral house mice at the sub-Antarctic Prince Edward Islands, *Oecologia* **96**, 508-516.

Chown, S.L., Gaston, K.J. and Gremmen, N.J.M. (2000), Including the Antarctic: Insights for ecologists everywhere, in W. Davison, C. Howard-Williams and P.A. Broady (eds.), *Antarctic Ecosystems: Models for Wider Ecological Understanding*, New Zealand Natural Sciences, Christchurch, pp. 1-15.

Chown, S.L., Gremmen, N.J.M. and Gaston, K.J. (1998) Ecological biogeography of Southern Ocean islands: Species-area relationships, human impacts, and conservation, *American Naturalist* **152**, 562-575.

Chown, S.L., Rodrigues, A.S., Gremmen, N.J.M. and Gaston, K.J. (2001) World Heritage status and the conservation of Southern Ocean islands, *Conservation Biology* **15**, 550-557.

Clarke, J.R. and Kerry, K.R. (1993) Diseases and parasites of penguins, *Korean Journal of Polar Research* **4**, 79-86.

Clayton, M.N., Wiencke, C. and Klöser, H. (1997) New records and sub-Antarctic marine benthic macroalgae from Antarctica, *Polar Biology* **17**, 141-149.

Convey, P. (1996a) The influence of environmental characteristics on life history attributes of Antarctic terrestrial biota, *Biological Reviews* **71**, 191-225.

Convey, P. (1996b) Reproduction of Antarctic flowering plants, *Antarctic Science* **8**, 127-134.

Convey, P. (2001) Antarctic Ecosystems, in S.A. Levin (ed.), *Encyclopaedia of Biodiversity*, Vol. 1, Academic Press, San Diego, pp. 171-184.

Convey, P. (2005) Recent lepidopteran records from sub-Antarctic South Georgia, *Polar Biology* **28**, 108-110.

Convey, P. (2006) Antarctic climate change and its influences on terrestrial ecosystems, in D.M. Bergstrom, P. Convey, and A.H.L. Huiskes (eds.), *Trends in Antarctic Terrestrial and Limnetic Ecosystems: Antarctica as a Global Indicator*, Springer, Dordrecht (this volume).

Convey, P. and McInnes, S.J. (2005) Exceptional, tardigrade dominated, ecosystems from Ellsworth Land, Antarctica, *Ecology* **86**, 519-527.

Convey, P., Greenslade, P. and Pugh, P.J.A. (2000a) Terrestrial fauna of the South Sandwich Islands, *Journal of Natural History* **34**, 597-609.

Convey, P., Smith, R.I.L., Hodgson, D.A and Peat, H.J. (2000b) The flora of the South Sandwich Islands, with particular reference to the influence of geothermal heating, *Journal of Biogeography* **27**, 1279-1295.

Convey, P., Pugh, P.J.A., Jackson, C., Murray, A.W., Ruhland, C.T., Xiong, F.S. and Day, T.A. (2002) Response of Antarctic terrestrial arthropods to multifactorial climate manipulation over a four year period, *Ecology* **83**, 3130-3140.

Cooper, J., Crafford, J.E. and Hecht, T. (1992) Introduction and extinction of brown trout (*Salmo trutta* L.) in an impoverished sub-Antarctic stream, *Antarctic Science* **4**, 9-14.

Copson, G.R. and Whinam, J. (1998) Response of vegetation on Subantarctic Macquarie Island to reduced rabbit grazing, *Australian Journal of Botany* **46**, 15-24.

Copson, G. and Whinam, J. (2001) Review of ecological restoration programme on subantarctic Macquarie Island: Pest management progress and future directions, *Ecological Management and Restoration* **2**, 129-138.

Courchamp F., Chapuis, J.-L. and Pascal, M. (2003) Mammal invaders on islands: impact, control and control impact, *Biological Reviews* **78**, 347-383.

Crafford, J.E., Scholtz, C.H. and Chown, S.L. (1986) The insects of subantarctic Marion and Prince Edward Islands, with a bibliography of entomology of the Kerguelen biogeographical province, *South African Journal of Antarctic Research* **16**, 41-84.

Cumpston, J.S. (1968) *Macquarie Island.* Antarctic Division, Department of External Affairs, Melbourne Australia. 380 pp.

Davaine, P. and Beall, E. (1997) Salmonid introductions into virgin ecosystems (Kerguelen Islands, Subantarctic): Stakes, results, prospects, *Bulletin Francais de la Pêche et de la Pisciculture* **344-345**, 93-110.

Derenne, P. (1976) Notes sur la biologie du chat haret de Kerguelen, *Mammalia* **40**, 531–595.

Dingwall, P.R. (Ed) (1995) *Progress in conservation of the Subantarctic Islands.* Proceedings of the SCAR/IUCN Workshop on Protection, Research and Management of Subantarctic Islands, Paimpont, France, 27-29 April, 1992. Gland: World Conservation Union.

Downes, J. (2003) Factors affecting the introduction and distribution of fungi in the Vestfold Hills, Antarctica, PhD Thesis, University of Nottingham, 213 pp.

Edwards, D.D., McFeters, G.A. and Venkatesan, M.I. (1998) Distribution of *Clostridium perfringens* and fecal sterols in a benthic coastal marine environment influenced by the sewage outfall from McMurdo Station, Antarctica, *Applied and Environmental Microbiology* **64**, 2596- 2600.

Ernsting, G. (1993) Observations on life cycle and feeding ecology of two recently introduced predatory beetle species at South Georgia, sub-Antarctic, *Polar Biology* **13**, 423-428.

Ernsting, G., Block, W., MacAlister, H., and Todd, C. (1995) The invasion of the carnivorous carabid beetle *Trechisibus antarcticus* on South Georgia (sub-Antarctic) and its effect on the endemic herbivorous beetle *Hydromedion sparsutum*, Oecologia **103** 34-42.

Ernsting, G., Brandjes, G.J., Block, W. and Isaaks, J.A. (1999) Life-history consequences of predation for a subantarctic beetle: evaluating the contribution of direct and indirect effects, *Journal of Animal Ecology* **68**, 741-752.

Finlay, B.J. and Clarke, K.J. (1999) Ubiquitous dispersal of microbial species, *Nature* **400**, 828.

Fowbert, J.A. and Smith, R.I.L. (1994) Rapid population increases in native vascular plants in the Argentine Islands, Antarctic Peninsula, *Arctic and Alpine Research* **26**, 290-296.

Franzmann, P.D. (1996) Examination of Antarctic prokaryotic diversity through molecular comparisons, *Biodiversity and Conservation* **5**, 1295-1305.

Freckman, D.W. and Virginia, R.A. (1997) Low-diversity Antarctic soil nematode communities: Distribution and response to disturbance, *Ecology* **78**, 363-369.

Frenot, Y. (1985) Etude de l'introduction accidentelle de *Dendrobaena rubida tenuis* (Oligochaeta, Lumbricidae) à l'Ile de la Possession, *Bulletin d' Ecologie* **16**, 47-53.

Frenot, Y. and Gloaguen, J.-C. (1994) Reproductive performance of native and alien colonising phanerogams on a glacier foreland, Îles Kerguelen, *Polar Biology* **14**, 473-481.

Frenot, Y., Gloaguen, J.C., Massé, L. and Lebouvier, M. (2001) Human activities, ecosystem disturbance and plant invasions in subantarctic Crozet, Kerguelen and Amsterdam Islands, *Biological Conservation* **101**, 33-50.

Frenot, Y., Chown, S.L., Whinam, J., Selkirk, P., Convey, P., Skotnicki, M. and Bergstrom, D. (2005) Biological invasions in the Antarctic: extent, impacts and implications, *Biological Reviews* **80**, 45-72.

Gabriel, A.G.A., Chown, S.L., Barendse, J., Marshall, D.J., Mercer, R.D., Pugh, P.J.A. and Smith, V.R. (2001) Biological invasions on Southern Ocean islands: the Collembola of Marion Island as a test of generalities, *Ecography* **24**, 421-430.

Gauthier-Clerc, M., Jiguet, F. and Lambert, N. (2002) Vagrant birds at Possession Island, Crozet Islands and Kerguelen Island from December 1995 to December 1997, *Marine Ornithology* **30**, 38-39.

Gibson, J.A.E., Wilmotte, A., Taton, A., Van De Vijver, B., Beyens, L. and Dartnall, H.J.G. (2006) Biogeographic trends in Antarctic lake communities, in D.M. Bergstrom, P. Convey, and A.H.L. Huiskes (eds.), *Trends in Antarctic Terrestrial and Limnetic Ecosystems: Antarctica as a Global Indicator*, Springer, Dordrecht (this volume).

Greenslade, P. (1995) Collembola from the Scotia Arc and Antarctic Peninsula including descriptions of two new species and notes on biogeography, *Polskie Pismo Entomologiczne* **64**, 305-319.

Greenslade, P. (2002) Assessing the risk of exotic Collembola invading subantarctic islands: prioritizing quarantine management, *Pedobiologia* **46**, 338-344.

Greenslade, P. and Wise, K.A.J. (1984) Additions to the collembolan fauna of the Antarctic, *Transactions of the Royal Society of South Australia* **108**, 203-205.

Gremmen, NJM (1997) Changes in the vegetation of sub-Antarctic Marion Island resulting from introduced vascular plants, in B. Bataglia, J. Valencia and D.W.H. Walton (eds.), *Antarctic communities: species, structure and survival*, Cambridge, Cambridge University Press, pp. 417-423.

Gremmen, N.J.M. and Smith, V.R. (1999) New records of alien vascular plants from Marion and Prince Edward Islands, sub-Antarctic, *Polar Biology* 21, 401-409.

Gremmen, N.J.M., S.L. Chown and D.M. Marshal (1998) Impact of the introduced grass Agrostis stolonifera on vegetation and soil fauna communities at Marion Island, sub-Antarctic, *Biological Conservation* 85, 223-231

Gressitt, J.L. (ed.) (1970) Subantarctic entomology, particularly of South Georgia and Heard Island, *Pacific Insects Monograph* 23, 1-374.

Hänel, C. and Chown, S.L. (1998) The impact of a small, alien invertebrate on a sub-Antarctic terrestrial ecosystem: *Limnophyes minimus* (Diptera, Chironomidae) at Marion Island, *Polar Biology* 20, 99-106.

Headland, R.K. (1984). *The Island of South Georgia*. Cambridge University Press, Cambridge, 293 pp.

Heywood, V.H. (1989) Patterns, extents and modes of invasions by terrestrial plants, in J.A. Drake, H.A. Mooney, F. di Castri, R.H. Groves, F.J. Kruger, M. Rejmanek, and M. Williamson (eds.), *Biological Invasions: A Global Perspective*, John Wiley, Chichester, UK, pp. 31-60.

Hughes, K.A. (2003a) Aerial dispersal and survival of sewage-derived faecal coliforms in Antarctica, *Atmospheric Environment* 37, 3147-3155.

Hughes, K.A. (2003b) The influence of seasonal environmental variables on the distribution of presumptive fecal coliforms around an Antarctic research station, *Applied and Environmental Microbiology* 5, 555-565.

Hughes, K.A. and Blenkharn, N. (2003) A simple method to reduce discharge of sewage microorganisms from an Antarctic research station, *Marine Pollution Bulletin* 46, 353-357.

Hughes, K.A., Ott, S., Bölter, M. and Convey, P. (2006) Colonisation processes, in D.M. Bergstrom, P. Convey, and A.H.L. Huiskes (eds.), *Trends in Antarctic Terrestrial and Limnetic Ecosystems: Antarctica as a Global Indicator*, Springer, Dordrecht (this volume).

Hughes, K.A., Walsh, S., Convey, P., Richards, S. and Bergstrom, D. (2005) Alien fly populations established at two Antarctic research stations, *Polar Biology* 28, 568-570.

Huiskes, A.H.L., Convey, P. and Bergstrom, D.M. (2006) Trends in Antarctic terrestrial and limnetic ecosystems: Antarctica as a global indicator, in D.M. Bergstrom, P. Convey, and A.H.L. Huiskes (eds.), *Trends in Antarctic Terrestrial and Limnetic Ecosystems: Antarctica as a Global Indicator*, Springer, Dordrecht (this volume).

Hull, B.B. and Bergstrom, D.M. (2006) Antarctic terrestrial and limnetic conservation and management, in D.M. Bergstrom, P. Convey, and A.H.L. Huiskes (eds.), *Trends in Antarctic Terrestrial and Limnetic Ecosystems: Antarctica as a Global Indicator*, Springer, Dordrecht (this volume).

Huyser, O., Ryan, P.G. and Cooper, J. (2000) Changes in population size, habitat use and breeding biology of lesser sheathbills (*Chionis minor*) at Marion Island: impacts of cats, mice and climate change? *Biological Conservation* 92, 299-310.

Jones, A.G., Chown, S.L. and Gaston, K.J. (2003) The free living pterygote insects of Gough Island, *Systematics and Biodiversity* 1, 213-273.

Jouventin, P., Stahl, J.C., Weimerskirch, H. and Mougin, J.L. (1984) The seabirds of the French subantarctic islands and Adelie land, their status and conservation, in J.P. Croxall, P.G.H. Evans and R.W. Schreiber (eds.), *Status and Conservation of the World's Seabirds*, International Council for Bird Preservation, Cambridge, Technical Publication No. 2, pp. 609-625.

Kennedy, A.D. (1995) Antarctic terrestrial ecosystem response to global environmental change, *Annual Review of Ecology and Systematics* 26, 683-704.

Kerry, E. (1990) Microorganisms colonising plants and soil subjected to different degrees of human activity, including petroleum contamination, in the Vestfold Hills and Mac.Robertson land, *Polar Biology* 10, 423-430.

Kerry, K., Riddle, M. and Clarke, J. (1999) *Diseases of Antarctic Wildlife*, A Report for The Scientific Committee on Antarctic Research (SCAR) and The Council of Managers of National Antarctic Programs (COMNAP), 104pp.

Kloppers, F.J. and Smith, V.R. (1998) First report of *Botryotinia fuckeliana* on Kerguelen Cabbage on the Sub-Antarctic Marion Island, *Plant Disease* 82, 710.

Lawley, B., Ripley, S., Bridge, P. and Convey, P. (2004) Molecular analysis of geographic patterns of eukaryotic diversity in Antarctic soils, *Applied and Environmental Microbiology* 70, 5963-5972.

Leader-Williams, N. (1988) *Reindeer on South Georgia: The Ecology of an Introduced Population*, Cambridge University Press, Cambridge, 319 pp.

Leader-Williams, N., Smith, R.I.L. and Rothery, P. (1987) Influence of introduced reindeer on the vegetation of South Georgia: results from a long-term exclusion experiment, *Journal of Applied Ecology* **24**, 801-822.

Lee, C.E. (2002) Evolutionary genetics of invasive species, *Trends in Ecology and Evolution* **17**, 386-391.

Le Roux, V., Chapuis, J.L., Frenot, Y. and Vernon, P. (2002) Diet of the House Mouse (*Mus musculus* L.) at Guillou Island, Kerguelen archipelago, subantarctic, *Polar Biology* **25**, 49-57.

Lésel, R. (1967) Contribution à l'étude écologique de quelques mammifères importés aux îles Kerguelen, *Terres Australes et Antarctiques Françaises* **38**, 3-40.

Lewis, P.N., Bergstrom, D.M. and Whinam, J. (in press) Barging in: a temperate marine community travels to Macquarie Island, *Biological Invasions*.

Lewis, P.N., Hewitt, C.L., Riddle, M. and McMinn, A. (2003) Marine introductions in the Southern Ocean: an unrecognised hazard to biodiversity, *Marine Pollution Bulletin* **46**, 213-223.

Lindsay D.C. (1973) Probable introductions of lichens to South Georgia, *British Antarctic Survey Bulletin* **33 and 34**, 169-172.

Lister, A., Block, W. and Usher, M.B. (1988) Arthropod predation in an Antarctic terrestrial community, *Journal of Animal Ecology* **57**, 957-971.

Lockwood, J.L. (1999) Using taxonomy to predict success among introduced avifauna: relative importance of transport and establishment, *Conservation Biology* **13**, 560-567.

Lyons, W.B., Laybourn-Parry, J., Welch, K.A. and Priscu, J.C. (2006) Antarctic lake systems and climate change, in D.M. Bergstrom, P. Convey, and A.H.L. Huiskes (eds.), *Trends in Antarctic Terrestrial and Limnetic Ecosystems: Antarctica as a Global Indicator*, Springer, Dordrecht (this volume).

Mack, R.N., Simberloff, D., Lonsdale, W.M., Evans, H., Clout, M. and Bazzaz, F.A. (2000) Biotic invasions: causes, epidemiology, global consequences, and control, *Ecological Applications* **10**, 689-710.

McKinney, M.L. and Lockwood, J. (1999) Biotic homogenization: a few winners replacing many losers in the next mass extinction, *Trends in Ecology and Evolution* **14**, 450-453.

Micol, T. and Jouventin P. (1995) Restoration of Amsterdam Island, South Indian Ocean, following control of feral cattle, *Biological Conservation* **73**, 199-206.

Micol, T. and Jouventin, P. (2002) Eradication of rats and rabbits from Saint-Paul Island, French Southern Territories, in C.R. Veitch and M.N. Clout (eds.), *Turning the tide: the eradication of invasive species*, IUCN SSC Invasive Species Specialist Group. IUCN, Gland, Switzerland, pp. 199-205.

Minasaki, R., Farrell, R.L., Duncan, S., Held, B.W. and Blanchette, R.A. (2001) Mycological biodiversity associated with historic huts and artefacts of the heroic period in the Ross Sea region. *Antarctic Biology in a Global Context*, Abstracts of VIII SCAR International Biology Symposium, Amsterdam. Abstract S5P28.

Myers, J.H., Simberloff, D., Kuris, A.M. and Carey, J.R. (2000) Eradication revisited: dealing with exotic species, *Trends in Ecology and Evolution* **15**, 216-320.

Nicolai, V. and Droste, M. (1984) The ecology of *Lancetes claussi* (Müller) (Coleoptera, Dytiscidae), the subantarctic water beetle of South Georgia, *Polar Biology* **3**, 39-44.

Ochyra, R., Smith, V.R. and Gremmen, N.J.M. (2003) *Thuidium delicatulum* (Hedw.) Schimp. (Thuidiaceae) - another bipolar moss disjunct from Subantarctic Marion Island, *Cryptogamie, Bryologie* **24**, 253-263

Pascal, M. (1980) Structure et dynamique de la population de chats harets de l'archipel des Kerguelen, *Mammalia* **44**, 171-182

Pascal, M. (1982) Les espèces mammaliennes introduites dans l'archipel de Kerguelen (Territoire des TAAF). Bilan des recherches entreprises sur ces espèces, *Comité National Français des Recherches Antarctiques* **51**, 333-343.

Pontier D., Say L., Debias F., Bried J., Thioulouse J., Micol T. and Natoli E. (2002) The diet of feral cats (*Felis catus* L.) at five sites on the Grande Terre, Kerguelen archipelago, *Polar Biology* **25**, 833–837.

Prat P., Charrier M., Deleporte S. and Frenot Y. (2002) Digestive carbohydrases in two epigeic earthworm of the Kerguelen Islands (Subantarctic), *Pedobiologia* **46**, 417-427.

Pugh, P.J.A. (1994) Non-indigenous Acari of Antarctica and the sub-Antarctic islands, *Zoological Journal of the Linnean Society* **110**, 207-217.

Pugh, P.J.A., Dartnall, H.J.G. and McInnes, S.J. (2002) The non-marine Crustacea of Antarctica and the islands of the Southern Ocean: biodiversity and biogeography, *Journal of Natural History* **36**, 1047-1103.

Pye, T. and Bonner, W.N. (1980) Feral brown rats, *Rattus norvegicus*, in South Georgia (South Atlantic Ocean), *Journal of Zoology* **192**, 237-255.

Pye, T., Swain, R. and Seppelt, R.D. (1999) Distribution and habitat use of the feral black rat (*Rattus rattus*) on subantarctic Macquarie Island, *Journal of Zoology* **247**, 429-438.

Pyšek, P. (1998) Is there a taxonomic pattern to plant invasions? *Oikos* **82**, 282-294.

Ryan, P.G., Smith, V.R. and Gremmen, N.J.M. (2003) The distribution and spread of alien vascular plants on Prince Edward Island, *African Journal of Marine Science* **25**, 555-562.

Sala, O.E., Chapin, F.S., Armesto, J.J., Berlow, E., Bloomfield, J., Dirzo, R., Huber-Sanwald, E., Huenneke, L.F., Jackson, R.B., Kinzig, A., Leemans, R., Lodge, D.M., Mooney, H.A., Oesterheld, M., Poff, N.L., Sykes, M.T., Walker, B.H., Walker, M. and Wall, D.H. (2000) Global biodiversity scenarios for the year 2100, *Science* **287**, 1770-1774.

Say L., Gaillard J.-M. and Pontier D. (2002) Spatio-temporal variation in cat population size in a sub-antarctic environment, *Polar Biology* **25**, 90-95.

Selmi, S. and Boulinier, T. (2001), Ecological biogeography of Southern Ocean Islands: The importance of considering spatial issues, *American Naturalist* **158**, 426-437.

Scott, J.J. and Kirkpatrick, J.B. (2005) Changes in subantarctic Heard Island vegetation at sites occupied by *Poa annua*, 1987-2000, *Arctic, Antarctic and Alpine Research* **37**, 366-371.

Shaw J.D, Bergstrom, D.M and Hovenden, M. (2005) The impact of feral rats (*Rattus rattus*) on populations of subantarctic megaherb (*Pleurophylum hookeri*), *Austral Ecology* **30**, 118-125.

Slabber, S. and Chown, S.L. (2002) The first record of a terrestrial crustacean, *Porcellio scaber* (Isopoda, Porcellionidae), from sub-Antarctic Marion Island, *Polar Biology* **25**, 855-858.

Smith, J.J. and McFeters, G.A. (1999) Microbial issues of sewage disposal from Antarctic bases: dispersion, persistence, pathogens and "genetic pollution", in K.M. Kerry, M. Riddle and J. Clarke (eds.), *Disease of Antarctic Wildlife. A report for the Scientific Committee on Antarctic Research (SCAR) and the Council of Managers of National Antarctic Programs (CONMAP)*, Australian Antarctic Division, Kingston, pp. 56-57.

Smith, R.I.L. (1994) Vascular plants as indicators of regional warming in Antarctica, *Oecologia* **99**, 322-328.

Smith, R.I.L. (1996) Introduced plants in Antarctica: potential impacts and conservation issues, *Biological Conservation* **76**, 135-146.

Smith V.R. and Steenkamp M (1990) Climate change and its ecological implications at a sub-Antarctic island. *Oecologia* **85**, 14-24

Smith, V.R. and Steenkamp, M. (1992*a*) Macroinvertebrates and litter nutrient release on a sub-Antarctic island, *South African Journal of Botany* **58**, 105-116.

Smith, V.R. and Steenkamp, M. (1992*b*) Soil macrofauna and nitrogen on a sub-Antarctic island, *Oecologia* **92**, 201-206.

Smith, V.R. and Steenkamp, M. (2001) Classification of the terrestrial habitats on sub-Antarctic Marion Island based on vegetation and soil chemistry, *Journal of Vegetation Science* **12**, 181-198.

Smith, V.R., Avenant, N.L. and Chown, S.L. (2002) The diet and impact of house mice on a sub-Antarctic island, *Polar Biology* **25**, 703-715

Tavares, M. and de Melo, G.A.S. (2004) Discovery of the first known benthic invasive species in the Southern Ocean: the North American spider crab *Hyas araneus* found in the Antarctic Peninsula, *Antarctic Science* **16**, 129-131.

Taylor, R.H. (1979) How the Macquarie Island Parakeet became extinct, *New Zealand Journal of Ecology* **2**, 42-45.

Todd, C.M. (1996) Body size, prey size and herbivory in Coleoptera from the sub-Antarctic island of South Georgia, *Pedobiologia* **40**, 557-569.

Torr, N. (2002) Eradication of rabbits and mice from subantarctic Enderby and Rose Islands, in C.R. Veitch and M.N. Clout (eds.), *Turning the Tide: The Eradication of Invasive Species*, Occasional Paper IUCN Species Survival Commission **27**, 319-328.

Turner P.A.M., Scott J.J. and Rozefelds A. (2006) Probable long distance dispersal of Leptinella plumosa Hook. f. to Heard Island: habitat, status and discussion of its arrival, *Polar Biology*, **29**, 160-168.

Vermeij, G.J. (1996) An agenda for invasion biology, *Biological Conservation* **78**, 3-9.

Vernon, P., Vannier, G. and Tréhen, P. (1998) A comparative approach to the entomological diversity of polar regions, *Acta Oecologica* **19**, 303-308.

Vogel, M., Remmert, H. and Smith, R.I.L. (1984) Introduced reindeer and their effects on the vegetation and the epigeic invertebrate fauna of South Georgia (subantarctic), *Oecologia* **62**, 102-109.

Weimerskirch, H., Zotier, R. and Jouventin, P. (1989) The avifauna of the Kerguelen Islands, *Emu* **89**, 15-29.

Williamson, M. (1996) *Biological Invasions*. Chapman and Hall, London, UK.

Wise, K.A.J. (1971) The Collembola of Antarctica, *Pacific Insects Monograph* **25**, 57-74.

Whinam J., Chilcott, N., Bergstrom, D.M. (2005) Subantarctic hitchhikers: expeditioners as vectors for the introduction of alien organisms, *Biological Conservation* **121**, 207-219.

Whinam, J. Copson, G. and Chapuis, J.-L. (2006) Subantarctic terrestrial conservation and management, in D.M. Bergstrom, P. Convey, and A.H.L. Huiskes (eds.), *Trends in Antarctic Terrestrial and Limnetic Ecosystems: Antarctica as a Global Indicator*, Springer, Dordrecht (this volume).

Wynn-Williams, D.D. (1996a) Antarctic microbial diversity: the basis of polar ecosystem processes, *Biodiversity and Conservation* **5**, 1271-1293.

Wynn-Williams, D.D. (1996b) Response of pioneer soil microalgal colonists to environmental change in Antarctica, *Microbial Ecology* **31**, 177-188.

11. LANDSCAPE CONTROL OF HIGH LATITUDE LAKES IN A CHANGING CLIMATE

A. QUESADA
Departamento de Biología, Universidad Autónoma de Madrid,
28049 Madrid, Spain
antonio.quesada@uam.es

W. F. VINCENT
Département de Biologie and Centre d'études nordiques,
Université Laval, Sainte-Foy, Québec, G1K 7P4, Canada
warwick.vincent@bio.ulaval.ca

E. KAUP
Institute of Geology at Tallinn University of Technology
Estonia pst 7, 10143 Tallinn, Estonia
kaup@gi.ee

J. E. HOBBIE
The Ecosystems Center, Marine Biological Laboratory Woods Hole,
MA 02543, USA
jhobbie@mbl.edu

I. LAURION
Institut national de la recherche scientifique, Centre Eau, Terre et
Environnement and Centre d'études nordiques
490 de la Couronne, Québec, G1K 9A9 Canada
Isabelle_Laurion@ete.inrs.ca

221

D.M. Bergstrom et al. (eds.), Trends in Antarctic Terrestrial and Limnetic Ecosystems, 221–252.
© *2006 Springer.*

R. PIENITZ

*Département de Géographie and Centre d'études nordiques
Université Laval, Sainte-Foy, Québec, G1K 7P4, Canada
reinhard.pienitz@cen.ulaval.ca*

J. LÓPEZ-MARTÍNEZ

*Departamento de Geología y Geoquímica,
Universidad Autónoma de Madrid, 28049 Madrid, Spain
jeronimo.lopez@uam.es*

J.-J. DURÁN

*Instituto Geológico y Minero de España,
Ríos Rosas, 23, 28003 Madrid, Spain
jj.duran@igme.es*

Introduction

Lakes are the downstream integrators of their surrounding catchments and are therefore highly responsive to variations in landscape properties. High latitude lakes share many characteristics with those of temperate latitudes and are subject to many of the same landscape controls. However, polar lakes and their catchments also experience persistent low temperatures, extreme seasonality and severe freeze-thaw cycles and these distinguishing features are likely to amplify their responsiveness to landscape and climate change.

General circulation models vary in their prediction of the future magnitude of regional climate change, but almost all converge on the conclusion that the polar regions will experience greater temperature increases than elsewhere and that these changes are likely to occur ever faster because of the positive feedback effects of melting snow and ice. Observations in the north polar region have shown that there have been significant rises in temperature throughout much of the area in recent decades, with effects on permafrost, lake ice cover, glacial extent and ice shelf break-up (Serreze et al. 2000, Mueller et al. 2003, ACIA 2004). Recent analysis of long term monitoring data from maritime Antarctic Signy Island (South Orkney Islands) and the McMurdo Dry Valleys in continental Antarctica has provided evidence of large regional variations in climate change and has shown that south polar lakes respond strongly to warming and cooling trends (Doran et al. 2002, Quayle et al. 2002, 2003).

In this review, we take a two-step approach towards examining climate-landscape-lake interactions in high latitude environments. First, we examine the general effects of landscapes on lake ecosystems through factors such as geomorphology, solute transport, vegetation and hydrology. We then examine how these properties are linked to climate, resulting in a set of mechanisms whereby climate change can have pronounced impacts on lakes. Throughout this review, we have drawn on examples from both polar regions. Lake ecosystems are an important part of the Arctic landscape and there are limnological similarities to Antarctica. In both regions permanently frozen soils exert a strong influence on catchment properties such as hydrological processes and geochemical interactions. Similarly in both regions, snow and ice cover are major controls on the structure and functioning of aquatic ecosystems. There is a long history of limnological research in high northern latitudes and much of this information is directly relevant to Antarctica. Current observations and model predictions indicate that the Arctic is much more sensitive to climate change and will continue to experience more rapid shifts in its temperature regime than Antarctica (Overland et al. 2004a, b). Much of the Subarctic and Arctic became ice-free in the early Holocene and the limnology and paleolimnology of northern lakes therefore provide a window into the potential future states of Antarctic lakes undergoing climate change.

Landscape Controls on Lakes

In this section we describe and illustrate the variety of ways in which landscape can exert control on the structure and dynamics of lake ecosystems. These effects take place at multiple time-scales: from pre-Holocene and Holocene (thousands of years), to recent changes observed over the last few decades. We have placed emphasis on examples from lakes in the polar regions to illustrate their distinctive properties and features that may make them particularly sensitive to climate change.

GEOMORPHOLOGY

Geomorphology exerts the most fundamental control on lakes and their surrounding catchments by dictating the shape of the landscape and the size and morphometry of basins that can receive, transport and store water. These properties also influence the geochemical weathering of substrates, the relative importance of hydrological flow pathways above and below soils and the transfer of solutes from land to water. A wide range of limnological characteristics are affected by the ratio of catchment to lake area, and by the area-volume and volume-depth relationships of lakes, including temperature, light availability for photobiological and photochemical processes, oxygen levels, nutrient retention times and habitat availability (Wetzel 2001, Kalff 2001). The geomorphological controls on lakes include not only surface effects, but also underground processes where lithology and geological structure play an important role in the water and solute supply. Lakes can be classified according to their geomorphological origins such as tectonic (eg folds, fractures), lithological (eg volcanics, karst), differential erosion (eg rock bars, dykes), glacial erosion (eg excavation, moraines), ice dynamics (eg thermokarst), coastal processes (eg lagoons) and slope processes (eg rotational slumps, avalanches).

In the polar regions, many geomorphological features of the landscape have been shaped by glacial processes and continue to respond to ice dynamics. In the most extreme systems, ice dams the outlet of lakes. Some of the most impressive of these proglacial lakes (eg lakes Agassiz and Barlow-Ojibway) formed during the last deglaciation in the Northern Hemisphere, especially in areas surrounding Hudson Bay, Canada. Glacial retreat beyond certain threshold areas often resulted in catastrophic floods (eg the 8.2 ka event, Barber et al. 1999) that had global impacts on climate. There are many examples of proglacial lakes today in Antarctica (eg Lake Hoare, McMurdo Dry Valleys, Lake Untersee, Dronning Maud Land). Ice and snowpack can also act as a temporary dam to lakes (eg Lake Limnopolar on Byers Peninsula, Livingston Island and Boeckella Lake on Hope Bay, both in the Antarctic Peninsula region) and such effects are also well known for northern rivers (Prowse and Culp 2003).

Epishelf lakes are a form of proglacial lake in which ice dams retain freshwater over the sea. This lake type is now rare in the Arctic, and one of the few examples was recently lost by climate change and break-up of the ice shelf at its mouth, with subsequent draining of the freshwater into the sea (Mueller et al. 2003). Many examples are known from Antarctica, including Moutonnée and Ablation lakes on the coast of Alexander Island (Heywood 1977, Hodgson et al. 2004), epishelf lakes in the Schirmacher Oasis (Korotkevich 1960) and Bunger Hills (Korotkevich 1972, Doran et al. 2000, Gibson and Andersen 2002, Verkulich et al. 2002) and Beaver Lake associated with the Amery ice shelf (Adamson et al. 1997, Laybourn-Parry et al. 2001). In some cases, the sea ice can be folded by the thrusting of glaciers and forms a series of synclinal troughs and anticlinal ridges roughly parallel to the edge of the glacier. Freshwater lakes of a few hundreds of meters long, tens of meters wide and a few meters deep can occupy the synclinal troughs during the summer months before the break-out of the sea ice, as in the Marguerite Bay area in Antarctica (Nichols 1960). Most of these systems formed during earlier periods of relative sea-level rise, lifting grounded ice masses and allowing sea water to penetrate into the lower part of the water column, or by connection to marine waters during periods of reduced freshwater input (Gibson and Andersen 2002, Gibson et al. this volume). A review of the formation and dynamics of Antarctic lake ecosystems (including epishelf lakes, epiglacial, subglacial and supra-glacial lakes) and their paleolimnology is provided by Hodgson et al. (2004).

Extensive flat terrains corresponding to erosive raised platforms of marine origin are relatively common in Antarctic coastal areas, due to glacioisostatic uplift and often contain numerous lakes. Examples include many ice-free peninsulas of the South Shetland Islands such as Fildes Peninsula on King George Island and Byers Peninsula on Livingston Island. In the Byers Peninsula, raised surfaces are extensive and contain more than 60 lakes and 50 pools, some of them up to $50\,000m^2$ in surface area and 9m deep. The flat, raised platforms show recent glacial erosion and absence of well-defined divides. There are diffuse boundaries and small differences in altitude between drainage basins, and this favours recent capture of meltwaters and precipitation, and sporadic connections between basins (López-Martínez et al. 1996).

Lithology and the geological structure exert a strong influence on high latitude lakes. Some lakes are located in craters (eg on Deception and Penguin Islands, South Shetland Islands) or in tectonic troughs. In many cases, faults and diaclases control the location of the lakes and the characteristics of the basins and the drainage network. Many lakes are located in glacial over-deepened, tectonically controlled sites. Lakes most strongly influenced by fractures are, in general, more elongate and commonly aligned parallel to each other, for example on Byers Peninsula, Livingston Island (López-Martínez et al. 1996).

Coastal lakes of lagoon origin are very common in the lower beaches of the maritime Antarctica. They have gently sloping edges and are either subcircular or elongate in plan (Jones et al. 1993, Cuchi et al. 2004). Their salinities tend to be high because of intermittent exchange with the sea (see below).

Igneous rocks often show abundant 'roches moutonnées' and the development of 'rock bars' and 'riegels' that act as dams for lakes in many cases. Lakes of karstic origin are rare in high latitudes due to the scarcity of limestone and other rocks suitable for karst development, in addition to the lack of water. However, thermokarst lakes of permafrost origin, are common in the polar regions (see below).

Under certain circumstances, liquid water masses can be maintained beneath the Antarctic ice sheet by geothermal heat fluxes to produce subglacial lakes. These lakes are likely to have been isolated for very long periods of time from the above-ice communities (over 15 million years, Bulat et al. 2004) and the potential for life in these waters is presently an intriguing frontier of research in polar microbial ecology.

Ice dams or other geomorphological features such as moraines or rock bars and riegels may produce arheic lakes in which there is no flushing and outflow, with the only loss of water via evaporation or ice ablation. This type of landscape effect results in hypersaline environments in which the solutes in the inflowing waters are concentrated by evaporation or freeze-concentration. This produces a stratified dense brine at the bottom layer of deeper lakes, or extreme salinities throughout the water column of shallower lakes. Winter freeze-up of shallow waters may impose a severe constraint on biodiversity in some high latitude environments by exposing organisms to extreme osmotic stress and sub-zero liquid water temperatures (Hawes et al. 1999).

A well studied example of geomorphological evolution and its effects on lakes is in the Toolik Long Term Ecological Research (LTER) site, Alaska. In this region, the broad retreat of ice sheets has left behind landscapes in which the glacial activity in the past has a strong effect on modern-day surface waters. In fact, three different glacial advances have left a legacy of different ecologies of lakes and streams. During the Pleistocene, most of northern Alaska was unglaciated except for mountain glaciers that flowed north from the peaks and valleys of the Brooks Range and spread out over the foothills. In the Toolik region, the till of the earliest glaciation, from the early Pleistocene, has been overlapped by till from three advances (Hamilton 2003). Lakes and streams on these three glacial surfaces have been studied: a ~10 000 year old surface (10ka), a ~60 000 year old surface (60ka) and a greater than 300 000 year old surface (300ka). The youngest surface (~10ka) has numerous kettle lakes, many isolated from streams. Flanking slopes are steep and may have slumps along the banks. Soils are well drained but surface drainage is poorly integrated. On the next surface (60ka) kettle lakes are still present but many have grassy slopes and marshy shores where solifluction has deposited silt. Drainage networks are well integrated and lakes are connected by small streams. The oldest surface (>300ka) is a subdued landscape that reflects a long period of postglacial modification. Moraines have broad crests and gentle flanking slopes. Crests and upper slopes have a cover of windblown silt (loess). Solifluction has redistributed much of the loess from upper to lower parts of moraine flanks. Soils are fine-grained and hold more moisture than those on the younger surfaces. Slopes are drained primarily by water tracks rather than by incised stream channels. Kettle lakes are rare but small thaw ponds have developed in swales. Drainages have silty channels with beaded thaw ponds. Even the hydrology is somewhat different in older terrestrial landscapes as there is better development of water tracks, narrow troughs that run downslope every few 10s of metres. The differences among landscapes are shown in Table 1. Similar studies are required in Antarctica

Table 1. Lake and stream averages for landscape of different ages in Alaska. Conductivity includes ranges. Assembled by W.J. O'Brien.

Landscape age	Lake depth (m)	Number of lakes (#/km^2)	Lake area (ha)	Stream length (km)	Conductivity (μS/cm) and range
Young <10ka	16	1.24	3.1	0.65	80 (45-105)
Intermediate ~60ka	11.3	0.77	5.7	0.74	40 (28-70)
Old >300ka	9.1	0.18	1.6	1.03	20 (8-35)

Hobbie et al. (2003) hypothesized that younger surfaces support more varied biotic communities because there are more types of habitat. On the younger surface, more lakes lead to greater buffering of stream hydrology, longer residence times of water in the aquatic systems and less stream disturbance. More lakes lead to more trapping of nutrients in sediments and to accumulation of nutrients in photosynthetic algae and secondary consumers. Downstream, the stream biota are more abundant because of the material from lakes. Isolated lakes in the youngest surfaces may lack fish, which drastically changes the food web, and have a very small drainage basin. The evolution of stream network structure is important because watershed disturbance interacts with stream network structure to create a distribution of habitats that affects the function of the entire stream network (Benda et al. 2004).

HYDROLOGY

The hydrologic properties of high latitude catchments are especially complex given the importance of freeze-thaw cycles and the influence of permafrost on surface and subsurface flow. Seasonal freezing and thawing processes play a major role by altering the amount of liquid water in the catchment and the pathways of flow to downstream lakes.

The heterogeneity of the permafrost and the active layer makes it very difficult to predict hydrological pathways in high latitude landscapes. Permafrost acts as a non-permeable layer, restricting the water exchange or movement to the active layer that thaws on a seasonal basis. Although groundwater is mostly frozen and immobile within permafrost, some groundwater movement can occur there by two mechanisms. First, ice within permafrost retains a small percentage of liquid water in films that can potentially exist at temperatures down to -50°C. This water may creep through the tortuous pore channels that pervade all frozen ground. There may be also salt brines in soils that are not flushed by precipitation. These brines are sufficiently concentrated to remain liquid to very low temperatures and can migrate as groundwater. These reservoirs of liquid water occur at different levels and can be interconnected (Woo 2000).

In spring, the soil is still frozen and infiltration from snowmelt only takes place through cracks or interstices through coarse elements of the soil. In Arctic soils, infiltration represents only 5-20% of the snowmelt (Marsh and Hey 1989). This infiltrated water may refreeze, further increasing the impervious nature of the soils. The spring flows may be very intense (up to 70% of the snow cover, Landals and Gill 1972) and largely restricted to the surface, washing the first centimetres of the soil and transporting moderate amounts of solutes. This spring flow is especially important for organisms inhabiting the soil surface, for example cyanobacterial mats and moss turfs that commonly grow on gentle slopes without permanent water supply. The active layer starts thawing after disappearance of the snow-cover and subsurface infiltration and flow pathways become more important at that time. Some groundwater flow may occur into lakes via taliks, unfrozen substrata that underlie lakes in permafrost regions. At some sites in the polar regions, groundwater fed by snowmelt and rainfall comes to the surface as springs that then flow into streams and lakes (Andersen et al. 2002).

The low albedo of barren high latitude soils causes surface heating and evaporation to be important features in the hydrological cycle of the polar regions (Woo 2000). Evaporative losses are highly variable, for example accounting for 15% of the 300-350mm precipitation in northern Alaska (Hobbie 1973, cited by Schindler et al. 1974) and a potential 165% of the 350-400mm precipitation in the Schirmacher Oasis, Antarctica (Haendel 1995). In polar desert catchments where precipitation is extremely low, the scarce snowmelt water is readily absorbed by the uppermost centimetres of unfrozen soil and then largely evaporated because of the dryness of the air and high winds. This results in an extreme negative water balance, for example an annual loss of ice of 305mm at Lake Vanda compared with only 10mm annual precipitation (Chinn 1993). Thus, groundwater contributes only sporadically to surface water bodies in permafrost regions (Fig. 1).

In certain areas of low-lying ground, snowmelt or precipitation can saturate the active layer and accumulate at the surface, producing extensive wetlands. This type of landscape is found throughout both polar regions, for example on Livingston Island in Antarctica and in the Canadian High Arctic. These polar wetlands provide persistent liquid water 'oases' in summer, and they are typically highly productive, well developed ecosystems with a diverse biota (Woo and Young 2000). The thawing of permafrost and the resulting decrease of soil volume and even slumping of shorelines gives rise to thermokarst lakes, small, shallow water bodies that are especially abundant throughout the Subarctic and Arctic (see below).

In continuous permafrost areas, the streams and rivers have little subsurface exchange and therefore respond rapidly to variations in precipitation (Woo 2000). Maximum runoff is recorded during the spring snowpack and glacial melt period and diurnal variations in discharge may occur associated with the daily melt cycle. The interannual variations in flow in the Onyx River, feeding Lake Vanda in the McMurdo Dry Valleys, have been related to the number of positive °C days on the surface of the glaciers that feed it with meltwater (Chinn 1993). In discontinuous permafrost areas, the streams and rivers are fed by subsurface water via the active layer, taliks or subpermafrost springs and so there is much less of a direct coupling with precipitation and melt cycles.

The nature and length of the transport pathway, from the water source to the lake, also influences the extent of solute accumulation and transfer. For example, glaciers and snow in the McMurdo Dry Valleys provide most of the hydrological input to the lakes, but contain only low solute concentrations (Lyons et al. 1998). Much of the dissolved content of these inflows is derived from the inflows picking up salts that have been deposited by wind or solid precipitation, or released by weathering. Thus, stream length has a strong influence on the ultimate downstream concentrations of solutes and hence chemical loading into the receiving lakewaters.

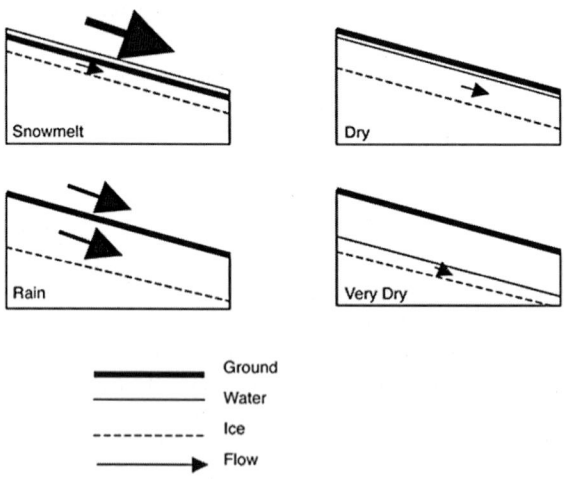

Figure 1. Water flows and relative position of the ice and water tables in a typical slope, under different meteorological conditions. The size of the arrows represents the difference in magnitude of the flow. The arrows above and inside the diagram represent the surface and the subsurface flow, respectively (redrawn from Woo 2000).

Subsurface flows are known in Antarctica, for example of increased salinity groundwaters in the Larsemann Hills. Widely fluctuating concentrations of subsurface nutrients have indicated patchily distributed soil brines and decaying buried organic matter as sources of salts (Kaup and Burgess 2002). Enriched by these sources, inflows have substantially increased the nutrient levels of lakes that they discharge into (Kaup et al. 2001). In dry soils, however, the subsurface transport of elements may be minimal. For example, in the McMurdo Dry Valleys region, Campbell et al. (1998) found that solute mobility ranged from 1m in very arid soils to 5m in wetter soils. With less soluble metals (Pb, Cu and Zn), movement was 0.5m or less in 30 years.

LITHOLOGY

Differences in geological substrates affect the extent of rock weathering and the chemical composition of soil water that ultimately discharges into lakes. This can be seen in lakes distributed across major geological boundaries, for example from highly alkaline sedimentary to acidic granite-gneiss formations in the region northeast of Yellowknife in the Canadian Arctic (Pienitz et al. 1997). As noted above, differences in glacial history can lead to different substrate types that in turn affect the solute chemistry of river and lake waters.

Geological differences also affect the type and productivity of vegetation that in turn will influence lakes through a variety of mechanisms (see below). For example, in their analysis of vegetation on the Arctic Slope, Alaska, Walker et al. (2003) found a significant statistical relationship between above ground phytobiomass and a summer temperature index (sum of mean monthly temperatures above 0°C). One conspicuous outlier, however, was at Atqasuk, where biomass was lower than at Barrow despite a more than two-fold higher temperature index. This anomaly was attributed to the sandy, nutrient-poor soils at Atqasuk. Similarly there are major differences in soil characteristics including pH and nutrients among glacial landscapes of different ages in the Toolik Lake LTER site, Alaska. This translates into differences in soil mineralization rates, foliar nutrient levels and plant community structure and biomass (Hobbie and Gough 2002).

On the Byers Peninsula, as an example of maritime Antarctic conditions, water conductivity shows a broad range of values from 4.8 - 1441µS cm^{-1}. Lowest values of conductivity are related to snow melt (4µS cm^{-1}) while rainfall varies between 17.3 and 53.7µS cm^{-1}, indicating the influence of marine aerosols. Groundwaters have conductivities from 246 to 715µS cm^{-1}, indicating rock weathering reactions, with differences according to lithology. Lakes and pools have higher conductivity values than snow/rain and underground water. Coastline lagoons have particularly high values due to overwash of brackish water, probably of marine origin. An inverse relationship between discharge and conductivity has been suggested (Cuchí et al. 2004). The hydrologic cycle during the end of the summer at Byers Peninsula indicates favourable conditions for rock-water interactions that increase conductivity of water according to rock type, temperature and interaction time. Permafrost plays an important role in the recharge and water flow, especially in interior areas of the peninsula.

Moisture and frequency of freeze-thaw cycles are key factors affecting chemical weathering (Haendel 1995). However, the relative importance of freeze-thaw cycles is under discussion. Hall (1992) indicated that in Byers Peninsula (Livingston Island) wetting and drying, salt weathering and chemical weathering could be more important than freeze-thaw, but Navas et al. (2006) suggested that the main processes in the same area seem to be related to cryogenic effects resulting from freeze-thaw cycles that lead to rock disintegration and the supply of solutes to the lakes. The migration of sulphatic, carbonatic and bicarbonatic solutions in soils and rocks has been documented by ionic content of water and in some cases by salt efflorescences in many ice-free areas (Simonov 1971, Wand et al. 1985). There is evidence of chemical weathering in the McMurdo Dry Valleys (Lyons and Mayewski 1993), although some authors argue that chemical weathering is reduced to negligible rates under conditions of extreme cold and dryness (Matsuoka 1995, Campbell and Claridge 1987). Chemical weathering is accelerated in aqueous environments, for example, silicate-based rocks in the stream beds of the McMurdo Dry Valleys (Lyons et al. 1998). Several studies (eg Conca and Malin 1986, Conca and Wright 1987) have shown that cations such as Fe^{3+}, Sr^{2+}, K^+ and Ca^{2+} are being weathered from the bedrock in Southern Victoria Land. However, it has been demonstrated that only a small proportion of the Mg^{2+} entering into the lakes in Taylor Valley, McMurdo Dry Valleys, is of weathering origin (Green et al. 1988), and that most of the Na^+, Cl^- and SO_4^{2-} enters as marine-derived aerosols. CO_3^{2-} and HCO_3^- may derive from dissolution of carbonate rocks in the drainage basins or via hydrolysis of silicate minerals. In less extreme areas, under higher humidity such as the Schirmacher Oasis, free water and temperatures above freezing are dominant in the soil for 2 to 3 months, causing release of K^+, Ca^{2+} and SO_4^{2-}. The residence time of water in the thawing zone has a strong influence on the extent of solute accumulation. As in the Taylor Valley, anorthosites contribute to a low Mg^{2+} and high Ca^{2+} content of waters in Antarctic Lakes Untersee and Obersee (Haendel 1995). Alkaline solutes, released during weathering of anorthosites are also the most probable reason for pH values of up to 12 in these lakes (Haendel et al. 1995), combined with the low buffering capacity of the receiving waters.

VEGETATION

Plant biomass, productivity and community structure are features of the landscape that have a wide-ranging influence on rivers and lakes. Vegetation alters the wind regime, the extent of snow cover and albedo, and the hydrological balance between evaporation and precipitation. The root structures of the vegetation in subantarctic and Arctic environments greatly influence the physical stability of the catchment and its susceptibility to soil erosion and particle transfer into waterways. The extent of permafrost development is also affected by vegetation cover, as are leaf litter, soil properties and terrestrial biota, including microbiota that can be eventually washed into streams and lakes.

One of the greatest controlling influences of terrestrial plants on aquatic ecosystems is through its effect on solutes in the waters that percolate through the leaf litter and root zones and eventually make their way into lakes via overland flow, streams and groundwater. Some nutrients and ions may be stripped out by biological uptake processes in the soil or by interactions with soil particles, while others may increase in concentration through decomposition, cation exchange and weathering. Dissolved organic matter (DOM) derived from vegetation breakdown in the soil has a particularly broad range of effects on lake and river ecosystems. Landscape processes control the quantity and quality of DOM entering the aquatic system (allochthonous DOM, ie that derived from outside the lake) with regional effects of climate, vegetation type and pH (acid rain) and strong gradients as a function of latitude and altitude (Schindler et al. 1990, Pienitz and Smol 1993, Vincent and Pienitz 1996, Lotter 1999, Laurion et al. 2000).

Even in Antarctic polar desert lakes, there is transfer of solutes from their surrounding catchments, including organic materials from mosses, lichens, algae and their associated decomposition products. Given the absence of vascular plants in such environments, the inputs will contain little complex humic material, but may be rich in other organic compounds. On Signy Island for example, polyols (sugar alcohols) from the catchment mosses may enter the lakes and be a substrate for certain bacteria (Wynn-Williams 1980). In the McMurdo Dry Valleys, more extensive lakes in the past have left residuals of organic material derived from previously submerged microbial mats. These occur in large quantities in the soils of some parts of the valleys and they constitute a legacy of organic carbon that today provides a supplemental input to the present lakes (Priscu 1998).

DOM derived from the breakdown of higher plants absorbs strongly in the ultraviolet as well visible (especially blue) wavebands. This coloured DOM (CDOM) therefore plays a central role in light availability and the underwater spectral regime of aquatic ecosystems (Williamson et al. 1996, Laurion et al. 2000, Markager and Vincent 2000), hence on photochemistry (Molot and Dillon 1997, Bertilsson and Tranvik 2000, Gibson et al. 2001) and photosynthesis (Carpenter et al. 1998, Williamson et al. 1999, Lehman et al. 2004). Absorption of solar energy by CDOM also has consequences for heat transfer in the water column and thermal stratification (Mazumder and Taylor 1994, Snucins and Gunn 2000). DOM is also a major energy source that can be used by bacteria and some of the energy of the bacteria may be eventually transferred to higher trophic levels (Schell 1983, Kirchman et al. 2004, del Giorgio and Davis 2003). DOM alters nutrient and micronutrient availability, for example phosphorus and iron, for autotrophic and heterotrophic communities and can thereby affect biological carbon and energy fluxes (Hessen 1992, Hobbie 1992, De Haan 1993).

ANTHROPOGENIC EFFECTS

Human activities are increasingly a major factor to consider in the landscape ecology of both polar regions, with consequences for downstream receiving waters. For example, construction and operational activities in addition to vehicle use at and near scientific stations in Antarctica are bringing about significant impacts upon the active layer and permafrost in the catchments of lakes. The damage to soils and vegetation from tracked vehicles has been evident or severe on King George Island,

South Shetland Islands, with slopes eroded and tracks that penetrate to a depth of 0.5m. Drainage patterns have been altered and quagmires formed (Harris 1991). Land disturbances have resulted in considerable loss of water from permafrost soils and have caused channelled flows, soil shrinkage, land slumping and salinization in the McMurdo Sound region (Campbell et al. 1994). In the Larsemann Hills, Antarctica, four stations and a network of roads have been established in a relatively small area. The roads run predominantly through gneiss that breaks down to fine sand and silt and this substrate is readily mobilized by water. Meltwaters from snowpacks and exposed permafrost have resulted in the roads becoming watercourses that channel the flow and alter lakes by changes in water input, salt loading and turbidity (Burgess and Kaup 1997, Kaup and Burgess 2003). These physically disturbed landscapes may also be more vulnerable to increased thaw and runoff associated with climate warming.

Human impacts are resulting also in chemical changes in surface and subsurface (active layer) waters of lake catchments. In the Larsemann Hills, these waters affected by anthropogenic inputs have order-of-magnitude higher conductivities than those in natural catchments. The origin of salts has been attributed to direct salt inputs from station activities (wastewater and urine, chemicals, building materials) and to intensive rock crushing by tracked vehicles and subsequent increased weathering, indicated by considerable silt increases in certain areas. The latter changes may be translated into increased nutrient loading on lakes (Kaup et al. 2001), eutrophication and associated changes in limnological properties such as decreased lake water transparency (Ellis-Evans et al. 1997). There is evidence for changed trophic status of lakes brought about by human-generated nutrients in other Antarctic lake districts, including the Schirmacher Oasis (Haendel and Kaup 1995) and Thala Hills (Kaup 1998). These enriched systems may respond quite differently from pristine high latitude lakes to climate-dependent changes in their surrounding landscape.

Effects of climate on landscape and lakes

Lakes are transient features of the landscape and experience continuous evolution, from the first geological origins of a basin to its eventual infilling by biotic and abiotic sediments. Climate change has the potential to affect lake evolution through a variety of processes, especially in the polar regions where small changes in temperature can have a large impact on landscape properties such as snowpack, glacier melt, hydrological inputs, vegetation and soil stability. In this section, we illustrate some of the mechanisms whereby climate affects landscapes and lakes, at various timescales.

GEOMORPHOLOGY

The climate-induced recession of glaciers and ice caps has a wide ranging effect on landscape geomorphology that in turn affects the presence, distribution and form of lake basins. Climate change can result in modifications of pre-existing geomorphological patterns, for example through the thawing of permafrost or changes in precipitation and water-induced erosion. Such erosion processes also affect the transport of sediment from land to lakes and therefore the extent of infilling. As noted in Geomorphology (above), the exposure of moraines of various ages after glacial retreat also has an influence on lake water chemistry. The isostatic adjustment of land following glacial retreat can also radically alter the influence of the sea on coastal lakes and lagoons (see below).

Ice bound lakes are especially sensitive to small variations in climate. For example, the extensive Ellesmere Ice Shelf that once dammed the northern fiords of Ellesmere Island underwent considerable break-up and contraction during the 20th Century (Vincent et al. 2001). This has resulted in the loss of ice-dammed epishelf lakes, most recently that of Disraeli Fiord in which the freshwater layer completely drained away as a result of the break-up of the Ward Hunt Ice Shelf in 2002 (Mueller et al. 2003).

MARINE CONNECTIONS

Many Antarctic and Arctic lakes show a connection to some degree to the sea (see Gibson et al. this volume). For example, in the Vestfold Hills (Antarctica) some ends of paleoceanic bays became lakes (eg Peterson et al. 1988, Cromer et al. 2005) and the coastal aquatic ecosystems include meltwater influenced fjords (eg Ellis Fjord and Crooked Fjord) and lakes that are intermittently connected to the sea with seasonal marine inputs, including microbiota and fish. The latter includes Burton Lake that has an ice dam across its entrance sill, isolating it from the sea for up to 8 months of the year (Bayly 1986). These connections are strongly affected by climate over long timescales (hundreds to thousand of years).

More than 20 lakes in the Vestfold Hills contain relict seawater that was isolated by isostatic uplift of the land after the last deglaciation, about 10 000 years ago (Gibson 1999). This climate-dependant process has resulted in meromictic lakes (lakes that never fully mix) in which dilute, oxygenated meltwaters lie over the ancient, saline, anoxic bottom waters. In the early stages of evolution of such lakes the formation of sea ice and associated brine drainage is likely to have contributed to the high salinities of their present-day bottom waters (Gallagher et al. 1989). Some of the lakes (eg Watts Lake (Pickard et al. 1986), Lake Druzby at the end of Ellis Fjord) have been gradually flushed out with meltwater and are now well-mixed freshwater systems. Sediment cores from Ace Lake in this region show three distinct phases of salinity as indicated by analysis of the fossil diatom flora (Roberts and McMinn 1999): a freshwater phase, followed by a period of inundation resulting in marine and sea ice biota, followed by the present-day hypersaline conditions. In this process, the local sea level, as the balance between the eustatic and isostatic components, is relevant. In the Holocene the sea level rose and then fell resulting in flooding of initially freshwater lakes and subsequent re-isolation (Zwartz et al. 1998)

Analogous effects are also known from the north polar region, again associated with climate change at millennial timescales, deglaciation and isostatic rebound. One set of these systems lies along the northern coastline of Ellesmere Island in the Canadian High Arctic where a series of lakes and fjords forms a chronosequence illustrating the various stages in landscape evolution and degrees of connectivity to the sea: well mixed fjords, ice-dammed, meltwater-influenced, stratified fjords, saline, meromictic lakes and lakes that have been flushed out by meltwater and are entirely fresh (Fig. 2).

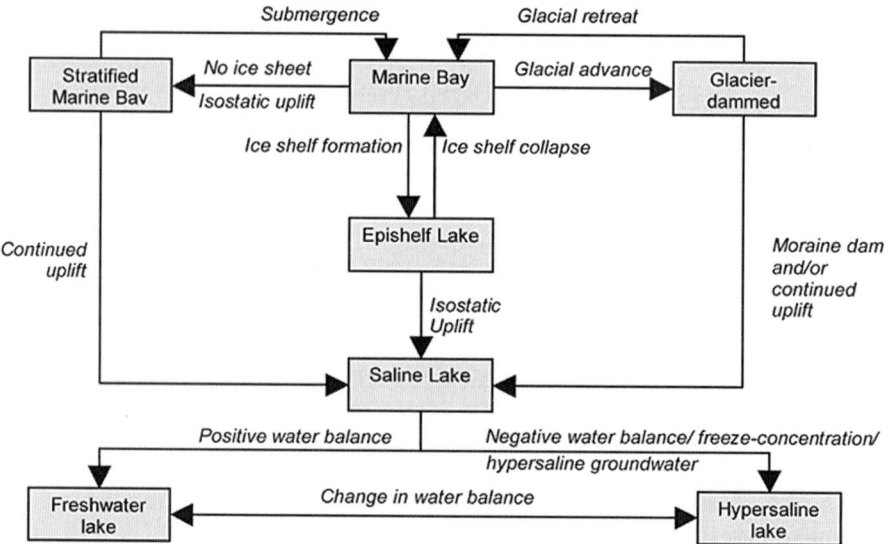

Figure 2. Postulated evolutionary sequence for coastal, high latitude landscapes, embayments and lakes. Reproduced by permission from van Hove et al. *(2006).*

Paleolimnological analysis of coastal Arctic and Antarctic lakes further to the south also show that their evolution began with a submerged phase beneath the sea. For example, analyses of the sediments of ultra-oligotrophic, freshwater Char Lake at Resolute Bay in the Canadian High Arctic show that it emerged from the sea ca. 6ka ago and subsequently became completely fresh ca. 4ka ago (Michelutti et al. 2003). Similarly, subarctic Lake Kachishayoot in coastal Hudson Bay has sediments indicating a marine then brackish water phase before isolation from the sea and its shift to present-day freshwater conditions (Saulnier-Talbot et al. 2003).

Small changes in sea level can produce important variations in the properties of small coastal lakes that are separated from sea level by barrier-beaches. Moreover, the fresh water/marine water interface will change over time, finding a new equilibrium with increased freshwater flow and sea level rise.

PERMAFROST EFFECTS

Permafrost is overlain by the active layer and the depth of this layer is a reflection of the dynamic equilibrium between hydrological and thermal properties of the soil and atmospheric conditions (Hinzman et al. 1991). Climate-related changes in this layer will have a wide-ranging influence on the transport of water, solutes and particulate materials to downstream receiving water bodies. No long-term (several decades) records of the depth of the permafrost active layer are available in the polar regions, however, several models (Frauenfeld et al. 2004) based on air temperatures identify potentially large variations in its thickness, deepening by 20cm in the period 1956-1990 in Russia. There are predictions for a 20-30% increase in the active layer thickness by the year 2050 in the Northern Hemisphere (Anisimov et al. 1997). This will expose upper permafrost to leaching with meltwater and also will produce shifts in the interconnections among subpermafrost, intrapermafrost and above-permafrost liquid water masses of unpredictable ecological consequences, since in many cases these groundwater masses are rich in solutes and only discharge on surface waters through permafrost discontinuities.

Permafrost may contain considerable stocks of ancient organic matter that can be liberated during melting. For example, in the Larsemann Hills and on Signy Island extensive vegetation beds (mosses, cyanobacterial mats) occurred over the upper permafrost during warmer climates of the past (Burgess et al. 1994, Smith et al. 2004). These can, by climate warming or other disturbances, increasingly contribute to the nutrients and DOM loads in the subsurface waters and, subsequently, in the lake inflows. In one case, gravel mining for road construction caused a drastic increase in permafrost thawing. Soil frozen for thousands of years began to weather and nutrients were released. The result was spring waters extremely rich in nutrients, mainly phosphorus, (Hobbie et al. 1999). Increased nutrient inflows will have a potential to alter trophic status in natural Antarctic lakes and such evidence has been indeed documented (eg Laybourn-Parry 2003).

The active layer in the Dry Valleys is mostly dry because evaporation and sublimation far exceed the annual recharge into it (Campbell et al. 1998). The result is that the soils are barely leached and weathering products accumulate in the soil profile. In contrast, in oases such as Schirmacher, Thala, Larsemann and Bunger (coastal continental Antarctica) with more intensive recharge from increased meltwater supply, considerable subsurface flow of water occurs (Haendel 1995, Burgess and Kaup 1997). Each summer, the depth of the thaw layer depends on temperature, solar radiation conditions and the extent of snow cover. With a snow cover more than 10cm thick, the solar radiation reaching the ground surface became almost negligible on King George Island (maritime Antarctic) where the active layer thickness varied between 0.5 - 3.5m. These conditions are probably representative for the entire maritime Antarctic region (Cannone and Guglielmin 2003) and also for the coastal continental Antarctic to a lesser extent.

The occurrence of many old soils at high inland elevation indicates that little response to global climatic change would be expected there. For the much younger soils in East Antarctica and the Antarctic Peninsula, when mean annual summer temperatures are higher, responses to global change and change in sea level may be significant (Bockheim et al. 1999). In these regions, the process of increasing the temperature due to global warming can, as in the Arctic, result in thawing of permafrost and an increase in active layer thickness. For example, a rise of ca. 1°C in summer air temperatures during the last 50 years due to local climate change on Signy Island (maritime Antarctic) has markedly increased chlorophyll a and nutrient levels in lakes. This response may be linked to deglaciation and also reductions in lake snow and ice cover (Quayle et al. 2002). However, if the climate warming in Antarctica is accompanied by increased precipitation then its effect on the active layer and on the lakes can be complex. If the summer snow cover in the oases is extended by increased precipitation, then chemical weathering and nutrient release from the catchments may also decrease.

Climate warming is likely to have pronounced effects on thermokarst landscapes. In these regions, surface melting and slumping affects the morphology of the landscape that in turn influences the hydrologic cycle and aquatic biota and is related to the movement of the tree line in the arctic tundra (Hinzman et al. 2004). These regions are abundant in parts of the Arctic and contain lakes and ponds in vast complexes that variously exchange water with each other and with rivers. The water bodies are typically shallow (< 1m depth) with low phytoplankton and stocks, but often with a productive, nutrient-rich benthic layer of cyanobacteria and other organisms. These luxuriant benthic mats appear to be a major food source for zooplankton that in turn supports other wildlife such as ducks and shorebirds. The formation processes and dynamics of the landscape in the thermokarst regions are illustrated in Fig. 3. The process of thermokarst formation starts by degradation of the ice wedge, then the subsidence of the surface and the presence of ice beneath the sunken area allow pond formation. The pond then increases the speed of thermokarst formation since the surface underneath does not freeze during winter. Eventually the reduction of permafrost can drain the pond, which may allow the colonization by terrestrial vegetation (shrubs in the Arctic). If the permafrost is completely destroyed the pond can dry up and the vegetation may disappear. If the atmospheric conditions are favourable, palsas may form after refreezing of the surface and new permafrost is formed.

Variations in permafrost, and particularly in the active layer thickness, can be important for man-made constructions. The variation in the thickness of the active layer also affects the stability of the structures since much of the existing infrastructure erected in northern regions is located in areas of high hazard potential and could be affected by thaw subsidence under conditions of global warming (Nelson et al. 2001). These activities are in turn likely to influence the sediment transport to lakes.

Human impacts are resulting also in chemical changes in surface and subsurface (active layer) waters of lake catchments. In the Larsemann Hills, waters subject to human impacts have up to an order of magnitude higher conductivity than those in natural catchments, the origin of salts being attributed to direct salt inputs from station activities (wastewater and urine, chemicals, building materials) and to intensive rock crushing by tracked vehicles and subsequent increased weathering, indicated by considerable silt increases in certain areas. The latter changes may result in the eutrophication, loss of transparency in lakes and major shifts in their ecosystem properties (Ellis-Evans et al. 1997).

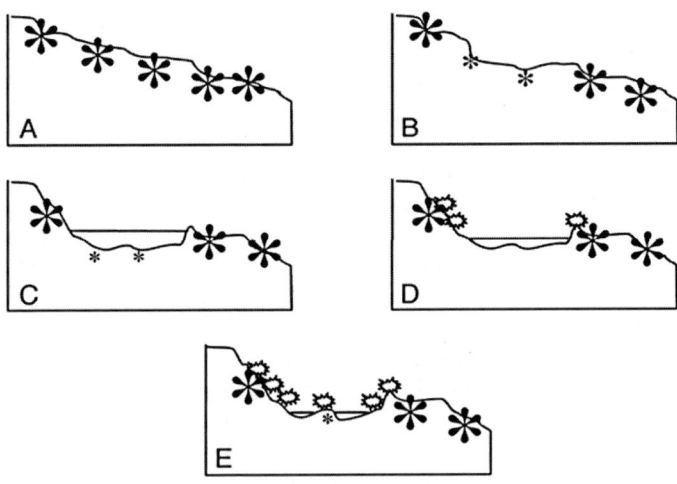

Figure 3. Hypothetical landscape evolution through thermokarst formation. A. Initial situation, permafrost is continuous. B. Some disturbance weakens the permafrost that becomes thinner leading to surface subsidence. C. Permafrost becomes very much reduced, the surface sinks considerably and ponds are formed. D. Vegetation colonizes this more humid and protected area. The pond drains because permafrost degrades more deeply. E. Plant colonization, almost complete drainage of the ponds, refreezing of some areas (palsa formation). The size of the symbol represents the permafrost thickness (adapted from Hinzman et al. 2004).

GLACIERS AND HYDROLOGY

Increased temperatures in the polar regions will be accompanied by a reduction of ice thickness and snowpack extent, the retreat of glaciers and also a shift in precipitation from snow to rain in some months (see Convey this volume). These variations are likely to induce modifications in the hydrology of the catchment and so in receiving lakes and wetlands.

Glacial retreat typically produces new lakes, when land depressions that originated from ice erosion are inundated with glacial meltwater. The glacial deposits left behind during retreat are prone to be eroded, weathered and transported by running waters. Moreover, the new land surfaces are also left open for running waters and wind erosion. Sediments are eventually carried to lakes that act as sediment traps. The amount and the size of the sediments in suspension will depend, apart from the water availability, upon the available energy in the ecosystem. In frozen systems, the available energy for transportation is very low and probably not enough to mobilise the sediments since most particles will be glued and/or protected by ice and snow. In such situations sediment reaching the lake will be small in size and low in quantity. On the other hand, where increased temperature results in ice melt and glacial retreat, the available energy for transportation will be higher and the sediment transported will cover a range from fine to larger in size and will be more abundant. The sediment cores collected from lakes in polar areas demonstrate a shift from coarse to fine size of the particles, indicating different stages of available energy and thus probably different climatic conditions (Björck et al. 1991, 1993), such as alternating warm and cold periods.

In shallow lakes fine sediments will remain in suspension during periods of open water since wind mixing will not allow the sedimentation of these inorganic particles. The presence of fine particles under open water conditions will increase the turbidity of the water reducing the available light for the photosynthetic organisms, both planktonic and benthic, likely limiting total autotrophic production in the ecosystem (see below).

An important effect of age on the biogeochemistry of the landscapes arises from the weathering of the glacial till. In the Alaska LTER, the till from all glaciations originated in a limited area of the Brooks Range (Hamilton 2003) so all the parent tills have similar chemistry. The effect of weathering and other important soil processes connected to vegetation and moisture results in a soil chemistry that reflects a similarity of origin and an evolution with age on all three glacial surfaces. The basic till in the region contains carbonates, apatite (calcium phosphate) and some calcium sulphate. Over time, the carbonates leach out and soil waters become more acidic (eg pH of the 10ka = 5.5-7, 60ka = 3.5-5.0). In streams and lakes, the pH and base cation content vary across the region in a pattern reflecting landscape age. For example, conductivity varies strikingly with landscape age (Table 1) and calcium and bicarbonate are more abundant in the lakes and streams of the younger surfaces. Within a drainage basin, the headwater streams and lakes receive the lowest concentrations of major ions while downstream freshwaters receive higher concentrations (Kling et al. 2000). Soil mineral phosphorus content is much greater in soils on younger landscapes (Walker et al. 1989, Giblin et al. 1991), while N availability is greater in older soils. Hobbie and Gough (2004) compared nutrient availability on the 10ka and the 60ka soils while Gebauer et al. (1996) documented the nutrients in the 300ka soils in the detailed study of Imnavait Creek. However, nutrients do not move as freely downslope as water and major cations. The uptake of nutrients by plants is increased by the shallow runoff due to permafrost thereby limiting nutrient inputs to freshwaters.

How do differences in biogeochemistry affect the functional aspects of aquatic ecology? Levine and Whalen (2001) found that the planktonic chlorophyll content (a proxy for algal biomass) was very similar in all of the numerous lakes they sampled in this region. Thus, pelagic production in lakes may not show a strong connection to landscape age but a stronger linkage appears to exist with benthic production (Fig. 4, Gettel 2006).

In shallow oligotrophic lakes benthic processes are especially important, light reaches the bottom and water column production is low (Ramlal et al. 1994). In lakes around Toolik Lake, shallow benthic primary production (< 6m) is similar to, or greater than water column production. The benthos also is important in nutrient cycling; benthic N_2 fixation can contribute up to 1 mg N m^{-2} d^{-1} during the summer months (Gettel 2006). In a survey of 15 lakes, Gettel (2006) found that shallow benthic N fixation showed a pattern of decreasing N_2 fixation among lakes on older landscapes (Fig. 4). However, Gettel (2006) also showed that fixation is controlled by a number of processes, including light and snail grazing, which co-vary with landscape age. In calcium rich lakes on the younger landscape, snail grazing can suppress N_2 fixation, altering the expected pattern of decreasing lake N_2 fixation with increasing landscape age. Benthic N_2 fixation is clearly important in shallow arctic lakes where benthic primary productivity can dominate.

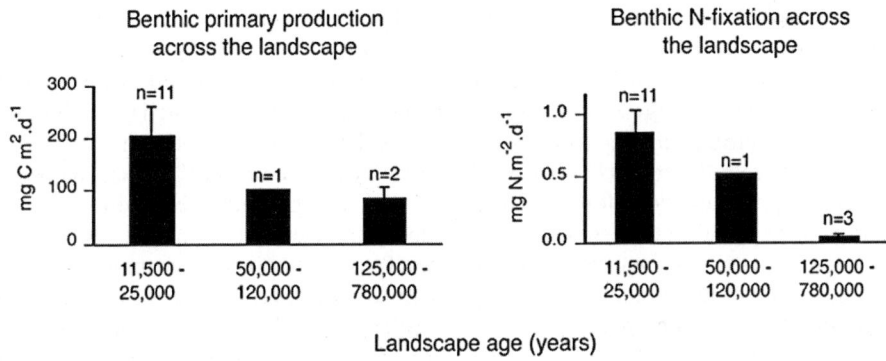

Figure 4. Benthic N_2 fixation and primary productivity on different landscape ages near Toolik Lake.

While soil dissolved organic carbon (DOC) increases with landscape age, DOC concentrations in lakes are under more complex controls because of the intensive processing of DOC in streams and lakes (Bowden and Group 1999). There are more snails and clams in lakes on the younger landscapes because the calcium concentrations are high. There is also evidence of more N_2 fixation in the younger networks possibly because of increased P (Fig. 4, Gettel 2006) and calcium. Position and morphometry also exert an important influence on primary production and these factors also vary with age. The older landscape has fewer, shallower lakes, which are less likely to be part of a network of lakes (Table 1). Lakes in a network tend to act as nutrient filters, so we expect that lake productivity will tend to decrease in long stream/lake sequences. For example, lakes at the headwaters of the Toolik inlet series of lakes (I-1 and I-2) have low primary productivity, but not as low as two lakes lower in the drainage (Kling et al. 2000). Lake I-8, which receives its water directly from streams, has higher productivity. Toolik Lake, which receives its water through nine intervening lakes, has the lowest primary productivity in the chain. This illustrates that the greater development of the stream-lake network on younger surfaces may serve to limit primary production in downstream freshwaters.

Lakes in both polar regions have undergone considerable changes in size associated with expansion and contraction of ice in their catchments. For example, Lake Vanda in the McMurdo Dry Valleys appears to have undergone major expansion 2000 - 3000 years ago, with evidence of ancient shorelines 64m above 1970s lake levels. This has been attributed to a brief period of climatic warming that increased the volume of meltwater from glaciers, and raised the snowline, causing the tributary glaciers to move to higher elevations. This meant that when conditions subsequently cooled, the higher glaciers were unable to provide substantial meltwater and Lake Vanda almost evaporated to dryness. This desiccation event is believed to have been the primary cause of the hypersaline bottom waters that now characterize this meromictic lake (Smith and Friedman 1993).

The long term water chemistry record at Toolik Lake has shown a doubling of the average acid neutralizing capacity (alkalinity) due primarily to changes in calcium and magnesium concentrations (Hobbie et al. 2003). There are no corresponding changes in the chemistry or amount of the precipitation that would account for these changes. It is suggested that dust from the road is causing the changes, but similar changes in alkalinity have been found in streams and lakes quite distant from the road (Hobbie et al. 2003). The most reasonable explanation is that alkalinity is an indicator of changes in soil and groundwater chemistry. For instance, it is possible that small, climate-dependent increases in active layer depth have exposed new soil material to weathering. Temperature however may not be the sole or even dominant factor affecting chemical weathering rates. For example, Lyons et al. (1997) compared weathering in Taylor Valley streams versus warm rivers in Alabama and found more than three times faster weathering rates in the former. In this comparison, lithology and water availability were likely to be the overriding controls on weathering.

Variations in temperature also affect the precipitation regime in polar regions, causing the increasing likelihood of rain rather than snow in the summer period. This change will have a major effect on the soil and geomorphology of the polar landscapes, since precipitation in solid form (snow) typically shows more gentle erosion effects because of the slower water release during thawing. However, rain produces an immediate effect eroding unstable soils, as those typical from polar regions. This also will also affect soil organisms, since the slow water release from snow allows the biota to maintain an appropriate water content for survival during longer periods. However, the same amount of precipitation in the form of rain saturates the active layer of soil but disappears quickly because of flow or evaporation (see above), limiting water availability for biological processes.

Temperature rise itself has direct effects on the lakes through several landscape processes. An increase in air temperature will produce a direct increase in water temperature but also will warm the rocks and soils of the catchment. These will transfer more heat to the lake via warmer water from the catchment entering the lake ecosystem. For instance, a physical model of Toolik Lake, Alaska, uses daily weather data as the input and simulates the annual thermal cycle (Hobbie et al. 1999). Under a scenario of a 5°C air temperature increase, lake temperatures increased by 3°C while the ice-free season increased by seven weeks. This reduced effect on water temperature could be critical, because in polar regions lake water temperature typically ranges between 0 and 6°C. If stream water that is warmer (less dense) and richer in solutes enters the lake after the spring runoff, it will float over the colder (more dense) lake water, triggering a stratification of the water column that will last until wind mixes the different density layers. In this potential stratification period surface waters will have higher growth rates of photosynthetic organisms due to higher nutrient concentrations. This change will have profound consequences on the lakes ecology.

DISSOLVED ORGANIC MATTER

As summarized in Vegetation (above), the transport of dissolved organic matter (DOM) is a primary mechanism coupling lakes to their surrounding landscapes, and it exerts a wide-ranging physical, chemical and biological influence on aquatic ecosystems. The export of DOM from land to water is affected by vegetation type, biomass and productivity, and by the mobilization of stored reserves of organic matter in the catchment. Each of these in turn is likely to be highly susceptible to climate change.

The effect of climate on vegetation and DOM export has been illustrated in a series of paleoecological studies in northern high latitudes (reviewed in Pienitz et al. 2004 and Vincent et al. 2005) including limnological changes accompanying the shift of the northern tree line during the Holocene (ca. last 10 ka) in the Canadian Northwest Territories (MacDonald et al. 1993, Pienitz et al. 1999, Pienitz and Vincent 2000, Lehmann et al. 2004), the succession of vegetation types during ice retreat and soil development at Glacier Bay Alaska (Engstrom et al. 2000, Williamson et al. 2001), and the isostatic uplift, isolation from the sea and plant colonisation of catchments in coastal subarctic Québec following deglaciation (Saulnier-Talbot et al. 2003). In the latter paleo-optical study (past underwater light regimes), the authors found that there had been abrupt increases in diatom-inferred DOC concentrations and water colour that coincided with the retreat of postglacial marine waters and the arrival of spruce trees within the surrounding landscape. Their investigation also revealed large changes in the underwater irradiance environment over the course of the postglacial period, from extremely high UV exposure following the initial formation of the lake and its isolation from the sea, to an order-of-magnitude lower exposure associated with the development of spruce forests in the catchment. The use of additional macrofossil markers revealed that underwater UV penetration remained low even following forest retreat due to the development of alternate DOC sources in the catchment such as *Sphagnum* wetlands (Saulnier-Talbot et al. 2003).

These effects of vegetation on DOM loading in the past provide insights into how climate change may affect landscapes and lakes in the future. Regions such as the Canadian Arctic contain latitudinal bands of ecozones that differ in vegetation type and standing stocks (Gould et al. 2003), and there is evidence that the DOM content of lakes generally follow this vegetation gradient, with highest concentrations in waters of the boreal forest and lowest concentrations in lakes fed by polar desert catchments with sparse, discontinuous plant communities (Pienitz and Smol 1993, Vincent and Pienitz 1996). However, there are exceptions to this general pattern associated with local oases of vegetation. Overall, there appears to be a statistical relationship between the density of phytobiomass and temperature, specifically the extent of summer warming (Walker et al. 2003) and thus in the future it is likely that these ecozones will move northwards. Already such effects are noticeable at some locations. There is evidence of increased densities of shrubs associated with recent warming of the Alaskan tundra (Sturm et al. 2001) and on the Seward Peninsula, Alaska, there is evidence of tree line advance (Lloyd et al. 2002). This arrival of larger plants with more lignin-containing tissues is likely to increase the amount of CDOM transferred to lakes, as has been inferred from the paleorecord.

In addition to affecting the above ground, living stocks of carbon and associated plant detritus, climate change may also cause the mobilization of stored organic carbon reserves in the catchment. This aspect is of major concern in the Arctic and Subarctic. Northern high latitude regions contain the largest peat-bog systems on Earth with as much as one third of global organic carbon stocks (Gorham 1991, McGuire et al. 2002, Smith et al. 2004). Changes in air temperature and precipitation could potentially mobilize part of the sequestered carbon of these systems (Agafonov et al. 2004, Christensen et al. 2004) through physical (hydrology, freeze-thaw cycles), chemical (photoreactions, oxidation) and biological mechanisms (respiration, biodegradation of DOM, energy transfer through the food web). The stability of the soil carbon pool appears sensitive to the depth and duration of thaw (Goulden et al. 1998). It has been shown that rising temperatures can stimulate the export of DOC from peatlands (Freeman et al. 2001). Since the Little Ice Age (ca. AD 1550-1850), permafrost degradation has resulted in the loss of forested lands and an increase in long-term net accumulation of organic matter (Vitt et al. 2000, Turetsky et al. 2002) that may eventually be mobilized.

There is considerable interest at present in the influence of DOM on the inorganic carbon balance of high latitude aquatic ecosystems, and the potential role of rivers and lakes as systems that may decompose and ventilate organic matter from the tundra into the atmosphere. Work by Kling et al. (1991) illustrates how carbon movement through streams and lakes can change the carbon balance of an entire watershed. These authors found that Alaskan tundra lakes and streams are typically supersaturated in carbon dioxide. The CO_2 concentration in 25 lakes averaged 1162ppmv, with calculated positive fluxes of CO_2 from water to the atmosphere averaging 20.9mmol m^{-2} d^{-1}. These high amounts of CO_2 came from groundwater that contained up to 46 500ppmv. Groundwater is likely confined by permafrost to shallow organic-rich soil layers where CO_2 produced by plant and microbial respiration can accumulate. Carbon loss by this aquatic pathway could equal the terrestrial carbon accumulation. They estimated the global C loss from tundra lakes and rivers to be between 7 - 20% of the current estimated C sink for arctic tundra, this estimation being conservative since part of the carbon exported to the ocean (POC and DOC) can also be respired and lost to the atmosphere. Kling et al. (1991) also measured CH_4 concentration in eight lakes and one river averaging 270ppmv, some 150 times the concentration at atmospheric saturation. Supersaturation of methane, and thus net efflux of this potent greenhouse gas, has also been reported by Whalen and Reeburg (1990) in Alaskan freshwaters.

BIOLOGICAL CYCLES AND BIODIVERSITY

Most models predict that climate change will lead to an increase in temperature and precipitation regime in polar regions. Experimental increases of temperature in non-aquatic ecosystems as Antarctic terrestrial soils have shown large variations such as an increase in cyanobacterial colonizers, and the size of the micro-arthropod and nematode populations (Kennedy 1994, Wynn-Williams 1996, Convey 2003, Convey this volume). These variations are already being detected in polar regions under natural conditions, since they are more sensitive to these changes than other ecosystems. The onset of these effects is now being observed, such as an increase in the distribution of the higher plant *Deschampsia antarctica* on King George Island from 1984 to 2001 (Gerighausen et al. 2003). Variations in the vegetation and edaphic communities of catchments are likely to have a large effect on the downstream receiving waters, shifting the amount and the type of DOC and POC entering into the lake (see above).

In terms of non-marine aquatic ecosystems the predicted changes include variations in the physical and chemical characteristics of the ecosystems. An increase in the open water period is expected, making light available during more time to deeper layers. An expanded open waters period will also allow more active mixing and through this will make available, in shallow lakes, more nutrients released from the lake sediments, therefore increasing nutrient availability for primary producers in these typically ultra-oligotrophic waters. The consequences of the exponential increase of primary producers are unpredictable but most probably will imply important variations in further steps of the food web. Apparently simple food webs, such as those found in polar freshwaters, can be strongly impacted by variations in organism numbers in any of the trophic levels (A. Camacho et al. unpubl. data). It is also likely that there will be an increase in the solute concentration of waters incoming into the lakes due to hydrological variation (see above). Both processes will most probably result in increased eutrophication. Typically eutrophication processes are characterised by an increase in biomass but also a decrease in biodiversity.

Another important factor is the increase of solids in suspension entering the lakes due to an increase of erosion in the catchments, or to higher discharge of streams into the lake. The effects of such an increase are unpredictable but most likely will reduce light availability in the water column shifting the biomass composition towards low light adapted primary producers (eg planktonic cyanobacteria). This increase in solids in suspension would also affect the benthic communities, which in some cases is the dominant biomass in some polar lakes (Imura et al. 1999). Most benthic communities are dominated by slow growing organisms so that rain of sediment on them may limit their survival (Imura et al. 1999). Any shifts in the underwater light regime will be a function of the balance between reduction caused by solids in suspension and increase due to reduction in ice cover duration and thickness. How these two factors will counteract is unknown at the moment, but most likely they will affect aquatic biodiversity.

After glacial retreat, milder conditions will make the new lands or aquatic ecosystems more suitable for invader species from non-polar latitudes. Clarke et al. (2005), as other authors suggested before (eg Ellis-Evans 1996), cast doubts on the biological isolation of Antarctica. In fact, redistribution and expansion of native populations and invaders have been documented for microalgal, invertebrate and plant taxa (Karentz 2003, Convey et al. this volume, Hughes et al. this volume). Several recent invasions have been documented, such as a carabid beetle at South Georgia (Ernsting et al. 1995). The subantarctic islands have been subject to the majority of recent introductions (Frenot et al. 2005), including taxa from many different phylogenetic groups. This topic is extensively treated by Gibson et al. (2006), in relation to the biogeographical distribution of the Antarctic organisms.

Biodiversity change and species redistribution will be more conspicuous in the Arctic since there are no boundaries to limit the free distribution of organisms. However, landscape variations due to climate change could also strongly affect aquatic ecosystems if new connections are formed among different water masses. Fish are one group of organisms that would quickly move into the new habitats. Evidence from the Arctic indicates that any invasions triggered by climate change in Antarctica will have substantial impacts and will profoundly modify these simple ecosystems in terms of biodiversity and ecosystem functioning.

Conclusions

Landscape exerts a broad range of controls on the properties and dynamics of lakes in both polar regions. The primary effect over long time scales is on basin and catchment geomorphology, hydrological flow patterns and connections with the sea. Over shorter timescales, climate can have a strong effect on permafrost degradation, rock weathering, soil formation, erosion and vegetation development. Each of these processes in turn affects the chemical, physical and biological properties of downstream receiving waters. Most of the variations that have been observed over the last few decades are the result of slow or continuous change in climatic conditions. In the future, however, more rapid and pronounced alterations are likely. In high latitude landscapes where permafrost, glaciers, snowpack and other cryosperic features are so dependent upon persistent cold, small temperature changes are likely to have abrupt, threshold-dependent effects that in turn impact strongly on lakes.

Acknowledgements

This study was supported by grant REN2000-0435-ANT from the Ministerio de Ciencia y Tecnología (Spain), grant SAB2003-0025 from Ministerio de Educación y

Ciencia (Spain) and the Networks of Centres of Excellence program ArcticNet (Canada). The work of E Kaup was supported by the grant SF0332089s02 from the Ministry of Education and Science (Estonia).

References

ACIA (2004) Impacts of a Warming Arctic: Arctic Climate Impact Assessment. Cambridge University Press, 140 pp.

Adamson, D.A., Mabin, M.C.G. and Luly, J.G. (1997) Holocene isostasy and late Cenozoic development of landforms including Beaver and Radok Lake basins in the Amery Oasis, Prince Charles Mountains, Antarctica, *Antarctic Science* **9**, 299-306.

Agafonov, L., Strunk, H. and Nuber, T. (2004) Thermokarst dynamics in Western Siberia: insights from dendrochronological research, *Palaeogeography Palaeoclimatology Palaeoecology*, **209**, 183-196.

Andersen, D.T., Pollard, W.H., McKay, C.P. and Heldmann, J. (2002) Cold springs in permafrost on Earth and Mars. *Journal of Geophysical Research-Planets*, **107**, 10.1029.

Anisimov, O.A., Shiklomanov, N.I. and Nelson, F.E. (1997) Global warming and active layer thickness: results from transient general circulation models. *Global and Planetary Change*, **15**, 61-77.

Barber, D.C., Dyke, A., Hillaire-Marcel, C., Jennings, A.E., Andrews, J.T., Kerwin, M.W., Bilodeau, G,. McNeely, R. Southon, J., Morehead, M.D. and Gagnon, J.-M. (1999) Forcing of the cold event 8200 years ago by catastrophic drainage of Laurentide lakes, *Nature*, **400**, 344-348.

Bayly, I.A.E. (1986) Ecology of the zooplankton of a meromictic Antarctic lagoon with special reference to *Drepanopus bispinosus* (Copepoda: Calanoida), *Hydrobiologia*, **140**, 199-231.

Benda, L., Poff, N.L., Miller, D., Dunne, T., Reeves, G., Pess, G., and Pollock M. (2004) The network dynamics hypothesis: How channel networks structure riverine habitats. *BioScience* **54**, 413-427.

Bertilsson, S., Tranvik, L.J. (2000) Photochemical transformation of dissolved organic matter in lakes. *Limnology and Oceanography* **45**, 753-762.

Björck, S., Hakansson, H., Olsson, S., Barnekow, L., and Janssens, J. (1993) Paleoclimatic studies in South Shetland Islands, Antarctica, based on numerous stratigraphic variables in lake sediments, *Journal of Paleolimnology*, **8**, 233-272.

Björck, S., Hakansson, H., Zale, R., Karlen, W., and Jönsson, B.L. (1991) A late Holocene lake sediment sequence from Livingston Island, South Shetland Islands, with paleoclimatic implications, *Antarctic Science*, **3**, 61-72.

Bockheim, J.G., Everett, L.R., Hinkel, K.M., Nelson, F.E. and Brown, J. (1999) Soil organic carbon storage and distribution in Arctic Tundra, Barrow, Alaska. *Soil Science Society of America Journal* **63**, 934-940.

Bowden, W.B. and Group, S.B. (1999) Roles of bryophytes in stream ecosystems. *Journal of the North American Benthological Society* **18**, 151-184.

Bulat, S.A., Alekhina, I.A., Blot, M., Petit, J-R., Angelis, M., Wagenbach, D., Lipenkov, V.Y., Vasilyeva, L.P. Wloch. D.M., Raynaud, D. and Lukin, V.V. (2004) DNA signature of thermophilic bacteria from the aged accretion ice of Lake Vostok, Antarctica: implications for searching for life in extreme icy environments, *International Journal of Astrobiology*, **3**, 1-12.

Burgess, J.S., Kaup, E. (1997) Some aspects of human impact on lakes in the Larsemann Hills, Princess Elizabeth Land, Eastern Antarctica, in W.B. Lyons, C. Howard-Williams and I. Hawes (eds.), *Ecosystem Processes in Antarctic Ice-free Landscapes*. Balkema, Rotterdam, pp. 259-264.

Burgess, J.S., Spate, A.P. and Shevlin, J. (1994) The onset of deglaciation in the Larsemann Hills, Eastern Antarctica. *Antarctic Science*, **6**, 491-495.

Campbell, I.B. and Claridge, G.G.C. (1987) Antarctica: Soils weathering processes and environment, Elsevier, Amsterdam, The Netherlands.

Campbell, I.B., Claridge, G.G.C. and Balks, M.R. (1994) The effect of human activities on moisture content of soils and underlying permafrost from the McMurdo Sound region, Antarctica. *Antarctic Science*, **6**, 307-316.

Campbell I.B., Claridge G.G.C., Campbell D.I. and Balks M.R. (1998) The soil environment of the McMurdo Dry Valleys, Antarctica, in J.C. Priscu (ed.), *Ecosystem dynamics in a polar desert: The McMurdo Dry Valleys, Antarctica.* American Geophysical Union, Washington, USA, pp. 297-320.

Cannone, N. and Guglielmin, M. (2003) Vegetation and permafrost: sensitive systems for the development of a monitoring program of climate change along an Antarctic transect, in, A.H.L. Huiskes, W.W.C. Gieskes, J. Rozema, R.M.L. Schorno, S.M. van der Vries, and W.J. Wolff (eds.) *Antarctic Biology in a Global Context*, Backhuys Publishers, Leiden, pp.31-36.

Carpenter, S.R., Cole, J.J., Kitchell, J.F. and Pace, M.L. (1998) Impact of dissolved organic carbon, phosphorus, and grazing on phytoplankton biomass and production in experimental lakes. *Limnology and Oceanography*, **43**, 73–80.

Chinn, T.J. (1993) Physical hydrology of the dry valley lakes, in W.J. Green and E.I. Friedmann (eds.), *Physical and Biogeochemical Processes in Antarctic Lakes*, Antarctic Research Series, Vol. 59, American Geophysical Union, pp. 1-51.

Christensen, T.R., Johansson, T., Akerman, H.J., Mastepanov, M., Malmer, N., Friborg, T., Crill, P. and Svensson, B.H. (2004) Thawing sub-arctic permafrost: Effects on vegetation and methane emissions, *Geophysical Research Letters,* **31,** L04501.

Clarke, A., Barnes, D.K.A. and Hodgson, D.A. (2005) How isolated is Antarctica? *Trends In Ecology and Evolution* **20**, 1-3.

Conca, J. and Malin, M. (1986) Solution etch pits in dolerite from the Allan Hills, *Antarctic Journal of the US* **21**, 18-19.

Conca, J. and Wright, J. (1987) The aqueous chemistry of weathering solutions in dolerite and the Allan Hills, Victoria Land, Antarctica. *Antarctic Journal of the US* **23**, 42-44.

Convey, P. (2003) Soil faunal community response to environmental manipulation on Alexander Island, southern maritime Antarctic, in, A.H.L. Huiskes, W.W.C. Gieskes, J. Rozema, R.M.L: Schorno, S.M. ven der Vies, W.J. Wolff (eds.). *Antarctic Biology in a global context.* Backhuys Publishers. Leiden pp. 74-78.

Convey, P. (2006) Antarctic climate change and its influences on terrestrial ecosystems, in D.M. Bergstrom, P. Convey, and A.H.L. Huiskes (eds.), *Trends in Antarctic Terrestrial and Limnetic Ecosystems: Antarctica as a Global Indicator*, Springer, Dordrecht (this volume).

Convey, P., Frenot, Y., Gremmen, N. and Bergstrom, D.M. (2006a) Biological invasions, in D.M. Bergstrom, P. Convey, and A.H.L. Huiskes (eds.), *Trends in Antarctic Terrestrial and Limnetic Ecosystems: Antarctica as a Global Indicator*, Springer, Dordrecht (this volume).

Cromer, L., Gibson, J.A.E., Swadling, K.M. and Ritz, D.A. (2005) Faunal microfossils: indicators of Holocene ecological change in a saline Antarctic lake. *Palaeogeography Palaeoclimatology Palaeoecology*, **221,** 83-97.

Cuchí, J.A., Durán, J.J., Alfaro, P., Serrano, E. and López-Martínez, J. (2004) Discriminación mediante parámetros fisicoquímicos in situ de diferentes tipos de agua presentes en un área con permafrost (Península Byers, Isla Livingston, Antártida Occidental). *Boletín de la Real Sociedad Española de Historia Natural (secc. Geol.)* **99**, 75-82.

De Haan, H. (1993) Solar UV-light penetration and photodegradation of humic substances in peaty lake water. *Limnology and Oceanography* **38**, 1072–1076.

del Giorgio, P.A., Davis, J. (2003) Patterns in dissolved organic matter lability and consumption across aquatic ecosystems, in S.E.G. Findlay, R.L. Sinsabaugh (eds.) *Aquatic Ecosystems: interactivity of dissolved organic matter.* Academic Press, San Diego, pp. 399-424.

Doran, P.T., Wharton, R.A., Lyons, W.B., DesMarais, D.J., and Andersen, D.T. (2000) Sedimentology and Geochemistry of a Perennially Ice-Covered Epishelf Lake in Bunger Hills Oasis, East Antarctica. *Antarctic Science* **12**, 131-140.

Doran, P.T., Priscu J.C., Lyons, W.B., Walsh, J.E., Fountain, A.G., McKnight, D.M., Moorhead, D.L., Virginia, R.A., Wall, D.H., Clow, G.D., Fritsen, C. H., McKay, C. P. and Parsons, A.N. (2002) Antarctic climate cooling and terrestrial ecosystem response. *Nature* **415**, 517-520.

Ellis-Evans, J.C. (1996) Microbial diversity and function in Antarctic freshwater ecosystems, *Biodiversity and Conservation* **5**, 1395-1431.

Ellis-Evans, J.C., Laybourn-Parry, J., Bayliss, P. and Perriss, S. (1997) Human impact on an oligotrophic lake in the Larsemann Hills, in B. Battaglia, J. Valencia, and D.W.H. Walton (eds.), *Antarctic Communities: Species, Structure and Survival*, Cambridge University Press, Cambridge, pp. 396-404.

Engstrom, D.R., Fritz, S.C., Almendinger, J.E. and Juggins, S. (2000) Chemical and biological trends during lake evolution in recent deglaciated terrain, *Nature* **408**, 161-166.

Ernsting, G., Block, W., MacAlister, H., and Todd, C. (1995) The invasion of the carnivorous carabid beetle *Trechisibus antarcticus* on South Georgia (sub-Antarctic) and its effect on the endemic herbivorous beetle *Hydromedion sparsutum*. Oecologia **103** 34-42.

Frauenfeld, O.W., Zhang, T.J., Barry, R.G. and Gilichinsky, D. (2004) Interdecadal changes in seasonal freeze and thaw depths in Russia, *Journal of Geophysical Research-Atmospheres* **109**, D5.

Freeman, C., Evans, C.D., Monteith, D.T., Reynolds, B. and Fenner, N. (2001) Export of organic carbon from peat soils, *Nature*, **412**, 785.

Frenot, Y., Chown, S., Whinam, J., Selkirk, P.M., Convey, P., Skotnicki, M., and Bergstrom D.M. (2005) Biological invasions in the Antarctic: extent, impacts and implications. *Biological Reviews* **80**, 45-72.

Gallagher, J.B., Burton H.R., and Calf, G.E. (1989) Meromixis in an Antarctic fjord: a precursor to meromictic lakes on an isostatically rising coastline, *Hydrobiologia* **172**, 235-54.

Gebauer, R., Grulke, N.E., Hahn, S.C., Lange, O.L., Oberbauer, S.F., Reynolds, J.F., Tenhunen, J.D. and Inhunen, J.D. (1996) Vegetation structure and aboveground carbon and nutrient pools in the Imnavait Creek watershed. *Landscape function and disturbance in Arctic tundra*, Berlin. Springer-Verlag, pp. 109-128.

Gerighausen, U., Bräutigam, K., Mustafa, O. and Peter, H.-U. (2003) Expansion of vascular plants on an Antarctic island – a consequence of climate change? In A.H.L. Huiskes, W.W.C. Gieskes, J. Rozema, R.M.L. Schorno, S.M. van der Vies and W.J. Wolff (eds.) *Antarctic Biology in a Global Context*, Backhuys, Leiden, pp. 79-83.

Gettel, G. (2006) Nitrogen cycling in lakes on different glacial surfaces in northern Alaska. PhD thesis, Cornell University, Ithaca, N.Y.

Giblin, A. E., Nadelhoffer, K., Shaver, G.R., and Laundre, J.A. (1991) Biogeochemical diversity along a riverside toposequence. *Ecological Monographs* **61**, 415-435.

Gibson, J.A.E. (1999) The meromictic lakes and stratified marine basins of the Vestfold Hills, East Antarctica. *Antarctic Science*, **11**, 175-192.

Gibson, J.A.E., and Andersen, D.T. (2002) Physical structure of epishelf lakes of the southern Bunger Hills, East Antarctica *Antarctic Science* **14**, 253-262.

Gibson, J.A.E., Vincent, W.F. and Pienitz, R. (2001) Hydrologic control and diurnal photobleaching of CDOM in a subarctic lake. *Archives for Hydrobiology* **152**, 143-159.

Gibson, J.A.E., Wilmotte, A., Taton, A., Van De Vijver, B., Beyens, L. and Dartnall, H.J.G. (2006) Biogeographic trends in Antarctic lake communities, in D.M. Bergstrom, P. Convey, and A.H.L. Huiskes (eds.), *Trends in Antarctic Terrestrial and Limnetic Ecosystems: Antarctica as a Global Indicator*, Springer, Dordrecht (this volume).

Gorham, E. (1991) Northern peatlands: role in the carbon-cycle and probable responses to climatic warming, *Ecological Applications*, **1**, 182-195.

Gould, W.A., Raynolds, M. and Walker, D.A. (2003) Vegetation, plant biomass, and net primary productivity patterns in the Canadian Arctic, *Journal of Geophysical Research-Atmospheres* **108**,

Goulden, M.L., Wofsy, S.C., Harden, J.W., Trumbore, S.E., Crill, P.M., Gower, S.T., Fries, T., Daube, B.C., Fan, S.M., Sutton, D.J., Bazzaz, A. and Munger, J.W. (1998) Sensitivity of boreal forest carbon balance to soil thaw *Science*, 279, 214-217.

Green, W.J., Angle, M.P. and Chave, K.E. (1988) The geochemistry of Antarctic streams and their role in the evolution of four lakes of the McMurdo Dry Valleys, *Geochimical and Cosmochimica Acta* **52**, 1265-1274.

Haendel, D. (1995) On the entry of weathering products into surface waters, in P. Bormann and D. Fritsche (eds.) *The Schirmacher Oasis, Queen Maud Land, East Antarctica*. Justus Perthes Verlag, Gotha, pp. 305-309.

Haendel, D., Kaup, E. (1995) Nutrients and primary production, in, P. Bormann and D. Fritsche (eds.) *The Schirmacher Oasis, Queen Maud Land, East Antarctica*. Justus Perthes Verlag, Gotha, pp. 312-319.

Haendel, D., Kaup, E., Loopmann, A. and Wand, U. (1995) Physical and hydrochemical properties of water bodies, in P. Bormann and D. Fritsche (eds.) *The Schirmacher Oasis, Queen Maud Land, East Antarctica. Justus Perthes Verlag, Gotha, pp. 279-295*.

Hall, K.J. (1992) Mechanical weathering on Livingston Island, South Shetland Islands, Antarctica, in, Y. Yoshida, K. Kaminuma and K. Shiraishi (eds.) *Recent Progress in Antarctic Earth Science*. Terra, Tokyo, pp 756-762.

Hamilton, T. D. (2003) Glacial Geology of the Toolik Lake and Upper Kuparuk River Regions. Institute of Arctic Biology, University of Alaska, Fairbanks, Series: *Biological papers of the University of Alaska*. No. **26**.

Harris, C.M. (1991) Environmental effects of human activities on King George Island, South Shetland Islands, Antarctica. *Polar Record*, **27**, 313-324.

Hawes, I., Smith, R., Howard-Williams, C., and Schwarz, A-M. (1999) Environmental conditions during freezing, and response of microbial mats in ponds of the McMurdo Ice Shelf, Antarctica, *Antarctic Science*, **11**, 198-208.

Hessen, D.O. (1992) Dissolved organic carbon in a humic lake: Effects on bacterial production and respiration. *Hydrobiologia* **229**, 115–123.

Heywood, R.B. (1977) A limnological survey of the Ablation Point area, Alexander Island, Antarctica, *Philosophical Transactions of the Royal Society, Series B*. **279**, 39-54.

Hinzman, L.D., Kane, D.L., Giek, R.E. and Everett, K. (1991) Hydrologic and thermal properties of the active layer in the Alaskan Arctic. *Cold Regions Science and Technology*, **192**, 95-110.

Hinzman, L.D., Toniolo, H.A., Yoshikawa, K., and Jones, J.B. (2004) Thermokarst development in a changing climate. ACIA International Symposium on Climate Change in the Arctic. Reykjavik.

Hobbie J.E. (1973) Arctic Limnology: A review, in M.E. Britton (ed.), *Alaskan Arctic Tundra*. Arctic Institute of North America Technical Paper 25. pp. 127-168.

Hobbie, J.E. (1992) Microbial control of dissolved organic carbon in lakes: Research for the future. *Hydrobiologia* **229**, 169–180.

Hobbie, S.E. and Gough, L. (2002) Foliar and soil nutrients in tundra on glacial landscapes of contrasting ages in northern Alaska. *Oecologia* **131**, 453-462

Hobbie, S.E. and Gough, L. (2004) Litter decomposition in moist acidic and non-acidic tundra with different glacial histories. *Oecologia*, **140**, 113-124.

Hobbie, J.E., Shaver, G. Laundre, J., Slavik, K., Deegan, L.A., O'Brien, J., Oberbauer, S., and MacIntyre, S. (2003) Climate forcing at the Arctic LTER Site, in, D. Greenland, D. Goodin and R. Smith, (eds.) *Climate Variability and Ecosystem Response at Long-Term Ecological Research (LTER) Sites*. Oxford University Press, New York. pp. 74-91.

Hobbie, J.E., Peterson, B.J., Bettez, N., Deegan, L., O'Brien, W.J., Kling, G.W., Kipphut, G.W., Bowden, W.B. and Hershey, A.E. (1999) Impact of global change on the biogeochemistry and ecology of an Arctic freshwater system, *Polar Research*, **18**, 207-214.

Hodgson, D.A., Doran, P.T., Roberts, D. and McMinn, A. (2004) Paleolimnological studies from the Antarctic and subantarctic islands, in, R. Pienitz, M.S.V. Douglas, and J.P. Smol (eds.) *Long-term Environmental Change in Arctic and Antarctic Lakes*. Springer, Berlin/New York. pp. 419-474.

Hughes, K.A., Ott, S., Bölter, M. and Convey, P. (2006) Colonisation processes, in D.M. Bergstrom, P. Convey, and A.H.L. Huiskes (eds.), *Trends in Antarctic Terrestrial and Limnetic Ecosystems: Antarctica as a Global Indicator*, Springer, Dordrecht (this volume).

Imura, S., Bando, T., Saito, S., Seto, K., and Kanda, H. (1999) Benthic moss pillars in Antarctic lakes, *Polar Biology* **22**, 137-140.

Jones, V.J., Juggins, S., and Ellis-Evans, J.C. (1993) The relationship between water chemistry and surface sediment diatom assemblages in maritime Antarctic lakes, *Antarctic Science*, **5**, 229-348.

Kalff, J. (2001) *Limnology*. Prentice Hall. 592pp.

Karentz, D. (2003) Environmental change in Antarctica: ecological impacts and responses, in, A.H.L. Huiskes, W.W.C. Gieskes, J. Rozema, R.M.L. Schorno, S.M. ven der Vies, W.J. Wolff (eds.). *Antarctic Biology in a global context*. Backhuys Publishers. Leiden pp. 45-55.

Kaup, E. (1998) Trophic status of lakes in Thala Hills -records from the years 1967 and 1988. *Proceedings NIPR Symposium Polar Biology*, **11**, 82-91.

Kaup, E. and Burgess, J.S. (2002) Surface and subsurface flows of nutrients in natural and human impacted lake catchments on Broknes, Larsemann Hills, Antarctica. *Antarctic Science*, **14**, 343-352.

Kaup, E. and Burgess, J.S. (2003) Natural and human impacted stratification in the shallow lakes of the Larsemann Hills, Antarctica, in: A.H.L. Huiskes, W.W.C. Gieskes, J. Rozema, R.M.L. Schorno, S.M. van der Vries, and W.J. Wolff (eds.) *Antarctic Biology in a Global Context*, Backhuys Publishers, Leiden, pp. 313-317.

Kaup, E., Ellis-Evans, J.C. and Burgess, J.S. (2001) Increased phosphorus levels in the surface waters of Broknes, Larsemann Hills, Antarctica. *Verhandlungen International Vereinigung fur Theoretische und Angewanote Limnology*, **27**, 3137-3140.

Kennedy, A.D. (1994) Simulated climate change: a field manipulation study of polar microarthropod community response to global warming, *Ecography* **17**, 131-140.

Kirchman, D.L, Dittel, A.I., Findlay, S.E.G. and Fischer, D. 2004. Changes in bacterial activity and community in response to dissolved organic matter in the Hudson River, New York. *Aquatic Microbial Ecology* **35**, 243-257.

Kling, G.W., Kipphut, G.W. and Miller, M.C. (1991) Arctic lakes and streams as gas conduits to the atmosphere: implications for tundra carbon budgets. *Science* **251**, 298-301.

Kling, G.W., Kipphut, G. W., Miller, M. M., and O'Brien, W. J. (2000) Integration of lakes and streams in a landscape perspective: the importance of material processing on spatial patterns and temporal cohesion. *Freshwater Biology* **4**, 477-497.

Korotkevich, E.S. (1960) Ocean bays in the Schirmacher Hills in Queen Maud Land. *Sovetskaia antarkticheskaia ekspeditsiia. Informatsionnyi biulleten.* **21**, 8-9

Korotkevich, E.S. (1972) *Polyarnye pustyni [Polar deserts].* Gidrometeoizdat, Leningrad, 420 pp.

Landals, A.L. and Gill, D. (1972) Differences in volume of surface runoff during the snowmelt period: Yellowknife, NWT. *International Association of Hydrological Sciences Publication* **107**, 927-942

Laurion, I., Ventura, M., Catalan, J., Psenner, R. and Sommaruga, R. (2000) Attenuation of ultraviolet radiation in mountain lakes: Factors controlling the among- and within-lake variability. *Limnology and Oceanography,* **45**, 1274-1288.

Laybourn-Parry, J. (2003) Polar limnology, the past, the present and the future, in A.H.L. Huiskes, W.W.C. Gieskes, J. Rozema, R.M.L. Schorno, S.M. van der Vies and W.J. Wolff (eds.) *Antarctic Biology in a Global Context*, Backhuys Publishers, Leiden, pp. 321-329.

Laybourn-Parry, J., Quayle, W.C., Henshaw, T., Ruddell, A., and Marchant, H.J. (2001) Life on the edge: the plankton and chemistry of Beaver Lake, an ultra-oligotrophic epishelf lake, Antarctica, *Freshwater Biology,* **46**, 1205-1217.

Lehman, M.K., Davis, R.F., Huot, Y. and Cullen, J.J. (2004) Spectrally weighted transparency in models of water-column photosynthesis and photoinhibition by ultraviolet radiation. *Marine Ecology Progress Series,* **269**, 101-110.

Levine, M.A. and Whalen, S.C. (2001) Nutrient limitation of phytoplankton production in Alaskan Arctic foothill lakes. *Hydrobiologia* **455**, 189-201.

Lloyd, A.H., Rupp, T.S., Fastie, C.L. and Starfield, A.M. (2002) Patterns and dynamics of treeline advance on the Seward Peninsula, Alaska, *Journal of Geophysical Research-Atmospheres* **108 D2**, Alt 2-1.

López-Martínez, J., Serrano. E. and Martínez de Pisón, E. (1996) Geomorphological features of the drainage system. In: Supplementary text of the Geomorphological Map of Byers Peninsula. *BAS Geomap Series*, 5-A, 15-19. Cambridge, British Antarctic Survey

Lotter, A.F. (1999) Late-glacial and Holocene vegetation history and dynamics as shown by pollen and plant macrofossil analyses in annually laminated sediments from Soppensee, central Switzerland. *Vegetation History and Archaeobotany* **8**, 165-184.

Lyons, W.B. and Mayewski, P.A. (1993) The geochemical evolution of terrestrial waters in the Antarctic: the role of rock-water interactions. In: W.J. Green and E.I. Friedmann (eds.), *Physical and Biogeochemical processes in Antarctic Lakes.* American Geophysical Union, Washington, USA, pp. 135-144.

Lyons, W.B., Welch, K.A., Nezat, C.A., Crick, K., Toxie, J.K., Mastrine, J.A., and McKnight, D.M. (1997) Chemical weathering rates and reactions in the Lake Fryxell Basin, Taylor Valley: Comparisons to temperate river basins, in W.B. Lyons, C. Howard-Williams and I. Hawes (eds.) *Ecosystem Processes in Antarctic Ice-free Landscapes.* Balkema, Rotterdam, pp. 147-179.

Lyons, B., Welch, K.A., Neumann, K., Toxey, J.K., McArthur, R., Williams, C., McKnight, D.M., MacDonald, and Moorhead D. (1998) Geochemical linkages among glaciers, streams and lakes within the Taylor Valley, Antarctica, in J.C. Priscu (ed.) *Ecosystem dynamics in a polar desert: The McMurdo Dry Valleys, Antarctica.* American Geophysical Union, Washington, USA, pp. 77-92

MacDonald, G.M., Edwards, T.W.D, Moser, K.A., Pienitz, R. and Smol, J.P. (1993) Rapid response of treeline vegetation and lakes to past climate warming, *Nature* **361**, 243-246.

Markager, S. and Vincent, W.F. (2000) Spectral light attenuation and the absorption of UV and blue light in natural waters, *Limnology and Oceanography* **45**, 642-650.

Marsh, P., and Hey, M. (1989) The flooding hydrology of Mackenzie Delta lakes near Inuvik, NWT. Canada, *Arctic* **42**, 41-49.

Matsuoka, N. (1995) Rock weathering processes and landform development in the Sor Rondane Mountains, Antarctica. *Geomorphology*, **12**, 323-339.

Mazumder, A., Taylor, W.D. (1994) Thermal structure of lakes varying in size and water clarity. *Limnology and Oceanography*, **39**, 968-976.

McGuire, A.D., Wirth, C., Apps, M., Beringer, J., Clein, J., Epstein, H., Kicklighter, D.W., Bhatti, J., Chapin, F.S., de Groot, B., Efremov, D., Eugster, W., Fukuda, M., Gower, T., Hinzman, L., Huntley, B., Jia, G.J., Kasischke, E., Melillo, J., Romanovsky, V., Shvidenko, A., Vaganov, E. and Walker, D. (2002) Environmental variation, vegetation distribution, carbon dynamics and water/energy exchange at high latitudes, *Journal of Vegetation Science* **13**, 301-314.

Michelutti, N, Douglas, M.S.V. and Smol, J.P. (2003) Diatom response to recent climatic change in a high arctic lake (Char Lake, Cornwallis Island, Nunavut). *Global and Planetary Change* **38**, 257-271.

Molot, L.A. and Dillon, P.J. (1997) Photolytic regulation of dissolved organic carbon in northern lakes. *Global Biogeochemical Cycles* **11**, 357-365.

Mueller, D.R., Vincent, W.F. and Jeffries, M.O. 2003. Break-up of the largest Arctic ice shelf and associated loss of an epishelf lake. Geophysical Research Letters 30: 2031, doi: 10.1029/2003 GL017931.

Navas, A., López-Martínez, J., Casas, J., Machín, J., Serrano, E., Durán, J.J. and Cuchí, J.A. (2006) Soil characteristics along a transect on raised marine surfaces on Byers Peninsula, South Shetland Islands, in D. Fütterer, D. Damaske, G. Kleinschmidt, H. Muller and F. Tessensohn (eds.) *Antarctic contributions to global earth science*. Berlin-Heildelberg, Springer.

Nelson, F.E., Anisimov, O.A. and Shiklomanov, N.I. (2001) Subsidence risk from thawing permafrost. The threat to man-made structures across regions in the far north can be monitored. *Nature*, **410**, 889-890.

Nichols, R.L. (1960) Geomorphology of Margarite Bay area, Palmer Peninsula, Antarctica. *Bulletin of the Geological Society of America*, **71**, 1421-1450.

Overland, J.E., Spillane, M.C. and Soreide, N.N. (2004b) Integrated analysis of physical and biological Pan-Arctic change. *Climate Change* **63**, 291-322.

Overland, J.E., Spillane, M.C., Percival, D.B., Wang, M. and Mofjeld (2004a) Seasonal and regional variation in Pan-Arctic surface air temperature over the instrumental record. *J. Climate* **17**, 3263-3282.

Peterson, J.A., Finlayson, B.L. and Zhang, Q.S. (1988) Changing distribution of late Quaternary terrestrial lacustrine and littoral environments in the Vestfold Hills, Antarctica. *Hydrobiologia,* **165**, 221-226

Pickard, J., Adamson, D.A. and Heath, C.W. (1986) The evolution of Watts Lake, Vestfold Hills, East Antarctica, from marine inlet to freshwater lake. *Palaeogeography, Palaeoclimatology, Palaeoecology,* **53**, 271-288.

Pienitz, R. and Smol, J.P. (1993) Diatom assemblages and their relationship to environmental variables in lakes from the boreal forest-tundra ecotone near Yellowknife, Northwest-Territories, Canada. *Hydrobiologia* 269, 391-404.

Pienitz, R. and Vincent, W.F. (2000) Effect of climate change relative to ozone depletion on UV exposure in subarctic lakes. *Nature* 404, 484-487.

Pienitz, R., Douglas, M.S.V. and Smol, J.P. (eds.) (2004) *Long-Term Environmental Change in Arctic and Antarctic Lakes*. Developments in Paleoenvironmental Research Series, vol. 8, Springer, Berlin/New York, 550 pp.

Pienitz, R., Smol, J.P. and Lean, D.R.S. (1997) Physical and chemical limnology of 24 lakes located between Yellowknife and Contwoyto Lake (Northwest Territories), arctic Canada. *Canadian Journal of Fisheries and Aquatic Sciences,* **54**, 347-358.

Pienitz, R., Smol, J.P. and MacDonald, G.M. (1999) Paleolimnological reconstruction of Holocene climatic trends from two boreal treeline lakes, Northwest Territories, Canada, *Arctic, Antarctic and Alpine Research*, **31**, 82-93.

Priscu, J.C. (ed.) (1998) *Ecosystem Dynamics in a Polar Desert: The McMurdo Dry Valleys, Antarctica*, Antarctic Research Series, Vol. 72, American Geophysical Union, Washington, D.C., 369pp.

Prowse, T.D. and Culp, J.M. (2003) Ice breakup: a neglected factor in river ecology. *Canadian Journal of Civil Engineering* **30**, 128-144.

Quayle, W.C., Peck, L.S., Peat, H., Ellis-Evans, J.C. and Harrigan, P.R. (2002) Extreme responses to climate change in Antarctic lakes, *Science* **295**, 645.

Quayle, W.C, Convey, P., Peck, L.S., Ellis-Evans, J.C., Butler, H.G., and Peat, H.J. (2003) Ecological responses of maritime Antarctic lakes to regional climate change, In E. Domack, A. Burnett, A. Leventer, P. Convey, M. Kirby and R. Bindschadler (eds.), *Antarctic Peninsula Climate Variability: Historical and Palaeoenvironmental Perspectives*, Antarctic Research Series, Vol. 79, American Geophysical Union, Washington, D.C. pp. 159-170.

Ramlal, P.S., Hesslein, R.H., Hecky, R.E., Fee, E.J., Rudd, J.W.M. and Guildford S.J. (1994) The organic carbon budget of a shallow arctic lake on the Tuktoyaktuk Peninsula, NWT, Canada. *Biogeochemistry* **24,** 145-172.

Roberts, D. and McMinn, A. (1999) A diatom based palaeosalinity history of Ace Lake, Vestfold Hills, Antarctica. *The Holocene* **9,** 401-408.

Saulnier-Talbot, É., Pienitz, R. and Vincent, W.F. (2003) Holocene lake succession and palaeo-optics of a subarctic lake, northern Québec (Canada). *The Holocene* **13,** 517-526.

Schell, D.M. (1983) C-13 and C-14 abundances in Alaskan aquatic organisms-delayed production from peat in arctic food webs *Science* **219,** 1068-1071.

Schindler, D.W., Welch, H.E., Kalff, J., Brunskil, G.J. and Kritsch, N. (1974) Physical and chemical limnology of Char Lake, Cornwallis Island (75° N lat). *Journal of the Fisheries Research Board of Canada,* **31,** 585-607.

Schindler, D.W., Beaty K.G., Fee, E.J., Cruikshank, D.R., Debruyn, E.R., Findlay D.L., Linsey G.A., Shearer, J.A., Stainton, M.P., Turner, M.A. (1990) Effects of climatic warming on lakes of the central boreal forest, *Science* **250,** 967-970.

Serreze M. C. J. E. Walsh, F. S. Chapin III, T. Osterkamp, M. Dyurgerov, V. Romanovsky, W.C. Oechel, J. Morison, T. Zhang and R.G. Barry (2000) Observational Evidence of Recent Change in the Northern High-Latitude Environment. *Climate Change* **46,** 159-207.

Simonov, I.M. (1971) *Oazisy Vostochnoy Antarktidy [The oases of East Antarctica].* Gidrometeoizdat, 176 pp.

Smith, G.I. and Friedman, I. (1993) Lithology and paleoclimatic implications of lacustrine deposits around Lake Vanda and Don Juan Pond, Antarctica. American Geophysical Union, Washington, D.C. *Antarctic Research Series,* **59,** 83-94.

Smith, L.C., MacDonald, G.M., Velichko, A.A., Beilman, D.W., Borisova, O.K., Frey, K.E., Kremenetski, K.V. and Sheng, Y. (2004) Siberian peatlands a net carbon sink and global methane source since the early Holocene, *Science* **303,** 353-356.

Snucins, E.D. and Gunn, J. (2000) Interannual variation in the thermal structure of clear and coloured lakes. *Limnology and Oceanography,* **45,** 1639-1646.

Sturm, M., Racine, C. and Tape, K. (2001) Increasing shrub abundance in the Arctic, *Nature,* **411,** 546-547.

Turetsky, M., Wieder, R.K., Halsey, L., Vitt, D.H. (2002) Current disturbance and the diminishing peatland carbon sink, *Geophysical Research Letters,* **29,** DOI:10.1029/2001GL014000.

Van Hove, P., Belzile, C., Gibson, J.A.E. and Vincent, W.F. (2006) Coupled landscape-lake evolution in the Canadian High Arctic. Canadian Journal of Earth Sciences (in press).

Verkulich, S.R., Melles, M., Hubberten, H.W., and Pushina, Z.V. (2002) Holocene environmental changes and development of Figurnoye Lake in the southern Bunger Hills, East Antarctica, *Journal of Paleolimnology* **28,** 253-267.

Vincent, W.F. and Pienitz, R. (1996) Sensitivity of high-latitude freshwater ecosystems to global change: Temperature and solar ultraviolet radiation. *Geoscience Canada* **23,** 231-236.

Vincent, W.F., Gibson, J.A.E. and Jeffries, M.O. (2001) Ice shelf collapse, climate change and habitat loss in the Canadian High Arctic. *Polar Record* **37,** 133-142.

Vincent, W.F., Rautio, M. and Pienitz, R. (2005) Climate control of underwater UV exposure in polar and alpine aquatic ecosystems, in, J.B. Orbaek (ed.). *Arctic Environmental Change.* Springer (in press).

Vitt, D.H., Halsey, L.A. and Zoltai, S.C. (2000) The changing landscape of Canada's western boreal forest: the current dynamics of permafrost. *Canadian Journal of Forest Research* **30,** 283-287.

Wand U., Fischer, L. and Schmitz, W. (1985) Salzausblühungen in der Schirmacher Oase, Ostantarktika. *Geod. Geophys. Veröff.,* Berlin, **R. I, 12,** 86-87.

Walker, D.A., Binnian, E., Evans, B.M., Lederer, N.D., Nordstrand, E. and Webber, P.J. (1989) Terrain, vegetation, and landscape evolution of the R4D research site, Brooks Range Foothills, Alaska. *Holarctic Ecology* **12,** 238-261.

Walker, D.A., Epstein, H.E., Jia, G.J., Balser, A., Copass, C., Edwards, E.J., Gould, W.A., Hollingsworth, J., Knudson, J., Maier, H.A., Moody, A. and Raynolds, M.K. (2003) Phytomass, LAI, and NDVI in northern Alaska: Relationships to summer warmth, soil pH, plant functional types, and extrapolation to the circumpolar Arctic. *Journal of Geophysical Research-Atmospheres* **108**, D2.

Wetzel, R.G. (2001) *Limnology: Lake and River Ecosystems*. 850 pp. Academic Press.

Whalen, S.C. and Reeburg, W.S. (1990) Consumption of atmospheric methane by tundra soils, *Nature*, **346**, 160-162.

Williamson, C.E., Morris, D.P., Pace, M.L. and Olson, A.G. (1999) Dissolved organic carbon and nutrients as regulators of lake ecosystems: Resurrection of a more integrated paradigm. *Limnology and Oceanography*, **44**, 795-803.

Williamson, C.E., Stemberger, R.S., Morris, D.P., Frost, T.M. and Paulsen, S.G. (1996) Ultraviolet radiation in North American lakes: attenuation estimates from DOC measurements and implications for plankton communities. *Limnology and Oceanography* **41**, 1024-1034.

Williamson, C.E., Olson, O.G., Lott, S.E., Walker, N.D., Engstrom, D.R. and Hargreaves, B.R. (2001) Ultraviolet radiation and zooplankton community structure following deglaciation in Glacier Bay, Alaska. *Ecology* **82**, 1748-60.

Woo, M. (2000) Permafrost and hydrology, in M. Nuttall and T.V. Callaghan (eds.), *The Arctic: environment, people, policy* Harwood Academic Publishers, Amsterdam, The Netherlands pp. 57-96.

Woo, M. and Young, K.L. (2000) Hydrological response of a patchy high Arctic wetland. 12th northern research basins symposium/workshop. *Nordic Hydrology*, Munksgaard, Copenhagen, Denmark, p. 317-338

Wynn-Williams, D.D. (1980) Seasonal fluctuations in microbial action in Antarctic moss peat. *Biological Journal of the Linnean Society*, **14**, 11-28.

Wynn-Williams, D.D. (1996) Response of pioneer soil microalgal colonists to environmental change in Antarctica. *Microbial Ecology*, **31**, 177-188.

Zwartz, D., Bird, M., Stone, J. and Lambeck, K. (1998) Holocene sea-level change and ice-sheet history in the Vestfold Hills, East Antarctica. *Earth and Planetary Science Letters*, **155**, 131-145.

12. ANTARCTIC CLIMATE CHANGE AND ITS INFLUENCES ON TERRESTRIAL ECOSYSTEMS

P. CONVEY

British Antarctic Survey, Natural Environment Research Council
High Cross, Madingley Road
Cambridge CB3 0ET, United Kingdom
p.convey@bas.ac.uk

Introduction

Most global circulation models (GCMs) predict enhanced rates of climate change, particularly temperature increase, at higher latitudes. In the Antarctic terrestrial environment, the severe stresses experienced combined with geographical isolation and the relative youth of most habitats are generally accepted to underlie the low contemporary terrestrial biodiversity that is found. Currently faced with rates of regional climate change that are amongst the most rapid seen on the planet (Quayle et al. 2002, Vaughan et al. 2003), Antarctic terrestrial and freshwater ecosystems are expected to show particular sensitivity and rapid responses (Freckman and Virginia 1997, Quayle et al. 2002, 2003). The Inter-governmental Panel on Climate Change (IPCC) Third Assessment (IPCC 2001) has singled out the polar regions as areas of special concern.

Climate change is not a new challenge for Antarctic biota. During the Pleistocene and Holocene, systematic variations in climate and ice cover are well documented (eg Hjort et al. 2003). During recent centuries there is evidence of glacial advances analogous to those of the Little Ice Age experienced in Europe while, over the last 1-10 000 years, there have also been considerable fluctuations in glacial and ice shelf extent and thickness (Clapperton and Sugden 1982, 1988, Lorius et al. 1985, Smith 1990, Pudsey and Evans 2001). Although widely held, the perception that Antarctica has remained under a permanent and stable icecap is clearly inaccurate.

D.M. Bergstrom et al. (eds.), Trends in Antarctic Terrestrial and Limnetic Ecosystems, 253–272.

Four aspects of changing climate will have a direct influence on Antarctic terrestrial organisms: temperature, water (precipitation and melt), solar radiation (photosynthetically active and UV radiation) and atmospheric CO_2 concentrations. The first three are considered here as, although increasing levels of CO_2 are recorded across Antarctica, any consequences for Antarctic biota have not been directly addressed. The remaining areas form the focus of separate sections of this chapter. Finally, an overview is given of recent studies of terrestrial biological responses that have been linked with these processes of contemporary climate change.

Temperature Trends

Low thermal energy input is a feature of Antarctic terrestrial habitats. This is highlighted by the broad generalised summary of air temperature characteristics for each of the continent's three major biogeographical zones presented in Table 1 (see also Convey 1996a, Danks 1999). While recognising that the link between standard air and small-scale microclimate temperature is complex and poorly characterised (see below), the relatively crude measure of physiological time (day degrees above 0°C) illustrates the lack of thermal energy even in comparison with the most extreme regions of the Arctic. The very low summer temperatures seen across the Antarctic are likely to result in microhabitat temperatures that are often near the minimum threshold for many physiological processes. A small temperature increment near to these threshold values may be expected to have a relatively greater biological impact than a similar increment well above the threshold, a factor underlying the prediction of extreme sensitivity to climate change in Antarctic biota. Furthermore it should also be noted that, other than for the very few meteorological stations in inland continental Antarctica, most temperature records are from coastal research stations at or very near to sea level. Macro and microclimatic conditions will become rapidly more extreme with increasing altitude relative to these values (eg Tweedie and Bergstrom 2000), as will climate-associated geomorphological processes such as periglacial activity. Finally, other features of temperature variation may also have considerable biological significance, including the upper and lower extremes and diurnal and annual ranges experienced and rates of change.

AIR TEMPERATURE TRENDS

Over the last half-century, temperature increases amongst the most rapid worldwide have been documented along the western Antarctic Peninsula and associated island archipelagos (Smith 1990, Fowbert and Smith 1994, King 1994, Harangozo et al. 1997, King and Harangozo 1998, Skvarca et al. 1998, King et al. 2003, Vaughan et al. 2003). Particularly strong warming is apparent in the detailed record available from Faraday/Vernadsky Station (Argentine Islands, c.65°S) and in records available from several other Stations on the Antarctic Peninsula and South Shetland and South Orkney Islands (Fig. 1). Here, mean annual temperature has risen at a rate of 5.7±2.0°C per century over this period (Vaughan et al. 2003). Although the most detailed records and strongest trends are obtained in the maritime Antarctic, warming trends have also been reported from a range of subantarctic sites (eg Marion Island - Smith and Steenkamp (1990), Smith (2002), Iles Kerguelen - Frenot et al. (1997), Macquarie Island - Tweedie and Bergstrom (2000), see also Jacka and Budd 1998) and at locations along the coast of continental Antarctica (Adamson and Adamson 1992, Jacka and Budd 1998, Vaughan et al. 2001).

Table 1. Generalised thermal characteristics of the major regions of Antarctica, in comparison with those of the High Arctic.

	Months with positive mean air temperatures	Air temperature range (°C)		Days above 0°C
		Mean winter to summer	Extreme range	
High Arctic	2 to 4	-34 to + 5	-60 to +20	50-350
Subantarctic	6 to 12	-2 to +8	-10 to +25	70-170
Maritime Antarctic	1 to 4	-12 to +2	-45 to +15	6-100
Continental Antarctic coast	0 to 1	-30 to -3	-40 to +10	0
Continental Antarctic inland	0	<-50 to -10	<-80 to -5	0

GCM simulations predict that some areas of continental Antarctica will experience cooling rather than warming trends, and there has been one report of general cooling, interpreted in the context of consequences for the ecosystems of the Dry Valleys region of Victoria Land (Doran et al. 2002). However, Turner et al. (2002) have suggested that this report is likely to be artefactual, being a consequence of the very low spatial density of data recording in this region.

The rapid temperature increases seen along the western coast of the Antarctic Peninsula and maritime Antarctic archipelagos appear to be linked to decreasing winter sea ice extent and, by teleconnection, to El Niño Southern Oscillation (ENSO) events in the southern Pacific Ocean (Cullather et al. 1996, Harangozo 2000). However, it must also be noted that, although many GCMs predict high rates of warming at polar latitudes, even the most recent and detailed spatial models (such as the Hadley Centre's HadCM3) do not accurately represent the observed patterns of warming seen in this region (King et al. 2003). This may be a model limitation, through a combination of insufficient resolution of the oceanic components and a lack of representation of the impact of the Antarctic Peninsula's topography on regional atmospheric circulation. Alternatively, Thompson and Solomon (2002), propose a link between Antarctic Peninsula warming and atmospheric circulation trends in the lower stratospheric polar vortex, and hence a link to photochemical ozone depletion.

TRENDS OVER THE ANNUAL CYCLE

At some locations, warming trends are not constant throughout the year. The annual trend on the western Antarctic Peninsula (Fig. 2) is formed from strong warming during the winter months (11±9°C per century), with much lower but still significant rates seen in summer (2.4±1.7°C per century: King and Harangozo 1998, King et al. 2003, Vaughan et al. 2003). These figures also illustrate the degree of between-year variability present in the data. The biological consequences of these differences in trend magnitude can be considerable, through extending the active season for terrestrial biota by shortening the winter period and releasing or maintaining liquid water in ecosystems through earlier spring thaws and later autumn freezing. At maritime Antarctic Signy Island (South Orkney Islands), a different pattern has been seen over the last three to four decades, with warming occurring during the summer months but little change noted in winter (Smith 1990, Convey et al. 2003).

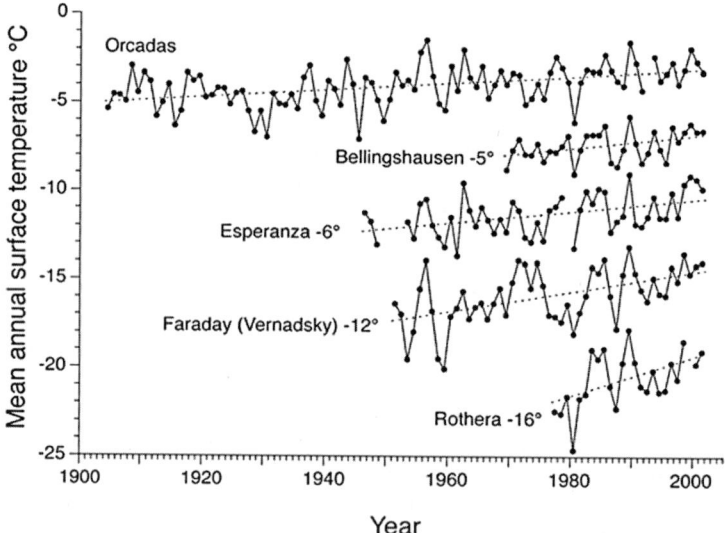

Figure 1. Time series of annual mean surface air temperature at selected stations in the vicinity of the Antarctic Peninsula, from Orcadas Station (Laurie Island, South Orkney Islands, c. 60°S to Rothera Station (Adelaide Island, c. 68°S). Records other than Orcadas have been offset by the amounts shown for clarity (from King et al. 2003).

In the subantarctic, the limited numbers of analyses available do not indicate consistent seasonal patterns across the different (and widely separated) islands. On Macquarie Island, the strongest warming trends have been reported during late summer and early autumn, although the island cools during severe ENSO events (Adamson et al. 1988) and Tweedie and Bergstrom (2000) considered that there was no overall seasonal bias to the warming trend. Smith and Steenkamp (1990) documented higher warming rates on Marion Island during later winter and early summer. On Iles Kerguelen, Frenot et al. (1997) reported no seasonal bias in data obtained between 1951 and 1993, while Allison and Keage (1986) found a bias towards summer warming in an analysis of a shorter dataset covering 1964 to 1984 at Heard Island.

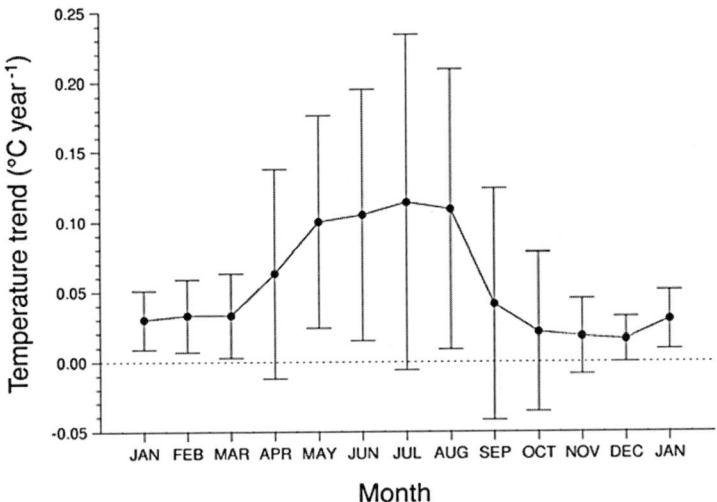

Figure 2. Trends of monthly mean temperature increase at Faraday Station (now Vernadsky Station) (Argentine Islands, c. 65°S), illustrating both greater magnitude and greater inter-annual variability during the austral winter months (from King et al. 2003).

TERRESTRIAL MICROCLIMATES

The difficulties in describing or modelling the relationship between coarse macroclimatic and fine-scale microclimatic data are well known (eg Walton 1984, Smith 1988), with microclimatic characteristics being influenced by a range of other potentially confounding variables (including insolation/cloud cover, snow cover, hydration, wind speed, physical aspect, substratum, biotic habitat structure). These difficulties are further compounded by a lack of long term datasets and monitoring studies of microclimate, meaning that there are currently no data available linking macro- and microclimatic trends. Several short-term microclimatic data sets have been obtained around the Antarctic, which serve to highlight the large and rapid changes possible in microclimatic conditions (eg Walton 1982, Smith 1988, Davey et al. 1992, Kennedy 1994), while also illustrating the inherent ecophysiological flexibility of many Antarctic terrestrial organisms that allows them to cope well with natural environmental variation, which is of much greater magnitude than the warming trends superimposed through regional climate change (Convey 1997a, 2001).

In this context, the use of a measure of physiological time (day degrees above 0°C), derived from macroclimatic records of air temperature, provides an illustration of the potential importance of relatively small increases in microhabitat temperature. Air temperature provides a poor indicator of microhabitat temperature in periods with significant insolation and in habitats under snow cover. However, it provides a more accurate estimate for shaded microhabitats and during periods of cloud cover (typical of much of the maritime and subantarctic). Trends of increase in the number of positive day degrees are again highly significant, with a 74% increase over the last 50 years recorded at Faraday/Vernadsky Station and the shorter records available from Rothera and Bellingshausen Stations generating trends of increase of 14% and 18.5% per decade respectively (D.G. Vaughan, unpubl. data). The potential biological significance of trends of this magnitude becomes apparent when it is realised that it has been calculated or demonstrated that increases of the order of 100-300 day degrees per season are sufficient to allow the completion of an annual rather than a biennial life cycle in a subantarctic beetle (Arnold and Convey 1998), and completion of an extra summer asexual generation in an Arctic aphid, resulting in an order of magnitude increase in the numbers of overwintering eggs (Strathdee et al. 1993).

FRESHWATER HABITATS

Limnetic ecosystems are discussed by Lyons et al. (this volume). There have been few extended or quantitative studies of the responses of Antarctic lakes or other freshwater ecosystems to climate change. A recent analysis of long-term macroclimatic (1947 – 1995) and limnological (1963 – 1996) datasets collected on maritime Antarctic Signy Island (South Orkney Islands) indicates that such ecosystems may magnify the macroclimatic trends already commented on above (Quayle et al. 2002, 2003). Thus, over the period 1980 – 1995, lake water temperatures were shown to have increased by 2-3 times the local air temperature increase (cf. Skvarca et al. 1998), while changes in the autumn freeze and spring ice break up dates have extended the open water period by up to four weeks. Given the significance of air temperature increases already discussed above, it is clear that increases of this magnitude may have a direct and large impact on lake biota, while the increased open water period has further knock-on impacts on diverse processes including water mixing, turbidity, nutrient concentrations, microbial population growth (primary and secondary productivity) and development of winter anoxia (Quayle et al. 2003).

Precipitation and Water Availability

The availability of liquid water is regarded as a key environmental variable controlling the distribution and activity of polar terrestrial biota, that may be even more important than temperature itself (Kennedy 1993, Sømme 1995, Block 1996). While GCMs lead to predictions of changes in precipitation patterns, as yet it is not possible to generate these at a fine enough spatial scale to be usefully applicable to biological systems.

PRECIPITATION

Precipitation patterns are linked with a range of other environmental variables including insolation, cloud cover and wind speed. Thus, there is likely to be a strong link with temperature, especially when considered at the microclimatic scale relevant to most Antarctic terrestrial biota. Modelling approaches have predicted a general increase in precipitation in the Antarctic coastal zone (Budd and Simmonds 1991) and data exist to support this prediction from the maritime Antarctic (Turner et al. 1997). This is another area of climate that may show teleconnection with ENSO events, in this case with a change in atmospheric circulation patterns leading to more depressions approaching the Antarctic Peninsula from the north. A further change in the detail of precipitation patterns has been noted in the maritime Antarctic, where it is increasingly likely that summer precipitation will fall as rain rather than snow (see also Noon et al. 2001, Quayle et al. 2003). An increase in precipitation is amongst a range of changes in environmental variables identified in analyses of 50 years of meteorological data from subantarctic Macquarie Island (C. Tweedie, D. Doley and D. Bergstrom, unpubl. data). Alternatively, decreases in precipitation are evident in some parts of Antarctica, including subantarctic Marion Island and Iles Kerguelen (Smith and Steenkamp 1990, Chown and Smith 1993, Frenot et al. 1997, Smith 2002, Chapuis et al. 2004). Winter precipitation has also decreased on maritime Antarctic Signy Island (Noon et al. 2001).

GLACIAL AND SNOW MELT

Over much of the Antarctic continent, precipitation rarely or never occurs as rain. Indeed, as measured by precipitation rates, large areas are described accurately as frigid deserts. In the maritime Antarctic, as noted above, regional warming has also been linked to an increasing probability of summer precipitation falling as rain rather than snow. On most subantarctic islands, the majority of precipitation falls as rain (snow at higher altitudes) and, with the exception of Iles Kerguelen, South Georgia and Heard Island, snow is short-lived and glaciers absent. In the absence of, or in addition to, direct precipitation, the pattern of availability of liquid water is, therefore, also governed by seasonal snow and glacial melt and is temporally and sometimes spatially, different to that of precipitation itself.

Rapid rates of glacial thinning and retreat and loss of 'permanent' snow cover observed at a range of maritime and subantarctic sites are particularly significant in this context (Smith 1990, Gordon and Timmis 1992, Fowbert and Smith 1994, Frenot et al. 1997, Pugh and Davenport 1997, Fox and Cooper 1998, Kiernan and McConnell 2002, Skvarca and De Angelis 2003). Melt will normally increase the availability of water to terrestrial ecosystems. However, as melt commences earlier and/or takes place at increased rates, there is also a possibility that spatially limited resources of ice or snow may become exhausted before the end of the summer season, imposing greater rather than reduced water stress on terrestrial organisms. A separate effect of warming may become apparent during winter as, through increasing the frequency of winter thaws, a sub-snow ice layer develops on the ground surface. On Arctic Svalbard, Coulson et al. (2000) have demonstrated that this ice layer has negative effects on some soil faunal communities. Comparable ice layers are known to develop at Antarctic locations during winter, but their biological impacts have not been determined (Davey et al. 1992, Arnold et al. 2003).

PERMAFROST

To date, there appear to be no data available relating to trends in permafrost development or loss in the Antarctic (Cannone and Guglielmin 2003). In the subantarctic, although appearing superficially similar to the tundra of the Arctic, there are a number of important biological and physical differences, one of the more significant being that permafrost is not present at low altitudes in the habitats underlying subantarctic vegetation. However, permafrost is much more widespread in the maritime and continental Antarctic and might be expected to follow analogous trends to those predicted or seen in the Arctic in response to climate change. Principally, the area underlain by continuous or discontinuous permafrost is expected to decrease significantly in the Arctic (Kettles et al. 1997), although this may be partially buffered by existing vegetation (Camill and Clark 2000) – a factor likely to be insignificant in the Antarctic. It is also possible for the melting of permafrost to have an important impact on hydrology and vegetation, through the development of thermokarst topography.

ICE SHEET FLUCTUATIONS

Over recent decades, much attention has been given to the catastrophic collapse of coastal ice shelves around the Antarctic Peninsula and their links with regional climate warming (Vaughan and Doake 1996, Morris and Vaughan 2003, Scambos et al. 2003). The immediate terrestrial biological consequences of these collapses are probably minimal. However, variation in ice shelf extent and thickness has considerably greater implications for the understanding of Antarctic terrestrial biogeography and biodiversity. Over the timescale of the Pleistocene and Holocene, it is also clear that sea level has varied (Hall 2003). In the continental and maritime Antarctic, the majority of ice-free terrestrial ground and contemporary biota are restricted to low altitude coastal regions. Thus, their precise distribution is likely to have been directly affected by sea level changes, both increases and decreases. Hall (2003) concludes that the marine limit during the Holocene (as indicated by raised beaches) was a maximum of c. 22m above that presently existing, with further raised beaches up to c. 60m ASL being related to previous interglacial periods. In contrast, during periods of glacial maxima, while sea level would have been lower (possibly by up to 120m), it is now thought that ice shelves expanded away from the coastline to the point of continental shelf drop-off (Larter and Vanneste 1995, Ó Cofaigh et al. 2002, COHIMAR/SEDANO Scientific Party 2003). Ice sheet depth over land was also considerably thicker than at present (eg Bentley and Anderson 1998), although there remains debate over the balance between thickness and flow rate of coastal ice shelves (Larter and Vanneste 1995). The combination of these findings is currently interpreted as indicating that most maritime or continental Antarctic terrestrial biota could not have survived in situ throughout the periods of climate change involved in glacial cycles, requiring the existence of as yet unknown refugia as many groups show levels of endemism or diversity inconsistent with recent (post-Pleistocene) colonisation (Convey 2003, Convey and McInnes 2005, Lawley et al. 2004, Stevens and Hogg this volume).

Ozone Depletion and Ultraviolet Radiation Climate

The final element of climate that has undergone considerable recent change in Antarctica is that of solar radiation. One aspect of this, direct insolation, is inextricably linked with the suite of environmental variables already discussed, ie cloud cover, precipitation, wind speed and temperature. Changes in any or all of these variables can be expected to have clear implications for primary production.

Second, over approximately the last 25 years, photochemical depletion during the austral spring of the protective ozone layer (the 'ozone hole', Farman et al. 1985) has led to changes in exposure to damaging shorter wavelength ultra-violet-B radiation. This phenomenon is separate from the processes of climate change already described, though the pattern of depletion over recent years has also been caused by anthropogenic chemical pollution of the atmosphere. It is in theory possible that release of chemicals into the atmosphere during certain natural events, particularly volcanic eruptions, could also have led to analogous episodes of ozone destruction in the past. Although, as yet, no evidence for this has been reported, there are biological proxies in the form of degradation resistant pigments that have been proposed as possible indicators of the history of environmental UV-B exposure (Rozema et al. 2001).

Spring ozone depletion allows greater penetration than would normally occur of short wavelength, potentially biologically damaging, UV-B radiation to terrestrial and shallow marine habitats (Fig. 3), but leaves levels of UV-A and photo-synthetically active (visible) radiation (PAR) unaffected. As this alters the ratio of PAR or UV-A to UV-B (Fig. 4), which can have important influence on intracellular repair processes (Santas et al. 1997), there is potential for disproportionate levels of ozone-mediated UV-B damage to occur. At maximum, approximately two-thirds of the protective ozone layer is lost over high southern latitudes.

Continental Antarctica is most consistently exposed to the ozone hole. However, the shape, extent and depth of the depleted zone change rapidly and, thus, much of the maritime Antarctic is routinely exposed to its effects and, as lobes circulate around the continent, the subantarctic islands, and southern South America, New Zealand and Australia may be exposed to depletion for shorter periods.

While the ozone hole exists, maximum exposure to UV-B is similar to typical mid-summer values (Fig. 3). Rather than intensity of exposure per se, its importance lies in the fact that maxima occur earlier in the season than normal (when many organisms may not have resumed normal physiological activity) and with shorter and more damaging wavelengths penetrating to ground level at this time of year.

The potential impact of increased UV-B radiation is also modulated by other environmental conditions such as cloud cover and type, albedo and solar angle (Lubin and Frederick 1991, Gautier et al. 1994, Sabburg and Wong 2000), which can have further subtle effects on the ratio of PAR to UV-A and UV-B (as the different wavelengths are differentially absorbed or scattered). Finally, at the level of the microhabitat, the importance of even thin layers of snow in the protection of (micro)biota from radiation damage has been demonstrated (Cockell et al. 2002), illustrating a further potentially complex interaction among the different changing climatic and radiative variables.

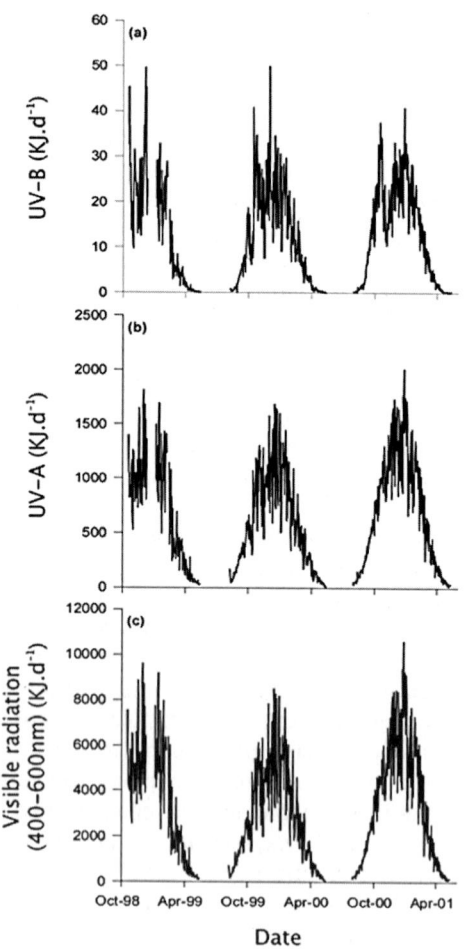

Figure 3. Daily doses of solar radiation measured using a Bentham scanning spectroradiometer at Rothera Station, Adelaide Island, October 1998 - April 2001. a) UV-B (280-315 nm). b) UV-A (315-400 nm). C) visible radiation (400-600 nm).

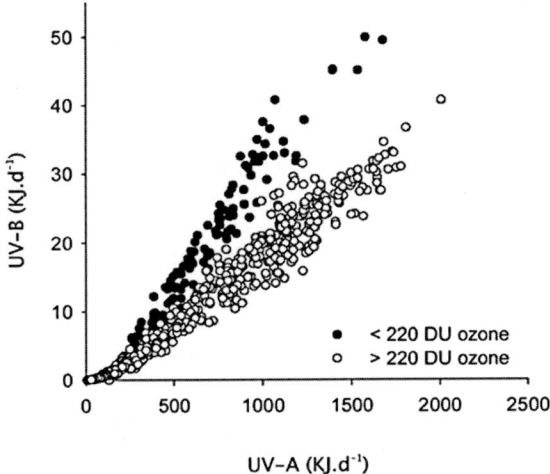

Figure 4. A plot of daily dose of UV-B against daily dose of UV-A using data obtained at Rothera Station, illustrating the alteration of the UV-A:UV-B ratio experienced under ozone depletion. Symbols indicate the minimum ozone level (as calculated from spectroradiometer data) for each day.

Terrestrial biological responses

It was quickly recognised that the combination of the magnitude of changes being experienced and the generally simple ecosystems of Antarctica led to an expectation of clear and identifiable consequences in this region (Roberts 1989, Smith and Steenkamp 1990, Voytek 1990). This led to the development of a predictive literature (eg Wynn-Williams 1994, 1996, Kennedy 1995a, Convey 1997b, Walton et al. 1997, Bergstrom and Chown 1999) and the initiation of observational and experimental studies aimed at testing some of these predictions. At the broad scale of the ecosystem, patterns of changes in two major environmental variables (temperature, water) have generally led to predictions of increases in the rates of both local colonization and of colonization by species new to the Antarctic. The overall consequences are generally expected to include increased terrestrial diversity, biomass and trophic complexity, all of which contribute to the development of more complex ecosystem structure and an increase in the importance of competitive interactions and abilities. The effects of changes in radiation patterns, while subtle, are expected to be negative, through requiring resource allocation to mitigation, while also ramifying throughout food webs.

Studies of Antarctic terrestrial biology that have been used to test or corroborate predictions associated with climate change separate into two classes – those that are simply descriptive or observational and those that involve more or less realistic manipulations, either in the laboratory or in natural habitats. A number of recent and extensive reviews of the findings of these studies are available (Kennedy 1996, Convey 2001, 2003, Walther et al. 2002) and will not be repeated in detail here. Other chapters within this volume deal primarily with studies of responses at the level of physiology and biochemistry (Hennion et al. this volume), life history strategy (Convey et al. this volume b) and biological invasion processes (Frenot et al. 2005, Convey et al. this volume a).

In the maritime and continental Antarctic, a key factor in predicting or understanding responses of terrestrial biota to climate change is the large degree of flexibility inherent to many life history and physiological characteristics (Convey 1996b, Convey et al. this volume b). This itself relates to the wide, sometimes rapid, and unpredictable variation in environmental parameters that is characteristic of the Antarctic terrestrial environment and far outweighs in magnitude many of the systematic changes in climate (particularly relating to temperature) that are being experienced. This life history flexibility allows immediate changes in development rate and shortening of life cycle duration in response to factors such as climate warming or extension of the active season. In the subantarctic, while flexibility remains an important factor, here it is driven by a rather different feature of the environment – the relative stability over the annual cycle afforded by the strong oceanic influence on island climate (Adamson et al. 1988, Convey 1996a). There is also an increased incidence of temporally defined events, such as obligate or cued diapause, in the life cycles of subantarctic species (eg Nicolai and Droste 1984), features that are poorly or not represented in the biology of maritime and continental Antarctic species (Convey 1996a,b).

Probably the most widely reported example of terrestrial organisms responding to climate change in the Antarctic has been that of the local colonization and striking recent increases in population numbers and extents of the two native Antarctic flowering plants (*Deschampsia antarctica* and *Colobanthus quitensis*) along the Antarctic Peninsula and associated islands and archipelagos (Fowbert and Smith 1994, Smith 1994, Grobe et al. 1997, Gerighausen et al. 2003). At some sites numbers of plants have increased by two orders of magnitude in as little as 30 years, although it is often overlooked that these increases have not involved any change in the species' overall geographic ranges. These increases have been interpreted in the context of increased temperature encouraging growth and vegetative spreading of established plants, in addition to increasing the probability of establishment of germinating seedlings. Additionally, warming is proposed to underlie a greater frequency of mature seed production (Convey 1996c), while it may also stimulate growth of seeds that have remained dormant in soil propagule banks (McGraw and Day 1997).

Strangely, given the attention focussed on these two higher plant species, there have been no similar quantitative field studies of the bryophytes that form the dominant vegetation of the Antarctic Peninsula and continent, or extended studies of the fauna of the habitats provided by the plants. A single long-term study of the water relations of the arthropod fauna of Signy Island (maritime Antarctic) (Block and Convey 2001, Convey et al. 2003) has generated evidence suggesting systematic changes in patterns of water availability in terrestrial habitats, interpreted as being consistent with local climate trends. On subantarctic Macquarie Island, Whinam and Copson (2006) have reported recent and considerable decreases in the occurrence of the boggy mire moss *Sphagnum falcatulum* over the last two decades. A combination of higher than average temperatures and wind speeds, and lower than average humidity and precipitation, are proposed to underlie this observation, through increasing the desiccation stress on this plant. Given the close linkage between occurrence of many Antarctic bryophyte species and habitat water availability, this example is likely to forewarn of many similar consequences in Antarctic terrestrial ecosystems.

Field manipulation studies at subantarctic (South Georgia) and maritime Antarctic sites have been used to demonstrate that bryophyte, microbiota and fauna respond rapidly to improved environmental conditions (Smith 1993, 2001, Kennedy 1994, Wynn-Williams 1996, Convey and Wynn-Williams 2002) with greatly increased populations. Even more so than higher plants, bryophytes, lichens and various microbiota produce spores and other propagules that may be transported easily and remain dormant in the soil propagule bank for long periods (Smith and Coupar 1986, Smith 1987, 1993, Hughes et al. this volume). Other than the above-mentioned manipulations at South Georgia (Smith 2001), all these studies have taken place at maritime Antarctic locations, with no comparable data available from elsewhere in the sub- or continental zones.

Long-term field manipulation methodologies, while subject to a range of valid criticisms and offering a less than perfect simulation of predicted environmental changes (Kennedy 1995b,c), remain the only practical option for field studies of biological responses to change in remote and severe environments such as the Antarctic. Their deployment over periods of years has led to large responses in terms of variables such as population number, structure, individual morphology or habitat structure in studies of all the major groups of Antarctic terrestrial fauna and flora (Convey 2003). Although many such studies were initiated with an (at least implicit) intention to examine responses either to warming or to selective exposure to particular wavelength ranges in the ultra-violet spectrum it has become increasingly clear that a more holistic approach is required in terms of experimental design, data analysis and interpretation, and in the elements of the ecosystem considered. Thus the most comprehensive studies completed to date anywhere in the Antarctic, those of Day and co-workers (Day et al. 1999, 2001, Convey et al. 2002) near Palmer Station (Anvers Island), employed experimental designs allowing a multifactorial analysis of the consequences of manipulation of thermal, hydration and radiation regimes, while examining responses at the levels of plant biochemistry, morphology, life history and ecology, and extending into studies of the decomposition cycle and wider food web consequences. It is becoming increasingly clear that biological responses to changing environmental variables are often likely to be subtle and hard to detect. However, because any responses will involve changes in resource allocation, a crucial element in the development of life history strategies, it is important to understand how they may integrate to give considerably greater impacts at the community or ecosystem levels. Thus, it is now recognised that a key to the understanding of ecological responses to climate change, in Antarctica or elsewhere, is to focus on the integration of responses across the entire food web (Searles et al. 2001, Day 2001, Convey 2003).

Summary

Climate changes are occurring throughout Antarctica, affecting four major groups of environmental variables each with considerable biological significance: temperature, water, solar radiation and carbon dioxide. The first three of these have been the focus of study. Parts of Antarctica, particularly the Antarctic Peninsula and Scotia Arc, are currently experiencing amongst the fastest rates of macroclimatic warming recorded worldwide, although our ability to link these with any microclimatic trends (the scale directly relevant to terrestrial biota) is limited by a lack of long term monitoring data. Likewise, although general changes in precipitation and melt patterns are predicted and reported, the scale of modelling and observation is currently far too great to allow useful application to specific terrestrial ecosystems. Patterns of change in radiation (UV-B) exposure in relation to the Antarctic ozone hole are better described though, again, few datasets have been obtained/applied at the microhabitat scale. Further adding to an already complex picture, it is also clear that variation in each of the major groups of variables is closely interlinked with the other groups.

In many respects, Antarctic terrestrial organisms are often well-adapted to the stresses of a highly variable environment, possessing features that should permit them to handle predicted levels of change that are often small compared with the natural variability already experienced. Indeed, with reference to temperature increase, resident biota will often be able to take advantage of reduced environmental stress, which will allow longer active periods/seasons, faster growth, shorter life cycles and population increase. Impacts of increased water availability are expected to be similar, although in both instances it is salient to note that exactly the reverse consequences can be experienced locally, either directly as a result of decreased water input, or as a result of interaction between increased temperature and water leading to greater evaporation and desiccation stress. Impacts of increased UV-B exposure associated with the spring ozone hole, while subtle, are expected to be negative. Finally, unlike most areas of the planet, Antarctic terrestrial habitats are protected from sources of alien colonisation by their very remoteness, meaning that in general the response of indigenous biota to changing climate can be considered separately from that to increased competition from colonizing species. The synergy between Antarctic climate change, reducing the barriers to establishment, and human activity, increasing import of exotic species, may soon act to destroy this protective barrier.

Acknowledgements

The ideas within this chapter have developed through many discussions with colleagues at the British Antarctic Survey and within the RiSCC community. Niek Gremmen, Marc Lebouvier and Ad Huiskes are thanked in particular for constructive and helpful comments on a previous version of the manuscript. Figures 1 and 2 are reproduced from King et al. (2003), with permission of the American Geophysical Union. Figures 3 and 4 were prepared by Paul Geissler and Helen Peat (BAS) using data collected as part of the BAS Long Term Monitoring and Survey Programme.

References

Adamson, H. and Adamson, E. (1992) Possible effects of climate change on Antarctic terrestrial vegetation, in P. Quilty (ed.), *Impact of Climate Change on Antarctica-Australia*, Australian Government Publishing Service, Canberra, Australia, pp. 52-61.

Adamson, D.A., Whetton, P. and Selkirk, P.M. (1988) An analysis of air temperature records for Macquarie Island: decadal warming, ENSO cooling and southern hemisphere circulation patterns, *Papers and Proceedings of the Royal Society of Tasmania* **122**, 107-112.

Allison, I.F. and Keage, P.L. (1986) Recent changes in the glaciers of Heard Island, *Polar Record* **23**, 255-271.

Arnold, R.J. and Convey, P. (1998) The life history of the diving beetle, *Lancetes angusticollis* (Curtis) (Coleoptera: Dytiscidae), on subantarctic South Georgia, *Polar Biology* **20**, 153-160.

Arnold, R.J., Convey, P., Hughes, K.A. and Wynn-Williams, D.D. (2003) Seasonal periodicity of physical factors, inorganic nutrients and microalgae in Antarctic fellfields, *Polar Biology* **26**, 396-403.

Bentley, M.J. and Anderson, J.B. (1998) Glacial and marine geological evidence for the ice sheet configuration in the Weddell Sea – Antarctic Peninsula region during the Last Glacial Maximum, *Antarctic Science* **10**, 309-325.

Bergstrom, D.M. and Chown, S.L. (1999) Life at the front: history, ecology and change on southern ocean islands, *Trends in Ecology and Evolution* **14**, 472-477.

Block W. (1996) Cold or drought - the lesser of two evils for terrestrial arthropods? *European Journal of Entomology* **93**, 325-339.

Block, W. and Convey, P. (2001) Seasonal and long-term variation in body water content of an Antarctic springtail - a response to climate change? *Polar Biology* **24**, 764-770.

Budd, W.F. and Simmonds, L. (1991) The impact of global warming on the Antarctic mass balance and global sea level, in G. Weller, C.L. Wilson and B.A.B. Severin (eds.), *Proceedings of the International Conference on the Role of Polar Regions in Global Change*, Geophysics Institute, University of Alaska, Fairbanks, pp. 489-494.

Camill, P. and Clark, J.S. (2000) Long-term perspectives on lagged ecosystem responses to climate change: permafrost in boreal peatlands and grassland/woodland boundary, *Ecosystems* **3**, 534-544.

Cannone, N. and Guglielmin, M. (2003) Vegetation and permafrost: sensitive systems for the development of a monitoring program of climate change along an Antarctic transect, in A.H.L. Huiskes, W.W.C. Gieskes, J. Rozema, R.M.L. Schorno, S.M. van der Vies and W.J. Wolff (eds.) *Antarctic Biology in a Global Context*, Backhuys, Leiden, pp. 31-36.

Chapuis, J.L., Frenot, Y. and Lebouvier, M. (2004) Recovery of native plant communities after eradication of rabbits from the subantarctic Kerguelen Islands, and influence of climate change, *Biological Conservation* **117**, 167–179.

Chown, S.L. and Smith, V.R. (1993) Climate change and the short-term impact of feral house mice at the sub-antarctic Prince Edward Islands, *Oecologica* **96**, 508-518.

Clapperton, C.M. and Sugden, D.E. (1982) Late quaternary glacial history of George VI Sound area, West Antarctica, *Quaternary Research* **18**, 243-267.

Clapperton, C.M. and Sugden, D.E. (1988) Holocene glacier fluctuations in South America and Antarctica, *Quaternary Science Reviews* 7, 185-198.

Cockell, C.S., Rettberg, P., Horneck, G., Wynn-Williams, D.D., Scherer, K. and Gugg-Helminger, A. (2002) Influence of ice and snow covers on the UV exposure of terrestrial microbial communities: dosimetric studies, *Journal of Photochemistry and Photobiology B: Biology* 68, 23-32.

COHIMAR/SEDANO Scientific Party. (2003) Uncovering the footprint of former ice streams off Antarctica, *Eos, Transactions, American Geophysical Union* 84(11), 97 and 102-103.

Convey P. (1996a) Overwintering strategies of terrestrial invertebrates in Antarctica - the significance of flexibility in extremely seasonal environments, *European Journal of Entomology* 93, 489-505.

Convey, P. (1996b) The influence of environmental characteristics on life history attributes of Antarctic terrestrial biota, *Biological Reviews* 71, 191-225.

Convey, P. (1996c) Reproduction of Antarctic flowering plants, *Antarctic Science* 8, 127-134.

Convey, P. (1997a) How are the life history strategies of Antarctic terrestrial invertebrates influenced by extreme environmental conditions? *Journal of Thermal Biology* 22, 429-440.

Convey, P. (1997b) Environmental change: possible consequences for the life histories of Antarctic terrestrial biota, *Korean Journal of Polar Research* 8, 127-144.

Convey, P. (2001) Terrestrial ecosystem response to climate changes in the Antarctic, in G.-R. Walther, C.A. Burga and P.J. Edwards (eds.), *"Fingerprints" of climate change - adapted behaviour and shifting species ranges*, Kluwer, New York, pp 17-42.

Convey, P. (2003) Maritime Antarctic climate change: signals from terrestrial biology, in E. Domack, A. Burnett, A. Leventer, P. Convey, M. Kirby and R. Bindschadler (eds.), *Antarctic Peninsula Climate Variability: Historical and Palaeoenvironmental Perspectives*, Antarctic Research Series, Vol. 79, American Geophysical Union, Washington, D.C., pp. 145-158.

Convey, P. and McInnes, S.J. (2005) Exceptional, tardigrade dominated, ecosystems from Ellsworth Land, Antarctica, *Ecology* 86, 519-527.

Convey, P. and Wynn-Williams, D.D. (2002) Antarctic soil nematode response to artificial environmental manipulation, *European Journal of Soil Biology* 38, 255-259.

Convey, P., Block, W. and Peat, H.J. (2003) Soil arthropods as indicators of water stress in Antarctic terrestrial habitats? *Global Change Biology* 9, 1718-1730.

Convey, P., Chown, S.L., Wasley, J. and Bergstrom, D.M. (2006b) Life history traits, in D.M. Bergstrom, P. Convey, and A.H.L. Huiskes (eds.), *Trends in Antarctic Terrestrial and Limnetic Ecosystems: Antarctica as a Global Indicator*, Springer, Dordrecht (this volume).

Convey, P., Frenot, Y., Gremmen, N. and Bergstrom, D.M. (2006a) Biological invasions, in D.M. Bergstrom, P. Convey, and A.H.L. Huiskes (eds.), *Trends in Antarctic Terrestrial and Limnetic Ecosystems: Antarctica as a Global Indicator*, Springer, Dordrecht (this volume).

Convey, P., Pugh, P. J. A., Jackson, C., Murray, A. W., Ruhland, C. T., Xiong, F. S. and Day, T. A. (2002) Response of Antarctic terrestrial arthropods to multifactorial climate manipulation over a four year period, *Ecology* 83, 3130-3140.

Coulson, S.J., Leinaas, H.P., Ims, R.A. and Søvik, G. (2000) Experimental manipulation of the winter surface ice layer: the effects on a High Arctic soil microarthropod community, *Ecography* 23, 299-306.

Cullather, R.I., Bromwich, D.H. and van Woert, M.L. (1996) Inter-annual variations in Antarctic precipitation related to El Niño - Southern Oscillation, *Journal of Geophysical Research* 101, 19109-19118.

Danks, H.V. (1999) Life cycles in polar arthropods - flexible or programmed? *European Journal of Entomology* 96, 83-102.

Davey, M.C., Pickup, J. and Block, W. (1992) Temperature variation and its biological significance in fellfield habitats on a maritime Antarctic island, *Antarctic Science* 4, 383-388.

Day, T.A. (2001) Multiple trophic levels in UV-B assessments - completing the ecosystem, *New Phytologist* 152, 183-186.

Day, T.A., Ruhland, C.T. and Xiong, F. (2001) Influence of solar UV-B radiation on Antarctic terrestrial plants: results from a 4-year field study, *Journal of Photochemistry and Photobiology B: Biology* 62, 78-87.

Day, T.A., Ruhland, C.T., Grobe, C.W. and Xiong, F. (1999) Growth and reproduction of Antarctic vascular plants in response to warming and UV radiation reductions in the field, *Oecologia* 119, 24-35.

Doran, P.T., Priscu J.C., Lyons, W.B., Walsh, J.E., Fountain, A.G., McKnight, D.M., Moorhead, D.L., Virginia, R.A., Wall, D.H., Clow, G.D., Fritsen, C. H., McKay, C. P., Parsons, A.N. (2002) Antarctic climate cooling and terrestrial ecosystem response, *Nature* **415**, 517-520.

Farman, J.C., Gardiner, B.G. and Shanklin, J.D. (1985) Large losses of total ozone in Antarctica reveal seasonal ClO_x/NO_x interaction, *Nature* **315**, 207-210.

Fowbert, J.A. and Smith, R.I.L. (1994) Rapid population increases in native vascular plants in the Argentine Islands, Antarctic Peninsula, *Arctic and Alpine Research* **26**, 290-296.

Fox, A.J. and Cooper, A.P.R. (1998) Climate-change indicators from archival aerial photography of the Antarctic Peninsula, *Annals of Glaciology* **27**, 636-642.

Freckman, D.W. and Virginia, R.A., (1997) Low-diversity Antarctic soil nematode communities: distribution and response to disturbance, *Ecology* **78**, 363-369.

Frenot, Y., Gloaguen, J.-C., and Tréhen, P. (1997) Climate change in Kerguelen Islands and colonization of recently deglaciated areas by *Poa kerguelensis* and *P. annua* in B. Battaglia, J. Valencia and D.W.H. Walton (eds.), *Antarctic Communities: Species, Structure and Survival*, Cambridge University Press, Cambridge, UK, pp. 358-366.

Frenot, Y., Chown, S.L., Whinam, J., Selkirk, P., Convey, P., Skotnicki, M. and Bergstrom, D. (2005) Biological invasions in the Antarctic: extent, impacts and implications, *Biological Reviews* **80**, 45-72.

Gautier, C., He, G. and Yang, S. (1994) Role of clouds and ozone on spectral ultraviolet-B radiation and biologically active UV dose over Antarctica, In C. Weiler and P. Penhale (eds.), *Ultraviolet radiation in Antarctica: measurement and biological effects*. Antarctic Research Series, Vol. 62, American Geophysical Union, Washington, DC, pp. 83-91.

Gerighausen, U., Bräutigam, K., Mustafa, O. and Peter, H.-U. (2003) Expansion of vascular plants on an Antarctic island – a consequence of climate change? In A.H.L. Huiskes, W.W.C. Gieskes, J. Rozema, R.M.L. Schorno, S.M. van der Vies and W.J. Wolff (eds.) *Antarctic Biology in a Global Context*, Backhuys, Leiden, pp. 79-83.

Gordon, J.E. and Timmis, R.J. (1992) Glacier fluctuations on South Georgia during the 1970s and early 1980s, *Antarctic Science* **4**, 215-226.

Grobe, C.W., Ruhland C.T. and Day T.A. (1997) A new population of *Colobanthus quitensis* near Arthur Harbor, Antarctica: correlating recruitment with warmer summer temperatures, *Arctic and Alpine Research* **29**, 217-221.

Hall, B.L. (2003) An overview of late Pleistocene glaciation in the South Shetland Islands, In E. Domack, A. Burnett, A. Leventer, P. Convey, M. Kirby and R. Bindschadler (eds.), *Antarctic Peninsula Climate Variability: Historical and Palaeoenvironmental Perspectives*, Antarctic Research Series, Vol. 79, American Geophysical Union, Washington, D.C., pp. 103-113.

Harangozo, S.A. (2000) A search for ENSO teleconnections in the west Antarctic Peninsula climate in austral winter, *International Journal of Climatology* **20**, 663-679.

Harangozo, S.A., Colwell, S.R. and King, J.C. (1997) An analysis of a 34-year air temperature record from Fossil Bluff (71°S 68°W), Antarctica, *Antarctic Science* **9**, 355-363.

Hennion, F., Huiskes, A.H.L., Robinson, S. and Convey, P. (2006) Physiological traits or organisms in a changing environment, in D.M. Bergstrom, P. Convey, and A.H.L. Huiskes (eds.), *Trends in Antarctic Terrestrial and Limnetic Ecosystems: Antarctica as a Global Indicator*, Springer, Dordrecht (this volume).

Hjort, C., Ingólfsson, Ó., Bentley, M.J. and Björck, S. (2003) The late Pleistocene and Holocene glacial and climate history of the Antarctic Peninsula region as documented by the land and lake sediment records – a review, In E. Domack, A. Burnett, A. Leventer, P. Convey, M. Kirby and R. Bindschadler (eds.) *Antarctic Peninsula Climate Variability: Historical and Palaeoenvironmental Perspectives*, Antarctic Research Series, Vol. 79, American Geophysical Union, Washington, D.C., pp. 95-102.

Hughes, K.A., Ott, S., Bölter, M. and Convey, P. (2006) Colonisation processes, in D.M. Bergstrom, P. Convey, and A.H.L. Huiskes (eds.), *Trends in Antarctic Terrestrial and Limnetic Ecosystems: Antarctica as a Global Indicator*, Springer, Dordrecht (this volume).

Inter-governmental Panel on Climate Change (IPCC) (2001) *Climate change 2001: impacts, adaptation and vulnerability*, In J.J. McCarthy, O.F. Canziani, N.A. Leary, D.J. Dokken and K.S. White (eds), Contribution of Working Group 2 to the Third Assessment Report of the IPCC, Cambridge University Press, Cambridge, UK.

Jacka, T.H. and Budd, W.F. (1998) Detection of temperature and sea-ice-extent changes in the Antarctic and Southern Ocean, 1949-96, *Annals of Glaciology* **27**, 553-559.

Kennedy, A.D. (1993) Water as a limiting factor in the Antarctic terrestrial environment: a biogeographical synthesis, *Arctic and Alpine Research* **25**, 308-315.

Kennedy, A.D. (1994) Simulated climate change: a field manipulation study of polar microarthropod community response to global warming, *Ecography* **17**, 131-140.

Kennedy, A.D. (1995a) Antarctic terrestrial ecosystem response to global environmental change, *Annual Review of Ecology and Systematics* **26**, 683-704.

Kennedy, A.D. (1995b) Temperature effects of passive greenhouse apparatus in high-latitude climate change experiments, *Functional Ecology* **9**, 340-350.

Kennedy, A.D. (1995c) Simulated climate change: are passive greenhouses a valid microcosm for testing the biological effects of environmental perturbations? *Global Change Biology* **1**, 29-42.

Kennedy, A.D. (1996) Antarctic fellfield response to climate change: a tripartite synthesis of experimental data, *Oecologia* **107**, 141-150.

Kettles, I.M., Tarnocai, C. and Bauke, S.D. (1997) Predicted permafrost distribution in Canada under a climate warming scenario, *Geological Survey of Canada, Current Research* **1997-E**, 109-115.

Kiernan, K. and McConnell, A. (2002) Glacier retreat and melt-lake expansion at Stephenson Glacier, Heard Island World Heritage Area, *Polar Record* **38 (207)**, 297-308.

King, J.C. (1994) Recent climate variability in the vicinity of the Antarctic Peninsula, *International Journal of Climatology* **14**, 357-369.

King, J.C. and Harangozo, S.A. (1998) Climate change in the western Antarctic Peninsula since 1945: observations and possible causes, *Annals of Glaciology* **27**, 571-575.

King, J.C., Turner, J., Marshall, G.J., Connally, W.M. and Lachlan-Cope, T.A. (2003) Antarctic Peninsula climate variability and its causes as revealed by analysis of instrumental records, In E. Domack, A. Burnett, A. Leventer, P. Convey, M. Kirby and R. Bindschadler (eds.), *Antarctic Peninsula Climate Variability: Historical and Palaeoenvironmental Perspectives*, Antarctic Research Series, Vol. 79, American Geophysical Union, Washington, D.C., pp. 17-30.

Larter, R.D. and Vanneste, L.E. (1995) Relict subglacial deltas on the Antarctic Peninsula outer shelf, *Geology* **23**, 33-36.

Lawley, B., Ripley, S., Bridge, P. and Convey, P. (2004) Molecular analysis of geographic patterns of eukaryotic diversity in Antarctic soils, *Applied and Environmental Microbiology* **70**, 5963-5972.

Lorius, C., Jouzel, J., Ritz, C., Merlivat, L. and Barkov, N.I. (1985) A 150,000-year climate record from Antarctic ice, *Nature* **316**, 591-596.

Lubin, D. and Frederick, J. (1991) The ultraviolet radiation environment of the Antarctic Peninsula: the roles of ozone and cloud cover, *Journal of Applied Meteorology* **30**, 478-493.

Lyons, W.B., Laybourn-Parry, J., Welch, K.A. and Priscu, J.C. (2006) Antarctic lake systems and climate change, in D.M. Bergstrom, P. Convey, and A.H.L. Huiskes (eds.), *Trends in Antarctic Terrestrial and Limnetic Ecosystems: Antarctica as a Global Indicator*, Springer, Dordrecht (this volume).

McGraw, J.B. and Day, T.A. (1997) Size and characteristics of a natural seed bank in Antarctica, *Arctic and Alpine Research* **29**, 213-216.

Morris, E.M. and Vaughan, D.G. (2003) Spatial and temporal variation of surface temperature on the Antarctic Peninsula and the limit of variability of ice shelves, In E. Domack, A. Burnett, A. Leventer, P. Convey, M. Kirby and R. Bindschadler (eds.), *Antarctic Peninsula Climate Variability: Historical and Palaeoenvironmental Perspectives*, Antarctic Research Series, Vol. 79, American Geophysical Union, Washington, D.C., pp. 61-68.

Nicolai, V., and Droste, M. (1984) The ecology of *Lancetes claussi* (Müller) (Coleoptera, Dytiscidae), the Subantarctic water beetle of South Georgia, *Polar Biology* **3**, 39-44.

Noon, P.E, Birks, H.J.B., Jones, V.J. and Ellis-Evans, J.C. (2001) Quantitative models for reconstructing catchment ice-extent using physical-chemical characteristics of lake sediments, *Journal of Palaeolimnology* **25**, 375-392.

Ó Cofaigh, C., Pudsey, C.J., Dowdeswell, J.A. and Morris, P. (2002) Evolution of subglacial bedforms along a paleo-ice stream, Antarctic Peninsula continental shelf, *Geophysical Research Letters* **29**, 1199, doi: 10.1029/2001GLO14488.

Pudsey, C.J. and Evans, J. (2001) First survey of Antarctic sub-ice shelf sediments reveals mid-Holocene ice shelf retreat, *Geology* **29**, 787-790.

Pugh, P.J.A. and Davenport, J. (1997) Colonisation vs. disturbance: the effects of sustained ice-scouring on intertidal communities, *Journal of Experimental Marine Biology and Ecology* **210**, 1-21.

Quayle, W.C., Peck, L.S., Peat, H., Ellis-Evans, J.C. and Harrigan, P.R. (2002) Extreme responses to climate change in Antarctic lakes, *Science* **295**, 645.

Quayle, W.C, Convey, P., Peck, L.S., Ellis-Evans, J.C., Butler, H.G., and Peat, H.J. (2003) Ecological responses of maritime Antarctic lakes to regional climate change, In E. Domack, A. Burnett, A. Leventer, P. Convey, M. Kirby and R. Bindschadler (eds.), *Antarctic Peninsula Climate Variability: Historical and Palaeoenvironmental Perspectives*, Antarctic Research Series, Vol. 79, American Geophysical Union, Washington, D.C. pp. 159-170.

Roberts, L. (1989) Does the ozone hole threaten antarctic life? *Science* **244**, 288-289.

Rozema, J., Noordijk, A.S.J., Broekman, R.A., van Beem, A., Meijkamp, B.M., de Bakker, N.V., van de Staaij, J.W.M., Stroetenga, M., Bohncke, S.J.P., Konert, M., Kars, S., Peat, H.J., Smith, R.I.L. and Convey, P. (2001) (Poly)phenolic compounds in pollen and spores of Antarctic plants as indicators of solar UV-B: a new proxy for the reconstruction of past solar UV-B? *Plant Ecology* **154**, 9-25.

Sabburg, J. and Wong, J. (2000) The effect of clouds on enhancing UVB irradiance at the earth's surface: a one year study, *Geophysical Research Letters* **27**, 3337-3340.

Santas, R., Koussoulaki, A. and Häder, D.-P. (1997) In assessing biological UV-B effects, natural fluctuations of solar radiation should be taken into account, *Plant Ecology* **128**, 93-97.

Scambos, T., Hulbe, C. and Fahnestock, M. (2003) Climate-induced ice shelf disintegration in the Antarctic Peninsula, In E. Domack, A. Burnett, A. Leventer, P. Convey, M. Kirby and R. Bindschadler (eds.), *Antarctic Peninsula Climate Variability: Historical and Palaeoenvironmental Perspectives*, Antarctic Research Series, Vol. 79, American Geophysical Union, Washington, D.C., pp. 79-92.

Searles, P.S., Kropp, B.R., Flint, S.D. and Caldwell, M.M. (2001) Influence of solar UV-B radiation on peatland microbial communities of southern Argentina, *New Phytologist* **152**, 213-221.

Skvarca, P. and De Angelis, H. (2003) Impact assessment of regional climatic warming on glaciers and ice shelves of the northwestern Antarctic Peninsula, In E. Domack, A. Burnett, A. Leventer, P. Convey, M. Kirby and R. Bindschadler (eds.), *Antarctic Peninsula Climate Variability: Historical and Palaeoenvironmental Perspectives*, Antarctic Research Series, Vol. 79, American Geophysical Union, Washington, D.C., pp. 69-78.

Skvarca, P., Rack, W., Rott, H. and Ibarzábal y Donángelo, T. (1998) Evidence of recent climatic warming on the eastern Antarctic Peninsula, *Annals of Glaciology* **27**, 628-632.

Smith R.I.L. (1987) The bryophyte propagule bank of Antarctic fellfield soils, *Symp. Biol. Hungarica* **35**, 233-245.

Smith, R.I.L. (1988) Recording bryophyte microclimate in remote and severe environments, In J.M. Glime (ed.), *Methods in Bryology*, Hattori Botanical Laboratory, Nichinan, Japan, pp. 275-284.

Smith, R.I.L. (1990) Signy Island as a paradigm of biological and environmental change in Antarctic terrestrial ecosystems, In K.R. Kerry and G. Hempel (eds.), *Antarctic Ecosystems, Ecological Change and Conservation*, Springer-Verlag, Berlin, pp. 32-50.

Smith, R.I.L. (1993) The role of bryophyte propagule banks in primary succession: case-study of an Antarctic fellfield soil, in: J. Miles and D.W.H. Walton, (eds), *Primary Succession on Land*, British Ecological Society Special Publication, Blackwell Scientific Publications, Oxford, UK, pp 55-78.

Smith, R.I.L. (1994) Vascular plants as indicators of regional warming in Antarctica, *Oecologia* **99**, 322-328.

Smith, R.I.L. (2001) Plant colonization response to climate change in the Antarctic, *Folia Facultatis Scientiarum Naturalium Universitatis Masarykiana Brunensis, Geographia*, **25**, 19-33.

Smith, R.I.L. and Coupar, A.M. (1986) The colonization potential of bryophyte propagules in Antarctic fellfield soils, *CNFRA* **58**, 189-204.

Smith, V.R. (2002) Climate change in the subantarctic: An illustration from Marion Island, *Climate Change* **52**, 345-357.

Smith, V.R. and Steenkamp, M. (1990) Climatic change and its ecological implications at a subantarctic island, *Oecologia* **85**, 14-24.

Sømme, L. (1995) *Invertebrates in Hot and Cold Arid Environments*, Springer-Verlag, Berlin.

Stevens, M.I. and Hogg, I.D. (2006) The molecular ecology of Antarctic terrestrial and limnetic invertebrates and microbes, in D.M. Bergstrom, P. Convey, and A.H.L. Huiskes (eds.), *Trends in Antarctic Terrestrial and Limnetic Ecosystems: Antarctica as a Global Indicator*, Springer, Dordrecht (this volume).

Strathdee, A.T., Bale, J.S., Block, W.C., Coulson, S.J., Hodkinson, I.D. and Webb, N.R. (1993) Effects of temperature elevation on a field population of *Acyrthosiphon svalbardicum* (Hemiptera: Aphididae) on Spitsbergen, *Oecologia* **96**, 457-465.

Thompson, D.W.J. and Solomon, S. (2002) Interpretation of recent Southern Hemisphere climate change, *Science* **296**, 895-899.

Turner, J., Colwell, S.R. and Harangozo, S. (1997) Variability of precipitation over the coastal western Antarctic Peninsula from synoptic observations, *Journal of Geophysical Research* **102**, 13999-14007.

Turner, J., King, J.C., Lachlan-Cope, T.A. and Jones, P.D. (2002) Recent temperature trends in the Antarctic, *Nature* **418**, 291-292.

Tweedie, C.E. and Bergstrom, D.M. (2000) A climate change scenario for surface air temperature at subantarctic Macquarie Island, in W. Davison, C. Howard-Williams and P.A. Broady (eds.), *Antarctic Ecosystems: Models for Wider Ecological Understanding*, New Zealand Natural Sciences, Christchurch, pp. 272-281.

Vaughan, D.G and Doake, C.S.M. (1996) Recent atmospheric warming and retreat of ice shelves on the Antarctic Peninsula, *Nature* **379**, 328-331.

Vaughan, D.G., Marshall, G. J., Connolley, W. C., King, J. C. and Mulvaney, R. (2001) Devil in the detail, *Science* **293**, 1777-1779.

Vaughan, D. G., Marshall, G.J., Connolley, W.M., Parkinson, C.L., Mulvaney, R., Hodgson, D.A., King, J.C., Pudsey, C.J. and Turner, J. (2003) Recent rapid regional climate warming on the Antarctic Peninsula, *Climate Change* **60**, 243-274.

Voytek, M.A. (1990) Addressing the biological effects of decreased ozone on the Antarctic environment, *Ambio* **19**, 52-61.

Walther, G.-R., Post, E., Convey, P., Parmesan, C., Menzel, M., Beebee, T.J.C., Fromentin, J.-M., Hoegh-Guldberg, O. and Bairlein, F. (2002) Ecological responses to recent climate change, *Nature* **416**, 389-395.

Walton, D.W.H. (1982) The Signy Island terrestrial reference sites: XV. Microclimate monitoring, 1972-74, *British Antarctic Survey Bulletin* **55**, 111-126.

Walton, D.W.H. (1984) The terrestrial environment, in R.M. Laws (ed), *Antarctic Ecology* Volume 1. Academic Press, London, pp 1-60.

Walton, D.W.H., Vincent, W.F., Timperley, M.H., Hawes, I. and Howard-Williams, C. (1997) Synthesis: polar deserts as indicators of change, in: W.B. Lyons, C. Howard-Williams and I. Hawes (eds.) *Ecosystem Processes in Antarctic Ice-Free Landscapes*. Balkema, Rotterdam, pp. 275-279.

Whinam, J. and Copson, G. (2006) *Sphagnum* moss: an indicator of climate change in the sub-Antarctic, *Polar Record* **42**, 43-49.

Wynn-Williams, D.D. (1994) Potential effects of ultraviolet radiation on Antarctic primary terrestrial colonizers: cyanobacteria, algae, and cryptogams, *Antarctic Research Series* **62**, 243-257.

Wynn-Williams, D.D. (1996) Response of pioneer soil microalgal colonists to environmental change in Antarctica, *Microbial Ecology* **31**, 177-188.

13. ANTARCTIC LAKE SYSTEMS AND CLIMATE CHANGE

W. B. LYONS
Byrd Polar Research Center
The Ohio State University
Columbus, OH 43210-1002 USA
lyons.142@osu.edu

J. LAYBOURN-PARRY
Natural Sciences, Keele University
Keele, Straffordshire ST5 5BG UK
j.laybourn-parry@natsci.keele.ac.uk

K. A. WELCH
Byrd Polar Research Center
The Ohio State University
Columbus, OH 43210-1002 USA
welch.189@osu.edu

J. C. PRISCU
Department of Land Resources and Environmental Sciences
Montana State University
Bozeman, MT 59717 USA
jpriscu@montana.edu

Introduction

Approximately 98% of the Antarctic continent is currently ice-covered and, except for subglacial lakes (Priscu and Christner 2004), possesses no liquid water environments. In contrast, the other 2% contains an extraordinary array of aquatic

273

D.M. Bergstrom et al. (eds.), Trends in Antarctic Terrestrial and Limnetic Ecosystems, 273–295.

environments including ice-covered freshwater and saline lakes and ephemeral streams. Lakes are found mostly in coastal, ice marginal regions in the Antarctic (Doran et al. 1994). Because of the great differences in mean-annual temperature related to the locations of these aquatic systems (Fig. 1), the temporal extent of ice covers on the lakes varies greatly. There are lakes that are ice-covered for part of each year, while there are perennially ice-covered systems in the dry valleys region of Southern Victoria Land where only 'moats' (ie ice-free littoral zones) form during the warmest summers. Because ice cover greatly influences both the physical and biological processes occurring within lakes, the extent and thickness of ice cover is an extremely important parameter in the biogeochemistry of Antarctic lakes (Wharton et al. 1993, Fritsen and Priscu 1999).

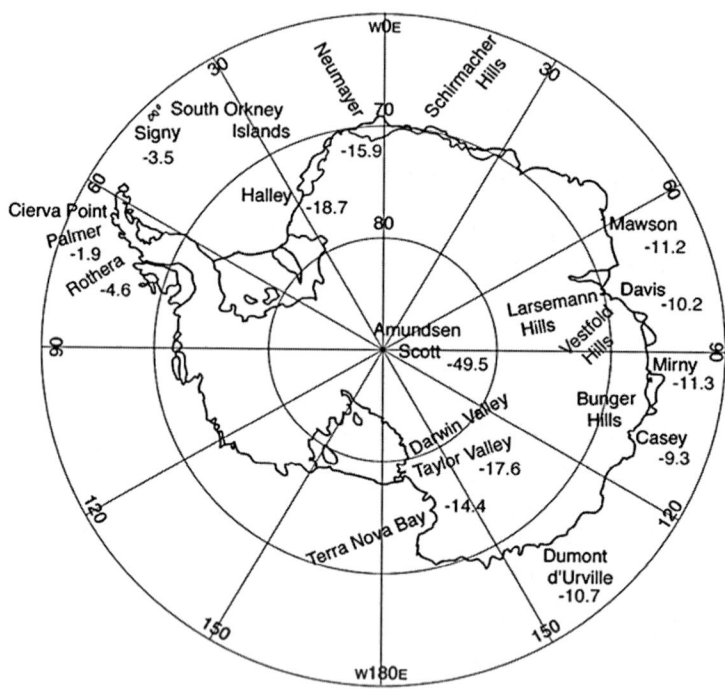

Figure 1. Mean annual temperatures (°C) for various stations in Antarctica.

Paleolimnological investigations demonstrate that climate variations have greatly impacted the extent of ice cover, in addition to the overall hydrology of Antarctic lakes throughout the Holocene back into the Last Glacial Maximum (LGM) (Wilson 1964, Bird et al. 1991, Björck et al. 1996, Fulford-Smith and Sikes 1996, Gore et al. 1996, Lyons et al. 1998b, Hendy 2000). Much of the paleolimnological work from Antarctic lakes has recently been summarized by Hodgson et al. (2004) and Doran et al. (2004) and will not be repeated here. Some of the earlier work on Antarctic lakes illustrated its value in delineating changes in the hydrologic response to climate (Wilson 1964). The integration of paleolimnological data with more recent observations has provided important insights into how these lakes respond to climate change (eg Gibson and Burton 1996, allied with the work of Roberts et al. 2001 in the Vestfold Hills and Poreda et al. 2004, allied with Hendy 2000 in Taylor Valley, Victoria Land). Understanding the impact of climate on the hydrologic balance of Antarctic lakes, and in turn, the influence of hydrologic changes on the overall ecology of these systems is a major challenge to Antarctic limnologists. This will be especially true in the future period of anthropogenically-induced climate change.

Kejna (2003) has recently reviewed the air temperature records from 34 stations around the Antarctic continent. The data span from 1958-2000 for 21 stations and for 1981-2000 for all the stations. In general, warming has occurred on the Antarctic Peninsula and in interior West Antarctica, with Faraday Station, on the west coast of the Peninsula having increased at a rate of 0.67°C per decade over the period 1958-2000. Many of the coastal stations in East Antarctica also demonstrated an increase in temperature over the longer time interval. However, since 1981, many regions of the continent, especially in East Antarctica, have shown a cooling. For example, Casey shows a 0.82°C per decade decrease (Kejna 2003). There has also been a weakening of the warming rate on the Peninsula during the last 20 years. Because Antarctic lakes are so sensitive to both increases and decreases in temperature (ie Wilson 1964, Gibson and Barton 1996, Foreman et al. 2004), the direct monitoring of Antarctic lakes provides an excellent sentinel of climate change, especially as climate impacts the local hydrologic cycle.

Climate variation and change are not the only factors affecting Antarctic lakes. The activities of mammal and bird populations also exert considerable influence on the physical, chemical and biological evolution of many Antarctic limnetic systems. Human activities can also be important in certain situations. For example, Heywood Lake, on Signy Island, has undergone eutrophication in the last 30 years because of input of nutrients from an expanding fur seal population within the catchment (Butler 1999). This has led to increased microbial abundance and changes in the structure of the ecosystem, with phytoplankton taxa more typical of polluted waters, as opposed to the more oligotrophic waters that existed in the 1970s and early 1980s (Butler 1999) before the fur seal population explosion. Eutrophication has led to longer periods of lake anoxia during winter thermal stratification and changes in seasonal biological patterns within the lake. Recent work in the Larsemann Hills has demonstrated that lakes impacted by human activities such as grey water and human waste discharge and even rock crushing by tracked vehicles have enhanced nutrient and total dissolved solid loads (Kaup and Burgess 2002). A comparison of human impacted catchments with catchments with little direct human activity indicates that the levels of dissolved nitrogen compounds were generally much higher in the human-influenced catchments. Salinities were also up to an order of magnitude higher in the human impacted catchments (Kaup and Burgess 2002).

Monitoring the response of lake dynamics to changing climate has long been recognized as an important task (Wilson 1981). The linkage between changing climate and lake dynamics becomes even more complicated in lakes that are influenced by direct human impacts. Most investigations of Antarctic lakes have been conducted over limited time periods (1-3 years). Although these short-term studies have been extremely important in establishing base-line conditions, determining the taxa that are present, and understanding biogeochemical processes in the lakes, they are not conducive to establishing long-term limnological trends. Our paper focuses on studies resulting in long-term data comparisons that have produced information about biological and/or physical and chemical trends thought to be driven by changing climatic parameters. This focus greatly limits the resources available to compare long-term trends that do exist and historic information on Antarctic limnology because it eliminates numerous one-time studies of specific limnetic systems in various parts of the continent. (We differentiate between long-term and historic data by defining 'long-term' as relatively continuous records through time and 'historic' as data collected with time gaps between collections). In addition, this paper will not address trends in the epishelf lakes that exist on the continent especially in the Bunger Hills and Schirmacher Oasis regions (Bormann and Fritzsche 1995, Doran et al. 2000, Gibson and Andersen 2002). For information on epishelf lakes, see Gibson et al. this volume.

Because of the important differences in the climate regimes within the Antarctic, we have separated our discussion into geographic regimes that have been used previously (Convey 2001).

Maritime Antarctica

This region includes the western side of the Antarctic Peninsula and island groups such as the South Orkney Islands.

SIGNY ISLAND, SOUTH ORKNEY ISLANDS

Signy Island (60°43'S) contains 17 lakes, some of which have been studied since 1961 and monitored since the 1980s (Butler 1999). Hence, the data from these lakes represent the longest continuous lacustrine records in the Antarctic (Quayle et al. 2002) and a significantly important resource for the assessment of the response of Antarctic limnetic systems to climate change. The Antarctic Peninsula region has warmed at a rate of 3.7±1.6°C per century over the last ~50 years (Vaughan et al. 2003). Other subantarctic islands are also exhibiting significant warming in conjunction with large variations in precipitation over the last 40-50 years (Bergstrom and Chown 1999).

The hydrologic patterns on Signy Island have also changed as the climate has warmed. The water budget of the lakes and streams in late summer is becoming increasingly dominated by direct precipitation rather than snowmelt and snow pack is presently lost earlier in the season (Noon et al. 2002). The monitoring record of these lakes going back to 1980 presents an important record on how these systems respond to warming. In the last 15 years, water temperature has increased 0.9°C in response to the atmospheric warming and the mean number of ice-free days has increased by 31 days from 1980 to the mid-1990s (Quayle et al. 2002). Historic photography indicates that the overall ice cover on the island has receded by about 45% since 1951, again in response to the 1°C temperature increase between 1950 and 1990 (Quayle et. al 2002). The reduction of the ice cover caused a subsequent increase in the fluxes of dissolved reactive phosphate, ammonium and inorganic carbon into the lakes that led to increased summer primary production and a four-fold increase in chlorophyll *a* during the last 15 years.

The geochemical evolution of the Signy Island lakes in response to climatic warming and local ice cap/glacier retreat can be compared to longer term trends exhibited by lakes in the Glacier Bay region, Alaska (Engstrom et al. 2000). The Alaskan lakes eventually became more dilute, more acidic and enriched in dissolved organic carbon with time as the glaciers retreated. These changes are thought to be due to the changes in watershed vegetation and soil conditions. However, the largest changes in the Alaskan lakes have been demonstrated to take place after approximately 100 years (Engstrom et al. 2000). Due to the differences in the evolutionary stage (ie time), the two systems cannot presently be adequately compared, but one might predict similar evolution of the Signy Island lakes over time. The Arctic data imply that in the next 50 years, if warming continues, the Signy Island lakes will evolve from Ca^{2+}/HCO_3^- rich systems to more dilute ones dominated by Na^+/Cl^- as chemical weathering of the watershed decreases (Engstrom et al. 2000). Like the Glacier Bay, Alaska lakes, the Signy Island lakes have also received higher nutrient inputs as the climate warmed (Quayle et al. 2002). The dissolved organic carbon concentrations in the Alaskan lakes also increased during this warming period, while more continental Antarctic lakes, such as those in the McMurdo Dry Valleys, have not (Lyons et al. 2001).

ANTARCTIC PENINSULA

A shorter lacustrine record from Otero Lake at Cierva Point on the Antarctic Peninsula (64°09'S) tells a similar story. Increases in lake temperature have led to increases in bacterial abundances and increases in conductivity (ie salinity) and nutrient loading in this shallow lake (maximum depth 3m) over a 5-year period between samplings (Mataloni et al. 1998, Izaguirre et al. 2001). Care must be taken to interpret these changes in terms of climate change alone, because the low sampling density could simply be recording inter-annual variability inherent to the lake itself.

Continental Antarctica

Continental Antarctica is defined to include the entire East Antarctic landmass and the east coast of the Antarctic Peninsula (Convey 2001). As shown in Fig. 1, this region has a much more extreme climate.

EAST ANTARCTICA

Schirmacher Oasis and the Gurber Mountains

The first scientific investigations of the Schirmacher Oasis occurred in 1938/39 by German scientists. This region has since been visited numerous times by both former Soviet Union and German scientists and a comprehensive investigation of the hydrology and limnology of the lakes was conducted in the 1980s. Aerial photographs revealed that the snow/ice cover in the eastern portion of the Schirmacher Oasis had been greatly reduced from 1939 to 1984 (Bormann and Fritzsche 1995). The ice covers of the lakes did not exceed 2m thick and many lakes were ice-free for a period in January/February (Hermichen et al. 1985). However, the comparison of the strandline of Lake Untersee, a large (10 km^2) freshwater lake in the area (71°20'S), using photographs from 1939, 1961 and 1982, showed no change in the shape of the lake, indicating there had been little change in the hydrologic balance of the lake over that time (Hermichen et al. 1985). Wand et al. (1997) revisited the lake in 1991/92 and observed that the lake was now physically and chemically stratified. The depth of Lake Untersee had also increased to 94m with a newly observed anoxic hypolimnion of 14m where methanogenesis was occurring and methane turnover rates have now been measured (V. Samarkin, pers. comm.). The increase in depth between 1981-1991 corresponds with similar increases of inflow into the McMurdo Dry Valleys lakes (Chinn 1993) and in Lake Wilson in Southern Victoria Land (see below), inferring a large geographical scale temperature warming, which increased melt water input over this time period.

Larsemann Hills

The lakes of the Larsemann Hills are thought to have formed by isostatic uplift following the last major glaciation some 9000 years ago (Gillieson 1991). The Larsemann Hills consist of two large peninsulas and a collection of offshore islands that support approximately 150 lakes and ponds. Geological evidence suggests that this area was largely ice-free during the Wisconsin Glaciation at 25 000 yr BP (Burgess et al. 1994). More recent work demonstrated that the Broknes Peninsula has been ice-free for at least 45 000 years (Hodgson et al. 2001). Changes in precipitation are reflected in the water chemistry of lakes. During dry periods, water levels fall and salinity increases resulting in biological changes in the lakewater. From an analysis of diatom frustules in the sedimentary record, it is possible to infer changes in salinity over time using transfer functions (Verleyen et al. 2003). The application of these diatom-based models has been applied to large saline and brackish lakes in addition to small freshwater lakes (Verleven et al. 2003). Diatom records reveal that the degree of species turnover in the oligosaline lakes is low when compared to much more saline lakes, indicating that changes in the diatoms assemblages were forced by environmental variables like depth and depth-related variables. This finding allows for the reconstruction of past climate conditions in this coastal oasis, parts of which were ice-free during the last glaciation. Based on an analysis of Pup Lagoon (salinity 0.5‰, maximum depth 4.6m), the climate on the Prydz Bay coast between 5800 and 5500 yr BP was typified by open water during spring and summer warm conditions, while between 5500 and 2750 yr BP, the extent of sea-ice duration increased, with only a few months when the sea was ice-free, much as it is today (Verleyen et al. 2004).

Vestfold Hills
The Vestfold Hills, 69°30'S, extend over an area of $410km^2$ and supports approximately 150 lakes, with fresh water lakes and saline lakes making up about 8% and 2% of the total area, respectively (Pickard 1986). These lakes have been extensively studied by the Australian Antarctic Program over the last 15-20 years. This suite of lakes offers an exciting prospect to limnologists because of their variety and abundance. Freshwater lakes include supraglacial and proglacial lakes, while the saline lakes range from brackish to hypersaline (6 x seawater) and include both permanently stratified (meromictic) and seasonally mixed (monomictic) lakes. As deglaciation occurred after the LGM, the Vestfold Hills landmass rebounded and marine inlets became cut off and isolated from the ocean. The Vestfold Hills and many of its lakes were formed by isostatic rebound after the last major glaciation (Adamson and Pickard 1986, Zwart et al. 1998). Many of the saline lakes arose from pockets of marine water that were trapped in closed basins. A review of a range of published palaeolimnological data indicates that eustatic sea level changes occurred more rapidly than isostatic rebound (Hodgson et al. 2004) resulting in marine incursions. A good example of this is Watts Lake, which has a surface area of $0.38km^2$ and a current salt content of 2.4%. Paleolimnological data indicate that the lake was a marine basin that was later flushed by freshwater, mainly from neighbouring Crooked Lake (Pickard et al. 1986, Gore et al. 1996). Evidence also indicates that a number of the saline lakes were initially freshwater which accumulated in closed basins formed by isostatic uplift. Following a rise in sea level, the basins were invaded by seawater, and subsequently when the sea level descended to below the edge of the lakes, they underwent evolution into the saline lakes we see today (Roberts and McMinn 1998, 1999, Cromer et al. 2005). Examples are Ace Lake, one of the most studied lakes in the Vestfold Hills and Anderson Lake, both of which are now saline meromictic lakes. Other lakes such as Crooked Lake, the largest ($9km^2$) freshwater lake in the area, were formed in deep basins by rivers fed by glacier melt behind a glacial impoundment that released incrementally, leaving four distinct shorelines (Gore et al. 1996). By applying transfer functions to the diatom paleolimnological record of the Vestfold Hills and other eastern Antarctic coastal oases, it is possible to determine changes in the climate during the Holocene, particularly changes in net precipitation (Roberts and McMinn 1998).

The saline lakes of the Vestfold Hills derived from relict seawater have become highly saline via cryoconcentration (Zwart et al. 1998). These lakes possess truncated marine food webs. Nutrient (N and P) levels tend to be higher in the saline lakes compared with the freshwater ones, the latter of which rank as ultra-oligotrophic in the global context. While none of the saline systems appears to be nutrient limited, the freshwater systems do periodically become nutrient depleted. Chlorophyll *a* concentrations in all of these lakes are not high, so others factors such as light and/or possibly trace elements may also limit primary production. Data from summer and early winter in Deep Lake, Crooked Lake, Ace Lake and Lake Druzhby show that both P and N concentrations rise during the winter as photosynthesis decreases (Kerry et al. 1977, Bayliss et al. 1997, Bell and Laybourn-Parry 1999, Henshaw and Laybourn-Parry 2002). There is considerable inter-annual variation in productivity and phytoplankton biomass in the lakes that have been subjected to long-term study. For example in Ace Lake, the primary production in January 1997 reached a maximum of 8.7mg $C.l^{-1}.hr^{-1}$ (Bell and Laybourn-Parry 1999) under ice-free conditions, while under ice-covered conditions in 1993 and 1999, the maximum rates were 0.68 and 7.0mg $C.l^{-1}.hr^{-1}$, respectively (Laybourn-Parry and Perriss 1995, Laybourn-Parry et al. 2002).

The light and temperature regime are greatly influenced by the extent of lake ice cover. Recent detailed remotely-sensed data taken every 5 min for an entire summer in the water column of Crooked Lake illustrate this extremely well (Palethorpe et al. 2004). Any climate change affecting lake ice retention could have a major impact on phytoplankton growth and nutrient dynamics in these lakes. For example, comparative work on the Vestfold Hills lakes reveals that differences in light regime and lake basin morphology have impact significantly the autotrophic organisms within the lakes (Laybourn-Parry and Bayliss 1996). Clearly, changing climatic parameters that affect the ice cover and/or the depth of these lakes can affect the distribution and diversity of the organisms within these lakes (Laybourn-Parry et al. 2002). In years when the ice cover is lost, the water temperatures of the lakes increase substantially (Laybourn-Parry et al. 2002). In the hypersaline lakes of the area, warming and increased melt can lead to dilution and changes in bacterial species composition (James et al. 1994). All of these studies show unequivocally that local climatic variability has an important role on the structure and function of these ecosystems. Data from Crooked Lake and Ace Lake spanning more than 10 years demonstrate that there are very significant inter-annual variations in biological activity related to local meteorological conditions that control the length of time ice cover occurs. These delicate lake ecosystems appear to respond very quickly to small climatic perturbations. While the paleolimnological record reveals long-term responses to cycles of precipitation and evaporation during the Holocene (Roberts and McMinn 1998, 1999, Hodgson et al. 2004), there are insufficient data to determine the fine scale response of these lakes to the present phase of climate warming as is possible from the continuous data set from the McMurdo Dry Valleys.

Northern Victoria Land
Recent research on streams and lakes in ice-free areas near Terra Nova Bay (~74°40'), when compared to similar data obtained 12 years earlier, indicates little geochemical variation and change in the biogeochemistry of these aquatic systems (Borghini and Bargagli 2004). Lake levels have decreased during this 12-year period despite a warming trend that has occurred in the region. This decrease in lake levels with increasing atmospheric temperatures has been argued to be related to an increase in the depth to the active layer in the permafrost and increased drainage and

evaporation loss (Borghini and Bargagli 2004). Although there was little change in the concentration of most major ions over the 12-year period, Borghini and Bargagli (2004) did observe a significant (as much as ~30-fold) increase in sulfate. The establishment of a long-term monitoring network in the region will enhance the overall understanding of climatic change on limnetic systems in this part of Antarctica (Borghini and Bargagli 2004), particularly since these data can be compared to McMurdo Dry Valleys data collected during the same time period.

McMurdo Dry Valleys/Southern Victoria Land

The lakes in the McMurdo Dry Valleys, Southern Victoria Land, have been investigated extensively since the 1957/58 IGY. There are important environmental data sets from these lakes dating back to the early 1960s (eg Angino et al 1962, Angino and Armitage 1963, Goldman et al. 1967). Long-term records of lake levels exist back to the early 1970s based on New Zealand investigations (Chinn 1993). In 1993, The U.S. National Science Foundation established a Long-Term Ecological Research (LTER) site in the McMurdo Dry Valleys, in which the primary area of scientific activity is Taylor Valley at ~78°S (Fig. 2). The LTER mandate requires long-term data collection (ie monitoring) from a series of locations in order to establish information on primary production, biomass concentration and species distribution (biodiversity) and how these parameters change over time (see http://www.lternet.edu/). The program was established to address ecological processes over long periods and across broad scales. As part of McMurdo Dry Valleys LTER, a suite of coordinated physical, chemical and biological parameters have been measured in Lakes Fryxell, Hoare and Bonney on a routine basis since 1993/94. General results of this study are summarised in two synthesis volumes (Priscu 1998, 1999) and the most recent work on the lakes is presented in Lyons et al. (2001). One of the major aspects of the McMurdo Dry Valleys LTER is the interpretation of long-term trends of lake biogeochemistry and ecology as they relate to climatic and, hence, hydrologic change. The reader is referred to the McMurdo Dry Valleys LTER website (http://www.mcmlter.org) as space here will not allow us to discuss every aspect of the data sets, however two aspects of the work are presented below.

Decadal Variation in Biogeochemistry of McMurdo Dry Valleys Lakes

Between 1986 and 2000, Taylor Valley (Fig. 2), in the McMurdo Dry Valleys cooled at the rate of 0.7°C per decade with the summer and autumn cooling being the most dramatic, 1.2° and 2.0°C per decade, respectively (Doran et al. 2002). As mentioned above, Taylor Valley contains three major lakes: Bonney, Fryxell and Hoare (Fig. 2). Lake Bonney contains two basins separated by a sill at ~13 m depth, which impedes exchange of deep saline water between the basins (Spigel and Priscu 1998). These lakes differ substantially in their salinities, geochemistry and physical stabilities (Lyons et al. 1998a, Spigel and Priscu 1998) with the hypolimnia of Lake Bonney being hypersaline, Lake Fryxell being brackish and Lake Hoare being fresh. The differences in chemistry (ie both salinity and ionic composition) of these lakes are related to their age and past responses to climate change (Matsubaya et al. 1979, Lyons et al. 1998b, Poreda et al. 2004).

As mentioned above, detailed multi-annual biogeochemical monitoring of these lakes began during the 1993/94 field season and continues to the present. Because Lake Hoare is the freshest of these lakes, its geochemical composition responds rapidly to climate changes. For example, during the decade-long period of relatively minimal glacier melt and low stream flow within Taylor Valley (McKnight et al. 1999), the Cl⁻ levels in Lake Hoare's surface water (top 12m) varied dramatically from 1993 to 2002 (Fig. 3). The decrease of freshwater input into the lake brought about by lower summer temperatures led to ~3x increase in Cl⁻ in the surface waters. This is more clearly seen in Fig. 4 where each set of connected data represents the changes in Cl⁻ concentrations (mean of 5 and 6m depth samples) in both Lakes Hoare and Fryxell during each summer. Note that the greatest variations are in Lake Fryxell. The hydrologic balance of Lake Fryxell may be most impacted by climatic variation because it has the highest number of streams and because of its greater surface area to depth ratio compared to the other lakes in Taylor Valley (Hood et al. 1998). Gibson and Burton (1996) have described in detail the increase in salinity in Antarctic lakes brought about by 'negative water balance'. Variations in the surface water salinities of the Taylor Valley lakes are clearly related to summer temperatures (Doran et al. 2002).

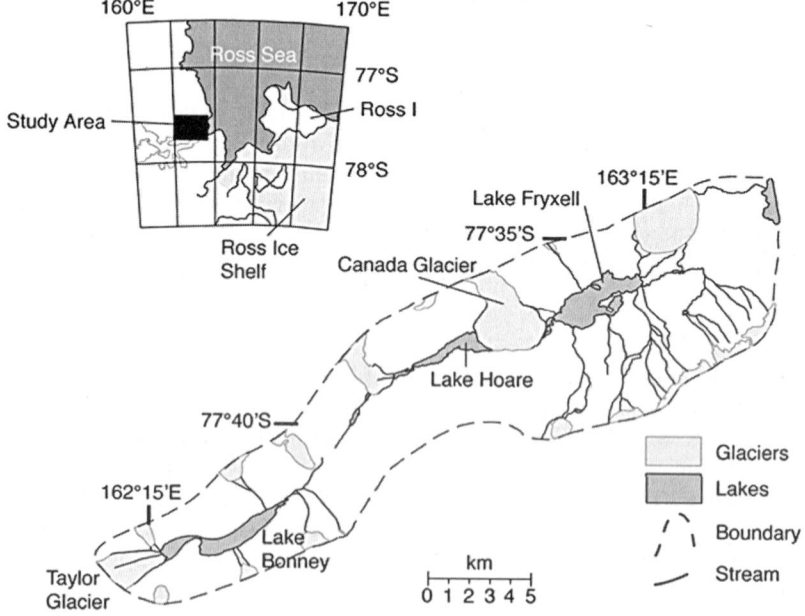

Figure 2. Map of Taylor Valley, Antarctica.

The 2001/02 summer was the warmest one on record in Taylor Valley since continuous records began in 1985 (P. Doran personal communication). Extremely high melt water flows were observed in late December 2001 into January 2002 when annual stream discharge for gauged sites in the McMurdo Dry Valleys was 33 times that of the previous year. Seeps or subsurface discharges also appeared where they had not been previously observed (Lyons et al. 2005). This increased melt water inflow resulted in drastic decreases in surface water Cl⁻ concentrations and increased lake levels (Figs. 3 and 4). The Cl⁻ concentration in the top 2m of water in Lake Hoare (5-6m depth) from the Spring 2002 are the lowest observed during the continuous decadal record (Fig. 3), while the values for Lake Fryxell approach the concentrations observed in the Spring of 1993 (Fig. 4). Chlorophyll *a* values in Lake Hoare in the surface waters do not appear to be greatly affected by the changes in salinity (Fig. 5). However, the year after the 2001/02 freshwater input, the highest chlorophyll *a* values measured over the previous decade were observed at depths below 15m (Fig. 5). These relatively high values may reflect the increase in irradiance resulting from a thinner ice cover and nutrient advection from elevated stream flows the previous year. Nutrient flux to the deeper water could also result from the freeze back of the large moat that formed during the warm year, a process that has been show to occur in Lake Fryxell (Miller and Aiken 1996).

Multi-decadal variation in biogeochemistry of McMurdo Dry Valleys lakes
From ~1970 until the documented cooling began in the McMurdo Dry Valleys starting in the late 1980s, lake levels had been rising and lake ice-covers had been thinning (Chinn 1993, Wharton et al. 1993, Spigel and Priscu 1998) indicating increased melt water input and suggesting local warming. An extrapolation by Chinn (1993) of his records to initial measurements made by Scott's party in 1903 suggested a rise of lake level in Lake Bonney of 13.5m from 1903 to 1990. The lake level increases were most pronounced in Lake Bonney and in Lake Vanda in Wright Valley to the north of Taylor Valley (Chinn 1993) and are probably related to the steep basins occupied by these lakes. In 1993/94, the water in Lake Fryxell, between the bottom of the ice cover and ~11m below the ice cover, was determined to be younger than 25 years using isotopic techniques (Hood et al. 1998). Both data sets (Chinn 1993, Hood et al. 1998) indicate that large volumes of water were emplaced into Lake Fryxell from the 1960s into the late 1980s. We can evaluate the impact of these long-term variations (the increase in lake level followed by the decrease in lake level in the 1990s discussed previously) on the biogeochemistry of Lake Fryxell by utilizing both LTER data and a series of other data collected as early as the 1960s.

Fig. 6 shows the dissolved SO_4^{2-} concentration vs. depth for Lake Fryxell. The data show a distinct decrease in SO_4^{2-} concentration at depth in the anoxic hypolimnion of the lake, especially between the early 1960s and the early 1980s (Fig. 6). The decrease continues from the early 1980s until the beginning of the McMurdo Dry Valleys LTER (1993), especially at the mid-depths of the water column. This decrease in SO_4^{2-} over time is interpreted to be due to microbial sulfate reduction, which we have now quantified using experimental isotopic measurements

(J. Priscu, unpubl. data). What is driving this process? We speculate that increased primary production in the upper water column of the lake may lead to an increased flux of organic matter to the hypolimnion of the lake, reducing oxygen levels and producing favourable redox conditions for sulfate reduction to occur (Lee et al. 2004). We have investigated the variation of stream flow into the lake to evaluate this contention. Since our gauged stream flow data in the Lake Fryxell basin only commence in the 1990/91 summer, we extended the Fryxell data using flows from another stream, which has a longer gauge record. McKnight et al. (1999) recognized a very strong relationship between the flow records from the Lake Fryxell basin streams to that from the nearby Onyx River, in Wright Valley. The Onyx River is the longest river in the Antarctic and has been monitored by New Zealand scientists since 1972, and became part of the McMurdo Dry Valleys LTER monitoring responsibilities in the mid 1990s. Using the relationship quantified by McKnight et al. (1999) for these two basins, we 'back casted' the Lake Fryxell basin melt water input into the lake itself (Figs. 7 and 8). We then related the mixolimnetic primary production rate in Lake Fryxell data to the extended stream flows. Where there are data available for the analysis (n=9), the R=0.68 and using a one-year lag the R=0.56. These data include McMurdo Dry Valleys LTER data and earlier work by Vincent (1981).

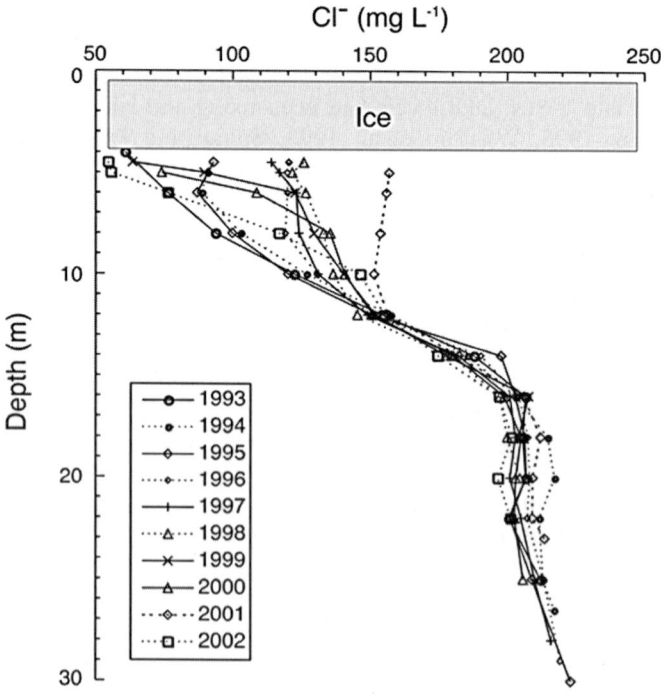

Figure 3. Cl⁻ concentrations in Lake Hoare for the November sampling event of each summer in the period 1993-2002. These data represent conditions in the lake before significant melting and stream flow have occurred.

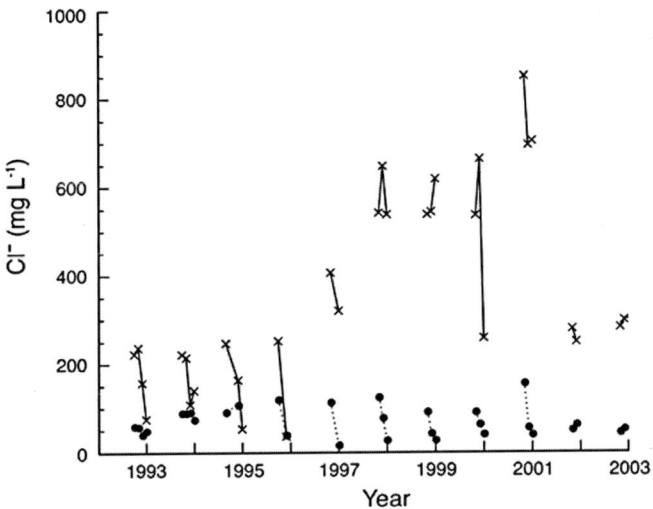

Figure 4. Each set of connected points represents the changes in Cl⁻ concentrations (average of 5 and 6m depth samples) in Lakes Hoare (circles) and Fryxell (crosses) during each summer.

Results from this exercise, in concert with the fact that Priscu (1995) has shown that primary production in Lake Fryxell is nutrient limited, implies that increased inflow enhances primary production in the lake, a contention that has been corroborated by the recent work of Foreman et al. (2004). The rapid sulfate reduction that occurred in the bottom of Lake Fryxell in the 1960s-1980s may result in part from an increase in freshwater input that increased the stability of the pycnocline reducing oxygen transport to deeper waters and the accompanying higher nutrient input to the photic zone that increased the rate of organic matter production by phytoplankton. At the 77mM Cl⁻ concentration in the water column, the SO_4^{2-} concentration in Lake Fryxell decreased from 4.77 to 1.40mM over a ~38yr period for a sulfate reduction rate of ~0.09mM.yr⁻¹. Howes and Smith (1990) first noted this decrease in sulfate in their studies during the mid-1980s. Using radiolabel techniques they measured values in the sediments as high as ~170nM.g⁻¹.yr (assuming a wet sediment density of 1.5g.cm⁻³). Using their water column data at 18m, we estimate an annual sulfate loss of 6.4kM at the sediment water interface. This should be considered a maximum value, however, as the rates in the shallower portion of the hypolimnion are possibly lower. These authors also argued that the present day stream input of SO_4^{2-} is currently balanced by microbial reduction and sediment burial.

These rates of reduction are within the same order of magnitude reported in the water column of Burton Lake in the Vestfold Hills (Franzmann et al. 1988). We contend that the subsequent decreased rate of sulfate reduction in Lake Fryxell in the 1990s (Fig. 6) is associated with a decrease in nutrient input and thickening of the ice cover, the latter of which decreased under-ice irradiance and reduced phytoplankton primary production (phytoplankton production that has been shown to be highly light dependent in this lake and all others in the McMurdo Dry Valleys (Lizotte and Priscu 1992, Priscu et al. 1999, Morgan-Kiss et al. 2006). This exercise establishes a clear linkage between climatic variability (both warming and cooling), the hydrologic cycle within the basin and the biogeochemistry of the lake.

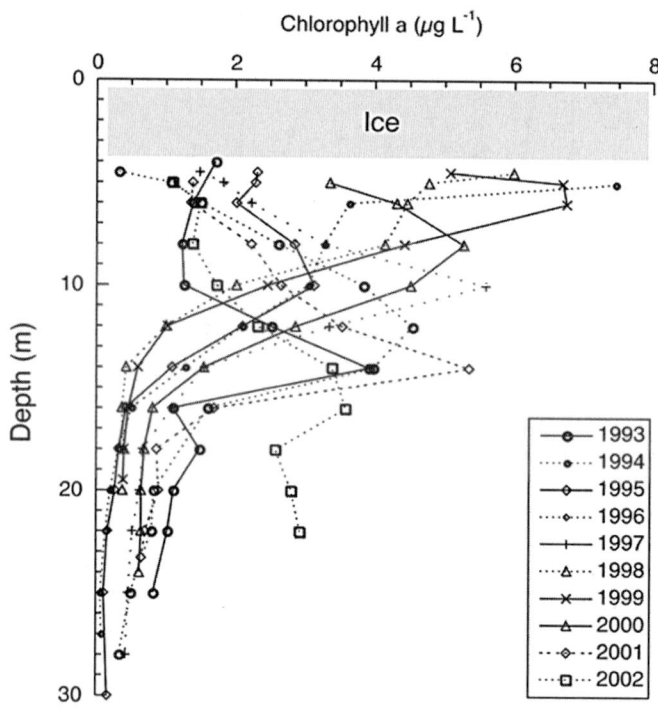

Figure 5. Chlorophyll a concentrations vs. depth in Lake Hoare from 1993-2002. These data are from the same samples as shown in Figure 3.

Subsequent investigations after the very high flow year of 2001/02, confirm the importance of increased stream inflow in enhancing primary production and increasing autotrophic biomass in the Taylor Valley lakes (Foreman et al. 2004). Depth integrated primary production increased by a factor of 5 in Lake Fryxell during the period of increased stream flows in late December 2001, and during the following spring, primary production increased in Lake Bonney and directly under the ice cover in Lake Fryxell (Foreman et al. 2004). The Lake Fryxell primary production was more than 6 times higher than the average value at this depth over the previous decade. Depth integrated chlorophyll-*a* values increased ~150% and

~50% in the lobes of Lake Bonney, but showed little change in Lake Fryxell (Foreman et al. 2004). The importance of warming and increased glacial melt with the enhanced nutrient input and the decreased ice cover thickness are extremely important in driving the production of these systems.

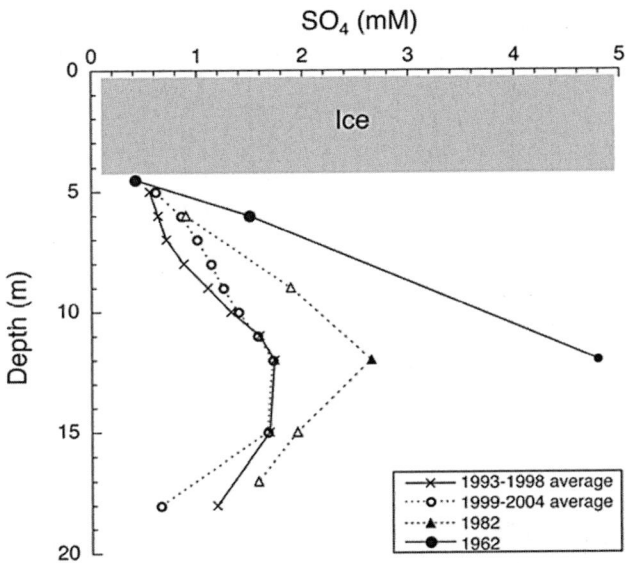

Figure 6. Dissolved SO_4^{2-} concentration vs. depth in Lake Fryxell at selected times. The 1962 and 1982 data are from Angino et al. (1962) and Green et al. (1989), respectively. The McMurdo Dry Valley LTER (MCM LTER) data have been averaged over the first six (MCM-I) years and then the next five (MCM-II) years of the LTER program. It should be noted that there is scatter in the LTER data at every depth, in part, because the depth of the lake is not fixed in time.

'Most Southern Lake'

Lake Wilson is a permanently ice-covered lake in the Darwin Valley at 80°S. It was first examined in 1975 by C. Hendy and then later in 1993 by Webster et al. (1996). These authors noted that the lake level had risen 25m between the two samplings, a 54% increase in volume, which is quasi-synchronous with the period of lake level rise and ice cover thinning noted for the McMurdo Dry Valleys lakes (eg Chinn 1993, Wharton et al. 1993). Calculations based on the Cl⁻ profiles of the lake suggest that the latest increased flux of meltwater into Lake Wilson began between ~1960-1970 (Webster et al. 1996). Water temperatures had slightly decreased in both surface layer and the deepest most saline portion of the lake over this period. Unfortunately there are no biological data from 1975 to compare to the most recent data, however, the on-going Latitudinal Gradient Project (LGP) (http://www.lgp.aq) undertaken by New Zealand and Italian scientist includes plans to return to the

Darwin Valley area in 2006 to 2008 field seasons. New data from Lake Wilson should be forthcoming at this time and extremely useful, when compared to the McMurdo Dry Valleys LTER and Terra Nova Bay data, in ascertaining the impact of climate variation on the aquatic systems along the Victoria Land Coast.

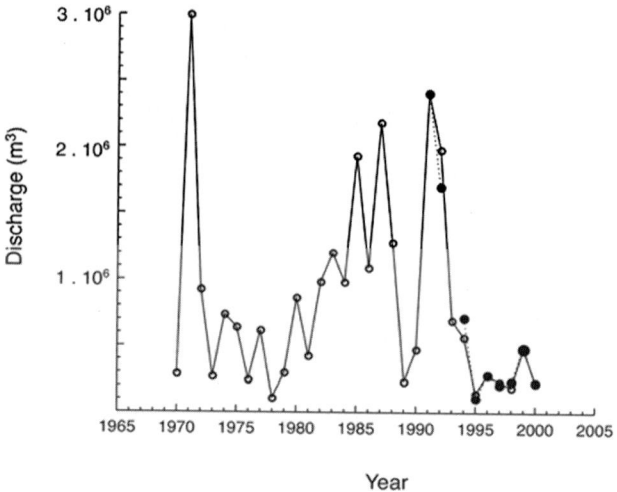

Figure 7. Stream discharge in the Lake Fryxell Basin. Measured values (dark circles) represent the sum of the volume of stream discharge from all of the gauged streams in the Lake Fryxell basin. The modelled discharge (open circles) in the Fryxell basin was back-cast by running a regression between the measured discharge in the Onyx River and the Fryxell basin.

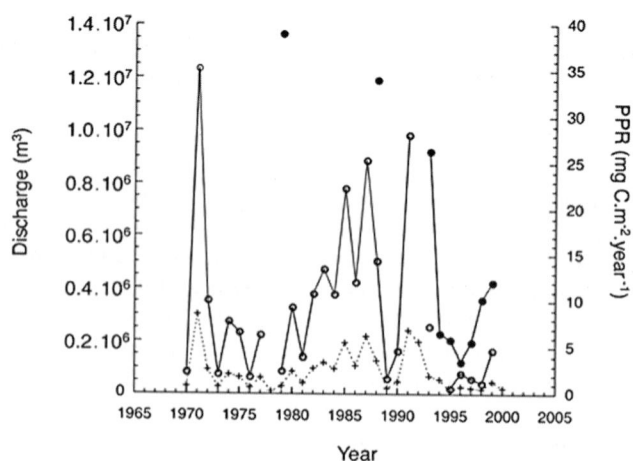

Figure 8. Mixolimnetic primary production rate in Lake Fryxell data (dark circles) compared to the measured Onyx River discharge (open circles) and modelled (crosses) Fryxell Basin stream discharge in the basin.

Conclusions

As pointed out recently by Convey (2001) for Antarctic terrestrial ecosystems, '*The identification of biological consequences of climate change in the Antarctic is still at a relatively early stage.*' However, it is evident that these delicate ecosystems appear to respond rapidly to climate change (Laybourn-Parry 2003, Doran et al. 2002). The causal response of aquatic ecosystems in the Antarctic to climate drivers can only be established, however, though the integration of meteorological, hydrological and geochemical information collected in concert with the biological observations. In addition, long-term data collection and monitoring are needed to better establish the linkages between climate and ecological change, with the best examples in the Antarctic to date being the BAS work on Signy Island and the LTER investigations in the McMurdo Dry Valleys (Quayle et al. 2002, Doran et al. 2002). Clearly there is a need to establish on-going, long-term lake investigations at other locations in the subantarctic and Antarctica, if a real understanding of how natural and anthropogenically induced climate changes affect ecosystems in these locations. Remote sensing technology offers polar limnologists a powerful tool to achieve this, allowing data to be gained at short-term intervals and accessed by satellite (Palethorpe et al. 2004). Although much more information is now available on the potential mechanisms that may influence Antarctic climate (eg Fischer et al. 2004, Goosse et al. 2004, Liu et al. 2004, Nielsen et al. 2004), the linkages between meteorological observations and hydrological and ecosystem changes over the short term are not always straightforward (McConchie and Hawke 2002, Gibson et al. 2002, Welch et al. 2003). In fact, recent work suggests that large-scale climate indices might predict ecological processes better than local weather variables (Hallett et al. 2004). Antarctic lakes offer ideal systems to study the impact of climate change on ecological structure and function because many of them are poised at the melting point of ice during the summer. We are only just beginning to understand the relationships between climate change and ecosystem function in these systems, mainly because of the long-term data sets that are not being collected in various regions of Antarctica. We must continue our long-term collection of a consortium of ecosystem parameters if we are to develop predictive tools of how ecosystems will respond in future climate scenarios.

Clearly, we are not at the stage of developing a quantitative, predictive model or even one that completely qualifies the response of Antarctic aquatic systems to climate change as has been accomplished in Arctic systems (eg Hobbie et al. 1999). However, based on the Signy Island and Taylor Valley results from long-term integrative physical, chemical and biological monitoring, we can begin to develop a scenario of what might transpire in these systems as global warming continues. Much of this is straightforward common sense, but it is substantiated by the work of Doran et al. (2002), Quayle et al. (2002) and Foreman et al. (2004).

1. As summer warming occurs there will be longer melt seasons with larger volumes of water flowing into the lakes. This in turn will increase lake volumes and lake ice covers will be lost completely for longer periods. (Note that albedo changes by precipitation increases and permafrost effects as noted by Borghini and Baragagli (2004) may complicate this).

2. The increased in-flow of ice/snow melt will increase the nutrient input into lakes and the loss of ice cover will increase exchange between the surface waters of the lake and the atmosphere.

3. Increased primary production due to enhanced nutrient input, increased CO_2 input via atmospheric exchange, increased light availability and higher water temperatures will ensue.

4. Fresh water advection into the upper water layers will enhance physical stability of the deeper lakes. Lack of ice cover in the shallow lakes will lead to increased wind induced mixing.

5. Increased primary production will lead to a higher organic carbon flux into the deeper waters of the lakes, which in the more hydrodynamically stable lakes may lead to anoxia and/or changes in the overall biogeochemical dynamics of the lake (eg increases in iron, nitrate or sulfate reduction).

6. Increased eutrophy in concert with changes in the physical mixing regime will influence the biological structure of the lakes and their biodiversity.

In short, small variations in climate, even those judged subtle by temperate region standards, will have a profound effect on Antarctic lake ecosystems. Continued monitoring together with experimental investigation is needed throughout the Antarctic continent to ascertain how variations in climate are driving ecosystem change.

Acknowledgements

We are deeply grateful to all our McMurdo Dry Valleys LTER colleagues of the last 12 years who have help collect and analyse lake samples, especially Rob Edwards, Craig Wolf and Christine Foreman. We especially thank Diane McKnight and Peter Doran for their thoughtful insights into the workings of the MCM lakes. We especially thank the limnological field teams for the collection and analysis of the chlorophyll data and Diane McKnight for the stream flow measurements presented within. We thank Tim Fitzgibbon for his editing skills. The paper benefited greatly from the reviews of Dr. W.F. Vincent and an anonymous reviewer. We thank them greatly. The data in Fig. 1 are from BAS (http://www.antarctica.ac.uk/index.php) and MCM-LTER (http://huey.colorado.edu/LTER/). This work was supported by NSF grants: OPP-9211773 and OPP-9813061.

References

Adamson, D.A. and Pickard, J. (1986) Cainozoic history of the Vestfold Hills, in J. Pickard (ed) *Antarctic Oasis*, Academic Press, Sydney pp 63-97.

Angino, E.E. and Armitage, K.B. (1963) A geochemical study of Lakes Bonney and Vanda, Victoria Land, Antarctica, *Journal of Geology* **71**, 89-95.

Angino, E.E., Armitage, K.B. and Tash, J.C. (1962) Chemical stratification in Lake Fryxell, Victoria Land, Antarctica, *Science* **138**, 34-36.

Bayliss, P., Ellis-Evans, J.C. and Laybourn-Parry, J. (1997) Temporal patterns of primary production in a large ultra-oligotrophic Antarctic freshwater lake, *Polar Biology* **18**, 363-370.

Bell, E.M. and Laybourn-Parry, J. (1999) Annual plankton dynamics in an Antarctic saline lake, *Freshwater Biology* **41**, 507-519.

Bergstrom, D.M. and Chown, S.L. (1999) Life at the front: History, ecology and change on southern ocean islands, *Trends in Ecology and Evolution* **14**, 472-477.

Bird, M.I., Chivas, A.R., Radnell, C.J. and Burton, H.R. (1991) Sedimentological and stable-isotope evolution of lakes in the Vestfold Hills, Antarctica, *Palaeogeography, Palaeoclimatology, Palaeoecology* **84**, 109-130.

Björck, S., Olsson, S., Ellis-Evans, C., Håkansson, H., Humlum, O. and de Lirio, J.M. (1996) Late Holocene palaeoclimatic records from lake sediments on James Ross Island, Antarctica, *Palaeogeography, Palaeoclimatology, Palaeoecology* **121**, 195-220.

Borghini, F. and Bargagli, R. (2004) Changes of major ion concentrations in melting snow and terrestrial waters from northern Victoria Land, Antarctica, *Antarctic Science* **16**, 107-115.

Bormann, P. and Fritzsche, D. (eds.) (1995) *The Schirmacher Oasis, Queen Maud Land, East Antarctica, and its surroundings*, Justus Perthes Verlag Gotha, Germany, 448 pp.

Burgess, J.S., Spate, A.P. and Shevlin, J. (1994) The onset of deglaciation in the Larsemann Hills, Eastern Antarctica. *Antarctic Science* **6**, 491-495.

Butler, H.G. (1999) Seasonal dynamics of the planktonic microbial community in a maritime Antarctic lake undergoing eutrophication, *Journal of Plankton Research* **21**, 2393-2419.

Chinn, T.J. (1993) Physical hydrology of the dry valley lakes, in W.J. Green and E.I. Friedmann (eds.), *Physical and Biogeochemical Processes in Antarctic Lakes*, Antarctic Research Series, Vol. 59, American Geophysical Union, pp. 1-51.

Convey, P. (2001) Terrestrial ecosystem response to climate changes in the Antarctic. In G.-R. Walther, C.A. Burga and P.J. Edwards (eds.), *"Fingerprints" of climate change - adapted behaviour and shifting species ranges*, Kluwer, New York, pp 17-42.

Cromer, L., Gibson, J.A.E., Swadling, K.M. and Ritz, D.A. (2005) Faunal microfossils: indicators of Holocene ecological change in a saline Antarctic lake, *Palaeogeography, Palaeoclimatology, Palaeoecology*, **221**, 83-97.

Doran, P.T. J. C. Priscu, W. Berry Lyons, R. D. Powell, D. T. Andersen and R.J. Poreda. (2004) Paleolimnology of Ice-covered Environments. pp. 475-507, In R. Pienitz, M. Douglas and John Smol (eds) *Long-term environmental change in Arctic and Antarctic lakes*, Kluwer Academic Publishers, Dordrecht, The Netherlands.

Doran, P.T., Priscu J.C., Lyons, W.B., Walsh, J.E., Fountain, A.G., McKnight, D.M., Moorhead, D.L., Virginia, R.A., Wall, D.H., Clow, G.D., Fritsen, C. H., McKay, C. P., Parsons, A.N. (2002) Antarctic climate cooling and terrestrial ecosystem response. *Nature* **415**, 517-520.

Doran, P.T., Wharton, Jr., R.A. and Lyons, W.B. (1994) Paleolimnology of the McMurdo Dry Valleys, Antarctica, *Journal of Paleolimnology* **10**, 85-114.

Doran, P.T, Wharton, Jr., R.A., Lyons, W.B., Des Marais, D.J. and Andersen, D.T. (2000) Sedimentology and geochemistry of a perennially ice-covered epishelf lake in Bunger Hills Oasis, East Antarctica, *Antarctic Science* **12**, 131-140.

Engstrom, D.R., Fritz, S.C., Almendinger, J.E. and Juggins, S. (2000) Chemical and biological trends during lake evolution in recent deglaciated terrain, *Nature* **408**, 161-166.

Fischer, H., Traufetter, F., Oerter, H., Weller, R. and Miller, H. (2004) Prevalence of the Antarctic Circumpolar Wave over the last two millennia recorded in Dronning Maud Land ice, *Geophysical Research Letters* **31**, L08202, doi:10.1029/2003GL019186.

Foreman, C.M., Wolf, C.F. and Priscu, J.C. (2004) Impact of episodic warming events on the physical, chemical and biological relationships of lakes in the McMurdo Dry Valleys, Antarctica, *Aquatic Geochemistry* **10**, 239-268.

Franzmann, P.D., Skyring, G.W., Burton, H.R. and Deprez, P.P. (1988) Sulfate reduction rates and some aspects of the limnology of four lakes and a fjord in the Vestfold Hills, Antarctica, *Hydrobiologia* **165**, 25-33.

Fritsen, C.H. and Priscu J.C. (1999) Seasonal change in the optical properties of the permanent ice cover on Lake Bonney, Antarctica: Consequences for lake productivity and dynamics. *Limnology and Oceanography* **44**, 447-454.

Fulford-Smith, S.P. and Sikes, E.L. (1996) The evolution of Ace Lake, Antarctica, determined from sedimentary diatom assemblages, *Palaeogeography, Palaeoclimatology, Palaeoecology* **124**, 73-86.

Gibson, J.A.E. and Andersen, D.T. (2002) Physical structure of epishelf lakes of the southern Bunger Hills, East Antarctica, *Antarctic Science* **14**, 253-262.

Gibson, J.A.E. and Burton, H.R. (1996) Meromictic Antarctic lakes as records of climate change: the structure of Ace and Organic Lakes, Vestfold Hills, Antarctica, *Proceedings of the Royal Society of Tasmania* **130**, 73-78.

Gibson, J.A.E., Gore, D.B. and Kaup, E. (2002) Algae River: an extensive drainage system in the Bunger Hills, East Antarctica, *Polar Record* **38**, 141-152.

Gibson, J.A.E., Wilmotte, A., Taton, A., Van De Vijver, B., Beyens, L. and Dartnall, H.J.G. (2006) Biogeographic trends in Antarctic lake communities, in D.M. Bergstrom, P. Convey, and A.H.L. Huiskes (eds.), *Trends in Antarctic Terrestrial and Limnetic Ecosystems: Antarctica as a Global Indicator*, Springer, Dordrecht (this volume).

Gillieson, D.S. (1991) An environmental history of two freshwater lakes in the Larsemann Hills, Antarctica. *Hydrobiologia* **214**, 327-331.

Goldman, C.R., Mason, D.T. and Hobbie, J.E. (1967) Two Antarctic desert lakes, *Limnology and Oceanography* **12**, 295-310.

Goosse, H., Masson-Delmotte, V., Renssen, H., Delmotte, M., Fichefet, T., Morgan, V., van Ommen, T., Khim, B.K. and Stenni, B. (2004) A late medieval warm period in the Southern Ocean as a delayed response to external forcing? *Geophysical Research Letters* **31**, L06203, doi:10.1029/2003GL019140.

Gore, D.B., Pickard, J., Baird, A.S. and Webb, J.A. (1996) Glacial Crooked Lake, Vestfold Hills, East Antarctica, *Polar Record* **32**, 19-24.

Green, W.J., Gardner, T.J., Ferdelman, T.G., Angle, M.P., Varner, L.C. and Nixon, P. (1989) Geochemical processes in the Lake Fryxell Basin (Victoria Land, Antarctica), *Hydrobiologia* **172**, 129-148.

Hallett, T.B., Coulson, T., Pilkington, J.G., Clutton-Brock, T.H., Pemberton, J.M. and Grenfell, B.T. (2004) Why large-scale climate indices seem to predict ecological processes better than local weather *Nature* **430**, 71-75.

Hendy, C.H. (2000) Late Quaternary lakes in the McMurdo Sound region of Antarctica, *Geografiska Annaler* **82A**, 411-432.

Henshaw, T. and Laybourn-Parry, J. (2002) The annual patterns of photosynthesis in two large, freshwater, ultra-oligotrophic Antarctic lakes, *Polar Biology* **25**, 744-752.

Hermichen, W.-D., Kowski, P. and Wand, U. (1985) Lake Untersee, a first isotope study of the largest freshwater lake in the interior of East Antarctica, *Nature* **315**, 131-133.

Hobbie, J.E., Peterson, B.J., Bettez, N., Deegan, L., O'Brien, W.J., Kling, G.W., Kipphut, G.W., Bowden, W.B. and Hershey, A.E. (1999) Impact of global change on the biogeochemistry and ecology of an Arctic freshwater system, *Polar Research* **18**, 207-214.

Hodgson, D.A., Noon, P.E. Vyverman, W., Bryant, C.L., Gore, D.B., Appleby, P, Gilmore, M., Verleyen, E. Sabbe, K., Jones, V.J., Ellis-Evans, J.C. and Wood, P.B. (2001) Were the Larsemann Hills ice-free through the Last Glacial Maximum? *Antarctic Science,* **13**, 440-454.

Hodgson, D.A., Doran, P.T., Roberts, D. and McMinn, A. (2004) Palaeolimnological studies from the Antarctic and sub-Antarctic islands, in R. Pienitz, M.S.V. Douglas, and J. P. Smol (eds) *Long-term Environmental Change in Arctic and Antarctic lakes.* Springer, Hague pp 419-474.

Hood, E.M., Howes, B.L. and Jenkins, W.J. (1998) Dissolved gas dynamics in perennially ice-covered Lake Fryxell, Antarctica, *Limnology and Oceanography* **43**, 265-272.

Howes, B.L. and Smith, R.L. (1990) Sulfur cycling in a permanently ice-covered amictic Antarctic lake, Lake Fryxell, *Antarctic Journal of the U.S.* **25**, 230-233.

Izaguirre, I., Mataloni, G., Allende, L., Vinocur, A. (2001) Summer fluctuations of microbial planktonic communities in a eutrophic lake-Cierva Point, Antarctica, *Journal of Plankton Research* **23**, 1095-1109.

James, S.R., Burton, H.R., McMeekin, T.A. and Mancuso, C.A. (1994) Seasonal abundance of *Halomonas merdiana, Halomonas subglaciescola, Flavobacterium gondwanense* and *Flavobacterium salegens* in four Antarctic lakes, *Antarctic Science* **6**, 352-332.

Kaup, E. and Burgess, J.S. (2002) Surface and subsurface flows of nutrients in natural and human impacted lake catchments on Broknes, Larsemann Hills, Antarctica, *Antarctic Science* **14**, 343-352.

Kejna, M. (2003) Trends of air temperature of the Antarctic during the period 1958-2000, *Polish Polar Research* **24**, 99-126.

Kerry, K.R., Grace, D.R., Williams, R. and Burton, H.R. (1977) Studies on some saline lakes of the Vestfold Hills, Antarctica, in G. Llano (ed.), *Adaptations within Antarctic Ecosystems*, Smithsonian Inst., Washington, pp. 839-858.

Laybourn-Parry, J. (2003) Polar limnology, the past, the present and the future, in A.H.L. Huiskes, W.W.C. Gieskes, J. Rozema, R.M.L. Schorno, S.M. van der Vies and W.J. Wolff (eds.) *Antarctic Biology in a Global Context*, Backhuys Publishers, Leiden, pp. 321-329.

Laybourn-Parry, J. and Bayliss, P. (1996) Seasonal dynamics of the planktonic community in Lake Druzhby, Princess Elizabeth Land, Eastern Antarctica, *Freshwater Biology* **35**, 57-67.

Laybourn-Parry, J. and Perriss, S.J. (1995) The role and distribution of the autotrophic ciliate *Mesodinium rubrum* (*Myrionecta rubra*) in three Antarctic saline lakes, *Archiv Fur Hydrobiologie* **135**, 179-194.

Laybourn-Parry, J., Quayle, W. and Henshaw, T. (2002) The biology and evolution of Antarctic saline lakes in relation to salinity and trophy, *Polar Biology* **25**, 542-552.

Lee P.A., Mikucki, Foreman, C.M., Priscu, J. C., DiTullio, G.R., Riseman, S.F., de Mora, S.J., Wolf C. F. and Kester L. (2004) Thermodynamic constraints on microbially mediated processes in lakes of the McMurdo Dry Valleys, Antarctica. *Geomicrobiology Journal,* **21**, 1-17.

Liu, J., Curry, J.A. and Martinson, D.G. (2004) Interpretation of recent Antarctic sea ice variability, *Geophysical Research Letters* **31**, L02205, doi:10.1029/2003GL018732.

Lizotte, M.P. and Priscu, J.C. (1992) Photosynthesis irradiance relationships in phytoplankton from the physically stable water column of a perennially ice-covered lake (Lake Bonney, Antarctica), *Journal of Phycology* **28**, 179-185.

Lyons, W.B., Tyler, S.W., Wharton Jr., R.A., Vaughn, B, McKnight, D.M. (1998b) A late Holocene desiccation of Lake Hoare, and Lake Fryxell, McMurdo Dry Valleys, Antarctica, *Antarctic Science* **10**, 245-254.

Lyons, W.B., Welch, K.A., Neumann, K., Toxey, J.K., McArthur, R. and Williams, C. (1998a) Geochemical linkages among glaciers, streams and lakes within the Taylor Valley, Antarctica, in J.C. Priscu (ed.) *Ecosystem Dynamics in a Polar Desert: The McMurdo Dry Valleys, Antarctica*, Antarctic Research Series, Vol. 72, American Geophysical Union, Washington, D.C. pp. 77-92.

Lyons, W.B., Welch, K.A., Priscu, J.C., Laybourn-Parry, J., Moorhead, D., McKnight, D.M., Doran, P.T. and Tranter, M. (2001) The McMurdo Dry Valleys Long-Term Ecological Research Program: A new understanding of the biogeochemistry of the Dry Valley lakes: A review, *Polar Geography* **25**, 202-217.

Lyons, W.B., Welch, K.A., Carey, A.E., Wall, D.H., Virginia, R.A., Fountain, A.G., Doran, P.T., Csathó, B.M. and Tremper, C.M, (2005) Groundwater seeps in Taylor Valley, Antarctica: An example of a subsurface melt event, *Annals of Glaciology* **40**, 200-206.

Mataloni, G., Tesolín, G. and Tell, G. (1998) Characterization of a small eutrophic Antarctic lake (Otero Lake, Cierva Point) on the basis of algal assemblages and water chemistry, *Polar Biology* **19**, 107-114.

Matsubaya, O., Sakai, H., Torii, T., Burton, H. and Kerry, K. (1979) Antarctic saline lakes-stable isotopic ratios, chemical compositions, and evolution, *Geochimica et Cosmochimica Acta* **43**, 7-25.

McConchie, J.A. and Hawke, R.M. (2002) Controls on streamflow generation in a meltwater stream, Miers Valley, Antarctica and implications for the debate on global warming, *Journal of Hydrology (NZ)* **41**, 77-103.

McKnight, D.M., Niyogi, D.K., Alger, A.S., Bomblies, A., Conovitz, P.A. and Tate, C.M. (1999) Dry Valley streams in Antarctica: Ecosystems waiting for water, *BioScience* **49**, 985-995.

Miller, L.G. and Aiken, G.R., 1996, Effects of glacial meltwater inflows and moat freezing on mixing in an ice-covered antarctic lake as interpreted from stable isotope and tritium distributions, *Limnology and Oceanography* **41**, 966-976.

Morgan-Kiss, R.M., Priscu, J.C., Pocock, T., Gudynaite-Savitch L., and Huner N.P.A (2006) Adaptation and acclimation of photosynthetic microorganisms to permanently cold environments, *Microbial and Molecular Biology Reviews* **70**, 222-252.

Nielsen, S.H.H., Koç, N. and Crosta, X. (2004) Holocene climate in the Atlantic sector of the Southern Ocean: Controlled by insolation or oceanic circulation? *Geology* **32**, 317-320.

Noon, P.E., Leng, M.J., Arrowsmith, C., Edworthy, M.G. and Strachan, R.J. (2002) Seasonal observations of stable isotope variations in a valley catchment, Signy Island, South Orkney Islands, *Antarctic Science* **14**, 333-342.

Palethorpe, B., Hayes-Gill, B., Crowe, J., Sumner, M., Crout, N., Foster, M., Reid, T., Benford, S., Greenhalgh, C. and Laybourn-Parry, J. (2004) Real-time physical data acquisition through a remote sensing platform on a polar lake, *Limnology and Oceanography: Methods* **2**, 191-201

Pickard, J. (1986) Antarctic oasis, Davis Station and the Vestfold Hills, in J. Pickard (ed.), *Antarctic Oasis: Terrestrial Environment and History of the Vestfold Hills*, Academic Press, Sydney, pp. 1-19.

Pickard, J., Adamson, D.A. and Heath, C.W. (1986) The evolution of Watts Lake, Vestfold Hills, East Antarctica, from marine inlet to freshwater lake, *Palaeogeography, Palaeoclimatology, Palaeoecology* **53**, 271-288.

Poreda, R.J., Hunt, A.G., Lyons, W.B. and Welch, K.A. (2004) The helium isotopic chemistry of Lake Bonney, Taylor Valley, Antarctica: Timing of Late Holocene climate change in Antarctica, *Aquatic Geochemistry* **10**, 353-371.

Priscu, J.C. 1995. Phytoplankton nutrient deficiency in lakes of the McMurdo dry valleys, Antarctica, *Freshwater Biology* **34**, 215-227.

Priscu, J.C. (ed.) (1998) *Ecosystem Dynamics in a Polar Desert: The McMurdo Dry Valleys, Antarctica*, Antarctic Research Series, Vol. 72, American Geophysical Union, Washington, D.C., 369pp.

Priscu, J.C. (1999) Life in the valley of the "dead", *BioScience* **49**, 959.

Priscu, J.C. and Christner B. (2004) Earth's Icy Biosphere, in A.T. Bull (ed.) *Microbial Diversity and Prospecting*, ASM Press, Washington, D.C. p. 130-145.

Priscu, J.C., Wolf, C. F., Takacs, C.D., Fritsen, C.H., Laybourn-Parry, J., Roberts E.C., and Lyons W.B. (1999) Carbon transformations in the water column of a perennially ice-covered Antarctic Lake, *Bioscience* **49**, 997-1008.

Quayle, W.C, Peck, L.S., Peat, H., Ellis-Evans, J.C. and Harrigan, P.R. (2002) Extreme responses to climate change in Antarctic lakes, *Science* **295**, 645.

Roberts, D. and McMinn, A. (1998) A weighted-averaged regression and calibration model for inferring lake water salinity from fossil diatoms assemblages in saline lakes of the Vestfold Hills: a new tool for interpreting Holocene lake histories in Antarctica, *Journal of Paleolimnology* **19**, 99-113.

Roberts, D. and McMinn, A. (1999) Palaeohydrological modeling of Ace Lake, Vestfold Hills, Antarctica. *The Holocene* **9**, 401-408.

Roberts, D., van Ommen, T.D., McMinn, A., Morgan, V., and Roberts, J.L. (2001) Late-Holocene East Antarctic climate trends from ice-core and lake-sediment proxies, *The Holocene* **11**, 117-120.

Spigel, R.H. and Priscu, J.C. (1998) Physical limnology of the McMurdo Dry Valleys lakes, in J.C. Priscu (ed.) *Ecosystem Dynamics in a Polar Desert: The McMurdo Dry Valleys, Antarctica*, Antarctic Research Series, Vol. 72, American Geophysical Union, Washington, D.C. pp. 153-188.

Vaughan, D.G., Marshall, G.J., Connolley, W.M., Parkinson, C., Mulvaney, R., Hodgson, D.A., King, J.C., Pudsey, C.J. and Turner, J. (2003) Recent rapid regional climate warming on the Antarctic peninsula, *Climate Change* **60**, 243-274.

Verleyen, E., Hodgson, D.A., Vyverman, W., Roberts, D., McMinn, A., Vanhotte, K. and Sabbe, K. (2003) Modelling diatom responses to climate induced fluctuations in the moisture balance in continental Antarctic lakes. *Journal of Paleolimnology*, **30**, 195-215.

Verleyen, E., Hodgson, D.A., Sabbe, K., Vanhotte, K. and Vyverman, W. (2004) Coastal oceanographic conditions in the Prydz Bay region (East Antarctic) during the Holocene recorded in an isolated basin. *The Holocene*, **14**, 246-257.

Vincent, W.F. (1981) Production strategies in Antarctic inland waters: phytoplankton eco-physiology in a permanently ice-covered lake, *Ecology* **62**, 1215-1224.

Wand, U., Schwarz, G., Brüggenmann, E., and Bräuer, K. (1997) Evidence for physical and chemical stratification in Lake Untersee (Central Dronning Maud Land, East Antarctica). *Antarctic Science* **9**, 43-45.

Webster, J., Hawes, I., Downes, M., Timperley, M. and Howard-Williams, C. (1996) Evidence for regional climate change in the recent evolution of a high latitude pro-glacial lake, *Antarctic Science* **8**, 49-59.

Welch, K.A., Lyons, W.B., McKnight, D.M., Doran, P.T., Fountain, A.G., Wall, D. Jaros, C., Nylen, T. and Howard-Williams, C. (2003) Climate and hydrological variation and implications for lake and stream ecological response in the McMurdo Dry Valleys, Antarctica, in D. Greenland, D.G. Goodin and R.C. Smith (eds.), *Climate Variability and Ecosystem Response at Long-Term Ecological Research Sites*, Oxford University Press, New York, pp. 174-195.

Wharton, R.A., Jr., McKay, C.P., Clow, G.D. and Andersen D.T. (1993) Perennial ice covers and their influence on Antarctic lake ecosystems, in W.J. Green and E.I. Friedmann (eds.) *Physical and Biogeochemical Processes in Antarctic Lakes*, Antarctic Research Series, Vol. 59, American Geophysical Union, Washington, D.C., pp 53-70.

Wilson, A.T. (1964) Evidence from chemical diffusion of a climate change in the McMurdo dry valleys 1200 years ago, *Nature* **201**, 176-177.

Wilson, A.T. (1981) A review of the geochemistry and lake physics of the Antarctic dry areas, in L.D. McGinnis (ed.) *Dry Valley Drilling Project*, Antarctic Research Series, Vol. 33, American Geophysical Union, Washington, D.C., pp. 185-192.

Zwartz, D., Bird, M. Stone, J. and Lambeck, K. (1998) Holocene sea-level change and ice-sheet history in the Vestfold Hills, East Antarctic. *Earth Planetary Science Letters*, **155**, 131-145.

14. SUBANTARCTIC TERRESTRIAL CONSERVATION AND MANAGEMENT

J. WHINAM
Biodiversity Conservation Branch
Department of Primary Industries and Water
GPO Box 44, Hobart, Tasmania
Jennie.Whinam@dpiw.tas.gov.au

G. COPSON
Wildlife Management Branch,
Department of Primary Industries and Water
GPO Box 44, Hobart, Tasmania
Geoff.Copson@dpiw.tas.gov.au

J.-L. CHAPUIS
Muséum National d'Histoire Naturelle
Département Ecologie et Gestion de la Biodiversité,
Paris, France
chapuis@mnhn.fr

Introduction

The subantarctic region (South Georgia, Marion and Prince Edward Islands, Iles Crozet, Iles Kerguelen, Heard and McDonald Islands and Macquarie Island) is faced with many conservation and management issues. From a scientific perspective, taxonomic surveys are a necessary pre-requisite to allowing biogeographic classification and/or comparisons between and among islands. This is particularly true for non-vascular plants and invertebrates. The main conservation issues in the subantarctic are global warming, introduced species and the management of human visitors. However, it is the combination of these conservation issues that poses the

D.M. Bergstrom et al. (eds.), Trends in Antarctic Terrestrial and Limnetic Ecosystems, 297–316.

biggest challenge for managers and the biggest threat to the integrity of the subantarctic terrestrial environment.

International legislative framework

There are numerous international treaties (eg Man and the Biosphere Reserve Convention and the World Heritage Area Convention) and national management plans (relating to National Parks and Reserves), that deal with various aspects of conservation management issues in the subantarctic. Details of these treaties and plans are outlined in Huiskes et al. (this volume). The manner in which these treaties and plans are implemented for conservation management varies from nation to nation. These treaties impose a statutory framework on issues of conservation management and are addressed in the sections below.

Conservation knowledge status

The current state of knowledge of the conservation values of the different regions of the subantarctic is highly variable. Whereas much conservation research has been undertaken in the different regions over the last few decades, the focus of this research varies. It is important to regularly update knowledge on the glaciology, geology, flora, fauna and climate of the region, if the impacts of threats to the conservation status of species or the impacts of climate change are to be identified and understood. Often the relationships among individual species are not simple, because each species forms part of a complex and integrated ecosystem.

GLACIOLOGY and GEOLOGY

The focus for glaciology has been on the information that glacial cores can provide on past climates and rates of deglaciation, especially on South Georgia, Heard Island, Iles Kerguelen and Marion Island (Frenot et al. 1997a, Anon 1999, see Bergstrom et al. this volume). Many of the subantarctic islands (eg Marion, Heard and McDonald Islands) are associated with active volcanic regions. The Iles Kerguelen (30 to 40 million years old) are considered the oldest oceanic archipelago in the Indian Ocean, with the presence of fossiliferous sites testifying to the emergence of these islands beginning in the Miocene (Giret et al. 2003). Much geological interest has been focused on Macquarie Island as the only known locality where oceanic crust formed in a normal submarine setting which is exposed above sea level in a major ocean basin (Department of the Environment, Sports and Territories/Parks and Wildlife Service Tasmania 1996), and this was the primary basis for listing the island as a World Heritage Area.

SOILS AND EROSION

On subantarctic islands, the coastal soils are usually organic and enriched by nutrient inputs from numerous seabirds and seals. Most of the energy in these ecosystems comes from the sea (Smith 1978). At lower altitudes, low temperatures and high levels of precipitation are responsible for a low rate of organic matter decomposition and peat deposits. Temperature dramatically decreases with increasing altitude and the closed vegetation cover present in coastal areas disappears. At higher altitudes, soils are generally more mineral with lower organic content and can form a surface desert pavement. The numerous freeze-thaw cycles can result in patterned ground. However, because of current climate changes, the patterned grounds located at low altitudes are no longer active in some areas (eg Iles Kerguelen, Y. Frenot pers. comm.).

On some subantarctic islands (eg the eastern part of the Iles Kerguelen and Macquarie Island), the combination of rabbit impacts and recent climatic changes is responsible for major changes in vegetation and some spectacular erosion (Copson and Whinam 1998, Chapuis et al. 2004, M. Lebouvier pers. comm., Fig. 1). Whereas rabbit grazing and burrowing reduces vegetation cover, summer dryness significantly affects the vegetation that has developed largely in the absence of aridity. This results in a significant reduction in plant cover, which in turn facilitates wind erosion, highlighting how ecosystem disturbances can lead to long-term or irreversible degradation of ecosystems.

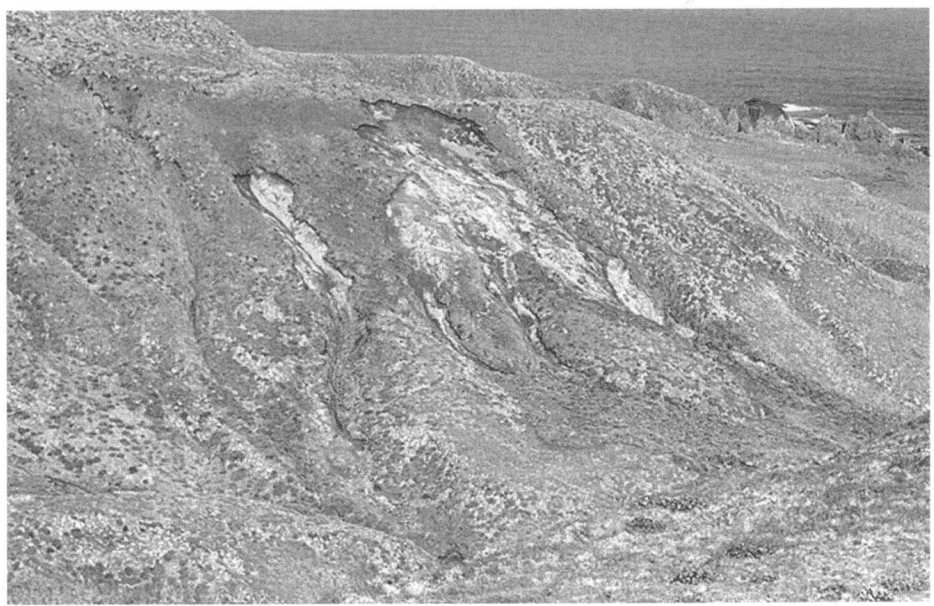

Figure 1. Grazing by introduced European rabbits has largely eliminated tall tussock grass communities from the steeper slopes of Macquarie Island, resulting in increased numbers of landslips. Photo: G. Copson.

Palynological studies carried out at Iles Kerguelen demonstrated that most of the subantarctic flora was present at the end of the last glacial event (Bellair-Roche 1972, Young and Schofield 1973). Vegetation changes on Marion Island over the last 16 000 years have been identified from pollen analysis in core samples (Schalke and van Zinderen Bakker 1971), whereas on Macquarie Island studies of palaeolake site deposits, up to 11 000yr BP have revealed vegetation composed of species growing on the island today (Selkirk et al.1988).

VEGETATION

Vegetation mapping has been undertaken on several subantarctic islands, although the degree of resolution and vegetation categories used necessarily varies. Complete taxonomic listings of vascular species are available for the subantarctic region and non-vascular flora has recently been the subject of taxonomic research. However, more work is required on taxonomic relationships among sites within the region (Selkirk et al. 1990) and genetic investigation is increasingly being used to undertake aspects of this research (eg Skotnicki et al. 2004, Skotnicki and Selkirk this volume, Stevens and Hogg this volume). Without this knowledge, the conservation status of species and communities cannot be evaluated properly (Chown et al. 1998). No plant species occurring in the subantarctic is currently listed on the IUCN endangered species list. Most of the terrestrial region is species-poor, with relatively few endemic species, variable numbers of alien taxa and with the noticeable absence of some structural groups (eg fernbrake at Heard Island). The isolation of these land areas has resulted in considerable interest in genetic evolution, methods of dispersal, colonisation processes and the ability of the plants to adapt to climate change.

In spite of their extreme remoteness, many alien species have been introduced to the subantarctic islands (Gremmen 1997, Gremmen and Smith 1999, Frenot et al. 2001, 2005). The current changes in climatic conditions might significantly alter the number of alien species, their competitive abilities and their status in the communities (Chapuis et al. 2004, Frenot et al. 2005). A detailed overview of the status of alien species in the Antarctic and subantarctic is given in Frenot et al. (2005) and Convey et al. (this volume).

VERTEBRATES

Good baseline data exists for some of the native vertebrate fauna (seals and birds) of the region. Much of the native fauna spends most of their time foraging in the oceans, often at great distances from their breeding areas. Little is known regarding the movements of individuals or of the genetic interchange among breeding grounds in different areas of the subantarctic. An example of this are southern elephant seals, a single species that is split into three main breeding populations based on the South Atlantic Ocean, Indian Ocean and Macquarie Island, with very little (documented) interchange among the populations (Gales et al. 1989, Olsen 2003). The taxonomic status of many subantarctic fauna species has been revised in the last decade, eg albatrosses (Robertson and Nunn 1998), which has affected the conservation status of several species. Changes in the conservation status of many of the native fauna species have been recorded in the last 50 years. Some species have decreased dramatically, eg southern elephant seals, rockhopper penguins and several albatross species (Cunningham and Moors 1994, Olsen 2003, BirdLife International 2004), whereas other species have experienced dramatic population increases, eg fur seals and king penguins (Rounsevell and Copson 1982, Boyd 1993, Guinet et al. 1995).

Knowledge of the biology of an introduced species at a site is essential for management purposes, as relationships to indigenous and/or other introduced flora and fauna species will greatly influence a management strategy, eg eradication, control or do nothing. For example on the Iles Kerguelen, an alien predatory carabid beetle significantly reduces the specific richness of native invertebrate communities (Chevrier et al. 1997). However, the introduced house mouse is the only predator of this beetle and in this situation, it was not considered appropriate to eradicate the rodent that has the role of a top predator (Le Roux et al. 2002, Courchamp et al. 1999, 2003). The knowledge is also important for prioritising actions in a pest management strategy, eg control of rabbits was the key to the integrated pest management program on Macquarie Island. In the 1970s, rabbits were the main food source for feral cats on the island. By controlling rabbit numbers, it was possible to eradicate wekas, an introduced flightless New Zealand bird when the cats turned to them as a secondary food source, and ultimately the feral cats (Copson 1995a, Copson and Whinam 2001).

INVERTEBRATES

The native terrestrial invertebrate fauna of the region as a whole and of individual locations is relatively poor in diversity and abundance compared to other regions. Our understanding of the native invertebrate fauna is much patchier than for the vertebrate fauna. Whereas taxonomic listings are good for some phyla in some locations (eg Crafford et al. 1986, Greenslade 1990, Pugh 1993), they are not so for all phyla in a single location - the terrestrial arthropod fauna usually being the best known. More comparative taxonomic work is required before biogeographical relationships and the status of species, eg the degree of endemism, are well understood.

Increasingly, studies are being undertaken into the physiological adaptations and ecological roles of the invertebrate fauna of Antarctica and the subantarctic islands (Block et al. 1984, Convey and Block 1996, Pugh 1996). Indigenous, macro-invertebrates fulfil a major role in the breakdown of vegetation matter and the nutrient cycle of the region (Pedersen and Bergstrom 1999) and the introduction of vertebrate herbivores and omnivores has significantly altered this on some subantarctic islands (Copson and Whinam 2001). There is also potential for introduced invertebrate species, such as those of predatory beetles on South Georgia, the diamond-backed moth on Marion Island (Chown and Avenant 1992, Ernsting 1993) to have long-term impacts. Because the removal of established introduced invertebrate species may not be possible, it is essential that management procedures be established to prevent further accidental introductions. This also applies to the movement of indigenous species, to avoid genetic mixing, and of soils, to locations within the region where they are not naturally found, eg between islands within groups and among island groups, during operations to move personnel and equipment among islands, or between sites on larger islands.

MICROBIOLOGY

The subantarctic, in common with any land surface with extensive plant cover, supports a rich and varied assemblage of soil and plant microorganisms. However, our knowledge of these organisms, together with the nature and focus of research on them, varies across the subantarctic region. Information is available from some areas on diatoms (McBride et al. 1999, Van de Vijver et al. 2002), soil arthropods (Greenslade 1992), kelp flies (McQuillan and Marker 1984), bacteria, algae and nematodes (Mawson 1958). Some information is available on seal and bird parasites (Murray 1967, Clarke and Kerry 1993) and helminth parasites of alien mammals (Pisanu and Chapuis 2003). Rabbit fleas (*Spilopsyllus cuniculi*) and the myxoma virus were introduced as a rabbit control measure on Macquarie Island in 1968 (Sobey et al. 1973, Skira et al. 1983) and on Ile du Cimetière (Iles Kerguelen) in 1998 (Chekchak et al. 2000). The composition and ecological roles of the microflora and microfauna of the subantarctic, both free-living and parasitic, are still poorly known.

ECOLOGICAL COMMUNITIES

Major factors which have influenced the development of each terrestrial ecological community in the subantarctic biogeographical regions include: the duration of a site's exposure above sea level, climate changes and founder effect influenced through location. Over the last 200 years, this process has been overwhelmingly changed by human impacts, mainly the exploitation of the indigenous fauna, the introduction of alien species and more recently, pollution and global warming. Currently, our knowledge of these ecological communities is far less developed than that of the taxa and their status, the latter having resulted in national recovery programs being implemented for several species. A better understanding of the ecological communities and processes is becoming increasing important as human visitation increases throughout the region and programs to manage introduced species are being implemented on more subantarctic islands. These two factors alone are likely to result in fast changes in regional ecosystems. Baseline knowledge is essential to evaluate and monitor their impacts on subantarctic environments (Walton and Shears 1994), especially in relation to assessing the efficacy of restoration programs.

Current Threats

Current threats to the conservation status of subantarctic fauna and flora can be classified into global (eg climate change and the ozone layer 'hole', Smith 1993), regional (eg commercial fishing practices and increasing human presence, Smith and Steenkamp 1990, Chown and Smith 1994) or local threats to specific areas (eg impacts of alien animal and plant species (Chapuis et al. 1994, Frenot et al. 2005), and oil spills (Pople et al. 1990, Smith and Simpson 1995). While research has enabled the implementation of adaptive management practices to mitigate some of the local and regional threats, this is not the case for the global threats where investigations are generally still in the research phase. The global threats will ultimately be more difficult to manage as they will need international cooperation and actions, as evidenced by the Kyoto Protocol.

The subantarctic region has long been recognised as an important area to monitor the impacts of climate change. Some of the greatest mean temperature changes in recent decades have been recorded in this region (Adamson et al. 1988, Frenot et al. 1997b, Tweedie and Bergstrom 1999, Convey this volume). There are direct implications of climate change for conservation management in the subantarctic, including:

- the possibility of greater success in the establishment of alien species (Frenot et al. 2005),
- increases in population sizes and distributions of existing alien species (eg development of *Taraxacum officinale* in the Iles Kerguelen, Chapuis et al. 2004),
- increase in wind erosion (Iles Kerguelen, M. Lebouvier, pers. comm.),
- impacts on the availability of food sources, especially marine resources (Cunningham and Moors 1994, BirdLife International 2004),
- breeding success (eg *Calliphora vicina* and alien plants on Iles Kerguelen, Chevrier et al. 1997, Frenot et al. 2001, and rabbits on Macquarie Island, Copson and Whinam 2001),
- confusing our understanding of the responses of native and introduced species to adaptive management practices (Chapuis et al. 2004),
- the need for reorientation of conservation research and management programs.

The management of alien species and their impacts on the indigenous species and ecosystems of the subantarctic terrestrial region has increasingly become a priority focus of conservation management programs. The availability of baseline data from the subantarctic, together with techniques that have been mainly developed elsewhere, are now being applied in alien species management programs. In recent years, the management of an individual alien species usually forms part of a broader restoration program. Besides political and public support, the prerequisites for a restoration program to be successful include:

- detailed knowledge of the damaged ecosystem(s) before intervention,
- an array of management techniques sufficient to deal with a wide variety of alien species, without having long-term adverse effects on indigenous species and natural ecosystems, and
- sufficient, long-term resources to enable successful completion of the program, (Copson 1995b).

An integral part of any restoration program is monitoring the responses of both target and non-target species. The generally simple nature of the subantarctic ecosystem enables an assessment of the outcomes to be applied to other areas in the subantarctic region (Copson and Whinam 2001, Frenot et al. 2005, Convey et al. this volume). Situations such as the Iles Kerguelen and Prince Edward Islands also provide opportunities to evaluate the impacts of introduced flora and fauna on indigenous species and ecosystems by comparing the biota on undisturbed islands with that of adjacent, disturbed islands (Chapuis et al. 2002). When eradication of exotic species is not possible, control programs could be implemented (Copson 1995b, Courchamp et al. 2003). Examples of widespread control include that of rabbits on Macquarie Island and the Iles Kerguelen (Copson and Whinam 2001, Courchamp et al. 2003).

The numbers of human visits, the levels of occupancy and the temperature of an island are known to strongly influence alien species richness (Chown et al. 1998, Frenot et al. 2005). Reasons for visitations to various parts of the subantarctic include all or some of the following: tourism, scientific research programs, construction programs associated with research and management and media visits. Visitations are likely to continue to increase, as is the intensities of visitations to certain popular areas, in addition to the range of activities undertaken (eg sightseeing, mountaineering and kayaking).

QUARANTINE AND CONSERVATION THREATS

A cornerstone of programs for the future management of these regions is effective prevention of further introductions of alien species. Throughout the region, this objective is being addressed through implementation of quarantine procedures and visitor education for both tourists and expeditioners, eg Prince Edward Islands Management Plan Working Group (1996). One aspect of quarantine is the need to recognise that isolated, ice-free areas are basically biological 'islands' (Bergstrom and Chown 1999) that also need to have effective protection in place to prevent intra-'island' spread of indigenous organisms or genetic interchange. This has become an increasingly important aspect of conservation management with both the retreat of glaciers opening up areas for colonisation, eg the expansion of reindeer on South Georgia (Moen and Macalister 1995) and the increased use of helicopters for remote field parties.

MANAGEMENT ACTIVITIES GUIDED BY MANAGEMENT PLANS

Increasingly, management activities on many of the subantarctic islands are governed by management plans, which are generally statutory documents. Only the French subantarctic islands have no statutory management plans at present, but have programs validated by the French Polar Institute and the Terres Australes et Antarctiques Françaises. The intention of these management plans is to establish rules and procedures, to protect the natural and cultural values of a site, for activities undertaken in the designated region. Whereas the legal status of the management plans varies, the intention is to establish what is acceptable in the specified area, with more prescriptive procedures often as an adjunct to these plans. Although the scope of these management plans varies, a number of activities are generally dealt with including:

- defining the designated area,
- the legislative framework for management,
- the terrestrial resources and values,
- in some cases the resources and values of territorial sea surrounding the islands (eg Heard and McDonald Islands, and Macquarie Island),
- management aims, goals and objectives with prescriptions to achieve these,
- zoning, which prescribes management activities on all or parts of these islands, and is used as a conservation protection measure to restrict access to sensitive sites,
- protection and management of geological features, flora, fauna and cultural heritage,
- issues associated with station activities, such as waste treatment (rubbish and sewage), building construction, visitor access (tourism, expeditioner and recreational), transportation (air, sea and vehicular), research programs, quarantine, fuel transfer (ship to shore), storage and usage, hazardous materials,
- recovery programs for species and/or communities listed as endangered,
- prevention of further introductions and management of existing alien species,
- research, including data collection and monitoring programs, including conservation management,
- interpretation and education,
- milestones to measure the progress and completion of prescribed management actions, and
- the timeframe that the management plan will cover.

Over recent decades, there has been increasing international cooperation among management authorities to develop best management practices for sites in the region

(Dingwall 1995). There is now a free exchange of ideas, data, draft management plans and personnel to efficiently address these management issues. This exchange has given management authorities the ability to adapt management practices developed elsewhere to their local situation, such as feral pest management programs (eg Copson and Whinam 2001, Bester et al. 2002). The various islands and sites in the region have different management needs and legislative frameworks, but in general there are land management guidelines ruling current practices for each of the above activities. We outline the common approaches to management of these activities and highlight some of the innovative responses of particular agencies.

ZONING

Although different terminologies are used, there are several identifiable zones, most of which require access permits. These include:

- service/main use zones: where the main infrastructure for management and research are located. This is generally the most disturbed zone and kept to a minimum size,
- natural zones: minimal infrastructure generally consisting of field huts, both permanent and temporary, tracks and helipads to support various conservation management and research programs,
- special management/restricted areas: established to provide extra protection to the natural and cultural values of specific sites. Access to these areas is usually restricted to scientists. Activities that would have long-term, adverse impacts on these values are not permitted and they generally contain little or no infrastructure,
- tourism zones: over the last decade it has become common for limited zones to be designated for tourist activities, including any infrastructure,
- marine reserves and/or buffer zones: these have been designated around some islands to limit resource extraction activities and to protect marine values.

PROTECTION AND MANAGEMENT OF GEOLOGICAL FEATURES, FLORA AND FAUNA AND CULTURAL HERITAGE

National and/or international legislation covers the protection and management of natural and cultural features of most subantarctic islands (Davis 1995). The legislation generally forbids exploitation of minerals, flora and fauna, and regulates activities that disturb, remove or endanger these values. Generally, the only activities allowed that directly impact on the natural and cultural values are approved scientific or management/support programs, often aimed at conservation management, which are regulated by permits. Cultural heritage sites are increasingly being evaluated, and appropriate provisions being made for them in management plans (eg Graham 1989, Davis 1995). Examples of this are the restoration of whaling stations at Iles Kerguelen and South Georgia.

RECOVERY PROGRAMS FOR SPECIES LISTED AS ENDANGERED

Endangered species may be listed under specific national and/or international legislative frameworks. Over the last decade, there has been an increase internationally in the number of programs aimed at recovery of species, most commonly vertebrate fauna, listed as endangered. These recovery programs may target a single species or several ecologically similar species, eg developing mitigation methods to lessen the risks of long-line fishing to albatrosses (Weimerskirch et al. 1997, Inchausti and Weimerskirch 2001). As many of the species involved have circumpolar distributions and/or major impacts occur in international waters, recovery programs can involve international cooperation, eg the Agreement on the Conservation of Albatrosses and Petrels (ACAP) that has been ratified by eight countries as of March 2006.

PREVENTION OF FURTHER INTRODUCTIONS AND MANAGEMENT OF ALIEN SPECIES

There has been a history of both accidental and deliberate introductions of alien species into the subantarctic region associated with various human activities, including early resource exploitation, scientific expeditions and tourism. Whereas the management of existing alien species has been undertaken on several sites since the 1970s, it is only in recent years that the prevention of further introductions has become a high priority management objective. For example, recent research has identified the main vectors associated with alien species transportation (Whinam et al. 2005). Quarantine programs are now in place for most areas of the region, although the prescriptions and management programs vary enormously. While most quarantine programs prescribe wash-down stations to prevent soil-borne pathogens and organisms, and restrictions on taking poultry produce into the field to reduce the risk of avian disease introduction are in place, the total prohibition on fresh food produce exists only at the Prince Edward Islands (Cooper et al. 2005). The transport of domestic stock to the subantarctic has ceased, although animals such as trained dogs may still be used in management programs (Copson 2003).

Control and/or eradication programs for alien species have been, or are, undertaken on the vast majority of subantarctic islands and some cold temperate islands. Details of alien species management programs are given in Convey et al. (this volume) and examples of eradication programs include:

- Auckland Is   cattle, goats, rabbits, house mice, tree daisy
- Campbell I   sheep, cattle, brown rats
- Ile Amsterdam   cattle
- Ile Saint-Paul   black rats, rabbits
- Iles Kerguelen   feral cats, rabbits, house mice, black rats
- Macquarie I   weka, feral cats, sweet vernal grass (*Anthoxantum odoratum*), curled dock (*Rumex crispus)*
- Marion I   feral cats, brown trout, thistle, *Agrostis gigantea*

In recent years, there has been an increased focus on diseases of wildlife in the region (eg Austin and Webster 1993, Clarke and Kerry 1993). An example of this was the deaths of Hooker's sea lions in the Auckland Islands caused by disease, with the potential to spread to Campbell and Macquarie Islands via tourist vessels (K. Kerry pers. comm.). Antibodies of avian diseases have been discovered in skuas and penguins in both the Antarctic and subantarctic (Alexander et al. 1989, Gauthier-Clerc et al. 2002) and a new plant virus has recently been identified on Macquarie Island (Skotnicki et al. 2003). Several research programs are currently investigating the status and possible sources of these diseases and the existence of other naturally occurring or introduced diseases. A recently identified source of marine introductions is marine species carried in the ballast waters of government expedition and tourist vessels (Lewis et al. 2003, 2006). Quarantine procedures are continually being developed as management tools to prevent further introductions of diseases into and out of the region (eg Whinam et al. 2005), with the potential of new strains of existing diseases developing in the isolation of the region. In addition, disease preparedness and response plans are being developed to deal with outbreaks of diseases in the region.

STATION ACTIVITIES

Most activities associated with subantarctic research stations are prescribed by the relevant management plan. Development of infrastructure generally requires approval from a managing authority and possibly an environmental impact assessment. Whereas on some subantarctic islands such as Marion Island, stringent measures have been adopted for waste disposal, this is not the case on many others. Special procedures are prescribed to cover the handling and storage of fuel for heating, electricity generation and transport, both at a station and in the field. Increasingly, Antarctic treaty nations are attempting to use more sustainable energy management practices, including the development of more energy efficient buildings, and alternative energy sources to reduce emission of greenhouse gases and to reduce the risks of oil spills, while reducing costs. In 1993, a joint Australian-French research project was established to investigate alternative energy options for Antarctic stations. Field trials of a 10 kW wind turbine were conducted at Casey and these have been followed by the installation of two 1mW turbines at Mawson.

TRANSPORT

Transportation within the region has implications for wildlife management, quarantine and expeditioner/tourism management. Prescriptions generally exist as to the type and use of various means of transport at a given site. For example, flight paths to minimise animal disturbance prescribe height, distances and routes for aircraft, minimum distances to be maintained from wildlife by inflatable rubber boats, aircraft and pedestrian traffic on animals (eg Giese and Riddle 1998, Giese et al. 2000). The type of transportation has a big influence on the risk that is posed to quarantine, eg helicopters allow rapid and easy access to multiple, isolated sites, increasing the potential of spread of organisms (both native and alien).

FIELD HUTS AND TRACKS

Management plans generally prescribe the location both of huts and tracks, and the management of their environmental footprints (such as waste disposal, fuel supplies, etc, Figs. 2 and 3). Recent monitoring work has indicated that many of the tracks on Macquarie Island are close to their environmental threshold (Dixon 2001). Recent technological advances have enabled field huts to be more energy efficient and to create smaller environmental footprints. Lightweight huts have been developed that can be placed and removed by helicopters at previously prepared sites, thereby reducing construction impacts (eg Australian and French territories). As they can be deployed at specific sites for the duration of a field program, they may have less long-term environmental impacts than huts which remain at the same site for many years. Careful siting can minimise the environmental impacts of field huts and tracks, eg locating huts at sea level on less sensitive vegetation.

Figure 2. Two remote field huts at La Mortadelle on Iles Kerguelen. The hut on the right is for accommodation while the hut on the left houses food and equipment. Photo: J. Whinam.

EXPEDITIONERS AND TOURIST IMPACTS

Higher standards of environmental awareness have been steadily developed as the number of visitors, both expeditioners and tourists, has increased to the region. With expeditioners, this is usually being achieved through selection of personnel and to a greater extent through their training and orientation before departure.

The development of adventure tourism, together with the steady increase in tourist numbers has resulted in the Antarctic Treaty nations currently investigating environmental procedures to minimise adverse tourism impacts in the subantarctic and Antarctica. Increasingly, tourist operators visiting the region are required to undertake environmental impact assessments before permits for visits are issued. These assessments generally cover prescriptions to minimise the impacts of the tourists on the natural and cultural values of the region, and to ensure that mitigation measures are in place in case of accident, oil spill, etc. Most subantarctic islands have specified contractual conditions and guidelines for tourist visits. These may include landing and visitation sites, duration of visits, safety procedures, numbers ashore and procedures to minimise impacts on flora and fauna. Observers are often carried on ships, or meet tourist parties as they arrive at sites, to oversee the compliance with permit conditions. In addition, the industry has developed its own guidelines for the conduct of tourist operations in the region through the International Association of Antarctica Tour Operators. However this does not cover all operators.

Figure 3. The barge 'L'Aventure II' which is used to move people and materials among islands at Iles Kerguelen. This barge is moving fencing and building materials to Ile Longue and transporting expeditioners. Photo: J. Whinam.

FISHING EXPLOITATION

Commercial fisheries pose the greatest danger to albatrosses through their incidental capture and death. Between the 1950s and 1980s, pelagic gillnets and driftnets ensnared tens of thousands of albatrosses, causing major drops in populations (Robertson and Gales 1998). Public condemnation of the unacceptable levels of by-catch in these fisheries resulted in a global moratorium on driftnet fishing in 1993. However, the global expansion of longline fisheries presented new threats to albatrosses, as they tend to follow fishing vessels (Robertson and Gales 1998). Albatrosses forage in waters favoured by fishing vessels, they commute great distances to foraging grounds (increasing the risk of contact with fishing vessels), and they are habitual ship followers and scavengers. It has been estimated that longline fishing kills more than 100 000 seabirds each year. Most of these deaths could be prevented if fishing vessels implemented an accepted set of mitigation measures (Gales 1998). However, regulation and enforcement are made more difficult with up to one half of all the birds caught as by-catch from long-lining being killed by illegal fishing vessels.

RESEARCH

A large part of the current focus of research in the region is on the conservation of its natural and cultural values. The rarity of the subantarctic resource combined with its environmental fragility has resulted in most management authorities stipulating minimal impact methods of research. International collaboration has increased the value of programs addressing issues of regional or global concerns, such as climate change, while minimising the impacts on any one area. Specific restrictions on research programs are generally addressed by scientific and/or access permits and, where appropriate, animal ethic guidelines.

INTERPRETATION/EDUCATION PROGRAMS

The use of interpretation and education programs as a valuable conservation tool in the region is being recognised. In addition to an interest in the many natural, cultural and historic values of the region, there has been a rise in interest in the conservation management programs being undertaken. The general public's understanding of the values of the region has been heightened by documentary film-makers who can graphically display the fragile nature of the subantarctic. Antarctic tourists now demand a high level of sophistication in interpretation/education programs.

Conclusions

Over the last 20 years, there have been significant changes in most areas of subantarctic management by most nations. However, the focus and degree of environmental protection afforded still varies among nations. Most importantly, international cooperation is occurring at various levels (such as information and staff exchanges) and is essential for efficient and successful management of conservation values and mitigation of conservation threats in the region. It is clear that statutory management plans prepared by nations for their territories are essential to ensure on-going management of the various aspects of station infrastructure, transport, research, tourism and quarantine. Subantarctic species provide a wonderful opportunity to monitor the health of the Southern Ocean. Changes in populations of animal species dependent on marine foraging areas that can be monitored on land are likely to provide vital information on marine resources. Similarly, changes in populations of both native and exotic plant species, combined with information on glacial retreat, are essential to understanding biotic responses to climate change. Baseline data (including updating taxonomic survey data to allow biogeographic analyses) and on-going monitoring are essential if there is to be effective feedback into management programs to address threats to the conservation values of subantarctic ecosystems.

Acknowledgements

We thank John Cooper for his useful comments on an earlier version of the manuscript.

References

Adamson, D.A., Whetton, P. and Selkirk, P.M. (1988) An analysis of air temperature records for Macquarie Island: decadal warming, ENSO cooling and southern hemisphere circulation patterns, *Papers and Proceedings of the Royal Society of Tasmania* **122**, 107-112.

Anonymous (1999) Environmental Management Plan for South Georgia: Public consultation paper. British Antarctic Survey.

Alexander, D. J., Manvell, R. J., Collins, M. S., Brockman, S. J., Westbury, H. A., Morgan, I. R. and Austin, F. J. (1989) Characterisation of paramyxoviruses isolated from penguins in Antarctica and Sub-Antarctica during 1976-1979, *Archives of Virology* 109, 135-144.

Austin, F. and Webster, R.G. (1993) Evidence of orthomyxoviruses and paramyxoviruses in fauna in Antarctica, *Journal of Wildlife Diseases* **29**, 568-571.

Bellair-Roche, N. (1972) Present knowledge of the Quaternary flora (Palynology) in the Southern Islands of the Indian Ocean, in R.J. Adie (Ed.), *Antarctic Geology and Geophysics*. Universitetsforlaget, Oslo, pp.789-791

Bergstrom, D.M. and Chown, S.L. (1999) Life at the front: history, ecology and change on southern ocean islands, *Trends in Ecology and Evolution* **14**, 472-477.

Bergstrom, D.M., Hodgson, D.A. and Convey, P. (2006) The physical setting of the Antarctic, in D.M. Bergstrom, P. Convey, and A.H.L. Huiskes (eds.), *Trends in Antarctic Terrestrial and Limnetic Ecosystems: Antarctica as a Global Indicator*, Springer, Dordrecht (this volume).

Bester, M.N., Bloomer, J.P., Van Aarde, R.J., Erasmus, B.H., Van Rensburg, P.J.J., Skinner, J.D., Howell, P.G. and Naude, T.W. (2002) A review of the successful eradication of feral cats from sub-Antarctic Marion Island, Southern Indian Ocean, *South African Journal of Wildlife Research* **32**, 65-73.

BirdLife International (2004) Threatened birds of the world 2004, CD-ROM. Cambridge, UK: BirdLife International.

Block, W., Burn A.J. and Richard, K.J. (1984) An insect introduction to the maritime Antarctic, *Biological Journal of the Linnean Society* **23**, 764-770.

Boyd, I.L. (1993) Pup production and distribution of breeding Antarctic fur seals *(Arctocephalus gazella)* at South Georgia, *Antarctic Science* **5**, 17-24.

Chapuis, J.-L., Boussès, P. and Barnaud, G. (1994) Alien mammals, impact and management in the French subantarctic islands, *Biological Conservation* **67**, 97-104.

Chapuis, J.-L., Frenot, Y. and Lebouvier, M. (2002) Une gamme d'îles de référence, un atout majeur pour l'évaluation de programmes de restauration dans l'archipel de Kerguelen, *Revue d'Ecologie (Terre Vie)* **Suppl. 9**, 121-130.

Chapuis, J.-L., Frenot, Y. and Lebouvier, M. (2004) Recovery of native plant communities after eradication of rabbits from the subantarctic Kerguelen islands, and influence of climate change, *Biological Conservation* **117**, 167-179.

Chekchak, T., Chapuis, J.-L., Pisanu, B. and Boussès, P. (2000) Introduction of the rabbit flea, *Spilopsyllus cuniculi* (Dale), to a subantarctic island (Kerguelen archipelago) and its assessment as a vector of myxomatosis, *Wildlife Research* **27**, 91-101.

Chevrier, M., Vernon, P. and Frenot, Y. (1997) Potential effects of two alien insects on a subantarctic wingless fly in the Kerguelen Islands, in B. Battaglia, J. Valencia and D.W.H. Walton (eds.) *Antarctic Communities: Species, Structure and Survival*, Cambridge University Press, Cambridge, UK, pp. 424-431.

Chown, S.L. and Avenant, N. (1992) Status of *Plutella xylostella* at Marion Island six years after its colonisation, *South African Journal of Antarctic Research* **22**, 37-40.

Chown, S.L. and Smith, V.R. (1993) Climate change and the short-term impact of feral house mice at the sub-antarctic Prince Edward Islands, *Oecologia* **96**, 508-518.

Chown, S.L., Gremmen, N.J.M. and Gaston, K.J. (1998) Ecological biogeography of Southern Ocean islands: Species-area relationships, human impacts, and conservation, *American Naturalist* **152**, 562-575.

Clarke, J.R. and Kerry, K.R. (1993) Diseases and parasites of penguins, *Korean Journal of Polar Research* **4**, 79-86.

Convey, P. (2006) Antarctic climate change and its influences on terrestrial ecosystems, in D.M. Bergstrom, P. Convey, and A.H.L. Huiskes (eds.), *Trends in Antarctic Terrestrial and Limnetic Ecosystems: Antarctica as a Global Indicator*, Springer, Dordrecht (this volume).

Convey, P. and Block, W. (1996) Antarctic Diptera: Ecology, physiology and distribution, *European Journal of Entomology* **93**, 1-13.

Convey, P., Frenot, Y., Gremmen, N. and Bergstrom, D.M. (2006) Biological invasions, in D.M. Bergstrom, P. Convey, and A.H.L. Huiskes (eds.), *Trends in Antarctic Terrestrial and Limnetic Ecosystems: Antarctica as a Global Indicator*, Springer, Dordrecht (this volume).

Cooper, J. de Villiers, M.S. and McGeoch, M.A. (2005) Quarantine measures to halt alien invasions of Southern Ocean islands: the South African experience (Prince Edward Islands Special Nature Reserve), *Aliens* **17**, 37-39.

Copson, G.R. (1995a) An integrated vertebrate pest strategy for subantarctic Macquarie Island, in *Proceeding of the 10th Vertebrate Pest Control Conference*, Hobart May 1995, pp. 29-33.

Copson, G.R. (1995b) Ecological restoration at Subantarctic islands: the issues, in P.R. Dingwall (ed.), *Progress in conservation of the Subantarctic Islands*. Proceedings of the SCAR/IUCN Workshop on Protection, Research and Management of Subantarctic Islands, Paimpont, France, 27-29 April 1992. Gland: World Conservation Union, pp. 153-156.

Copson, G. R. (2003) *Integrated vertebrate pest management on subantarctic Macquarie Island 1997 – 2002*. Final Report. Department of Primary Industries Water and Environment: Annual report to the Natural Heritage Trust.

Copson, G. and Whinam, J. (1998) Response of vegetation on subantarctic Macquarie Island to reduced rabbit grazing, *Australian Journal of Botany* **46**, 15-24.

Copson, G. and Whinam, J. (2001) Review of ecological restoration program at subantarctic Macquarie Island: Pest Management progress and future directions, *Ecological Management and Restoration* **2**, 129-138.

Courchamp, F., Chapuis, J.-L. and Pascal, M. (2003) Mammal invaders on islands: impact, control and control impact, *Biological Reviews* **78**, 347-383.

Courchamp, F., Langlais, M. and Sugihara, G. (1999) Cats protecting birds: modelling the mesopredator release effect, *Journal of Animal Ecology* **68**, 272-292.

Crafford, J.E., Scholtz, C H. and Chown, S.L. (1986) The insects of sub-Antarctic Marion and Prince Edward Islands, with a bibliography of entomology of the Kerguelen biogeographical province, *South African Journal of Antarctic Research* **16**, 41-84.

Cunningham, D. M. and Moors, P. J. (1994) The decline of the Rockhopper Penguin *Eudyptes chrysocome* at Campbell Island, Southern Ocean and the influence of rising sea temperature, *Emu* **94**: 27-36.

Davis, B. (1995) Problems in management of historical and cultural sites, in P.R. Dingwall (ed.), *Progress in conservation of the Subantarctic Islands*. Proceedings of the SCAR/IUCN Workshop on Protection, Research and Management of Subantarctic Islands, Paimpont, France, 27-29 April, 1992. Gland: World Conservation Union, pp.179-181.

Department of the Environment, Sports and Territories/Parks and Wildlife Service Tasmania (1996) *Nomination of Macquarie Island by the Government of Australia for Inscription on the World Heritage List.* Unpublished document, 96 pp.

Dingwall, P.R. (Ed) (1995) *Progress in conservation of the Subantarctic Islands*. Proceedings of the SCAR/IUCN Workshop on Protection, Research and Management of Subantarctic Islands, Paimpont, France, 27-29 April, 1992. Gland: World Conservation Union.

Dixon, G. (2001) *Management strategy for walking tracks and access corridors on Macquarie Island.* Unpublished Report for the Parks and Wildlife Service, Tasmania. 55p.

Ernsting, G. (1993) Observations on life cycle and feeding ecology of two recently introduced predatory beetle species at South Georgia, sub-Antarctic, *Polar Biology* **13**, 423-428.

Frenot, Y., Gloaguen, J.-C., and Tréhen, P. (1997b) Climate change in Kerguelen Islands and colonization of recently deglaciated areas by *Poa kerguelensis* and *P. annua* in B. Battaglia, J. Valencia and D.W.H. Walton (eds.), *Antarctic Communities: Species, Structure and Survival*, Cambridge University Press, Cambridge, UK, pp. 358-366.

Frenot, Y., Gloaguen, J.-C., Massé, L. and Lebouvier, M. (2001) Human activities, ecosystem disturbance and plant invasions in subantarctic Crozet, Kerguelen and Amsterdam Islands, *Biological Conservation* **101**, 33-50.

Frenot, Y., Gloaguen, J.-C., Van de Vijver, B. and Beyens, L. (1997a) Datation de quelques sédiments tourbeux holocènes et oscillations glaciaires aux Iles Kerguelen, *Comptes Rendus de l'Académie des Sciences, série III, Sciences de la vie* **320**, 567-573.

Frenot, Y., Chown, S.L., Whinam, J., Selkirk, P.M., Convey, P., Skotnicki, M. and Bergstrom, D.M. (2005) Biological invasions in the Antarctic: extent, impacts and implications, *Biological Reviews* **80**, 45-72.

Gales, R. (1998) Albatross populations: Status and threats, in G. Robertson and R. Gales (eds.), *Albatross Biology and Conservation*. Surrey Beatty and Sons, Australia, pp. 20-45.

Gales, N.J., Adams, M. and Burton, H.R. (1989) Genetic relatedness of two populations of the southern elephant seal, *Mirounga leonina*, *Marine Mammal Science* **5**, 57-67.

Gauthier-Clerc, M., Jiguet, F. and Lambert, N. (2002) Vagrant birds at Possession Island, Crozet Islands and Kerguelen Island from December 1995 to December 1997, *Marine Ornithology* **30**, 38-39.

Giese, M. and Riddle, M.J. (1998) Guidelines for people approaching breeding groups of Adélie penguins, in A. Hassan (ed.), *New Zealand Natural Science* **23**, 65.

Giese, M.A., Kerry, E.J. and Riddle, M.J. (2000) Managing interactions between people and Antarctic wildlife, in T. Hughson and C. Ruckstuhl, C. (eds.), *Proceedings of the Sixth International Symposium on Cold Region Development, Hobart, Tasmania, 31 Jan - 04 Feb 2000, pp.* 291-294.

Giret, A., Weis, D., Grégoire, M., Mattielli, N., Moine, B., Michon, G., Scoates, J., Tourpin, S., Delpech, G., Gerbe, M.C., Doucet, S., Ethien, R. and Cottin, J.-Y. (2003) L'archipel de Kerguelen : les plus vieilles îles dans le plus jeune océan, *Géologue* **137**, 23-29.

Graham, T. (1989) Cultural resource management at the Prince Edward Islands. Unpubl. BA Hons. Diss. Department of Archaeology, University of Cape Town, Rondebosch, pp. 85.

Greenslade, P.J. (1990) Annotated checklist of the free-living terrestrial invertebrate fauna of Macquarie Island with notes on biogeography, *Papers and Proceedings of the Royal Society of Tasmania* **124**, 35-50.

Greenslade, P.J. (1992) New records of Mesaphorura (Collembola: Onychiuridae, Tullbergiinae) species from Australia, Macquarie Island and the Antarctic, *Transactions of the Royal Society of South Australia* **116**, pp 141-143.

Gremmen, N.J.M. (1997) Changes in the vegetation of subantarctic Marion Island resulting from introduced vascular plants, in B. Battaglia, J. Valencia and D.W.H. Walton (eds.), *Antarctic Communities: Species, Structure and Survival*. Cambridge University Press, Cambridge, pp. 417-423.

Gremmen, N.J.M. and Smith, V.R. (1999) New records of alien vascular plants from Marion and Prince Edward Islands, sub-Antarctic, *Polar Biology* **21**, 401-409.

Guinet, C., Jouventin, P. and Malacamp, J. (1995) Satellite remote sensing in monitoring the changes of seabirds: Use of Spot image in king penguin population increase at Ile aux Cochons, Crozet Archipelago, *Polar Biology* **15**, 503-510.

Huiskes, A.H.L., Convey, P. and Bergstrom, D.M. (2006) Trends in Antarctic terrestrial and limnetic ecosystems: Antarctica as a global indicator in D.M. Bergstrom, P. Convey, and A.H.L. Huiskes (eds.), *Trends in Antarctic Terrestrial and Limnetic Ecosystems: Antarctica as a Global Indicator*, Springer, Dordrecht (this volume).

Inchausti, P. and Weimerskirch, H. (2001) Risks of decline and extinction of the endangered Amsterdam albatross and the projected impact of long-line fisheries, *Biological Conservation* **100**, 377-386

Le Roux, V., Chapuis, J.-L., Frenot, Y. and Vernon, P. (2002) Diet of the House Mouse (*Mus musculus* L.) at Guillou Island, Kerguelen archipelago, subantarctic, *Polar Biology* **25**, 49-57.

Lewis, P.N., Bergstrom, D.M. and Whinam, J. (2006) Barging in: a temperate marine community travels to the subantarctic, *Biological Invasions*

Lewis, P.N., Hewitt, C.L., Riddle, M. and McMinn, A. (2003) Marine introductions in the Southern Ocean: an unrecognised hazard to biodiversity, *Marine Pollution Bulletin* **46**, 213-223.

Mawson, P.M. (1958) Free-living nematodes. Section 3: Enopoidea from subantarctic stations, *British, Australian and New Zealand Antarctic Research Expedition 1929-1931. Reports, Series B, Zoology and Botany*, **6**, 307-358.

McBride, T.P., Selkirk, P.M. and Adamson, D.A. (1999) Present and past diatom communities on subantarctic Macquarie Island, in S. Mayama, M. Idei, and I. Koizumi (eds.), *Proceedings 14th International Diatom Symposium*, Tokyo Japan 2-8 Sept 1996, 353-365.

McQuillan, P.B. and Marker, P. (1984) The kelpflies, (Diptera: Coelopidae) of Macquarie Island, *Tasmanian Naturalist* **79**, 17-20.

Moen, J. and Macalister, H. (1995) Continued range expansion of introduced reindeer on South Georgia, *Polar Biology* **14**, 459-462.

Murray, M.D. (1967) Ectoparasites of Antarctic seals and birds, *JARE Scientific Reports Special Issue No. 1, Proceedings of the Symposium on Pacific-Antarctic Sciences*, pp. 185-191.

Olsen, P. (2003) *Sub-Antarctic Fur Seal and Southern Elephant Seal Recovery Plan 2004-2008*, Department of the Environment and Heritage, Canberra.

Pedersen, T.K. and Bergstrom, D.M. (1999) Role of invertebrates in nutrient cycling on subantarctic Macquarie Island, *Proceedings from the Ninth ITEX Meeting, January 5-9, 1999, Arctic Laboratory Report 1.* 1. 57.

Pisanu, B. and Chapuis, J.-L. (2003) Helminths from introduced mammals on sub-Antarctic islands, in A.H.L Huiskes, W.W.C. Gieskes, J. Rozema, R.M.L. Schorno, S.M. van der Vies and W.J. Wolf (eds.), *Antarctic Biology in a Global Context*, Backhuys Publishers, Leiden, The Netherlands, pp. 240-243.

Pople, A., Simpson, R.D. and Cairns, S.C. (1990) An incident of Southern Ocean oil pollution: Effects of a spillage of diesel fuel on the rocky shore of Macquarie Island (sub-Antarctic), *Australian Journal of Marine and Freshwater Research* **41**, 603-620.

Prince Edward Islands Management Plan Working Group (1996) *Prince Edward Islands Management Plan*, Department of Environmental Affairs and Tourism, Pretoria, 64 pp.

Pugh, P.J.A. (1993) A synonymic catalogue of the Acari from Antarctica, the sub-Antarctic islands and the Southern Ocean, *Journal of Natural History* **27**, 323-422.

Pugh, P.J.A. (1996) Edaphic oribatid mites (Cryptostigmata: Acarina) associated with an aquatic moss on sub-Antarctic South Georgia, *Pedobiologia* **40**, 113-117.

Robertson, G. and Gales, R. (1998) (eds.) *Albatross Biology and Conservation*. Surrey Beatty and Sons, Australia.

Robertson, C.J.R. and Nunn, G.B. (1998) Towards a new taxonomy for albatrosses. in G. Robertson and R. Gales (eds), *Albatross Biology and Conservation*. Surrey Beatty and Sons, Chipping Norton, pp. 13-18.

Robertson, G. and Gales, R. (1998) (eds.) *Albatross Biology and Conservation*. Surrey Beatty and Sons, Australia.

Rounsevell, D.E. and Copson, G.R. (1982) Growth rate and recovery of a King Penguin, *Aptenodytes patagonicus*, population after exploitation, *Australian Wildlife Research* **9**, 519-525.

Schalke, H.J.W.G. and Van Zinderen Bakker, E.M. Snr. (1971) History of the vegetation, in E.M. Van Zinderen Bakker Snr., J.M. Winterbottom and A. Dyer (eds.) *Marion and Prince Edward Islands*. A. A. Balkema, Cape Town, pp. 89-97.

Selkirk, D.R., Selkirk, P.M., Bergstrom, D.M. and Adamson, D.A. (1988) Ridge top peats and paleolake deposits on Macquarie Island, *Papers and Proceedings of the Royal Society of Tasmania* **122**, 83-90.

Selkirk, P.M., Seppelt, R.D. and Selkirk, D.R. (1990) *Subantarctic Macquarie Island. Environment and Biology*. Studies in Polar Research Cambridge University Press, Cambridge, 285 pp.

Skira, I. J, Brothers, N. P. and Copson, G. R. (1983) Establishment of the European rabbit flea *Spilopsyllus cuniculi* on Macquarie Island, Australia., *Australian Wildlife Research* **10**, 121-127.

Sobey, W. R., Adams, K. M., Johnston, G. C., Gould, L. R., Simpson, K. N. G. and Keith, K. (1973) Macquarie Island: the introduction of the European rabbit flea *Spilopsyllus cuniculi* (Dale) as a possible vector for myxomatosis, *Journal of Hygiene* (Cambridge) **71**, 299-308.

Skotnicki, M.L. and Selkirk, P.M. (2006) Plant biodiversity in an extreme environment: genetic studies of origins, diversity and evolution in the Antarctic, in D.M. Bergstrom, P. Convey, and A.H.L. Huiskes (eds.), *Trends in Antarctic Terrestrial and Limnetic Ecosystems: Antarctica as a Global Indicator*, Springer, Dordrecht (this volume).

Skotnicki, M.L., Mackenzie, A.M., Ninham, J.A. and Selkirk, P.M. (2004) High levels of genetic variability in the moss *Ceratodon purpureus* from continental Antarctica, subantarctic Heard and Macquarie Islands, and Australasia, *Polar Biology* **27**, 687-698.

Skotnicki, M.L., Selkirk, P.M., Kitajima, E., McBride, T.P., Shaw, J. and Mackenzie, A. (2003) The first subantarctic plant virus report: *Stilbocarpa* bacilliform badnavirus (SMBV) from Macquarie Island, *Polar Biology* **26**, 1-7.

Smith, S.D.A. and Simpson, R.D. (1995) Effects of the "Nella Dan" oil spill on the fauna of *Durvillaea antarctica* holdfasts, *Marine Ecology Progress Series* **121**, 73-89.

Smith, V.R. (1978) Animal-Plant-Soil Nutrient Relationships on Marion Island (Subantarctic), *Oecologia* **32**, 239-253

Smith, V. (1993) Climate change and ecosystem functioning - A focus for sub-Antarctic research in the 1990s, *South African Journal of Science* **89**, 69-70.

Smith, V.R. and Steenkamp, M. (1990) Climatic change and its ecological implications at a subantarctic island, *Oecologia* **85**, 14-24.

Stevens, M.I. and Hogg, I.D. (2006) The molecular ecology of Antarctic terrestrial and limnetic invertebrates and microbes, in D.M. Bergstrom, P. Convey, and A.H.L. Huiskes (eds.), *Trends in Antarctic Terrestrial and Limnetic Ecosystems: Antarctica as a Global Indicator*, Springer, Dordrecht (this volume).

Tweedie, C.E. and Bergstrom, D.M. (1999) Time series analysis of 50 years of surface air temperature records from subantarctic Macquarie Island: a baseline study for predicting changes to terrestrial ecosystem structure and function, in *Proceedings from the Ninth ITEX Meeting, January 5-9, 1999, Arctic Laboratory Report 1*, 69-73.

Van de Vijver B., Frenot, Y. and Beyens, L. (2002) Freshwater diatoms from Ile de la Possession (Crozet Archipelago, Subantarctica), *Bibliotheca Diatomologica*, Cramer Verlag, Berlin-Stuttgart.

Walton, D.W.H. and Shears, J. (1994) The need for environmental monitoring in Antarctica - baselines, environmental impact assessments, accidents and footprints, *International Journal of Environmental Analytical Chemistry* **55**, 77-90.

Whinam, J., Chilcott, N. and Bergstrom, D.M. (2005) Subantarctic hitchhikers: Expeditioners as vectors for the introduction of alien organisms, *Biological Conservation*, **121**, 207-219.

Weimerskirch, H., Brothers, N. and Jouventin, P. (1997) Population dynamics of wandering albatross *Diomedea exulans* and Amsterdam albatross *D. amsterdamensis* in the Indian Ocean and their relationships with long-line fisheries: Conservation implications, *Biological Conservation* **79**, 257-270.

Young, S.B. and Schofield, E.K. (1973) Pollen evidence for late quaternary climate changes on Kerguelen Islands, *Nature* **245**, 311-312.

15. ANTARCTIC TERRESTRIAL AND LIMNETIC ECOSYSTEM CONSERVATION AND MANAGEMENT

B.B. HULL

Department of Environment and Heritage
Australian Government Antarctic Division
203 Channel Highway
Kingston, Tasmania 7050, Australia
bruce.hull@agad.gov.au

D. M. BERGSTROM

Department of Environment and Heritage
Australian Government Antarctic Division
203 Channel Highway
Kingston, Tasmania 7050, Australia
dana.bergstrom@agad.gov.au

Introduction

Antarctica's natural environment is known globally for its attributes such as wilderness and aesthetic values and as being a unique and valuable locality to conduct science. This chapter explores the current framework that exists for conservation and management of terrestrial and limnetic ecosystems under the Antarctic Treaty System (ATS), ie applying to continental and maritime Antarctica south of 60°S. We provide a brief history of human activities in the region and discuss the formation and operation of the ATS and the Madrid Protocol. We then consider knowledge management and examine the Antarctic continent within the context of the four big conservation issues facing the world in the first decade of the 21st Century: (i) local impacts and habitat loss, (ii) homogenisation of biota, (iii) the impact of climate change, and (iv) harvesting and removal of resources. Finally we provide a critique of management of these issues within the ATS regime and consider possible directions for the future.

D.M. Bergstrom et al. (eds.), Trends in Antarctic Terrestrial and Limnetic Ecosystems, 317–340.
© *2006 Springer.*

Humans and Antarctica

Perhaps one of the first interactions between humans and the Antarctic region was the Polynesian voyage of Ui–te–Rngiora. Legend tells of him and other Polynesians sailing south to the frozen ocean, known as Te tai–uka–a–pia, in about 650CE (Headland 1989). However, it was not until the mid 16th Century that the first recorded extensive harvesting of Antarctic resources took place. Francis Pretty, accompanying Sir Francis Drake on his second circum–navigation from 1577 to 1580, when south of Tierra del Fuego, recorded that Drake and his crew found:

> *"great store of foule which could not fly with the bigness of geese, whereof we killed in*
> *lessee than one day three thousand and victualled ourselves thoroughly* (Pretty 1910)."

Drake's victualling was the precursor to the massive harvesting of Southern Ocean whales and seals from the late 18th to the early 20th Centuries, following the discoveries and publication of reports by Cook, Kerguelen and Bellingshausen. The extensive exploitation of seals led James Weddell, a sealer himself, to propose what might well be the first, although unheeded, Antarctic conservation measure:

> *"the fur seal, might, by a law similar to that which restrains fisherman in the size of the*
> *mesh of their net, have been spared to render annually 100,000 furs for many years to*
> *come. This would have followed from not killing the mothers till the young were able to*
> *take to the water; and even then, only those which appear to be old, together with a*
> *proportion of the males, thereby diminishing their numbers, but in a slow progression"*
> (Weddell 1825)."

The last decade of the 19th Century saw the major scientific investigations and explorations of the Heroic Era, also described by Headland as the era of 'continental penetration' (Headland 1989). Adrian Gerlache, leading the Belgian Antarctic Expedition in the *Belgica,* involuntarily spent the winter of 1898 drifting in the pack ice south of Peter I Øy. Carstens Borchgrevink of the British Antarctic Expedition led the first continental wintering expedition at Cape Adare in 1899. Other wintering expeditions followed: Drygalski's German South Polar Expedition (1901–03), the British National Antarctic Expedition under Scott (1901–04), Nordenskjold's Swedish South Polar Expedition (1901–03), the Scottish National Antarctic Expedition commanded by Bruce (1902–04), and Mawson's Australasian Antarctic Expedition (1911–14). These expeditions left evidence of their activities, with 11 of the remaining historic huts in Antarctica being from this period.

Following these exploratory campaigns, Argentina, Australia, Chile, France, New Zealand, Norway and the United Kingdom all claimed parts of Antarctica, covering approximately 85% of the continent. A number of expeditions were active during the period between World Wars I and II, with exploration often being associated with whaling activities. The expeditions of the Heroic Era were the predecessors of an unremitting and increasing flow of visitors: explorers, scientists, tourists and adventurers to Antarctica that continues to this day. The 1950s saw the establishment of large permanent bases in Antarctica by Argentina, Australia, France, Great Britain, Chile and the United States of America.

Genesis of the Antarctic Treaty 1959

Momentum for the development of a treaty on Antarctica emanated from the political and military conditions of the world following World War II. Conflict was developing in Antarctica in concert with the Cold War between the Western powers and the Soviet Union and its allies. Counter to the conflict of the period, the International Geophysical Year (IGY) took place from 1957 to 1958 in an atmosphere of unprecedented scientific cooperation (Vicuna 1985). These two counter points provided impetus for the creation of the Antarctic Treaty and in the year following the IGY, the USA convened a conference on the topic, inviting States with an interest in Antarctic affairs (comprising the 12 national IGY participants) to Washington in late 1959.

Foremost in the minds of the Countries that met in Washington in 1959 to agree the Antarctic Treaty was to recognise:

> *"...that it is in the interests of all mankind that Antarctica shall continue forever to be used exclusively for peaceful purposes and shall not become the scene or object of international discord..."*

The conference was *"a remarkable diplomatic achievement"* (Dodds 2002) which defused States' differing positions on sovereignty and territorial claims, and preserved and promoted the international cooperation in scientific research developed during the IGY (Scully 1985). Despite the challenging nature of a number of the issues, the Treaty was quickly concluded and signed on 1 December 1959. It was ratified by the 12 participating IGY nations and, additionally, Poland, and entered into force on 23 June 1961. There are now 45 signatories to the Treaty.

The preamble to the Treaty provided the principles by which Antarctica has been governed for the last 45 years: Antarctica is to be used exclusively for peaceful purposes, and for the promotion of international cooperation through freedom of scientific investigation. The Treaty prohibits activities of a military nature, nuclear explosions and the disposal of radioactive waste in Antarctica. Military personnel and equipment however, were (and still are) allowed to support scientific research.

Although the focus of the Treaty was primarily international cooperation and scientific research in Antarctica, Article IX recognised the need for "measures to conserve the living resources". The Agreed Measures for the Conservation of Antarctic Fauna and Flora were adopted in 1964, however it took more than 20 years for these measures to come into effect (Ministry of the Environment New Zealand 2006). Despite this delay, awakening environmental awareness of the Treaty System is demonstrated by that approximately 60% of the total 330 Recommendations, Measures and Decisions that have been made between 1961 and 2005 have been environmentally based. The incremental awareness of environmental matters has reflected changing attitudes within Treaty countries, and this trend led to the adoption of the Protocol on Environmental Protection to the Antarctic Treaty (the Madrid Protocol) in 1991.

The suite of instruments that now comprises the ATS includes the Antarctic Treaty, the Agreed Measures for the Conservation of Antarctic Fauna and Flora 1964, the Convention for the Conservation of Antarctic Seals 1972, the Convention on the Conservation of Antarctic Marine Living Resources 1980, the Convention on the regulation of Antarctic Mineral Resource Activities 1988, (although this has never been ratified), the Protocol on Environmental Protection to the Antarctic Treaty 1991 (the Madrid Protocol) and, most recently, the Agreement on the Conservation of Albatrosses and Petrels 2004. Subordinate instruments such as 'Measures', 'Decisions' and 'Resolutions' supplement the Treaty. These interrelated instruments and recommendations provide the framework for the management of activities and protecting environmental values in Antarctica. Consultative Parties are expected to implement the requirements through appropriate national processes. Legally binding Measures come into effect when all Parties have done so, unless reservations have been entered. Heap (1991) described the legal framework of the Treaty as one that is:

"hortatory rather than mandatory in character — it cajoles rather than orders"

Governance of Antarctica through the Treaty is provided by means of an annual forum, the Antarctic Treaty Consultative Meeting (ATCM). Parties meet to exchange information, consult on matters of common interest, and formulate, consider, and recommend to their respective Governments further measures. In addition, the Treaty provides a framework for national jurisdiction over visitors to the continent. A vote in the consensus–based ATCM is conferred on those of the 45 signatories of the Treaty who have demonstrated a substantial commitment to the Antarctic by the establishment of a research station or the undertaking of a significant research program.

The need to foster cooperation across a number of Antarctic issues was of concern to the original drafters of the Treaty (Kimball 1988). The ATCM adopted Recommendation I–IV that outlined the role and provision of scientific advice by the Scientific Committee on Antarctic Research (SCAR, see SCAR 2006) and Recommendation I–V that supported the establishment of bilateral relations between Treaty Governments, United Nations agencies and other organisations. The Parties also readily accepted an offer from the World Meteorological Organisation (WMO) to cooperate in meteorology and data collection, and recommended that Governments pursue cooperation through their representatives in the WMO. The Fourth Consultative Meeting, held in Santiago (Chile) in 1966, made provision for meetings of experts to provide advice on practical problems relating to Antarctic activities. Meetings could be called from time to time as the need arose, and be attended by experts of the Consultative Parties, who could also invite other experts with the agreement of all Consultative Parties. Meetings of experts have now been held covering issues such as environmental monitoring and tourism.

However, until the early 1980s, the ATS and its associated bodies remained to some degree insulated from the mainstream of international environmental governance and convention making. This was challenged by debates promoted by a group of non–Antarctic States on the "Question of Antarctica" in the United Nations General Assembly (Beck 2004), and the international debate surrounding the issue of mining in Antarctica. Thus from 1983, those countries that had ratified the Treaty but were not Consultative Parties were invited to take part in the ATCMs, although not allowed to participate in decision-making. In addition, the ATCM was opened up to a greater number of non–governmental environmental organisations. For example, UNEP (United Nations Environmental Program), IUCN (International Union for the Conservation of Nature) and ASOC (Antarctica and the Southern Ocean Coalition) attended as invited experts, as did (more recently) the International Association of Antarctica Tour Operators (IAATO). Since the 1980s, the trend within the ATS has been towards more open governance, greater participation and increased environmental regulation.

The Protocol on Environmental Protection to the Antarctic Treaty (Madrid Protocol)

The Madrid Protocol was very quickly developed and adopted after rejection of Convention on the Regulation of Antarctic Mineral Resource Activities (CRAMRA), which sought to regulate prospecting, exploration, and exploitation of minerals. The Madrid Protocol, which entered into force in January 1998, designates Antarctica as "*a natural reserve devoted to peace and science*" and provided for comprehensive environmental protection, the pursuit of scientific activity and a ban on mining in the Antarctic.

The Madrid Protocol established a fundamental set of Environmental Principles by which all activities in Antarctica are to be planned and conducted so as to limit adverse impacts on the environment. Matters identified as being worthy of protection included: the atmosphere, climate, weather patterns, air and water quality; terrestrial (including aquatic), glacial and marine environments; the distribution, abundance and productivity of species or populations of fauna and flora; endangered or threatened species or populations and areas of biological, scientific, historic, aesthetic or wilderness significance.

Article 11 of the Protocol established the Committee for Environmental Protection (CEP, see CEP 2006) that provides advice and formulates recommendations to the ATCM on the operation of the Protocol and its Annexes. The ATCM may also refer other issues to the CEP. In particular, the CEP is expected to provide advice, where appropriate, on the following areas: minimisation or mitigation of environmental impacts and the application and implementation of the environmental impact assessment; procedures for response action in environmental emergencies; operation and improvement of the Antarctic protected area system; procedures and conduct of inspections; collection and exchange of information relating to environmental protection; the state of the Antarctic environment; the need for scientific environmental research, including environmental monitoring.

The annexes to the Protocol deal with specific environmental management and protection concerns. The original Annexes make provision for environmental impact assessment, conservation of Antarctic fauna and flora, waste disposal and management, prevention of marine pollution, and for area protection and management. At the ATCM XXVIII in 2005, Parties adopted a sixth Annex to the

Protocol, which covers liability arising from environmental emergencies. Of immediate relevance to terrestrial and limnetic environments are the four annexes (I, II, III and V) dealing with environmental impact assessment, waste management, conservation of flora and fauna, and area protection and management.

Annex I established the procedures for prior environmental impact assessment of all activities or changes in activities undertaken in the Antarctic Treaty Area. Prior assessment of activities is undertaken to establish whether the activities are identified as having one of three levels of impact on the Antarctic environment and associated ecosystems: (1) a less than a minor or transitory impact, (2) a minor or transitory impact, and (3) more than a minor or transitory impact. The Parties have resolved however, in revised guidelines for the preparation of environmental impact assessments, that it no agreement can be reached on the definition of the term 'minor or transitory' as no suggested frameworks have been universally acceptable. The definition of impacts has therefore been based on national interpretations, on a site–specific and case-by-case basis (Antarctic Treaty Secretariat 2005).

Preliminary assessment of the impact of activities is undertaken in accordance with appropriate national procedures and, if the activity is determined to have "less than a minor or transitory impact", the activity may proceed. If the activity is determined to have a minor or transitory impact an Initial Environmental Evaluation (IEE) is prepared, and if the activity has more than a minor or transitory impact, a Comprehensive Environmental Evaluation (CEE) is prepared. The latter includes a description of the activity and condition of the site, possible alternative processes; estimation of the nature, extent, duration and intensity of immediate and cumulative impacts, monitoring activities, and mitigation and remediation measures. A draft CEE is made publicly available and circulated to all Parties and to the CEP before the ATCM at which it is to be considered. A final CEE is then prepared which addresses comments received from other Parties and is circulated to all parties before commencement of the activity. IEEs are not automatically made publicly available but are included in the required annual exchange of information among Treaty Parties made available on the Antarctic Secretariat's website (http://www.ats.aq).

Annex II sets out measures to protect the native fauna and flora of Antarctica. Taking or harmful interference is restricted to a limited number of circumstances, such as the collection for scientific study. The Annex also allows for the listing of any species of native mammal, bird or plant as a "Specially Protected Species", prohibits the importation of any non–indigenous species of plant or animal, and requires Parties to put in place precautions to prevent the introduction of non–indigenous micro–organisms.

General obligations for the disposal, management, and reduction minimisation of waste are imposed by Annex III. Parties are required to clean up past and present waste disposal sites (on land) and abandoned work sites, except where the site has been designated as an historic site or monument, or where its removal would result in greater adverse environmental impact than leaving the material in place. The Annex elaborates on the specific requirements for the removal of waste from the Treaty area, the disposal of waste by incineration, restrictions on disposal of waste onto ice–free areas or into fresh water systems, prohibition of the importation of PCBs, non–sterile soil, polystyrene beads or chips and similar packaging and pesticides, conditions for the discharge of sewage and domestic waste into the sea, and requirements that waste generated at field camps should be removed to base camps or ships when practicable. Finally, Parties are required to develop, exchange and review waste management plans, exchange information and appoint waste management officials to develop and monitor their plans.

Area protection and management is addressed in Annex V, which allows for the designation of Antarctic Specially Protected Areas (ASPAs) to protect outstanding environmental, scientific, historic, aesthetic or wilderness values. Antarctic Specially Managed Areas (ASMAs) are to assist in the planning and co–ordination of activities, avoid possible conflicts, improve co–operation among Parties or minimise environmental impacts. A permitting system is in place to control entry into and the undertaking of activities allowed for under the management plan for an ASPA. No permit is required to enter an ASMA, although activities are to be undertaken in accordance with a code of conduct contained in the ASMA's management plan. Parties may also nominate sites or monuments to the Historic Sites and Monuments List.

The current status of environmental knowledge, data and information exchange

In addition to relying on the ATS structure for the fundamentals of effective management of the Antarctic environment which is an entirely borderless biome, the ATCM requires effective mechanisms for gaining and exchanging environmental knowledge. Article III–c of the Antarctic Treaty states that Contracting Parties agree, where feasible and practicable that,

> *"scientific observations and results from Antarctica shall be exchanged and made freely available"*

Data capture and exchange has been one of the main advances underpinning Antarctic science over the last 20 years, and a major driver for enhanced international collaboration. This is partly due to substantial advances in all forms of digital technology, from the digital recording of data in the field using hand-held devices to substantial advances in Geographic Information Systems (GIS) technology and the establishment of global data repositories such as GenBank (GenBank 2006).

There has also been a shift in scientific outlook towards increased scientific data exchange, encouraged not least by developments in the attitudes and priorities of the funders of science towards integrative and interdisciplinary research programmes addressing 'big science issues'. Collaboration and exchange are a prerequisite for success in such ventures, and almost inevitably generate greater quantities of data requiring more complex analytical tools and hardware. National and international funding agencies are also typically requiring that all data collected during a project be made available to the global community after a nominal time.

Scientific data are exchanged in a number of ways within the Antarctic research community. Many national Antarctic programs now have national data centres that are linked though the Joint Committee on Antarctic Data Management (JCADM 2006) which promotes the establishment, coordination and support of the Antarctic Data Directory System (ADDS) and appropriate data management within the Antarctic scientific community. JCADM reports jointly to two bodies, SCAR and the Council of Managers of National Antarctic Programs (COMNAP 2006). The ADDS is a framework for linking metadata (indexes or records of data) and the data themselves, into Antarctic and global data directories such as NASA's Global Change Master Directory. Individual project data are managed at both the researcher level and at the national program level.

Globally, the number of Antarctic–wide databases is increasing, both in number and in their content. There are at least four topographical databases for the Antarctic continent. The Antarctic Digital Database (Fox and Cooper 1994, ADD–SCAR 2006), produced by the British Antarctic Survey on behalf of SCAR, is constructed from data at 50m contour intervals and with a horizontal accuracy of between 150m – 1km. However, this and the other databases only provide, at best, broad–brush perspectives and are still of insufficient spatial scale for many scientific research purposes. The development of a systematic environmental–geographical framework is also in its early stages (Antarctic Treaty Secretariat 2005). BAS also supports an extensive meteorological database on behalf of SCAR (READER 2006). Other databases contain physical and biological information. For example, SCAR's RiSCC Biodiversity Database (2006), housed at the Australian Antarctic Data Centre, holds over 80 000 terrestrial and limnetic biodiversity records. Increased awareness of the value of these databases with regard to their use in higher order scientific and environmental management questions is also increasing substantially (eg Peat et al. in press).

The current nature of Antarctic data, biological or otherwise, is extremely spatially disparate and fragmentary. Figure 1 demonstrates these characteristics for part of East Antarctica, showing the spatial coverage of the majority (78% of 590) of biological metadata from the Australian Antarctic program since 1988. This coverage generally reflects the pattern of human presence and research effort. Most of the major blocks of biological metadata in Fig. 1 correspond with the locations of permanently occupied research stations. Basic biodiversity information is still to be gained from many large ice–free areas in this region alone, including the Bunger Hills (66°10'S 100°52'E), the Prince Charles Mountains (71°25'S 67°14'E), far–northern Victoria Land (74°15'S 163°00'E), Edward VII Land (77°40'S 155°00'E) and Enderby Land (70°00'S 50°00'E). Incomplete geographical coverage is compounded by the taxonomic uncertainties that remain present for many taxa (Chown and Convey this volume), although the advent of molecular technology may deliver much–needed assistance (Stevens and Hogg this volume).

The transfer of terrestrial and limnetic environmental information within the ATS takes place principally through three pathways. First, SCAR may formally or informally advise the CEP and the ATCM of new information or understanding emerging from the scientific literature. Second, scientific knowledge may filter through national programs and become the basis of working or information papers submitted to the ATCM by Parties. Third, it may occur through individual scientists participating as either national representatives or in inter–sessional expert working groups or workshops. For example, in April 2006, New Zealand hosted an inter-sessional workshop on the impact of non–indigenous species on Antarctic ecosystems (http://www.anta.canterbury.ac.nz/resources/non-native species in the antarctic/non-native species.shtml). Outcomes of such meetings may be developed initially into a working or information paper (in this example ATCM–WP13 2006), eventually leading to a Recommendation or Measure.

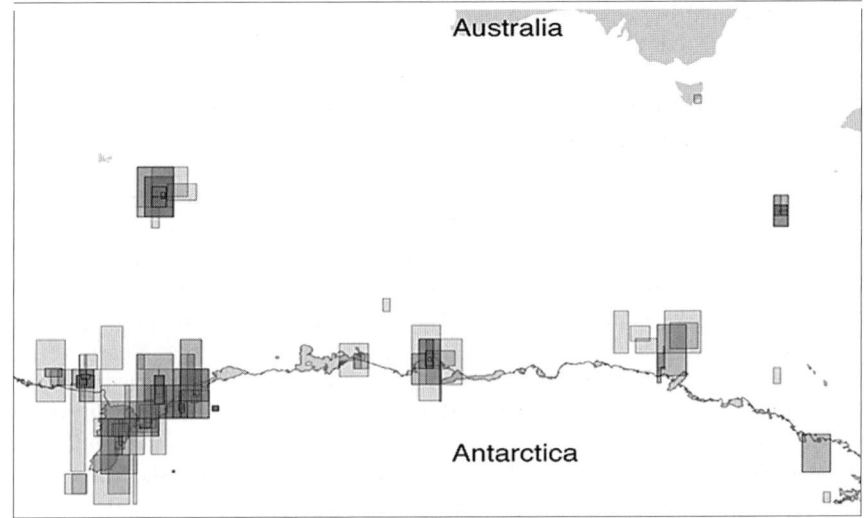

Figure 1. Spatial coverage of 590 metadata records of biological studies by the Australian Antarctic Program, mapped using 5° boxes. Darker intensities represent greater numbers of records. Five major concentrations are within the vicinity of Australian Antarctic or subantarctic stations. Data gained along ship transects are not included.

Current issues and threats

Globally, there a number of ubiquitous major environmental threats and issues. These are local impact of human activity and habitat loss, homogenisation of biodiversity, the impact of climate change and impacts associated with harvesting or resource extraction (although in terms of mining and agricultural practices these can equally be considered as habitat loss). Despite its isolation, Antarctica is not immune to such threats, however, the pressures from these threats are unequal.

LOCAL IMPACTS AND HABITAT LOSS

Local impacts can roughly be divided into habitat loss and disturbance of the environment and biota, and contamination. When considering habitat loss, it is important to understand just how little ice–free area there is in Antarctica. The Antarctic continent is approximately 14 million km^2 in area, of which the total ice–free area is only 46 000km^2 (0.33%, Fox and Cooper 1994), an area similar in size to Switzerland. It is easiest to consider the terrestrial and limnetic ecosystems of the Antarctic as a series of 'habitat islands' — either true islands or small areas of ice–free land isolated by surrounding ice or sea (Frenot et al. 2005). The degree of biological isolation or connectivity of these conceptual 'islands' is still one of the major questions to be addressed this century. The combination of limited ice–free land availability and its attractiveness to humans invites environmental pressure.

Most Antarctic research stations are typically sited within 2km of the coast, determined largely by ease of shipping and logistic accessibility. Totalling just 6000 km^2 (ca 0.05% of the total land area of the Antarctic), this ice–free coastal strip is also the only suitable habitat for most of the flora and fauna on the coast, excluding off–shore islands. Therefore, human habitation is inevitably concentrated in areas that have increased environmentally sensitivity (Poland et al. 2003) with subsequent loss of ice–free habitat. As of March 2006, 82 stations had been established in the Antarctic (including the maritime islands). Of these, 37 year–round stations and 12 significant seasonal stations, with a combined average winter population of 1030 and a peak summer population of scientists and support crew of 3760 people, were active during 2005 (COMNAP 2006). The Antarctic Peninsula has been disproportionately occupied by stations due to its high degree of accessibility from South America. The imminent International Polar Year (2007–09) may also encourage further infrastructure expansion to facilitate participation by national operators. Either in support of IPY or as part of Consultative Parties' on–going operations, numerous science support developments are planned, underway or have been built including six new stations, two station rebuilds, three station upgrades, and the development of four new transport links including air links (UN 2005).

The other significant source of human visitation to Antarctica is presently through commercial tourism, which is almost exclusively ship–based and overwhelmingly undertaken on islands of the Scotia Arc and on the Antarctic Peninsula, again due to the ease of accessibility from South American ports (Frenot et al. 2005). Tourist numbers have increased substantially since the early 1990s (Fig. 2). In the summer of 2004/05, approximately 27 950 tourists visited Antarctica, of which almost 23 000 were on voyages that made landings (IAATO 2006). Typically excluded in these annual figures is the number of support crew landings. In 2004/05, IAATO reported that staff and crew totalled around 18 000 and, of these, approximately 13 500 participated in voyages that made landings. Poland et al. (2003) noted that due to the very different nature of their operations, many more person–days are spent in Antarctica by those working for national programs based at established stations; however with the sustained increase in tourism, the difference is narrowing rapidly. In any event, much of the tourism is concentrated at sites characterised by abundant flora and fauna, and visitors come at the height of the breeding season when the melt is at its maximum and vegetation is most exposed.

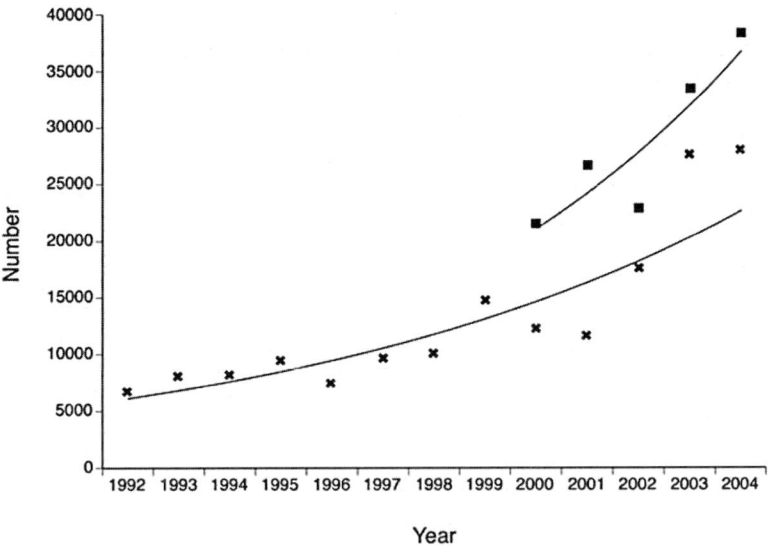

Figure 2. Trends in total tourist numbers, including voyages with and without landings. Crosses denote annual tourist number, and squares denote total numbers of tourists and crew. Exponential best–fit trendlines are shown for each series.

Trends in human usage and local impact vary between 'science and sovereignty' and 'tourism' modes. Most national programs are based at stations with a 'hub and spoke' footprint of their activities on their surrounding environment. Stations usually consist of permanent buildings close to a logistic facility – either a wharf or an airstrip (or both), with associated infrastructure including accommodation and operations buildings, roads, fuel storage facility, scientific laboratories and equipment, communication equipment (aerials, satellite dishes), garbage and sewage facilities and power generation facilities. Local areas can be contaminated or impacted by past and present activities such as fuel spills or construction, while the activities of personnel and their derived contamination (Table 1) can directly impact on the local flora and fauna.

Until the late 1980s, waste from Antarctic stations was routinely either dumped in local landfill sites or onto the winter sea–ice. Most waste disposal sites are undocumented (Stark et al. 2006), and it has been estimated "that contamination in the Antarctic is in the order of $1 - 10$ million m^3 of abandoned waste, with possibly a similar volume of petroleum–contaminated soil" (Snape 2003). Indeed, one of the major issues in the context of local impact is that of cumulative effects with contamination, in most cases, representing the effects of multiple activities over many years (Goldsworthy et al. 2003).

Stations are obvious areas of high impacts from human presence and usage (ie high human impact hours), but so too are areas accessible for fieldwork by researchers, and by support personnel for recreation. Impacts from research activities can include both long– and short–term physical damage, such as track formation, camping ground disturbance and point source pollutants from vehicles or equipment, while many scientific research activities inevitably disturb the flora or fauna. Recreational activities also risk local disturbance.

Table 1. Sources of contamination in the terrestrial environment generally associated with stations (derived from Poland et al. *2003, M.J. Riddle pers. comm.).*

Source of contamination	Notes
Long–range airborne contaminants	Broad range of organic and metal pollutants, acidifying compounds and radioactive contamination transported to Antarctica from remote sources (Bargagli et al. 2005). Often accumulated by top order predators such as South Polar skua *Catharacta maccormicki*.
Legacy waste and contaminants	Past waste disposal sites (garbage tips) and associated seepage and run–off, past spills and remains of past activities (eg buildings). Most common compounds from waste sites are hydrocarbons from fuel and lubricants, lead, copper and zinc.
Unavoidable operational emissions	Sewage effluent is often discharged in the form of macerate into the sea or into deep ice pits. Combustion products from generators incinerators and vehicles provide local point source pollutants. Hydrocarbon-based lubricants in ice and rock drilling activities.
Accidents	Contemporary spills of fuel, chemicals and lubricants.

Tourism activities are substantially different to those of national research programs, consisting generally of high numbers of people with low individual person-hours ashore. Trips typically involve multiple landings by small boats at several areas of scenic, natural or historic interest. Larger ships may land high numbers of people for a short period, repeating this pattern several times over a short period to allow all passengers the opportunity to land. Over a season, this can result in substantial numbers landing at particular sites. For example in the 2004/05 summer in the maritime Antarctic, 10 570 tourists landed at the historic whaling station at Whalers Bay, Deception Island; 10 523 people landed at Cuverville Island to view penguin colonies and 9452 landed at scenic Neko Harbour (northwest Antarctic Peninsula: IAATO 2006). Many of these sites have extensive ice–free habitats, with vegetated areas and freshwater bodies.

HOMOGENISATION OF BIOTA, INTRODUCTION OF NON–INDIGENOUS SPECIES

Homogenisation of biota, arises from the transportation and establishment of non–indigenous (alien) species. Frenot et al. (2005) and Convey et al. (this volume) have discussed this topic in detail, highlighting that as yet the Antarctic continent has not suffered the same pressure from introduced taxa as have the subantarctic islands. Chown and Convey (this volume) eloquently emphasise that, although warmer areas of the Antarctic (including the subantarctic islands) historically have incurred higher visitor frequencies, energy availability and the number of established alien species are also positively correlated (see also Chown et al. 2005). Thus, it is understandable that the number of non–indigenous species established is presently lower in the maritime and continental Antarctic than in the subantarctic. Although, at least for some locations, there is now a reasonable level of baseline knowledge of the presence and impacts of alien species (higher plants, vertebrates, larger invertebrates) in the Antarctic (Frenot et al. 2005, Convey et al. this volume), much less is known about much of the microfauna or microbial taxa. The recent workshop on the non–indigenous biota in Antarctica and the associated Treaty submissions (ATCM-WP13 2006) nevertheless demonstrates that this issue is now identified as a key threat and issue.

National programs transport a wide variety of material into terrestrial locations in the Antarctic including, but not limited to, vehicles, scientific equipment, fresh food and building supplies. Dartnall's (2005) and Whinam et al.'s (2005) studies both note that scientific equipment is often used at more than one location, potentially being the vector for intra–regional transfer of propagules between field sites. Inter– and intra–continental transport of propagules, both alien and regionally indigenous, has only recently been recognised as an important issue (Frenot et al. 2005). In recognition of this, in 2004 the RiSCC program developed a *Code of Conduct for Fieldwork* enabling scientists to minimise the chance of introduction of alien taxa, and to reduce the risk of accidental transfer of taxa between major ice–free localities (Appendix 1).

Tourist operations offer different threats with regard to the introduction of alien species. Being mainly ship–based, there are reduced risks associated with cargo and foodstuffs. However, the increasing numbers of tourists and crew participating in landings results collectively in a significant potential propagule load. Five patterns in tourist visits have been identified (Naveen et al. 2001, Frenot et al. 2005): (a) tourist visits are disproportionately located at sites of high/medium bio– and environmental diversity (b) the intensity of use by visitors is increasing; (c) the sites of high popularity are not consistent, thus the potential for human impact varies with time; (d) the spectrum of tourist activities is changing, to include more locations with a greater range of activities on offer; (e) typical visitor itineraries include successive visits to a number of sites within the Antarctic. In the Scotia Arc region, these often include a transect from the relatively diverse Tierra del Fuego and Falkland Islands, to the subantarctic island of South Georgia to progressively more extreme locations along the Scotia Arc archipelagos and Antarctic Peninsula, carrying a very clear risk of step–wise transport of likely pre–adapted biota between successive sites visited. Antarctic tourism, additionally attracts the 'luxury' element of the tourist market, meaning that expectations of a memorable, 'once–in–a– lifetime' travel event are high, and inevitably pressurises operators to offer that 'little bit more'.

CLIMATE CHANGE

Bergstrom et al. (this volume) place contemporary climate change into the perspective of the last million years, while Convey (this volume) and Lyons et al. (this volume) discuss the current and potential impacts of climate change on terrestrial and freshwater organisms and ecosystems. In the context of the last half million years, the Earth is experiencing a rare warming phase, while levels of atmospheric CO_2 and their rate of increase are unprecedented within this (and probably a far longer) timeframe. Many species present within the Antarctic region are at their distributional limit (mostly southern), while the entire distributions of regional endemic species are included. Four biologically significant parameters are currently demonstrating change in some areas of the Antarctic region (Convey this volume): temperature, free water availability, UV–B radiation and carbon dioxide levels. These parameters are closely interlinked. Much standard climatic observation and modelling activities have yet to be undertaken at spatial scales relevant to that of Antarctic life, while there are insufficient long–term field monitoring or holistic field manipulative experiments to allow accurate predictions of impacts (Convey this volume). However, one disturbing possibility identified is that climate change processes are likely to interact synergistically with the increasing direct human impacts in terms of alien species introductions, to give a greater chance of successful transport to and establishment of these aliens in the Antarctic environment (Frenot et al. 2005, Convey et al. this volume).

RESOURCE EXTRACTION

The final major environmental issue is that of harvesting or the removal of resources from Antarctic ecosystems. Currently, there is no major harvesting of any biota from terrestrial environments in the Antarctic region. Although most sealing occurred in the subantarctic with extensive destruction of fur and elephant seal populations, small–scale, harvesting of pack–ice seals (mainly for food for dog teams) took place up to 1985 (Kock 2006). Commercial fisheries in the Southern Ocean focus on a relatively small (in global terms) Antarctic krill (*Euphausia superba*) fishery based in the South Atlantic sector, and a more widely spread but still small fin–fish fishery (Kock 2006). Management of all legal fisheries, with an aim of sustainable harvesting, is the responsibility of the Commission for the Conservation of Antarctic Marine Living Resources (CCAMLR 2006). In the context of this chapter and volume, we are unaware of any study that has examined whether these fisheries have impacts on 'marine to terrestrial' ecosystem processes such as nutrient transfer (see Erskine et al. 1998 for an example of such processes).

Small–scale collection of biological material for 'bio-prospecting' is occurring in Antarctica, and the extraction of genetic resources for commercialisation is an emerging issue globally. Lohan and Johnston (2003) noted that genes and resultant compounds obtained from Antarctic organisms have the potential to be valuable in a range of applications including novel bioactives such as cold–adapted enzymes, novel pigments, and antifreeze use in biotechnological, transgenic and pharmaceutical endeavours. They further reported that the European Patent Office had 62 registered patents that relied upon Antarctic biodiversity and another 92 applications for patents in the US Patent Office, although search of the United States Patent and Trademark Office's (2006) database lists 500 patents with the word 'antarctica' mentioned within their descriptions. These figures do not differentiate between marine and terrestrial/freshwater biodiversity, while many refer to temperature–stable, enzyme–based processes (particularly lipidases) derived from the alkaline–tolerant yeast *Candida antarctica,* which was initially collected from Lake Vanda in the Victoria Land Dry Valleys (Lohan and Johnston 2003). The main issues currently arising from bio-prospecting within Antarctica are political or legal rather than environmental, including ownership of intellectual property, benefit sharing, and freedom of scientific information exchange (Antarctic Treaty Secretariat 2005). The CEP has concluded that environmental impacts of sampling for bio-prospecting are small (as opposed to harvesting), and sufficiently covered within the environmental impact assessment (EIA) procedures in Annex X of the Protocol (ATCM 2003).

The other major resource extraction issue in the Antarctic concerns mineral resources. Berkman and Woyach (1992) implied that there was a dramatic rate of increase in signatories to the Antarctic Treaty after the world oil crises of the early 1970s, however as stated above, the Convention on the Regulation of Antarctic Mineral Resource Activities of 1988 has never been ratified. Article 7 of the Madrid Protocol states:

> *"any activity relating to mineral resources, other than scientific research, shall be prohibited"*

The Madrid Protocol, including the mining prohibition, can be modified at any time if all parties agree. If there is less than consensus for modification, and if a party so requests, a review conference may be called after 2048 to review the Protocol. The Protocol may then be modified, provided that it is agreed by a majority of the Treaty Parties at the time of the review, including ¾ of the Consultative Parties at the time of the adoption of the Protocol (1991). In addition, it is also necessary to have in place a binding legal regime for managing mining and which protects the sovereign interest of parties for mineral resource activities to commence. With current increases in world oil prices and the rapid and substantial increase in the demand and exploitation of fossil fuels and mineral resources in both emerging and established economies, this current prohibition on minerals activity in the Antarctic may come under threat. Mining could, of course, happen at any time by nationals of a state not party to the Antarctic Treaty and the Madrid Protocol. However, as this situation does not, at present, pose any current environmental threat and that bio-prospecting is effectively covered by the Environmental Impact Assessment process, no further attention to this topic will be given here.

Management of current environmental change issues within the ATS

Traditionally Antarctica has been viewed as an isolated, insular and extreme continent carrying the appellation of being the highest, coldest, driest and windiest of all continents and cocooned from the rest of the world by ocean currents, deep water and distance (Clarke 2003). However, that view has radically changed in recent decades, with recognition that:

> "the Antarctic is no longer remote and disconnected from the rest of the world: global environmental processes are inextricably linked with Antarctica; technology and wealth ensure the Antarctica is no more immune from resource pressures than any other part of the planet; legally and politically, Antarctica is part of the thinking of decision makers and regulators" (Press 2001)

This more global view has not yet universally permeated environmental management and conservation in the Antarctic. Our thesis in this concluding section is that full acknowledgment that Antarctica is an important part of a complex global system, both environmentally and economically, will require Parties to look for more innovative approaches in responding to global environmental concerns and their impact in Antarctica. This is distinct from their previous concentration on the management of localised impact from activities in Antarctica itself. Such an approach calls for more external engagement than currently is seen in the Parties' consensual approach to decision making. This may be seen as a challenge, as some may fear fracturing what has come to be seen as a successful international regime that has survived intact for 45 years and provides a model for other international treaties.

LANGUAGE OF THE TREATY

International Treaties, by their very nature use ambiguous language and diluted terminology, and the Madrid Protocol is by no means an exception. While this helps to preserve and facilitate a consensus approach and allows party states to participate at a level that their national geopolitical environment permits, it also risks introducing ambiguity in standard setting and implementation. This is clearly demonstrated within the environmental impact assessment process (see below).

LOCAL IMPACTS

There are a number of major issues concerning local impacts: standard setting and cumulative impacts, waste management, protected sites and visitor pressure. The lack of agreement concerning the meaning of "minor or transitory impact" has resulted in applications of differing standards in assessment of environmental impacts across Consultative Parties' jurisdictions (Bush 2000, Berguño 2000). For example, between 1993 and 2005, a total of 22 proposals were prepared and submitted to the ATS relating to environmental impact assessments for projects based around ice or rock drilling activities. Of these, 13 were prepared as IEEs (Initial Environmental Evaluation) and nine as CEEs (Comprehensive Environmental Evaluation). This illustrates differing interpretations of the potential for local environmental impact among Parties for broadly similar activities. CEEs are subject to review by Parties that, in principle, leads to a higher standard of assessment of the activity. Hemmings and Roura (2003) have also noted that CEEs are generally of high quality but that the number produced was not commensurate with the actual level of activity in the Antarctic. They also found greater variability in the quality of IEEs. A review (COMNAP 2004) of the IEEs by the Antarctic Environment Officers Network within COMNAP concluded that the:

> "Treatment of cumulative impacts was limited or missing in most IEE documents reviewed"

The concept of cumulative impacts has yet to attain the high level of awareness needed for the adequate protection of Antarctica in both the scientific and tourism fraternities, despite the recent review of the Environmental Impact Assessment guidelines (Antarctic Treaty Secretariat 2005). This mainly arises through a lack of scientific studies and monitoring demonstrating cumulative impacts, and the lack of established criteria of assess cumulative impacts.

The scientific community can offer the ATS substantial support with regard to providing spatial, temporal and biological frameworks to define impact, and the baseline data and the means to monitor ecosystem health. Furthermore, applied scientific study can define criteria and monitoring systems for assessing cumulative impacts (Anon. 2005). State of Environment reporting is just one example in which scientific pursuits have assisted environmental management. This is a developing area within Antarctic science, with significant investment already by JCADM and COMNAP.

Waste management currently lacks clarity within the ATS. Annex III of the Protocol commits Parties to remediation of past and present work sites where practicable, and there is limited (although growing) research and remediation being undertaken. However there are no universally accepted guidelines or binding processes for assigning past liability for environmental damage (Snape 2003). Snape (2003) contrasts this with the widely perceived view of Antarctica being of global environmental significance and a symbol of good environmental stewardship. Stark et al. (2006) noted, however, that there has been a major paradigm shift by Parties towards recognition of the need to remediate sites, though also emphasising that while remediation is currently socially and culturally acceptable, it is economically unpalatable and often in direct budget competition with other Antarctic science. Nonetheless, major scientific advances have been made in contaminated site assessment, ecotoxicology, and site remediation (Snape 2003) with an ever–growing science network among polar remediation scientists, with Antarctic science playing a lead role. This is demonstrated by a bi–annual series of contaminants in freezing ground conferences (http://www.freezingground.org/Portal/index.htm).

A major advance in terms of local environmental damage is the recently agreed Liability Annex (Antarctic Treaty Secretariat 2005). This agreement provides a limited framework for responding to environmental emergencies. The annex provides for compensation if the operator responsible for an incident fails to respond and another party carries out the needed response.

The establishment of protected status for locations has improved significantly with the introduction of a more uniform system of ASMAs and ASPAs. The selection of Antarctic protected areas, however, has largely been an *ad hoc* process, inevitably therefore not ensuring systematic and representative coverage of the Antarctic environment (Harris and Woehler 2004). SCAR (SCAR 2005) noted that:

> *"whilst there is a wide range of measures in place for the conservation and sustainable use of biodiversity in the region, rational, spatially explicit conservation planning, which has been undertaken for many other parts of the globe, lags far behind in the Antarctic realm"*

SCAR further observed that, despite some spatially-explicit data being available for some groups such as seabirds and terrestrial biodiversity, as a consequence of the requirements for the establishment of ASPAs and ASMAs, these data are yet to be used to develop a formal, spatially-explicit conservation planning framework for Antarctica (but see Cooper et al. 2000).

There are positive moves within the ATS and SCAR towards developing a more systematic approach to conservation. These are both stimulated by, and building on, the construction of geographic and biodiversity databases, such as New Zealand's Antarctic Geographic Domains (ATCM–WP32 2006) and the SCAR RiSCC biodiversity database (http://www.aad.gov.au/default.asp?casid=3968). The application of such databases within frameworks of conservation theory, which has been aptly demonstrated for a subantarctic island (Chown et al. 2001), in concert with funding of biodiversity research, will provide a more solid scientific basis for more effective management decisions.

One of the recurrent themes identified throughout this volume is the concept of Antarctic isolation. The degree of connectivity or isolation of the various ice–free areas of the continent remains poorly known. It is clear that most parts of the Antarctic cannot simply be described as 'pristine', as a consequence of aerosol pollution such as PCBs and heavy metals (Bargagli et al. 2005). Some areas however, due to a combination of geography, geology and geomorphological history, and extremely limited or zero direct human impact through visitation, can aptly be described as the most isolated ecosystems on the planet. It is recognised that these areas must be identified and allocated the highest environmental protection possible. New scientific data at the species level (Gibson et al. this volume) and at the population and gene flow level (Skotnicki and Selkirk this volume, Stevens and Hogg this volume) are essential for such Antarctic conservation measures to be effective.

The final major issue at the local scale regards the rapid increase in visitor pressure on Antarctica, and links with the previous point in the context of the breakdown of isolation. Here, the ATCM is in the initial stages of grasping the implications of this growth. Hemmings and Rouse (2004) described a:

> *"conflict between the embedded Antarctic Treaty System priority of protecting the Antarctic environment and the priority to open, low–cost access to Antarctica that commercial ventures assume or argue for"*

Without mitigating measures, tourism will carry increased propagule loads to Antarctica, and sequential visits will have significant impacts at local sites. There is currently a lack of consideration of the cumulative impacts of multiple visits to the same location, and Hemmings and Roura (2004) concluded that there was no established procedure to assess the overall impact of tourist visits at any one site. Within the ATS, no tourism activity has been the subject of a CEE, and only limited use has been made of IEEs. The response of the ATS has been to begin to establish site guidelines (Antarctic Treaty Secretariat 2005), particularly with reference to mitigating impacts on charismatic megafauna. However, there appear as yet to be no monitoring strategies in place assessing the impact of tourism activities on terrestrial and limnetic ecosystems, and there is no 'collective testing' of the multiple predictions of minor and transitory impact as required in Protocol. This is an issue begging scientific investigation.

Effective management of tourism activities to minimise environmental harm, follow–up of the required prediction of impacts, establishment of standards of practice and the regulation of operators who do not belong to the self–regulating IAATO remain issues confronting the ATS and CEP. Furthermore, within the framework of the emerging patterns of visitation described above, discussion of the twin concepts of concentrating tourism and identifying 'sacrificial sites', and integrating these within the ASPA or ASMA systems is just beginning.

CLIMATE CHANGE

Predicted impacts of climate change vary from positive, plastic metabolic responses of individual species (as associated flow effects such as improved competitive ability) to negative loss in biodiversity associated with changes in environmental conditions such as changed water regime and invasion by non–indigenous species (Frenot et al. 2005, Convey et al. this volume). Although Antarctica has relatively low terrestrial and limnetic biodiversity in terms of species numbers when compared to warmer biomes, one should note that value in biodiversity is not just related to species numbers alone, but should also include a measure of uniqueness and rarity from the genomic through to ecosystem levels.

Chown and Convey (this volume) and Gibson et al. (this volume) emphasised that as we further develop our understanding of Antarctic biodiversity, more evidence is being uncovered of important ancient and vicariant elements in the Antarctic biota. Such findings should send out clear signals of substantial, global biodiversity significance. Furthermore, the widespread use of enzymes from the endemic fungi *Candida antarctica* throughout the biotechnological industries demonstrates the potential value to human endeavours of such biota.

The Treaty, as an entity, has not engaged in world discussion concerning the impacts of climate change. Further, it chose to actively avoid engagement with other Treaties and Agreements (ATCM 1994) reinforcing its bilateral approach, leaving this up to individual nation states. Scientific research in Antarctica generates a considerable contribution to the global knowledge base on climate change and its impacts on ecosystems. This knowledge filters into the ATS as described above but mechanisms are currently lacking for ensuring its further transmittance.

The environmental principles of the Madrid Protocol identify atmosphere, climate and weather patterns as worthy of protection, however this issue can not be mitigated locally in Antarctica by Parties, as it is indeed widely acknowledged as a global issue. The ATS could consider playing a greater role in the direct facilitation of the transfer of regionally–derived information into the wider community to provide input into a process that has a large human component and will directly impact on Antarctic biodiversity. The Treaty and the Protocol provides the framework for such engagement as it urges co–operation with international organisations. Such actions, however, provide a challenge to the *status quo*. Rothwell (2000) has noted the Protocol has correlations and overlaps, and no conflicts, with other environmental regimes and conventions but was established in isolation and without mutual recognition of those treaties, thus reflecting ATS reluctance to engage with non–ATS fora.

Conclusions

The Antarctic Treaty has been recognized as a successful regime for the last 45 years. However, as we have argued, there is greater complexity in the current environment than was ever envisaged at the Treaty's birth. Global forces such as climate change and pollution impact on the Antarctic environment, while the Antarctic continent itself is a significant driving force of the world's climate. Any substantial impact of climate change will not only affect Antarctic biota and habitats, but the entire Earth's biota, particularly if sea level is affected. Thus, there appear to be strong imperatives for the ATS to improve engagement with the wider world. Furthermore, there is much to be learned from other environment management systems, such as a more systematic approach to environmental protection and linking with conservation theory, and the development of criteria and standards of risk assessment, impact assessment and remediation. As we have described above, there are specific areas that could be improved with regard to environmental management and conservation. This volume details major advances in the last 20 years in terrestrial and limnetic science; however we are still just scraping the 'tip of the iceberg'. If there is one major 'take home message', then that is that increased scientific knowledge of the natural environment will undoubtedly improve the setting in which sound environmental management decisions are made.

References

ADD-SCAR (2006) Antarctic Digital Database, http://www.add.scar.org/.

Anonymous (2005) *Practical Biological Indicators of Human Impacts in Antarctica.* Report of workshop, 16-18 March 2005, College Station, Texas, 2 volumes, pp.39, http://vpr.tamu.edu/antarctic/workshop/workshop.php.

Antarctic Treaty Secretariat (2005) Final report of the Twenty-Eighth Antarctic Treaty Consultative Meeting. Buenos Aires, pp. 700.

ATCM (1994) Final Report of the Eighteenth Antarctic Treaty Meeting, Antarctic Treaty Consultative Meeting, Kyoto, 11-22 April 1994.

ATCM (2003). Final Report of the Twenty–Sixth Antarctic Treaty Consultative Meeting, Madrid, 9–20 June 2003.

ATCM–WP13 (2006) Working Paper 13, Non–native Species in the Antarctic. Report of a Workshop, Antarctic Treaty Consultative Meeting, submitted by New Zealand.

ATCM–WP32 (2006) Working Paper 32, Systematic Environmental Protection in Antarctica – refining and reviewing the "proof of concept" Environmental Domains of Antarctica classification for a systematic environmental geographic framework, Antarctic Treaty Consultative Meeting, submitted by New Zealand.

Bargagli, R., Agnorelli, C. Borghini, F. and Monaci F. (2005) Enhanced deposition and bioaccumulation of mercury in Antarctic terrestrial ecosystems facing a coastal polynya, *Environmental Science and Technology* **39**, 8150-8155.

Beck, P.J. (2004) Twenty years on: the UN and the 'Questions of Antarctica,' 1983–2003, *Polar Record* **40**, 205-212.

Bergstrom, D.M., Hodgson, D.A. and Convey, P. (2006) The physical setting of the Antarctic, in D.M. Bergstrom, P. Convey, and A.H.L. Huiskes (eds.), *Trends in Antarctic Terrestrial and Limnetic Ecosystems: Antarctica as a Global Indicator*, Springer, Dordrecht (this volume).

Berguño, J. (2000) Institutional issues for the Antarctic Treaty System with the Protocol in force: an overview, in D. Vidas (ed) *Implementing the Environmental Protection regime for the Antarctic*, Kluwer, Dordrecht, pp. 93-106.

Berkman, P.A. and Woyach, R.B. (1992) International transitions in the Antarctic Treaty System, *Polar Record* **28**, 244-245.

Bush, W. (2000) Means and methods of implementation of Antarctic environmental regimes and national environmental instruments: an exercise in comparison, in D. Vidas (ed) *Implementing the Environmental Protection regime for the Antarctic*, Kluwer, Dordrecht, pp. 21-43.

CCAMLR (2006) Commission for the Conservation of Antarctic Marine Living Resources, http://www.ccamlr.org.

CEP (2006) The Committee for Environmental Protection, http://www.cep.aq.

Chown, S.L. and Convey, P. (2006) Biogeography, in D.M. Bergstrom, P. Convey, and A.H.L. Huiskes (eds.), *Trends in Antarctic Terrestrial and Limnetic Ecosystems: Antarctica as a Global Indicator*, Springer, Dordrecht (this volume).

Chown, S. L., Hull, B.B. and Gaston, K.J. (2005) Human impacts, energy availability, and invasion across Southern Ocean Islands, *Global Ecology and Biogeography* **14**, 521-528.

Chown, S.L., Rodrigues, A.S.L., Gremmen, N. and Gaston, K.J. (2001) World Heritage status and the conservation of Southern Ocean islands, *Conservation Biology* **15**, 550-557.

Clarke, A. (2003) Evolution, adaptation and diversity: global ecology in an Antarctic context, in A.H.L. Huiskes, W.W.C. Gieskes, J. Rozema, R.M.L. Schorno, S.M. van der Vries and W.J. Wolff (eds.), *Antarctic Biology in a Global Context*, Backhuys Publishers, Leiden, The Netherlands, pp. 3-17.

COMNAP (2004) An Analysis of Initial Environmental Evaluations (IEEs), Antarctic Treaty Consultative Meeting 27, Information Paper 15, pp. 8.

COMNAP (2006) Council of Managers of National Antarctic Programs. http://www.comnap.aq.

Convey, P. (2006) Antarctic climate change and its influences on terrestrial ecosystems, in D.M. Bergstrom, P. Convey, and A.H.L. Huiskes (eds.), *Trends in Antarctic Terrestrial and Limnetic Ecosystems: Antarctica as a Global Indicator*, Springer, Dordrecht (this volume).

Convey, P., Frenot, Y., Gremmen, N.J.M. and Bergstrom, D.M. (2006) Biological invasions, in D.M. Bergstrom, P. Convey, and A.H.L. Huiskes (eds.), *Trends in Antarctic Terrestrial and Limnetic Ecosystems: Antarctica as a Global Indicator*, Springer, Dordrecht (this volume).

Cooper, J., Woehler, E. and Belbin, L. (2000) Selecting Antarctic Specially Protected Areas: Important Bird Areas can help, *Antarctic Science* **12**, 129.

Dartnall, H.J.G. (2005) A new species of *Keratella* (Rotifer, Monogonata: Brachionidae) from South Georgia and the Falkland Islands, *Quekett Journal of Microscopy* **40**, 41-46.

Dodds, K. (2002) Pink ice: Britain and the South Atlantic Empire, Tauris, London, pp.229.

Erskine, P.D., Bergstrom, D.M., Schmidt, S., Stewart, G.R., Tweedie, C.E., and Shaw, J.D. (1998) Subantarctic Macquarie Island — a model ecosystem for studying animal-derived nitrogen sources using 15 N natural abundance, *Oecologia* **117**, 187-193.

Fox, A.J. and Cooper, A.P.R. (1994) Measured properties of the Antarctic ice sheet derived from the SCAR Antarctic digital database, *Polar Record* **30**, 201-206.

Frenot, Y., Chown, S.L., Whinam, J., Selkirk, P., Convey, P., Skotnicki, M. and Bergstrom, D.M. (2005) Biological invasions in the Antarctic: extent, impacts and implications, *Biological Reviews* **80**, 45-72.

GenBank (2006) GeneBank, http://www.ncbi.nlm.nih.gov/Genbank.

Gibson, J.A.E., Wilmotte, A., Taton, A., Van De Vijver, B., Beyens, L. and Dartnall, H.J.G. (2006) Biogeographic trends in Antarctic lake communities, in D.M. Bergstrom, P. Convey, and A.H.L. Huiskes (eds.), *Trends in Antarctic Terrestrial and Limnetic Ecosystems: Antarctica as a Global Indicator*, Springer, Dordrecht (this volume).

Goldsworthy, P., Canning, E.A. and Riddle M.J. (2003) Soil and water contamination in the Larsemann Hills, East Antarctica, *Polar Record* **39**, 319–337.

Harris, J.W. and Woehler, E.J. (2004) Can the Important Bird Area approach improve the Antarctic Protected Area System? *Polar Record* **40,** 97-105.

Headland, R.K. (1989) Chronological list of Antarctic expeditions and related historical events. Cambridge University Press, Cambridge, United Kingdom, pp. 730.

Heap, J.A. (1991) Antarctic politics and Antarctic Science — are they at logger heads? *Antarctic Science* **3**, 1.

Hemmings A.D. and Roura R. (2003) A square peg in a round hole: fitting impact assessment under the Antarctic Environmental Protocol to Antarctic tourism, *Impact Assessment and Project Appraisal* **21,** 13-24.

IAATO (2006) International Association of Antarctica Tour Operators, http://www.iaato.org.

JCADM (2006) Joint Committee on Antarctic Data Management, http://www.jcadm.scar.org/).

Kimball, L. (1988) The Role of Non-Government Organizations in Antarctic Affairs, in C.C. Joyner and S.K. Chopra (eds), *Antarctic Legal Regime*, Martinus Nijhoff Publishers, Dordrecht, pp. 33-63.

Kock, H–D. (2006) Understanding CCAMLR's approach to management: An Introduction, http://www.ccamlr.org/pu/E/e_pubs/am/p2.htm#1.2_The_History.

Lohan, D. and Johnston, S. (2003) *The international regime for bioprospecting: Existing policies and emerging issues for Antarctica*, UNU/IAS Report, Tokyo, pp. 1-24.

Lyons, W.B., Laybourn-Parry, J., Welch, K.A. and Priscu, J.C. (2006) Antarctic lake systems and climate change, in D.M. Bergstrom, P. Convey, and A.H.L. Huiskes (eds.), *Trends in Antarctic Terrestrial and Limnetic Ecosystems: Antarctica as a Global Indicator*, Springer, Dordrecht (this volume).

Ministry of the Environment, New Zealand (2006) Antarctica, http://www.mfe.govt.nz/laws/meas/antarctica.html

Naveen, R., Forrest, S.C., Dagit, R.G., Blight, L.K., Trivelpiece, W.Z. and Trivelpiece, S.G (2001). Zodiac landings by tourist ships in the Antarctic Peninsula region, 189-99, *Polar Record* **37**, 121-132.

Peat H.J., Clarke, A. and Convey, P. (in press) Diversity and biogeography of the Antarctic flora, *Journal of Biogeography.*

Poland, J.S., Riddle, M.J., Zeeb, B.A. (2003) Contaminants in the Arctic and the Antarctic: a comparison of sources, impacts, and remediation options, *Polar Record* **39**, 369-383.

Pretty, F. (1910) Sir Francis Drake's Famous Voyage Around the World, in C.W. Elliott (ed), *Voyages and Travels: Ancient and Modern,* Vol. XXXIII, The Harvard Classics. New York: P.F. Collier and Son, 1909–14, New York.

Press, A.J. (2001) Antarctica and the future, in J. Jabour–Green and M. Haward (eds) *The Antarctic: Past Present and Future*, Antarctic CRC Report 28, 153-167.

READER (2006) REference Antarctic Data for Environmental Research, http://www.antarctica.ac.uk/met/READER/).

RiSCC Biodiversity Database (2006) RiSCC Biodiversity Database, Australian Antarctic Division Data Centre, http://www.aad.gov.au/default.asp?casid=3968.

Rothwell, D.R. (2000) Relationships between the Environmental Protocol and UNEP Instruments, in D Vidas (ed), *Implementing the Environmental Protection Regime for the Antarctic*, Kluwer Academic Publishers, Dordrecht, pp. 221-241.

SCAR (2005) Biodiversity in the Antarctic, ATCM 28 — IP85, pp. 5.

SCAR (2006) Scientific Committee on Antarctic Research. http://www.scar.org.

Scully, R.T. (1985) The Evolution of the Antarctic Treaty System – The International Perspective, in *Antarctic Treaty System: An Assessment. Proceedings of a workshop held at Beardmore, South Field Camp, Antarctica, January 7-13, 1985,* Polar Research Board, Commission on Physical Sciences, Mathematics, and Resources, National Research Council, National Academy Press, Washington D.C. pp. 391-411.

Skotnicki, M.L. and Selkirk, P.M. (2006) Plant biodiversity in an extreme environment: genetic studies of origins, diversity and evolution in the Antarctic, in D.M. Bergstrom, P. Convey, and A.H.L. Huiskes (eds.), *Trends in Antarctic Terrestrial and Limnetic Ecosystems: Antarctica as a Global Indicator*, Springer, Dordrecht (this volume).

Snape I. (2003) The Third International Conference on Contaminants in Freezing Ground, Hobart, Tasmania, 14–18 April 2002, *Polar Record* **39**, 289–290.

Stark, J.S, Snape, I. and Riddle, M.J. (2006) Abandoned Antarctic waste disposal sites: Monitoring remediation outcomes and limitations at Casey Station, *Ecological Management and Restoration* **7**, 21-31.

Stevens, M.I. and Hogg, I.D. (2006) The molecular ecology of Antarctic terrestrial and limnetic invertebrates and microbes, in D.M. Bergstrom, P. Convey, and A.H.L. Huiskes (eds.), *Trends in Antarctic Terrestrial and Limnetic Ecosystems: Antarctica as a Global Indicator*, Springer, Dordrecht (this volume).

UN (2005) United Nations General Assembly Records Sixtieth Session. Report of the Secretary–General on the Question of Antarctica, 11 August 2005, A/60/222, pp. 22.

United States Patent and Trademark Office (2006) http://www.uspto.gov/patft/index.html.

Vicuna, F.O. (1985) Antarctic Conflict and International Cooperation, in *Antarctic Treaty System: An Assessment. Proceedings of a workshop held at Beardmore, South Field Camp, Antarctica, January 7-13, 1985,* Polar Research Board, Commission on Physical Sciences, Mathematics, and Resources, National Research Council, National Academy Press, Washington D.C., pp. 55-64.

Weddell, J. (1825) *A Voyage Towards the South Pole, Performed in the Years 1822–24. Containing an examination of the Antarctic Sea, to the seventy–fourth degree of latitude: and a visit to Tierra del Fuego, with a particular account of the inhabitants.* Longman, Hurst, Rees, Orme, Brown and Green, London, pp. 276.

Whinam, J., Chilcott, N. and Bergstrom, D.M. (2005) Subantarctic hitchhikers: expeditioners as vectors for the introduction of alien organisms, *Biological Conservation* **121**, 207-219.

Appendix 1. RiSCC protocol for minimising potential for introductions at field research sites.

Risk assessment
As part of the fieldwork planning process, the following simple risk assessment is conducted:

Risk assessment questions:
- a. Has any equipment/ equipment cases/field clothing/boots, planned for use in the subantarctic/Antarctica been used in other natural environments, particularly alpine or polar environments?
- b. What are the means needed to clean this equipment/equipment cases/clothing/boots?
- c. Will the field party be visiting more than one major locality?
- d. If yes, how will the field party ensure that equipment/equipment cases/clothing/boots do not carry diaspores between sites?

Fieldwork
The following recommendations are made with regard to field–work:

Field planning
If fieldwork requires moving between major ice–free localities, aim to conduct fieldwork in low diversity localities before high diversity localities.

Equipment
- 1. When designing field equipment, reduce the capacity of the equipment to carry additional material and make the equipment easy to clean and sterilise.
- 2. If equipment cannot be cleaned effectively, do not use this equipment to multiple major localities, but instead take multiple sets of equipment (eg plankton nets).
- 3. Be aware of where equipment cases are stored and that these cases do not accumulate dust or invertebrate infestations.
- 4. When cleaning items, be particularly vigilant in removing soil, seeds and bryophyte propagules (including leaves).

Outdoor clothing and boots and packs
If clothing cannot be cleaned with bleach or a similar compound, take new clothing/boots and packs. Be aware that items with Velcro can collect seeds (Whinam et al. 2005). Chose items with a minimum, or preferably, no Velcro.

Clean field items between sites. Be particularly vigilant in removing soil, seeds and bryophyte propagules (including leaves).

16. THE ANTARCTIC: LOCAL SIGNALS, GLOBAL MESSAGES

D. M. BERGSTROM
Department of Environment and Heritage
Australian Government Antarctic Division
203 Channel Highway
Kingston, Tasmania 7050, Australia
dana.bergstrom@agad.gov.au

A.H.L. HUISKES
Unit for Polar Ecology
Netherlands Institute of Ecology (NIOO-KNAW)
POB 140, 4400 AC Yerseke, The Netherlands
a.huiskes@nioo.knaw.nl

P. CONVEY
British Antarctic Survey, Natural Environment Research Council
High Cross, Madingley Road
Cambridge CB3 0ET, United Kingdom
p.convey@bas.ac.uk

Introduction

"Antarctica is a difficult continent to interpret" (Pickard 1986).

Twenty years after John Pickard summed up the complexity of the Antarctic region in this simple yet deeply profound statement, it is appropriate to ponder whether the situation has changed: how has our understanding of the bottom end of the Earth improved? Have there been any 'earth-shifting' paradigms developed in the recent past?

Antarctica has now clearly been identified as having a major role in processes associated with the Earth System. The role of the formation of Antarctic bottom

341

D.M. Bergstrom et al. (eds.), Trends in Antarctic Terrestrial and Limnetic Ecosystems, 341–347.
© 2006 *Springer.*

water as a driver of the thermohaline current is now relatively well understood and its influence on the properties of the world's oceans widely accepted (Foldvik et al. 2004). Antarctica and the Southern Ocean are recognised as major elements in the equation controlling the Earth's climate and weather, and furthermore, the signals of change are acute in some Antarctic locations (Convey this volume).

Our understanding of the role of the supercontinent Gondwana and its subsequent breakup in shaping the Earth's biota has increased (McLoughlin 2001), and if anything, there is a developing understanding and realisation that long distance, over-ocean dispersal has actually played a more important role in shaping southern continental and island floras (Muñoz et al. 2004, Hughes et al. this volume) than previously thought during the era of recognition of plate tectonics. For example, using modern molecular methods it is now hypothesised that the major lineages in the Poales (sedges and grasses) evolved in separate southern continents (Bremmer 2001). As both groups are widely dispersed in all Gonwanan land masses, but Africa had rifted apart early during angiosperm evolutionary time, over-ocean dispersal must again be considered as a component to the Gondwanan-Antarctic equation. Those subantarctic islands with unbroken oceanic histories provide classic examples of over–ocean, long distance dispersal being responsible for their entire biotas.

Palaeostudies of various forms have increased our understanding of the past climate and biological history of the Antarctic region (Bergstrom et al. this volume) and recent ice cores in Antarctica have provided a much needed perspective on recent climatic events. The Antarctic continental ice–sheet developed far later than previously thought, involving a complicated process of ice advance and retreat, but, over the last 480kyr, ice has been the norm, with only 10% of this time spent in interglacials (EPICA 2004). Thus the climate change that Earth is experiencing now must be viewed as a rare event when compared with the last half a million years of the planet's history, and over the last 1Ma there has been greater percentage of icehouse than greenhouse conditions.

One of the remarkable features of the Antarctic continent and surrounding islands is that despite the repeated ice-ages, not all land was covered by substantial ice sheets. Thus, there were ice-free refugia. Some localities such as Macquarie Island were too low in elevation for the development of an extensive ice-cap (Selkirk et al. 1990), while other areas, including continental areas, escaped ice cover for other reasons. Regardless of reason, ice-free areas mean habitat available for biota. In fact, not even ice-free land is a prerequisite for life, as the emergence of the study of epishelf lakes, cryoconite holes and sub-glacial lakes over the last 20 years has demonstrated (Priscu et al. 2005, Gibson et al. this volume). The essence of life, as clearly demonstrated by the driest continent on Earth, is liquid water (Convey this volume, Hennion et al. this volume). The loss of liquid water at some sites in Antarctica with current climate change will undoubtedly result in loss of local biodiversity (Convey this volume).

Liquid water and ice-free refugia during ice ages has meant long availability of habitat, even extending back to the Gondwana era (Bergstrom et al. this volume). Understanding of colonisation processes and biogeographic patterns has increased markedly over the last 20 years (eg Hughes et al. this volume, Chown and Convey, this volume). The degree of vagility has been identified as an important feature in influencing the distribution patterns of species in the Antarctic region (Greve et al. 2005). Indeed, it has been clearly shown that standard biogeographic factors such as distance between land masses, age, local environmental variables and human impact are as much relevant in the Antarctic region as elsewhere in the world (Chown et al. 1998, Chown and Convey this volume), giving strength to the universality of these principles. Aerobiological (Marshall 1996), cloche (Smith 2001) and molecular genetic studies (Hughes et al. this volume, Skotnicki and Selkirk this volume, Stevens and Hogg this volume) have provided evidence of propagule transport of vagile species to and/or within the Antarctic regions and their subsequent colonisation patterns.

A tantalising picture of the survival of non–vagile species in ice-free habitats is emerging however. Pugh (2003) has concluded that some groups of mites are not capable of long–distance dispersal and Bayly et al. (2003) have provided strong argument for the presence of vicariant invertebrates in Antarctica (Gibson et al. this volume). The recent successful extraction for ancient copepod DNA from Antarctic palaeo–samples (Bissett et al. 2005) highlights new and exciting directions for future research. The identification of vicariant Gondwana species in the Antarctic region will allow us to examine what features of species allow survival through multiple 'icehouse/ greenhouse' cycles (Bergstrom et al. this volume).

The resounding message of current Antarctic biota, is that of isolation and life at the end of a planetary spectrum of conditions with regard to many variables (Bergstrom et al. this volume, Convey et al. this volume b, Peck et al. 2006). 'Simple' and 'extreme', words commonly used in the past to describe the Antarctic have clearly been shown to be relative terms and perhaps often inappropriate under new terms of reference that have been established. Where there is life in the Antarctic, even under the most extreme of conditions there is a community of sorts, although one of the intriguing components is that, both in terrestrial/freshwater environments and marine, often elements, functional groups or guilds are missing (Bergstrom and Chown 1999, Clarke and Johnston 2003). This fact highlights that Antarctica demonstrates a simple but fundamental concept — that not all ecosystem processes are of equal impact. At this end of the spectrum, environmental factors (mainly chronically low temperatures, availability of liquid water and the various components of solar radiation) generally outweigh the impact of biotic interactions such that selection pressures from these physical environmental factors are strong (Bergstrom et al. this volume, Convey et al. this volume b). Results of these selection pressures include the presentation of life history strategies with features such as stress tolerance, lack of competitive ability or investment in dispersal strategies, reduced reproductive investment and output in some groups, and extended lifespans and life cycles. But more generally, a key element in the response of many Antarctic terrestrial and freshwater biota to environmental variation is the possession of considerable flexibility in many life history and physiological characteristics (Convey et al. this volume b, Hennion et al. this volume).

With increases in the temperature component of current climate change in many locations of the Antarctic, many terrestrial species may respond positively by faster metabolic rates, shorter life cycles and local expansion of populations. But subtle negative impacts can also be predicted (and are perhaps being observed) with regard to increased exposure to UV-B (Convey this volume, Hennion et al. this volume, Robinson et al. 2005). Changes in water availability will also impact on both terrestrial and the more the stable, limnetic environments. Local reduction in water availability in terrestrial habitats can lead to desiccation stress (Convey this volume) and subsequent changes in ecosystem structure, as has been reported from Marion Island where there have been dramatic changes in mire communities associated with a substantial decrease in rainfall (Smith 2002).

Changes in both temperature and precipitation, even changes judged as subtle by climate scientists, will probably have profound effects on limnetic ecosystems through the alteration of the surrounding landscape and of the time, depth and extent of surface ice cover, water body volume and lake chemistry (with increased solute transport from the land in areas of increased melt) (Quesada et al. this volume, Lyons et al. this volume, Quayle et al. 2002, 2003). Predicted impacts of such changes will be varied. In shallow lakes, lack of surface ice cover will lead to increased wind–induced mixing. Input of freshwater into the mixolimna of deeper lakes will increase stability and this, associated with increased primary production, will lead to higher organic carbon flux. Such a change will have flow–on effects including potential anoxia, shifts in overall biogeochemical cycles and alterations in the biological structure and diversity of ecosystems (Lyons et al. this volume).

The predictions of Lyons et al. (this volume) also serve to illustrate a profound paradigm shift in Antarctic science that has occurred in the last 20 years. Although they and Convey (this volume) state that we are not yet in a situation where we can develop a quantitative predictive model or even models that completely qualifies the response of Antarctic systems to climate change, much of the predictions are based on a foundation of long-term studies and monitoring, such as those at the McMurdo Dry Valleys LTER or British Antarctic Survey sites on Signy Island. Furthermore, digital data capture and exchange has been one of the main recent advances in Antarctic science, partly due to major advances in all forms of digital technology and in shift in altitude towards more scientific data exchange and collective approaches to 'big science issues' – indeed bringing to fruition one of the founding principles of the Antarctic Treaty system itself (Hull and Bergstrom this volume). This means that a more holistic approach can be taken to complex Antarctic science questions and increasingly larger datasets made available for these analyses.

Finally, if the selection pressures of the past in the Antarctic have resulted in adaptations with emphases in stress tolerance, plasticity and variation in life histories but reduced competitive ability, then Antarctic ecosystems are vulnerable to the impact of colonisation by better competitors, that may be at more advantage under changed climatic conditions (Bergstrom and Chown 1999, Convey and Chown this volume, Convey et al. this volume a). These competitors may be either naturally dispersed or have 'hitch-hiked' with humans (Frenot et al. 2005, Whinam et al. 2005, Convey this volume).

With drastic increases in the number of humans visiting the subantarctic and Antarctic region (Whinam et al. this volume, Hull and Bergstrom this volume), the looming threat that recent research has gained insight into, is the loss of isolation of the region and actual and potential changes to subantarctic and Antarctic ecosystems caused by alien species. That there is a correlation between energy availability and success rate of alien establishment (Chown and Convey this volume) has given us a 'lucky break'. Some subantarctic islands such as Iles Kerguelen and Marion Island have provided us with salient examples of the impact of alien species in isolation of other impacts found elsewhere on the planet such as land clearance, salination, acid–rain and pollution. Some subantarctic islands such as the Heard and McDonald Islands and much of maritime and continental Antarctica, however, still remain relatively free of the impact of alien species. At locations where aliens have been recently been introduced attempts should be made to remove or at least curb the expansion of populations and mitigation methods employed such described in Whinam et al. (2005). Antarctica has been declared a place of 'peace and science'. If we fail to stop the transfer of alien biota to and between regions in the Antarctic, we greatly degrade the scientific value of the continent. Currently parts of Antarctica are the only places on the planet where we can study natural biological phenomena in the knowledge that humans have not altered most of the processes. This legacy is far too valuable to lose.

References

Bayly, I.A.E., Gibson, J.A.E., Wagner, B. and Swadling, K.M. (2003) Taxonomy, ecology and zoogeography of two Antarctic freshwater calanoid copepod species: *Boeckella poppei* and *Gladioferens antarcticus, Antarctic Science* **15**, 439-448.

Bergstrom, D.M. and Chown, S.L. (1999) Life at the front: history, ecology and change on southern ocean islands, *Trends in Ecology and Evolution* **14**, 472-477.

Bergstrom, D.M., Hodgson, D.A. and Convey, P. (2006) The physical setting of the Antarctic, in D.M. Bergstrom, P. Convey, and A.H.L. Huiskes (eds.), *Trends in Antarctic Terrestrial and Limnetic Ecosystems: Antarctica as a Global Indicator*, Springer, Dordrecht (this volume).

Bissett, A., Gibson, J.A.E., Jarmen, S.N., Swadling, K.M. and Cromer, L. (2005) Isolation, amplification, and identification of ancient copepod DNA from lake sediments, *Limnology and Oceanography: Methods* **3**, 533-542.

Bremmer, K. (2002) Gondwanan evolution of the grass alliance of families (Poales), *Evolution* **56**, 1374-1387.

Chown, S.L. and Convey, P. (2006) Biogeography, in D.M. Bergstrom, P. Convey, and A.H.L. Huiskes (eds.), *Trends in Antarctic Terrestrial and Limnetic Ecosystems: Antarctica as a Global Indicator*, Springer, Dordrecht (this volume).

Chown, S.L., Gremmen, N.J.M. and Gaston, K.J. (1998) Ecological biogeography of Southern Ocean islands: species-area relationships, human impacts and conservation, *American Naturalist* **152**, 562-575.

Clarke, A. and Johnston, N.M. (2003) Antarctic marine benthic diversity, *Ocean Marine Biology Annual Review* **41**, 47–114.

Convey, P. (2006) Antarctic climate change and its influences on terrestrial ecosystems, in D.M. Bergstrom, P. Convey, and A.H.L. Huiskes (eds.), *Trends in Antarctic Terrestrial and Limnetic Ecosystems: Antarctica as a Global Indicator*, Springer, Dordrecht (this volume).

Convey, P., Chown, S.L., Wasley, J. and Bergstrom, D.M. (2006b) Life history traits, in D.M. Bergstrom, P. Convey, and A.H.L. Huiskes (eds.), *Trends in Antarctic Terrestrial and Limnetic Ecosystems: Antarctica as a Global Indicator*, Springer, Dordrecht (this volume).

Convey, P., Frenot, Y., Gremmen, N. and Bergstrom, D.M. (2006a) Biological invasions, in D.M. Bergstrom, P. Convey, and A.H.L. Huiskes (eds.), *Trends in Antarctic Terrestrial and Limnetic Ecosystems: Antarctica as a Global Indicator*, Springer, Dordrecht (this volume).

EPICA (2004) Eight glacial cycles from an Antarctic ice core, *Nature* **429**, 623-628.

Foldvik, A.,Gammelsrød, T., Østerhus,S., Fahrbach, E., Rohardt, G., Schröder, M., Nicholls, K.W., Padman, L. and Woodgate, R.A. (2004) Ice shelf water overflow and bottom water formation in the southern Weddell Sea, *Journal of Geophysical Research* **109**, C02015, doi:10.1029/2003JC002008.

Frenot, Y., Chown, S.L., Whinam, J., Selkirk, P., Convey, P., Skotnicki, M. and Bergstrom, D. (2005) Biological invasions in the Antarctic: extent, impacts and implications, *Biological Reviews* **80**, 45-72.

Gibson, J.A.E., Wilmotte, A., Taton, A., Van De Vijver, B., Beyens, L. and Dartnall, H.J.G. (2006) Biogeographic trends in Antarctic lake communities, in D.M. Bergstrom, P. Convey, and A.H.L. Huiskes (eds.), *Trends in Antarctic Terrestrial and Limnetic Ecosystems: Antarctica as a Global Indicator*, Springer, Dordrecht (this volume).

Greve, M., Gremmen, N.J.M., Gaston, K.J. and Chown, S.L. (2005) Nestedness of Southern Ocean island biotas: ecological perspectives on a biogeographical conundrum, *Journal of Biogeography* **32**, 155-168

Hennion, F., Huiskes, A.H.L., Robinson, S. and Convey, P. (2006) Physiological traits or organisms in a changing environment, in D.M. Bergstrom, P. Convey, and A.H.L. Huiskes (eds.), *Trends in Antarctic Terrestrial and Limnetic Ecosystems: Antarctica as a Global Indicator*, Springer, Dordrecht (this volume).

Hughes, K.A., Ott, S., Bölter, M. and Convey, P. (2006) Colonisation processes, in D.M. Bergstrom, P. Convey, and A.H.L. Huiskes (eds.), *Trends in Antarctic Terrestrial and Limnetic Ecosystems: Antarctica as a Global Indicator*, Springer, Dordrecht (this volume).

Hull, B.B. and Bergstrom, D.M. (2006) Antarctic terrestrial and limnetic ecosystem conservation and management, in D.M. Bergstrom, P. Convey, and A.H.L. Huiskes (eds.), *Trends in Antarctic Terrestrial and Limnetic Ecosystems: Antarctica as a Global Indicator*, Springer, Dordrecht (this volume).

Lyons, W.B., Laybourn-Parry, J., Welch, K.A. and Priscu, J.C. (2006) Antarctic lake systems and climate change, in D.M. Bergstrom, P. Convey, and A.H.L. Huiskes (eds.), *Trends in Antarctic Terrestrial and Limnetic Ecosystems: Antarctica as a Global Indicator*, Springer, Dordrecht (this volume).

Marshall, W.A. (1996) Biological particles over Antarctica, *Nature* **383**, 680.

McLoughlin, S. (2001) The breakup history of Gondwana and its impact on pre-Cenozoic floristic provincialism, *Australian Journal of Botany* **49**, 271-300

Muñoz, J., Felicísimo, A. M., Cabezas, F., Burgaz, A. R. and Martínez, I. (2004) Wind as a long-distance dispersal vehicle in the Southern Hemisphere, *Science* **304**, 1144-1147.

Pickard, J. (1986) The Vestfold Hills: A window on Antarctica, in J. Pickard (ed), *Antarctic Oasis, Terrestrial Environments and History of the Vestfold Hills*, Academic Press, Sydney.

Peck, L.S., Convey, P. and Barnes, D.K.A. (2006) Environmental constraints on life histories in Antarctic ecosystems: tempos, timings and predictability, *Biological Reviews* **81**, 75-109.

Priscu, J.C., Kennicutt, M.C., Bell, R.E., Bulat, S.A., Ellis-Evans, J.C., Lukin, V.V., Petit, J–R., Powell, R.D., Siegert, M.J., and Tabacco, I. (2005) Exploring subgalcial lake environments, *EOS* **86**, 193-197.

Pugh, P.J.A. (2003) Have mites (Acarina: Arachnida) colonised Antarctica and the islands of the Southern Ocean via air currents? *Polar Record* **39**, 239-244.

Quayle, W.C, Peck, L.S., Peat, H., Ellis-Evans, J.C. and Harrigan, P.R. (2002) Extreme responses to climate change in Antarctic lakes, *Science* **295**, 645.

Quayle, W.C, Convey, P., Peck, L.S., Ellis-Evans, J.C., Butler, H.G., and Peat, H.J. (2003), Ecological responses of maritime Antarctic lakes to regional climate change, in E. Domack, A. Burnett, A. Leventer, P. Convey, M. Kirby and R. Bindschadler (eds.), *Antarctic Peninsula Climate Variability: Historical and Palaeoenvironmental Perspectives*, Antarctic Research Series, Vol. 79, American Geophysical Union, Washington, D.C. pp. 159-170.

Quesada, A., Vincent, W.F. Kaup, E., Hobbie, J.E., Laurion, I., Pienitz, R., López-Martínez, J. and Durán, J.-J. (2006) Landscape control of high-latitude lakes in a changing climate, in D.M. Bergstrom, P. Convey, and A.H.L. Huiskes (eds.), *Trends in Antarctic Terrestrial and Limnetic Ecosystems: Antarctica as a Global Indicator*, Springer, Dordrecht (this volume).

Robinson, S.A., Turnbull, J.A., and Lovelock, C.E (2005) Impact of changes in natural ultraviolet radiation on pigment composition, physiological and morphological characteristics of the Antarctic moss, *Grimmia antarcticii, Global Change Biology* **11**, 476–489.

Selkirk, P.M., Seppelt, R.D. and Selkirk, D.R. (1990) *Subantarctic Macquarie Island. Environment and Biology,* Studies in Polar Research Cambridge University Press, Cambridge, pp. 285.

Skotnicki, M.L. and Selkirk, P.M. (2006) Plant biodiversity in an extreme environment: genetic studies of origins, diversity and evolution in the Antarctic, in D.M. Bergstrom, P. Convey, and A.H.L. Huiskes (eds.), *Trends in Antarctic Terrestrial and Limnetic Ecosystems: Antarctica as a Global Indicator,* Springer, Dordrecht (this volume).

Smith R.I.L. (2001) Plant colonization response to climate change in the Antarctic, *Folia Facultatis Scientiarum Naturalium Universitatis Masarykianae Brunensis, Geographia* 25, 19–33.

Smith, V.R. (2002) Climate change in the subantarctic: an illustration from Marion Island, *Climatic Change* 52, 345-357.

Stevens, M.I. and Hogg, I.D. (2006) The molecular ecology of Antarctic terrestrial and limnetic invertebrates and microbes, in D.M. Bergstrom, P. Convey, and A.H.L. Huiskes (eds.), *Trends in Antarctic Terrestrial and Limnetic Ecosystems: Antarctica as a Global Indicator*, Springer, Dordrecht (this volume).

Whinam, J. Copson, G. and Chapuis, J.-L. (2006) Subantarctic terrestrial conservation and management, in D.M. Bergstrom, P. Convey, and A.H.L. Huiskes (eds.), *Trends in Antarctic Terrestrial and Limnetic Ecosystems: Antarctica as a Global Indicator*, Springer, Dordrecht (this volume).

Whinam, J., Chilcott, N. and Bergstrom, D.M. (2005) Subantarctic hitchhikers: expeditioners as vectors for the introduction of alien organisms, *Biological Conservation* 121, 207-219.

INDEX